ESSENTIALS OF FUNCTIONAL FOODS

Mary K. Schmidl, PhD

National Food and Nutrition Consultants
Adjunct Assistant Professor
Department of Food Science and Nutrition
University of Minnesota
St. Paul, Minnesota

Theodore P. Labuza, PhD

Professor of Food Science and Technology
Department of Food Science and Nutrition
University of Minnesota
St. Paul, Minnesota

AN ASPEN PUBLICATION®
Aspen Publishers, Inc.
Gaithersburg, Maryland
2000

FLORIDA GULF COAST
UNIVERSITY LIBRARY

Library of Congress Cataloging-in-Publication Data

Schmidl, Mary.
Essentials of functional foods / Mary Schmidl, Theodore P. Labuza.
p. cm.
Includes bibliographical references and index.
ISBN 0-8342-1261-7
1. Functional foods.   2. Food industry and trade.   I. Labuza, Theodore Peter, 1940–   II. Title.
RM217.S284   2000
613.2—dc21   00-020858

Orders: (800) 638-8437
Customer Service: (800) 234-1660

About Aspen Publishers • For more than 40 years, Aspen has been a leading professional
publisher in a variety of disciplines. Aspen's vast information resources are available in both
print and electronic formats. We are committed to providing the highest quality information
available in the most appropriate format for our customers. Visit Aspen's Internet site for more
information resources, directories, articles, and a searchable version of Aspen's full catalog,
including the most recent publications: **www.aspenpublishers.com**
Aspen Publishers, Inc. • The hallmark of quality in publishing
Member of the worldwide Wolters Kluwer group.

Editorial Services: Joan Sesma
Library of Congress Catalog Card Number: 00-020858
ISBN: 0-8342-1261-7

Printed in the United States of America

1 2 3 4 5

This book is dedicated to the K(C)atherines. Over 150 years ago
great-great-great grandmother Catherine McNaboe, sister of Michael,
a great priest involved in the California Gold Rush, forged a new
life leaving her beloved family and Ireland to come to the United States.
Later, Katherine Schmidl and Catherine Labuza, our respective mothers,
gave so much to us. Finally our daughter Katherine, who at seven is
desirous of learning and is facing a vast and difficult world.
This book describes one facet of that world.

Table of Contents

Contributors

Jorma Ahokas
Professor
Director
Key Centre for Applied & Nutritional
  Toxicology
RMIT University
Melbourne, Victoria
Australia

John J.B. Anderson, PhD
Professor of Nutrition
Department of Nutrition
Schools of Public Health and Medicine
University of North Carolina
Chapel Hill, North Carolina

John D. Benson, PhD
Director
Pediatric Nutrition Research &
  Development
Ross Products Division of Abbott
  Laboratories
Columbus, Ohio

Dominique Brassart, PhD
Nestle Research Center
Lausanne, Switzerland

Mary Ellen Camire, PhD
Associate Professor
Department of Food Science & Human
  Nutrition
University of Maine
Orono, Maine

Jonathan W. DeVries, PhD
Principal Scientist, Technical Manager
Medallion Laboratories
Minneapolis, Minnesota

L. Lotte Dock, MS
Department of Food Science
Purdue University
West Lafayette, Indiana

Lorraine Eve, BSc
Senior Regulatory Advisor
Leatherhead Food Research Association
Leatherhead, Surrey
United Kingdom

David Firestone, PhD
Scientist Emeritus
Food and Drug Administration
Washington, DC

Ann Marie Flammang, PhD, RD
Medical Nutrition, Medical Affairs
Ross Products Division of Abbott
    Laboratories
Columbus, Ohio

John D. Floros, PhD
Professor
Department of Food Science
Purdue University
West Lafayette, Indiana

Daniel D. Gallaher, PhD
Associate Professor
Department of Food Science and Nutrition
University of Minnesota
St. Paul, Minnesota

Sanford C. Garner, PhD
Assistant Research Professor
Department of Surgery
Duke University Medical Center
Durham, North Carolina

Peter Barton Hutt
Covington & Burling
Washington, DC

Theodore P. Labuza, PhD
Professor of Food Science and Technology
Department of Food Science and Nutrition
University of Minnesota
St. Paul, Minnesota

Paul A. Lachance, PhD, DSc, FACN
Executive Director
The Nutraceuticals Institute
Professor of Food Science and Nutrition
Rutgers University
New Brunswick, New Jersey

Lillian Langseth, DrPH, FACN, CNS
Fellow
American College of Nutrition
Adjunct Professor
Columbia University
New York, New York

Gary A. Reineccius, PhD
Professor
Department of Food Science and
    Nutrition
University of Minnesota
St. Paul, Minnesota

Alan S. Ryan, PhD
Senior Group Leader
Pediatric Nutrition Research and
    Development
Ross Products Division of Abbott
    Laboratories
Columbus, Ohio

Seppo Salminen, PhD
Professor
Department of Biochemistry and Food
    Chemistry
University of Turku
Turku, Finland

Gertjan Schaafsma, PhD
DMV International
Nutrition Expertise Center
Wageningen
Netherlands

Eduardo J. Schiffrin, MD
Nestle Research Center
Lausanne, Switzerland

Barbara O. Schneeman, PhD
Professor
Department of Nutrition
University of California
Davis, California

Karlene R. Silvera, PhD
Senior Research Analytical Chemist
General Mills
Minneapolis, Minnesota

J.M. Steijns, PhD
DMV International
Nutrition Expertise Center
Veghel
Netherlands

Foreword

The area of functional foods is exploding worldwide and now represents one of the most dynamic growth categories that has emerged from food science and nutrition. Defined as foods that provide a health benefit beyond inherent nutrition, this new category represents an increasingly diverse collection of foods, including infant formulas, medical foods, dietary supplements, performance foods, probiotics, and foods specially designed for delivery of specialized nutrients and components, such as fiber, vitamins, minerals, antioxidants, prebiotics, dairy proteins, soy proteins, fats and oils.

Essentials of Functional Foods has assembled an impressive collection of expert authors to overview each of these food and nutrient categories and discuss product formulations, functional components, and their associated structure-function claims. The text provides key information on the history of functional foods, legal definitions, and the environment regulating product labeling and structure-function claims. The text also presents detailed sections on the technological aspects of processing, preservation, safety, and packaging of functional foods.

Integration of the core principles of food science and technology into the analysis and processing of the major functional food categories and components makes *Essentials of Functional Foods* a platform reference for any student or professional working in food science and functional foods.

Todd R. Klaenhammer
William Neal Reynolds Professor of Food Science and Microbiology
Director of the Southeast Dairy Foods Research Center
North Carolina State University

PART I

Introduction

CHAPTER 1

Relationship of Food, Nutrition, and Health

Barbara O. Schneeman

The concept of functional foods is often cited as a newly emerging area in food science and nutrition; however, this concept is based on advances in knowledge and evolution of the field of nutrition that have occurred throughout the 20th century. During the first half of the 20th century, the focus was on undernutrition and strategies to modify foods to correct nutritional deficiencies. In the second half, the focus shifted to issues of overnutrition and an interest in modifying foods to correct associated public health problems. But in both situations, food scientists and nutritionists have asked how foods can be modified or formulated to have specific physiological or nutritional effects that improve health. Even before this modern era of defining functional foods, many cultures have identified foods or natural materials that could correct disease states or maintain health. The origins of modern pharmaceutical sciences can be traced to this study of natural materials.[1]

BACKGROUND

At the beginning of the 20th century, the discovery of vitamins established that foods contained compounds essential for health over and beyond the well-known components of protein, available energy, and ash. In 1912, Casimir Funk synthesized this understanding in a landmark paper that presented the "vitamine theory." Funk proposed four different "vitamines" based on four diseases that had empirically been associated with food. These "vitamines" would cure scurvy, pellagra, beri-beri, and rickets. His theory and the shift in thinking about disease are pivotal in our understanding of the modern development of functional foods. Funk proposed that the absence of substances from foods, rather than presence of germs or other disease-causing factors, could be a cause of illness and that the substances could be isolated and fed to correct the disorder. This era of modern nutrition was dramatic in terms of its impact on the prevalent public health issues of the times. Within less than five decades, the vitamins had been identified and characterized.[2] The vitamins were recognized as associated with specific metabolic functions and, in their absence, specific deficiency syndromes. During this era of discovery, compounds were also identified as meeting the definition of a vitamin under certain conditions or in certain species. Such compounds included choline, lipoic acid, myo-inositol, ubiquinones, and carnitine. While these compounds are not considered vitamins for humans, the research raised questions about the ability of certain compounds to maintain health, even though it was not possible to define a specific deficiency syndrome if the compounds were absent from the diet. Thus, during the first half of the century the discovery of vitamins and their relationship to disease made us aware that certain compounds present in only minute amounts in foods were important to health. This understanding also enabled us to realize that certain diseases could be corrected by adding these compounds to foods. One outcome of this awareness was the development of enriched and fortified foods. Enrichment of flour originated as a government program to correct nutrient deficiency problems—perhaps one of the first modern attempts at designing a food for functional purposes related to a nutritional outcome. The incidence of pellagra, a significant public health problem in the southern United States, was virtually eliminated once enrichment of flour began. The Food and Drug Administration (FDA) has recently included folic acid in the enrichment program with the intent of lowering the incidence of neural tube defects in pregnancy. For more detailed information, visit the FDA Website, http://vm.cfsan.fda.gov/.

RELATIONSHIP OF DIET, NUTRITION, AND HEALTH

Midway through the 20th century, the emphasis in diet, nutrition, and health began to change. In preparing the *Surgeon General's Report on Nutrition and Health*, Dr C Everett Koop captured the essence of the shift that began in the United States when he noted that nutrient deficiency diseases were no longer major public health problems in the United States; rather, the report noted, "overconsumption of certain dietary components is now a major concern for Americans. While many food factors are involved, chief among them is the disproportionate consumption of foods high in fats, often at the expense of foods high in complex carbohydrates and fiber—such as vegetables, fruits and whole grain products—that may be more conducive to health."[3] This report, published in 1988, along with the National Research Council's report *Diet and Health* in 1989,[4] brought focus to the dietary pattern that is associated with the major causes of death in the United States. These reports had been preceded by several reports and recommendations[5–12] and have formed the scientific basis for several subsequent reports and recommendations[13–19] (see Table 1–1 for a chronology of key reports on food, nutrition, and health). The dietary pattern that has been associated with the major causes of disease in developed countries like the United States can be characterized by its components as relatively high in saturated fatty acids, total fat, cholesterol, salt, and sugars and relatively low in unsaturated fatty acids, complex carbohydrates, and fiber. Alternatively, this dietary pattern can be characterized by food choices in that it is relatively high in animal products and foods that have a high-energy–low-nutrient density and is relatively low in fruits, vegetables, and whole-grain products. A consequence of the first approach (dietary components) is to address primarily the energy distribution of the diet (ie, the percentage of energy derived from fat versus carbohydrate), whereas a consequence of the second approach is to ask the additional question of whether plant foods have compounds with physiologic or metabolic functions that affect health. Again, we see that the process of addressing nutrition research in relation to health influences strategies for designing functional foods: should we modify the energy distribution of foods, modify the fatty acid composition, or consider the presence/absence of compounds or specific food items that may have physiological functions related to health?[20]

If we use a working definition of functional foods as foods that have been modified or formulated to have specific physiological or nutritional effects, then both strategies, modification of the energy distribution and addition of compounds with physiological activity, are relevant. However, a shift in our thinking about the outcome of consuming such functional foods is needed. The addition of an essential nutrient that is deficient in the food supply will correct the specific deficiency syndrome associated with that compound. However, as pointed out by the 1988 *Surgeon General's Report on Nutrition and Health*,[3] the diseases that have emerged as major public health issues in the latter half of the 20th century are multifactorial and are related to dietary pattern, lifestyle, and heredity. Consequently, a compound may have physiological activity, but the overall impact of consumption relative to disease risk will depend upon numerous other factors, including total dietary pattern. This relationship suggests that functional foods should be evaluated, not by concepts such as optimal

Table 1–1 Key Reports on the Relationship between Diet and Health

Report	Agency	Year
White House Conference on Food, Nutrition, and Health[5]	Executive branch, US government	1969
Diet Related to Killer Diseases[6]	Senate Select Committee on Nutrition and Human Needs	1977
Dietary Goals for the United States[7]	Senate Select Committee on Nutrition and Human Needs	1977
Toward Healthful Diets[8]	National Academy of Sciences/ National Research Council	1980
Nutrition and Your Health: Dietary Guidelines for Americans[9]	US Department of Agriculture and US Department of Health and Human Services	1980
Promoting Health/Preventing Disease: Objectives for the Nation[10]	US Department of Health and Human Services	1980
Diet, Nutrition, and Cancer[11]	National Research Council	1983
Nutrition and Your Health: Dietary Guidelines for Americans, 2nd ed.[12]	US Department of Agriculture and US Department of Health and Human Services	1985
Surgeon General's Report on Nutrition and Health[3]	US Public Health Service	1988
Diet and Health: Implications for Reducing Risk of Chronic Disease[4]	National Research Council	1989
Nutrition and Your Health: Dietary Guidelines for Americans, 3rd ed.[13]	US Department of Agriculture and US Department of Health and Human Services	1990
Food Pyramid: A Guide to Daily Food Choices[14]	US Department of Agriculture	1992
Nutrition Labeling and Education Act of 1990[15]	Food and Drug Administration	1990
Nutrition and Your Health: Dietary Guidelines for Americans, 4th ed.[16]	US Department of Agriculture and US Department of Health and Human Services	1996
Healthy People 2000 Review, 1995–96[17]	US Public Health Service	1996
Surgeon General's Report of Physical Activity and Health[18]	US Public Health Service	1996
Preparation and Use of Food-Based Dietary Guidelines[19]	FAO and WHO	1996

health, well-being, or disease prevention, but in terms of physiological functions that affect the risk of disease.[21] Some examples help illustrate this point.

Consumption of diets high in dietary fiber have been associated with lower risk of gastrointestinal disease, including large-bowel cancer and cardiovascular disease. This epidemiological observation has spawned a variety of clinical and experimental studies to understand the impact of dietary fiber on gastrointestinal function and the relationship to metabolism. These studies have revealed that viscous polysaccharides lower low-density lipoprotein (LDL) and total plasma cholesterol via effects on bile acid turnover and lipid absorption.[22,23] With respect to large-bowel function, dietary fiber is the primary substrate for the microorganisms in the large bowel and the main dietary component that increases stool bulk. These data illustrate physiological responses due to consumption of fiber: reduction of plasma and LDL cholesterol, slowing of the rate of lipid and carbohydrate absorption, increase in stool bulk, and production of short-chain fatty acids in the large intestine. However, prevention of disease such as cardiovascular disease or colon cancer is associated with a dietary pattern that is high in plant foods that provide fiber but also features other compounds or has a composition that may influence health (eg, antioxidants, low fat levels, high monounsaturated fatty acids levels). Thus, a physiologic function may be achieved by a certain type of polysaccharide, but disease prevention is the result of many factors working in conjunction with specific physiological functions. Similarly, several compounds found in plant foods that include phenolics, carotenoids, and flavonoids have been shown to have antioxidant activity. Again, a specific physiological function, such as reducing the rate of oxidative change to LDL particles or DNA, can be demonstrated by these compounds and may be one of many factors that are associated with a reduced risk of cardiovascular disease and cancer when diets high in plant foods are consumed. Another type of example concerns a protein, trypsin inhibitor (TI), found in legumes, that has primarily been studied for its antinutritional properties.[24] TIs are known to interfere with protein digestion and when protein is limited in the diet may result in malnutrition. However, when protein intake is adequate, they serve primarily as a potent stimulant of cholecystokinin (CCK) release. Burton-Freeman et al[25] have demonstrated that release of CCK serves as a potent satiety signal and is related to the dose of fat infusion into the intestine. Thus, the TI naturally occurring in dry beans or as a compound added to foods is an example of a compound with physiological function that under certain circumstances (ie, inadequate protein) has negative consequences for nutritional status. However, under conditions of nutritional adequacy it may help to curb appetite, even when low-fat foods are consumed, which may aid in managing obesity. Again, the functional component by itself may not prevent disease, but its physiological function will affect the overall risk pattern.

Another concept in the development of functional foods has been the recognition of new functions for known essential nutrients.[26] For example, folic acid has been associated with lower risk for neural tube defects and with a lower risk of heart disease, carotenoids have been associated with lower risk of cancer, and vitamin E has been associated with reduced risk of cardiovascular disease. These associations have typically been based on consumption of the vitamin at levels well above the amount needed to prevent the classic deficiency syndrome associated with lack of the nutrient in the diet. In addition, the role of these vitamins in disease prevention may contribute to the protective effects of diets high in fruits, vegetables, and whole-grain products.

Recommendations regarding diet, nutrition, and health have evolved to reflect our growth in knowledge about the components of food that are essential for health as well as of the dietary pattern that lowers risk of chronic disease and maintains health. In the first half of the century, most nutritional advice emphasized the food groups and our growing knowledge of the vitamins in foods. As nutrient deficiency diseases began to be managed, the emphasis shifted to the macronutrients or energy distribution of the diet associated with increasing risk of cardiovascular disease and cancer and then shifted again to a food-based approach that includes an awareness of dietary pattern associated with health as well as meeting micronutrient requirements. In 1977, the Senate Select Committee on Nutrition published a report titled *Diet Related to Killer Diseases* that focused on diets high in fat, saturated fatty acids, and cholesterol as culprits in the increasing incidence of cardiovascular disease and cancer.[6] In addition, this committee established *dietary goals* that were seen as setting targets for change in the macronutrient composition of the diet. By establishing dietary goals, the committee also recognized the importance of gathering food intake data at both the household and the individual level, in addition to food disappearance data, to assess changes in dietary pattern among the American public. Subsequently, the National Academy of Sciences released *Toward Healthful Diets*,[8] and the US Department of Agriculture (USDA) began the process of setting *dietary guidelines* in 1980. Dietary guidelines were seen as an educational tool to aid the public in achieving dietary goals. The evolution of the recommendations on dietary fiber in the *Dietary Guidelines for Americans*[9,12,13,16] illustrates the evolution in our thinking and understanding of dietary factors that affect risk of disease. In 1980 and 1985, the dietary guidelines' advice on fiber was to "eat foods with adequate starch and fiber."[9,12] These reports emphasized intake of fiber for gastrointestinal function and intake of starch in the form of complex carbohydrate, which was preferred to sugars or fat as an energy source. In 1990, this guideline was changed to read, "Choose a diet with plenty of fruits, vegetables, and grain products."[13] This change is a significant shift from recommending a component of food as an energy source to encouraging the consumption of certain types of foods that reflect a dietary pattern associated with health. While the food-based guideline was emphasized, the focus in the 1990 report[13] was still on starch, complex carbohydrate, and fiber in these foods. In 1995, the text of the guidelines was changed significantly to indicate that many components in fruits, vegetables, and grains, not just starch and fiber, are associated with reducing risk of chronic diseases.[16] The 1995 guidelines highlight plant foods as sources of antioxidant nutrients; folates; certain minerals; vitamins E, C, and B_6; and compounds that are not traditionally viewed as nutrients but that have physiological activity that may be associated with health. Taken as a whole, the 1995 dietary guidelines promote a food-based approach, noting that consumption of fruits, vegetables, and grains can contribute to dietary strategies that help lower blood pressure and intake of total fat, saturated fatty acids, and cholesterol.

In 1995, an expert consultation on food-based dietary guidelines was convened by the Food and Agriculture Organization (FAO) and the World Health Organization (WHO) in Cyprus. This consultation defined food-based dietary guidelines as expressing the principles of nutrition in terms of foods.[19] The importance of food-based strategies, as defined by this group, is shown in Table 1–2.

The emphasis on food-based strategies to improve health and maintain nutritional status provides an opportunity to consider the role of functional foods—foods modified or formulated to have specific physiological or nutritional effects that affect health—in the total dietary pattern associated with health. A key part of this opportunity is the recognition, as stated in the 1995 *Dietary Guidelines for Americans*, that "foods also contain other components . . . that are important to health. Although each of these food components has a specific function in the body, all of them together are required for overall health."[16] At both the national and international level, foods are viewed as a critical element of a healthful lifestyle, and understanding the role of the components of food will enable us to ensure their adequacy in the food supply.

The Food and Nutrition Board has initiated the process of developing the Dietary Reference Intakes (DRIs), and a DRI committee was established in 1995 to oversee and conduct this project. The DRIs are a set of reference values that include the Estimated Average Requirement (EAR), the Recommended Dietary Allowances (RDAs), the Adequate Intake (AI), and the Tolerable Upper Intake Level (UL).[27] The definitions given to these terms are shown in Table 1–3. The DRI process

---

Table 1–2 Key Principles for Food-Based Dietary Guidelines (FBDGs)

- Relevant FBDGs are determined by public health issues.
- FBDGs need to recognize the social, economic, agricultural, and environmental aspects of foods and eating patterns.
- FBDGs reflect food patterns rather than numeric goals and express nutrition education in terms of foods to be included in the dietary pattern.
- FBDGs are positive and encourage enjoyment of appropriate dietary intakes.
- A variety of dietary patterns can be consistent with good health.

Source: Data from Food and Agricultural Organization and World Health Organization, *Preparation and Use of Food-Based Dietary Guidelines*, 1996, WHO Geneva.

Table 1–3 Dietary Reference Intakes (DRIs)

Term	Definition
Estimated average requirement (EAR)	The intake that meets 50% of the requirement of the individuals in that group
Recommended dietary allowance (RDA)	The intake that meets the nutrient need of almost all (97–98%) individuals in the group
Adequate intake (AI)	Average observed or experimentally derived intake by a defined population or subgroup that appears to sustain a defined nutritional state, such as normal circulating nutrient values, growth, or other functional indicators of health
Tolerable upper intake level (UL)	The maximum intake by an individual that is unlikely to pose risks of adverse health effects in almost all individuals

Source: Adapted from Institute of Medicine-Food and Nutrition Board, *Dietary Reference Intakes: Calcium, Phosphorus, Magnesium, Vitamin D and Fluoride*, 1997, National Academy of Sciences Press, Washington, D.C.

has several implications for the development of functional foods. In the past, we have had only the one value, the RDA, for evaluation and planning of diets and assessment of adequacy of intake. The more complete set of reference values will provide a better approach to determining adequacy, and the DRI committee has established a subcommittee on the uses and interpretation of DRIs to expand upon the appropriate uses of these values. While the need to establish recommendations for the intake of essential nutrients to prevent deficiency disease continues to be an important focus of the DRI committee and its subcommittees, the committees have also been instructed to consider the role of essential nutrients in reducing the risk of chronic disease and other diseases and conditions. Included in the DRIs are estimates of the UL, which provide reference values for the maximum daily nutrient intake level that is not likely to pose a health risk to almost all individuals within a specified life stage and gender group. The need for establishing ULs is based on the increased availability of nutrient-fortified foods as well as supplements in the food supply. For certain groups of consumers, the risk of exceeding the UL of intake has increased because of the use of such products. It is important to keep in context that the UL does not imply a beneficial effect but refers to the level that is likely to be tolerated biologically. Additionally, the use of AIs will allow recommendations for adequate intake to be made on functional indicators of health rather than the more classic indicators, which were based primarily on prevention of deficiency diseases.[28] Finally, the DRI committee has proposed a subcommittee on other food components, which, if convened, will provide an opportunity to examine the scientific evidence regarding the relationship of these compounds to health and disease prevention. Taken together, this approach toward the DRI is likely to provide a new and more scientifically valid framework in which to evaluate the formulation and modification of foods for nutritional and physiological functions.

GOVERNMENT POLICIES

Currently, labeling and health claims on food packaging are regulated under FDA through the Federal Food, Drug and Cosmetic Act (FFDCA),[29] the Nutrition Labeling and Education Act of 1990 (NLEA),[15] and, for foods marketed as supplements, the Dietary Supplement Health and Education Act of 1994 (DSHEA).[30] Current information on regulations for labeling can be found at the FDA Web site (http://vm.cfsan.fda.gov/list.html). The NLEA changed the format of nutrient labeling on packaged foods and allowed statements on the relationship between a nutrient or a food and the risk of disease or a health-related condition. The original legislation approved seven health claims: those related to (1) calcium and osteoporosis; (2) fat and cancer; (3) saturated fat and cholesterol and coronary heart disease; (4) fiber-containing grain products, fruits, vegetables, and cancer; (5) fruits, vegetables, and grain products that contain fiber and risk of coronary heart disease; (6) sodium and hypertension; and (7) fruits and vegetables and cancer. Subsequently, additional claims were approved for sugar alcohols and dental caries, folic acid and neural tube defects, oat bran and cardiovascular disease, psyllium and cardiovascular disease, and soy protein and cardiovascular disease. The claims allowed under the NLEA are based on significant scientific agreement and are required to be phrased so that consumers can understand the relationship between the nutrient or food and the disease and the role of total diet in disease risk. In addition, there are specific requirements for foods to be eligible for each of the health claims. For example, to carry a claim about fat and cancer, a food must meet the descriptor requirements for low-fat foods; and to carry a claim about fruits, vegetables, and grain products that contain fiber and the risk of coronary heart disease, a food must also meet the descriptor requirements for "low saturated fat," "low cholesterol," and "low fat," must contain, without fortification, at least 0.6 g of soluble fiber per serving, and must contain 10% of Daily Values (DV) of at least one of six nutrients.

The FDA Modernization Act of 1997 (FDAMA)[31] changed the process, essentially allowing for a different notification process to approve the use of health claims. Prior to FDAMA, a claim could not be used unless FDA published a regulation authorizing such a claim. FDA will now permit use of a claim if the claim is based on a current, published, authoritative statement from certain federal scientific bodies, as well as the National Academy of Sciences. Notifications for a health claim or nutrient content claim based on an authoritative statement must still be submitted to the FDA Center for Food Safety and Applied Nutrition, which will be responsible for determining if the statement complies with the relevant provisions of the FDAMA. These changes were proposed with the intent to expedite the process by which health claims are allowed.

The DSHEA of 1994 created a provision for "structure-function" claims as well as health claims on products defined as "dietary supplements" under DSHEA. These products are labeled as dietary supplements on the principal display panel (PDP) and are intended to supplement the diet of humans. Ingredients in dietary supplements include vitamins, minerals, amino acids, herbs, or botanicals as well as mixtures of these materials, concentrates, or extracts. They are intended for ingestion in the form of pills, capsules, tablets, or liquids and are not conventional foods or meal replacers (see the FDA Web site for more complete descriptive information). These products can bear health claims that have been authorized by FDA. If a product qualifies to carry such claims, any references to treatment or cure of specific diseases must be approved under the drug provisions of the FFDCA, and if these are approved, the product qualifies as a drug and not a dietary supplement. Manufacturers of a supplement can make statements about its effects on the structure or function of the body or well-being. To make such a claim, they are responsible for substantiating that the statement is truthful and not misleading; this type of claim need not be approved by FDA, although the agency must be notified within 30 days after a product bearing such a claim is marketed. These types of claims cannot refer to treatment, mitigation, prevention, diagnosis, or cure of specific diseases and must bear the statement on the label that "this statement has not been evaluated by the Food and Drug Administration. This product is not intended to diagnose, treat, cure, or prevent any disease."[30]

Depending on how a product is formulated and marketed, items considered within the broad category of functional foods fall under either the NLEA or the DSHEA regulations for labeling requirements. A key factor will be intended use of the product and whether it is labeled a dietary supplement.

Dietary advice such as the *Dietary Guidelines for Americans* emphasize the importance of a food-based approach to meeting our nutrient requirements. Foods provide the essential nutrients to prevent deficiency diseases as well as the other compounds that are important for health. The development of functional foods has value if they enable us to reinforce this food-based strategy for meeting nutrient needs. Perhaps some of the most exciting opportunities exist for the development of products that are modified or formulated to address the nutritional impact of specific lifestyle issues—for example, high levels of physical activity or sedentary behavior—as well as specific genetic predisposition for disease risk.[32]

The shift in government policy to allow health claims on food products and the new approach to allow authoritative statements as substantiation for health claims are likely to expand the number of claims and make the approval process more rapid. This shift will stimulate interest in formulating or modifying products for nutritional or health reasons. Likewise, it illustrates the need for research that will create significant scientific agreement on the health attributes of nutrients and food components to comply with the regulatory process. Presumably, the ability to state valid health claims will provide an opportunity in the marketplace for these products. As our knowledge expands regarding the association of food components and nutrients, lifestyle choices, and genetic factors or disease states, new opportunities for product development will emerge.

REFERENCES

1. Consejo Social de la Universidad Complutense de Madrid. *El Museo de la Farmacia Hispana.* Madrid, Spain: Grafica Internacional; 1993.

2. Combs GF Jr. *The Vitamins: Fundamental Aspects in Nutrition and Health.* San Diego, CA: Academic Press; 1992.

3. *Surgeon General's Report on Nutrition and Health.* Washington, DC: US Public Health Service; 1988.

4. National Research Council, Food and Nutrition Board. *Diet and Health: Implications for Reducing Risk of Chronic Disease.* Washington, DC: National Academy Press; 1989.

5. *White House Conference on Food, Nutrition, and Health.* Washington, DC: US Government Printing Office, 1969.

6. Senate Select Committee on Nutrition and Human Needs. *Diet Related to Killer Diseases.* Washington, DC: US Government Printing Office; 1977.

7. Senate Select Committee on Nutrition and Human Needs. *Dietary Goals for the United States.* Washington, DC: US Government Printing Office; 1977.

8. Food and Nutrition Board, National Research Council. *Toward Healthful Diets.* National Academy Press, Washington, DC; 1980.

9. US Dept of Agriculture and US Dept of Health and Human Services. *Nutrition and Your Health: Dietary Guidelines for Americans.* Washington, DC: US Government Printing Office; 1980.

10. US Dept of Health and Human Services. *Promoting Health/Preventing Disease: Objectives for the Nation.* Washington, DC: US Government Printing Office; 1980.

11. Food and Nutrition Board, National Research Council. *Diet, Nutrition, and Cancer.* Washington, DC: National Academy Press; 1983.

12. US Dept of Agriculture and US Dept of Health and Human Services. *Nutrition and Your Health: Dietary Guidelines for Americans.* 2nd ed. Washington, DC: US Government Printing Office; 1985.

13. US Dept of Agriculture and US Dept of Health and Human Services. *Nutrition and Your Health: Dietary Guidelines for Americans.* 3rd ed. Washington, DC: US Government Printing Office; 1990.

14. US Dept of Agriculture. *Food Pyramid: A Guide to Daily Food Choices.* Washington, DC: US Government Printing Office; 1996.

15. US Food and Drug Administration. *Nutrition Labeling and Education Act of 1990.* Public Law 101-535.

16. US Dept of Agriculture and US Dept of Health and Human Services. *Nutrition and Your Health: Dietary Guidelines for Americans.* 4th ed. Washington, DC: US Government Printing Office; 1995.

17. Centers for Disease Control, Dept of Health and Human Services. Healthy People 2000: National Health Promotion and Disease Prevention Objectives for the Year 2000. *MMWR.* 1990;39:689–690, 695–697.

18. Centers for Disease Control, Dept of Health and Human Services. *Physical Activity and Health: A Report of the Surgeon General.* Washington, DC: US Government Printing Office; 1996.

19. Food and Agricultural Organization, World Health Organization. *Preparation and Use of Food-Based Dietary Guidelines.* Geneva, Switzerland: World Health Organization; 1996.

20. Schneeman BO. Dietary influences on health. *Prev Med.* 1996;25:38–40.

21. Clydesdale FM. A proposal for the establishment of scientific criteria for health claims for functional foods. *Nutr Rev.* 1997;55:413–422.

22. Kritchevsky D, Bonfield C. *Dietary Fiber in Health and Disease: Advances in Experimental Medicine and Biology.* New York, NY: Plenum Press; 1997.

23. Schneeman BO. Dietary fiber and gastrointestinal health. *Nutr Res.* 1998;18:625–632.

24. Gallaher D, Schneeman BO. Nutritional and metabolic response to plant inhibitors of digestive enzymes. In: Friedman M, ed. *Nutritional and Biological Aspects of Food Safety.* New York, NY: Plenum Press; 1984:299–320.

25. Burton-Freeman B, Gietzen DW, Schneeman BO. Cholecystokinin and serotonin receptors in the regulation of fat-induced satiety in rats. *Am J Physiol.* 1999;276:R429–434.

26. Institute of Medicine, Food and Nutrition Board. *Dietary Reference Intakes: Proposed Definition and Plan for Review of Dietary Antioxidants and Related Compounds.* Washington, DC: National Academy of Sciences Press; 1998.

27. Institute of Medicine, Food and Nutrition Board. *Dietary Reference Intakes: Calcium, Phosphorous, Magnesium, Vitamin D and Fluoride.* Washington, DC: National Academy of Sciences Press; 1997.

28. Turnland J, Schneeman BO, eds. New approaches to define nutrient requirements. *Am J Clin Nutr.* 1996;63:983S–1001S.

29. 21 CFR § 301.

30. US Food and Drug Administration. *Dietary Supplement Health and Education Act of 1994.* Public Law 103-417.

31. Food and Drug Administration. *Food and Drug Modernization Act of 1997.* Public Law 105-115.

32. Whitehead RG. Lowered energy intake and dietary macronutrient balance: potential consequences for micronutrient status. *Nutr Rev.* 1995;53:S2–S8.

PART II

Technological Aspects

Functional Foods and Dietary Supplements: Product Safety, Good Manufacturing Practice Regulations, and Stability Testing

Theodore P. Labuza

This chapter will address the legal and scientific issues involved in the manufacturing of nutraceuticals/functional foods/dietary supplements with respect to safety, Good Manufacturing Practice regulations (GMPs), and shelf life testing. Since there are many definitions of these products, we need to define them here. Nutraceuticals were originally defined by Dr Steve DeFelice of the Foundation for Innovation in Medicine in 1989 as "any substance that is a food or part of a food that provides medical and/or heath benefits, including prevention and treatment of disease."[1] This is an all-encompassing definition, including foods, medical foods, and dietary supplements, but it has no regulatory basis. This chapter will concentrate on two categories of nutraceuticals, dietary supplements and functional foods, or foods on which a structure-function claim related to health benefits is made because of the presence of some added active ingredient. It should be noted that it is not a legal definition, since none exists in the United States, although functional foods defined as "foods of special dietary use" (FOSHUs) are a legal entity in Japan.

Dietary supplements by law[2] are products in pill, capsule, soft gel, tablet, powder, or liquid form. Dietary supplements may also be in food form but must be labeled as dietary supplements and cannot constitute a whole-meal replacement. There are now many dietary supplements in baked or extruded form resembling real foods such as chips or health bars that are labeled as dietary supplements. Dietary supplements can carry a structure-function claim that relates the value of some active ingredient, hereafter indicated as "X," in terms of its usefulness for health, where "X" can be a metabolite or an herb/botanical product. More importantly, the structure-function claim cannot be a drug claim and must include the following Food and Drug Administration (FDA) disclaimer: "This statement has not been evaluated by the Food and Drug Administration. This product is not intended to diagnose, treat, cure or prevent any disease."[3]

As noted, functional foods are a new and undefined regulatory category in the United States. In essence, they are foods that also contain an active ingredient "X" that is either generally recognized as safe (GRAS) or a food additive and for which a structure-function claim is made, similar to one on a dietary supplement. Theoretically and legally, a functional food may also make a health claim as regulated in 21 CFR §101.71. Since a functional food is not a dietary supplement, there seems to be no requirement for an FDA disclaimer statement as there is with a dietary supplement if a structure-function claim is made. Of importance is that the functional food, if marketed as a food, cannot contain a "new dietary ingredient" unless that ingredient is also either GRAS or an approved food additive. The process for establishing a "new dietary ingredient will be covered later in this chapter. Since dietary supplements and functional foods are regulated under the Federal Food, Drug and Cosmetic Act (FFDCA),[4] all parts apply unless exempted. Of critical importance is the fact that these products must be processed and marketed under the condition that they be neither adulterated nor misbranded and that they carry facts material to the product.

In terms of processing either a functional food or a dietary supplement, factors that must be considered include the legal status of what can be added, what can be claimed, the safety of the product, the quality of the product with respect to the efficacy of active ingredient "X," including any potential losses in processing and distribution, and, obviously, the cost of manufacturing and distribution. From a manufacturing and distribution standpoint, these products must meet the standard GMP regulations.[5] The GMPs can be interpreted to require the knowledge of the composition of the active ingredient and its safety for the population group for which it is intended, the choice of proper processing technologies to ensure that the finished product is not adulterated, and required controls in distribution to ensure that the amount of "X" is at a level that would be efficacious for the claim made. This chapter will focus on these points with respect to several key questions:

1. What is the legal status of the active ingredient "X"?
2. How do we ensure the identity of the active ingredient "X"?
3. What is the safety of the active ingredient "X"?
4. What losses of the active ingredient "X" take place during processing, distribution, and storage, and what is the impact on the claims made?

THE OVERRIDING LEGAL PRINCIPLES

The overriding laws that control foods, functional foods, and dietary supplements concern products that are either adulterated (unsafe or unfit for food) or misbranded (false or misleading label). Those laws passed by the US Congress are

- Federal Food Drug and Cosmetic Act of 1938 as Amended (FFDCA)[4]
- Nutritional Labeling and Education Act of 1990 (NLEA)[6]
- Dietary Supplement and Health Education Act of 1994 (DSHEA)[2]
- Federal Food Quality Protection Act of 1996 (FQPA)[7]
- Food and Drug Administration Modernization Act of 1997 (FDAMA)[8]

Like all federal laws, these laws are broad in scope and are extended through the promulgation of regulations under the Administrative Procedures Act.[9] These regulations are published in the *Federal Register*, undergo a notice and comment process, and, if agreed upon, take on the power of law when finalized in the Code of Federal Regulations. Those pertaining to foods and dietary supplements are published in 21 CFR pts 1 through 199. These are under the regulatory authority of FDA, as administered by the Secretary of Health and Human Services (HHS). In some instances, because of legal processes, some general rules that FDA expects the industry to comply with but that have not undergone the same scrutiny as finalized regulations are published in the *Compliance Policy Guides Manual* (CPGM).[10] All of the above legal documentation is available in electronic format on the World Wide Web. A useful portal to it is the FDA home page at http://www.fda.gov.

With respect to violations of these laws, FDA has several routes of action. In some cases, FDA can seize the product. This is done by a federal marshall of the US Department of Justice who has to charge the actual product face to face with the violation. A notice of such action must then be sent to the owner of the seized goods, who has the right to defend the product in federal court. If not defended, the product is destroyed. FDA uses seizure only when the violations are most serious with respect to product use. In general, since seizure is a long drawn-out process, especially if the company decides to defend the product in court, FDA generally would rather have the product withdrawn from the marketplace. This "recall process" is not mandatory but is usually carried out by the manufacturer. If carried out, it must be performed as outlined in 21 CFR part 7.

If FDA feels that the company needs to change its manufacturing practices, FDA may seek a court injunction that would require the company to shut down and clean up its facility. The court will listen to the arguments and decide the route to take; any violation of that is considered contempt of court.

As a last resort, and only in the most serious situations, FDA may go the route of a criminal action against the company. In fact, the action is also against the chief operational officers of the company, as has been held up by the US Supreme Court in *United States v Park*.[11] This means the officers can be criminally charged and there is no need to prove intent to violate the applicable laws; the mere act of adulterating or misbranding is prima facie evidence that does not need to be proved, as it would in general criminal law. Thus, the defense must rest solely on proving that the law was applied wrongly or that the law is in fact unconstitutional.

In *United States v Hata*,[12] the defense pleaded that the reason for the adulteration of foods held in a warehouse that had not been completely finished in construction was that the company could not get the building materials in time, a legal defense called *objective impossibility*. The court disagreed and stated that if a food-processing or storage facility cannot do it right, they should not do it at all; that is, food processors owe a duty to the public to use all means to ensure safe and wholesome foods and objective impossibility does not apply. Similarly, in *United States v Starr*,[13] intentional disregard of instructions to clean up a processing area by an employee still resulted in criminal action against the CEO of the company; thus, even sabotage does not remove a company's legal responsibility.

ADULTERATION DUE TO FILTH

The simplest way to understand the impact of federal laws on dietary supplements and functional foods is to look at each major component of the law that may lead to violations. In this section, the major tenets of adulteration in the FFDCA will be reviewed with respect to fitness as food and safety.

Section 402(a)(3): A Food Is Adulterated If It Consists in Whole or Part of Filthy, Putrid or Decomposed Substances or Is Otherwise Unfit for Food

This section of the law deals with "unfitness as food," which means that the food should not contain objectionable material (filth) from an aesthetic standpoint and that the raw material used in processing should not be decomposed—for example should have minimal rot spots and sliminess. In the case *United States v 133 Cases of Tomato Paste*,[14] the court ruled that *filth* has a common meaning and that even if it is processed to become invisible to the naked eye, the product is still adulterated if the levels of filth exceed what is reasonably set as the maximum, legally a level called *de minimus*. From a legal standpoint, the term *de minimus* means that there are certain violations that are so trivial that the courts do not want to be bothered with them. In this respect, since there is always going to be some filth present, FDA had to decide what was *de minimus*. This is especially relevant to herbs and botanicals, which are plants that are mostly grown outside the United States (OUS), harvested by hand, and dried in the sun. Often there is little knowledge of the growing, harvesting, and processing (mostly drying) procedures. In the early days of regulation of the food supply in the United States, it became obvious that foods could be contaminated with extraneous matter (filth), some of which we do not necessarily want in there but that is to a degree unavoidable. Filth includes insects (whole and parts), animal hairs/feathers and parts thereof, animal droppings, and nonpathogenic mold spots that affect the aesthetic sensibilities of the consumer.[15] There is also an implication that such filth could be a vector for human pathogens carried by birds, insects, and rodents.[16,17] In most cases, the pieces are so small so as to be observable only with a microscope. Over the years, both industry and the government have perfected methods for detection of this filth. The FFDCA of 1938, as written, would have made most plant foods illegal, so there was general agreement about what levels of filth would be too much. In 1971, after passage of the Freedom of Information Act,[18] FDA had to publish these levels. Rather than going through a laborious *Federal Register* process to set limits, FDA established defect action levels (DALs) for unavoidable and natural defects in foods that technically would cover both functional foods and dietary supplements.[19] Essentially, it stated:

1. There is a problem of contamination of foods with filth, and this filth will always be present even under GMPs.
2. The level that is set is necessary and feasible to attain under normal practices but is subject to change.
3. Any new DAL or change will be published in the CPGM, and a notice of its availability will be put in the *Federal Register*.
4. Mixing is not allowed to reduce the level to below the published DAL.

DALs are currently available electronically on the FDA Web site, allowing for easier searches.

Many dietary supplement ingredients are new in our food supply, although they may have been used as foods or in traditional medical practice outside the United States. Consequently, no regulatory DALs exist for them, although some trade associations and manufacturers have established some internal standards. Manufacturers in this area will have to begin the process of measuring filth and then agreeing on some voluntary standards or work with FDA to establish a DAL in the CPGM. This can best be facilitated by some of the trade and professional associations involved in this arena. Of help to the current situation are the DALs for spices, since many of these products are handled in the same manner as herbs/botanicals. Examples of DALs for some spices are shown in Table 2–1.

From the quite high allowable levels of filth in cinnamon, the dietary supplement industry should take solace that the levels set will probably not be unreasonable. One fear, however, is that use of

---

Table 2–1 Sample DALs for Spices

Ground Paprika (CPGM 7109.03, §525.200)
<75 insect fragments or 11 rodent hairs per 25-g composite sample

Ground Capsicum Pepper (CPGM 7109.03, §525.200)
<20% mold count
<50 insect fragments or 6 rodent hairs per 25 g

Cinnamon CPGM (71904.4, §525.225)
<5% insect-infested pieces by weight
<5% moldy pieces by weight
<1 mg of mammalian excreta per pound
<400 insect fragments in 50 g of ground product
<11 rodent hairs per 50 g of ground product

Source: Reprinted from U.S. Food and Drug Administration, *Compliance Policy Guides Manual.*

dietary supplement ingredients may be in much higher amounts than that of spices and by a more immunocompromised public, so that FDA may take a more conservative stance in setting a DAL. It must be remembered that if no level is set for a particular product, and FDA seizes it because of filth, the company will have to prove that the product in particular is not adulterated under section 402(a)(3) of the FFDCA, since the law itself sets no allowable level.

Section (402)(a)(4): A Food Is Adulterated If It Is Prepared or Held under Conditions Whereby It May Become Contaminated with Filth

Section 402(a)(4) makes a food adulterated and thus illegal if it is prepared or held under conditions where filth is present in the environment such that the food may be contaminated by this filth. This includes the harvesting and drying operations. The product itself does not have to be examined; the fact that the conditions surrounding the product are filthy implicates the food itself and presumably any dietary ingredient or active ingredient "X." In the last several years, because of foodborne disease outbreaks from fruits and vegetables of foreign origin, there has been a move by the US Congress to more formally mandate that OUS foods and ingredients meet the same stringent requirements as products made inside the United States. This could have great implications for the availability of ingredients such as herbs/botanicals and suggests that manufacturers should begin to work with suppliers to ensure that their products meet US standards. In addition, as noted above, the industry must work with FDA to set additional DALs.

ADULTERATION DUE TO SAFETY FACTORS

Section 402(a)(1): A Food Is Adulterated If It Contains Any Poisonous or Deleterious Substance Which May Render the Food Injurious to Health

This section of the FFDCA deals with the safety of the food supply with respect to unintentionally added substances or naturally occurring substances. The first clause above deals with the unintentional addition of harmful substances either by humans or from the environment and includes microbial pathogens, pathogenic mold toxins such as aflatoxin, viruses, parasites, banned pesticides such as DDT and dieldrin, and environmental pollutants such as lead, mercury, dioxin, and polychlorinated biphenyls (PCBs).

Section 406 of the FFDCA allows the FDA to set tolerances for poisonous and unavoidable contaminants in foods, and 21 CFR §109 indicates the process used to promulgate the tolerance. The regulations in 21 CFR §109.6(b) allow for a tolerance in food of a poisonous or deleterious substance as long as

1. it is required in production
2. or it cannot be avoided in GMPs
3. there is no new technology that can lower the level

Surprisingly, the only action under 21 CFR §109.6(b) has been for setting a tolerance for PCBs, which took 10 years to finalize. For all other environmental toxins, including mold toxins, FDA has taken a similar route as for filth. Thus, under 21 CFR §109.6(c), the regulation states that if there is foreseeable technology in the future that can be used to lower the levels of such a poisonous or deleterious substance, FDA will not set a tolerance but rather will publish an "action level" in the CPGM. This is somewhat confusing, since the filth standards are called "defect action levels"; the two terms should not be confused. Table 2–2 contains examples of such action levels, which includes those for banned pesticides such as DDT as well as for pathogen-derived toxins such as aflatoxin and vomit

Table 2–2 Action Levels for Poisonous or Deleterious Substances

Aflatoxin
<25 μg/Kg food
DDT
<0.5 to 5 ppm, depending on food type
<50 insect fragments or 6 rodent hairs per 25 g

Dieldrin
<0.03 to 0.3 ppm, depending on food type

Lead
<0.5 to 7 mg/kg leached, depending on food product or packaging material

Source: Reprinted from U.S. Food and Drug Administration, *Compliance Policy Guides Manual.*

---

toxin. It should be noted that pathogens present in food are also considered added (by humans) poisonous or deleterious substances and thus make a product illegal at the level of detection (~1 cfu/25 g). In this case, FDA has chosen not to set a tolerance or action level.

One key problem facing the dietary supplement industry is that of herbs and botanicals manufactured OUS. Often there is very little knowledge of the natural or unintentionally added toxicant levels present, especially in countries where environmental pollution control is nonexistent or unregulated. In many OUS countries, use of pesticides that are banned in the United States presents a particular problem. This suggests that importers should have a good history of soil use and farm practices in these growing areas and conduct routine analyses of incoming materials. Several US vegetable processors who obtain raw material through import today may analyze for more than 350 compounds before accepting a batch. Although one can ask for a guarantee that the product meets all aspects of the law, manufacturers should still protect themselves since a guarantee is no protection against FDA action. For example, in *United States v Walsch*,[20] the company that obtained the guarantee is still responsible under the FFDCA for introducing an adulterated product. Of course, they may later file civil suit against the supplier, but their business may be ruined. In addition to use of chemicals, certain practices such as the use of human and animal fecal matter for fertilizer enhance the chance that such botanical products will become contaminated and that the organisms will survive the manufacturing process. No data exist to determine if this is a real problem.

As with filth, Congress realized that section 402(a)(3) could make all food illegal because there are many naturally occurring toxins. Thus, the FFDCA was amended to make a further statement in what is called the second clause, discussed next.

Section 402(a)(1): However, If Such Substance Is Not an Added Substance, Then the Food Is Not Adulterated If the Quantity Would Not Ordinarily Render the Food Injurious to Health

This is the old principle that "the dose makes the poison," but it applies only to naturally occurring substances, that is, to naturally present metabolites in a food or herb. Several court actions were undertaken by some segments of the food industry to force FDA to apply this concept to microbial pathogens (other than how aflatoxin is regulated), but the courts have consistently denied such action and allowed FDA to regulate the level of pathogens at the level of detection, generally 1 cfu/25 g.[21] In that case, the court ruled that pathogens like salmonella species are "added" poisonous or deleterious substances rather than being naturally present. Thus, the first clause of 401(a)(1) applies rather than the second clause, and FDA has discretion to use an action level that can be at the level of detection and thus need not be published in the CPGM. Although one could petition for a regulatory tolerance, in the current climate for food safety this is not a feasible route.

Where the second clause of section 402(a)(1) has come into play is in the regulation of naturally occurring metabolites in foods. For example, potatoes, especially near the growing eyes, produce large amounts of the compound solanine, which is highly toxic. But since the level of consumption on a daily basis is *de minimus* (of such a low quantity as not to be of concern to the court), application of the above law allows potatoes to be used in the US diet, and, in fact, since they are GRAS substances, they cannot be banned. Certain unknown substances in dietary ingredients such as herbs and botanicals, if consumed on a regular basis, may in fact present such a similar problem. A good example is the common *Agaricaceae* mushroom species. It has been shown in several studies that such mushrooms contain the compound agaratine, which is a known DNA breaker.[22] On the basis of toxicological data, Sheppard et al[23] suggested that the safe consumption rate would be less than 4 g per day, or one mushroom meal every 100 days. Exceeding this results in a risk of about 2 cancer cases per 100,000 population per year. However, since mushrooms are considered GRAS foods and are not food additives, they are not banned.

Section 402(a)(2): A Food Is Adulterated If It Contains Any Added Poisonous or Deleterious Substance Unless It is an Approved Food Additive, Approved Food Color, GRAS Substance, Regulated Pesticide on a Raw Agricultural Commodity, [or] New Animal Drug

This section of the law specifically bans the intentional addition of any substances to a food unless there is a law or regulation allowing such introduction. This was added in 1958 under the Food Additives Amendment to help clarify the confusion under section 402(a)(1). Up until recently, any substance going through studies aimed toward its use in food had to be designed to meet one of the five exempted categories listed above. Of importance, if approval as a food additive is sought, the petitioner must provide evidence that the compound is not a carcinogen, as required under section 409(c)(3) of the FFDCA, which bans any substance that may induce cancer in humans or animals. This is the Delaney Clause. Up until 1996, pesticides in processed foods were considered food additives and thus were subject to the Delaney Clause. Current US law allows pesticides that we have banned because of their toxicity to still be manufactured in the United States and sold for use in foreign countries. Thus, they present a hazard, as we may import and use food or dietary supplement ingredients, including herbs/botanicals, that are treated with such pesticides.

The Environmental Protection Agency (EPA) has regulatory control over pesticides on raw agricultural commodities, which should include any herbs/botanicals before they are processed. There must be a specific clearance for each chemical for each specific crop, and there is no broad use. This became clear in 1996 when a major cereal company had to withdraw more than $160 MM of finished product from the marketplace because a silo operator had substituted a different pesticide for the one that was allowed in order to cut his operating costs. Although it was safe and was allowed to be used on several cereal grains for insect control, it was not cleared for use on the particular oat grain used in the finished processed cereal product. Thus, one must be very diligent in getting information about pesticide usage on herbs/botanicals from the growers or face a similar consequence.

In the past, once the raw agricultural commodity was processed—for example, into a food or a dietary supplement in the form of a tablet—the pesticide became an official food additive controlled by FDA and subject to the FFDCA. FDA promulgated regulations allowing the presence of the pesticide in the finished product at no more than the level on the raw agricultural product. Thus, if the product was dried and tableted, it would become illegal because of the concentration effect unless some means was used to remove the pesticide during the processing.

In 1996, the US Congress promulgated the FQPA. This act has been criticized for deleting pesticides from the food additive definition. As noted above, prior to this act, a pesticide present in or on a processed food was a food additive and could not be present in foods at a level greater than in the raw commodity. In addition, since it was defined as a food additive in the FFDCA, a pesticide could not be used if it was found to induce cancer in humans or animals; that is, the Delaney Clause of the FFDCA applied. However, FDA never invoked the Delaney Clause for some pesticides that were considered to be carcinogens, since EPA felt that the benefit to the farmer was greater than the risk to the consumer. The Federal Insecticide, Fungicide and Rodenticide Act (FIFRA)[24] allows this under the aegis of the meaning of an economic poison. Cancer could be one of the allowable risks, generally calculated at no more than one case per million people per lifetime.

Because of this inaction by FDA and EPA, several environmental groups in California moved to force their hand. In a series of lawsuits,[25,26] the courts ruled that some 81 current pesticides would have to be banned because they were carcinogens. In the spirit of helping the farmer, Congress passed the FQPA to partially delete pesticides from the definition of a food additive. Thus, they can be regulated only by the use of 21 CFR §109 to set an action level that does not preclude carcinogens. The action-level setting would thus apply to any pesticide use on any herb/botanical as well. Although the FQPA was established to maintain the then-current use of pesticides, FIFRA requires they be evaluated every 5 years. The FQPA in fact set stricter standards for pesticide evaluation and use, requiring greater analysis, including the totality of consumption from all sources (air, water, foods, environment), and using children, the most susceptible age group, as the target for risk calculations. As a result, EPA in 1999 has banned a universal pesticide, methyl parathion, a widely used organophosphate compound previously approved for many plant crops.[27] The implication for dietary supplements that use herbs and botanicals is that their increased use may inspire regulatory agencies to look at the potential consumption levels in the elderly, who may be greater consumers of such products as well as being more immunocompromised. Thus, the industry needs to carefully examine pesticide use on its ingredients.

As noted in section 402(a)(2), there is no mention of dietary ingredients specifically. Many of them could fall under the GRAS exemption if they were products used in foods prior to 1958. The definition of a food additive under section 201(s) of the FFDCA (which was added in 1958 as part of the Food Additive Amendment) specifically states that "a food additive does not include such substance that is generally recognized as safe (GRAS) among experts qualified by scientific training and experience to evaluate the safety through adequately shown scientific procedures or in the case of a substance used for food prior to Jan. 1, 1958, through scientific procedures or common use in food to be safe under conditions of its common use." Essentially, this means that plant materials used for foods prior to 1958 are self-declared as GRAS and not subject to the food additive definition and the ensuing Delaney Clause.

Since 1958, FDA has held that foods, herbs, and botanicals not consumed in the United States prior to passage of the 1958 Food Additives Amendment either were unapproved food additives or required the company introducing them to get a GRAS declaration or go through the food additive process. Thus, in 1973, FDA prevented a sassafras herb tea from being marketed because it contained safrole, an unapproved food additive that was a carcinogen.[28] In 1982, FDA had the Bureau of Customs blocklist (prevent from entering the United States at an official port of entry) the importation of an herb from Korea, *renshan-fenwang-jian*, on the basis that it was not GRAS and therefore would become an unapproved food additive. FDA had no evidence of its common use in the United States prior to 1958. The importer objected and filed suit in the US District Court in order to lift the ban.[29] The court ruled that the definition of GRAS, given above, did not exclude a compound from GRAS status if there was common use for food outside the United States prior to 1958 because Congress at the time of the passage of the Food Additive Amendment had no intent to restrict GRAS to consumption in the United States. Thus, the blocklist entry was removed, and the company was allowed to import and sell the product. On the basis of this ruling, substances used for food purposes OUS can be brought into the United States and incorporated into foods or dietary supplements. To further validate this, FDA published a regulation almost six years later[30] saying that *substantial use* meant use of the ingredient as food by a significant number of people. This would, however, preclude one from finding a new plant in the jungle that had never been used before for food and importing it as a GRAS substance or from bringing in an herb as a GRAS substance that was used by a relatively small group of natives who lived in a remote location in the jungle. In either case, the new product would technically have to undergo safety testing for approval as either a GRAS substance or a food additive unless it was declared a "new dietary ingredient" under the DSHEA and thus was restricted to dietary supplements. This new regulatory category and process will be reviewed subsequently.

It should also be noted that FDA instituted a new self-affirmation process for GRAS declaration[31] that has not been finalized. Thus, some new, never previously used herb would have to be petitioned as GRAS before it could be used in food, or it could be more easily introduced as a new dietary ingredient under the DSHEA. In 1999, two companies introduced margarine products that contained sterol- or stanol-like compounds as a food rather than as a dietary supplement. These compounds have been shown to inhibit cholesterol absorption and thus may be useful in maintaining a healthy cholesterol level in the body. In one case, the plant sterol was extracted from pine tree resin, obviously not a food, while in the other case it was extracted from a waste stream in soybean oil manufacturing. The latter would be present in foods containing whole soybeans but not in commercial salad oils or margarines, where it has been removed by the process. Rather than going the route of marketing the product as a dietary supplement with a new dietary ingredient, both companies chose to use the GRAS self-affirmation process noted above for their sterols. This action took many lawyers who specialize in foods by surprise because (1) the GRAS self-affirmation process had not been finalized and (2) FDA has apparently allowed the companies to keep the data private rather than, as usual, requiring them to have the data on display in a federal office.[32]

Congress in 1994, in the DSHEA, added a means to declare a substance as a new dietary ingredient so that it could be approved for use specifically in dietary supplements. As is typical in law language, they created some caveats to restrain FDA from acting in too powerful a way. In 21 USC § 321(ff)(3)(B)(1) of the DSHEA, it declared that a new dietary ingredient is not "an article that is approved for as a new drug under Section 355 . . . which was not before such approval . . . marketed as a dietary supplement or as a food." This negative wording means that if the material in question was used somewhere in the world as a food or dietary supplement, a manufacturer can declare that it is a new dietary ingredient under 21 CFR §190.6 et seq. even if it is now marketed in some form in the United States as a drug.[33] To get approval for usage of a new substance as a "new dietary ingredient," a manufacturer must notify FDA 75 days prior to introduction into the marketplace that it is introducing it as such. Thus, any plant, animal product, or metabolite of any organism can supposedly be introduced as a new dietary ingredient, even if it has never used before in the diet anywhere in the world. The information submitted must include the manufacturer's name; the Latin name of the herb or botanical; a description of use, including level and conditions; and the history of use and other safety data.[34] There is no description, however, of the required depth of the safety data, although FDA in the *Federal Register* notice stated that it will look for reasonable assurance that the substance does not present a significant or unreasonable risk of injury or illness. FDA also importantly stated, that if it gives no response after 75 days to the party initiating the process, the party may go ahead and introduce the product but is under risk, since no response is not an endorsement of the safety of the product. Also, it must be understood that any substance used as a dietary supplement before passage of DSHEA occurred (October 15, 1994) is grandfathered in.

On the basis of this new regulation, there is a new category of substances that can be added to food, specifically new dietary ingredients. These ingredients are exempted from the GRAS and the food additive definition, including the Delaney Clause,[35] which in itself prevents use of potential carcinogens as ingredients.[36] Furthermore, the DSHEA requires that the burden of proof of safety be on FDA[37]: that is, the manufacturers are not required to prove safety before introduction into the marketplace, and, as noted above, the level of safety data they submit seems to be minimal. Thus, ingredients that are not approved for use as food additives according to section 409 of the FFDCA can be added to a dietary supplement as long as the manufacturers have some data in their files as to its safety and toxicity, but how much data is needed is not known. For example, the compound stevioside, a sweetener extracted from the leaves of chrysanthemum species and sold over the counter in Japan as an artificial sweetener, although petitioned for, has not been approved for use in foods as a food additive because FDA felt it to be unsafe. It is now a component of several dietary supplements, especially herbal teas, since it was introduced as a new dietary ingredient using the new method of FDA premarket notice under 21 CFR §190.6 instead of going the GRAS route, which would require much more safety information to be gathered, even if the substance was self-declared as GRAS. Further, on the basis of the case *Fmali Herb v Heckler* discussed earlier in this chapter, the leaves of the stevia plant from which stevioside is extracted could be imported and used in a dietary supplement as a GRAS substance, since they have been consumed for centuries by Peruvian Indians.

In several instances, FDA has been concerned about the safety of several new dietary ingredients and has taken action. For example, FDA moved against ephedra products on June 2, 1997, with the publication of a *Federal Register* notice[38] on the basis that the dietary supplement products containing it have been responsible for more than 800 reports of illness, including some deaths. They proposed to limit the consumption level (it is widely available in most food/fuel stores and is used as a stimulant) by requiring some package warning labels. For example, they suggested a maximum dose of 8 mg per serving or 24 mg/d, and they wanted the label to state that the product should not be used for more than 7 days as well as to have a warning statement that included the words "may cause death." The dietary supplement industry strongly objected to this proposed regulation, and it has never been implemented. As another example, on September 14, 1998, FDA, in response to a new dietary ingredient submission, sent a letter asking that the introduction of a pokeweed preparation as a new dietary supplement be stopped on the basis that the plant naturally contained some highly toxic lectins. In this case, the industry complied with the FDA decision.

In a more complex action, on May 20, 1998, FDA issued a notice to Pharmanex Inc (Siam Valley, CA) that their dietary supplement product Cholestin was a drug and therefore misbranded. At that time, the product label stated that the product could reduce both total cholesterol and low-density lipoprotein cholesterol by 10%. Cholestin is a fermented red yeast rice product that was imported from China, where it was used both as a food coloring (use as a food) and in traditional Chinese medicine for centuries (ie, before 1958). The yeast fermentation produces a compound, mevinolin, that is exactly the same compound as in a prescription drug called Mevacor (Lovastatin), manufactured by Merck and approved as a drug by FDA in 1987 to inhibit cholesterol synthesis in the liver and thereby reduce cholesterol levels in the blood. In 1998, FDA asked the Bureau of Customs to blocklist the dried fermented rice at all ports of entry into the United States, thereby preventing Pharmanex from getting the raw material to manufacture the dietary supplement in the form of tablets. Subsequently, Pharmanex sued the United States to overturn that decision.[39] On February 16, 1999, the district court agreed and overturned the FDA decision, thereby declaring that Cholestin is a dietary supplement on the basis of its prior use in China before the DSHEA was enacted. What this says is that a dietary supplement may contain a substance with druglike activity but if that substance was being used or marketed as a supplement by one party prior to the DSHEA, the fact that it was also marketed as a drug in the United States by another party prior to that time does not make the supplement a drug. This case is somewhat similar to the previously discussed standard set by *Fmali Herb v. Heckler* in that prior use in the legal category can take precedent.

ADULTERATION DURING PROCESSING, DISTRIBUTION, AND STORAGE

Section 402(a)(4): A Food Is Adulterated If It Is Prepared or Held under Conditions Whereby It May Be Rendered Injurious to Health

As with filth, section 402 (a)(4) allows FDA to declare a food adulterated if the conditions of the plant, warehouse, or distribution are such that they could render the food (or dietary supplement) unsafe. This means that FDA does not have to prove the product itself is unsafe, only that it has a chance to become unsafe. With foods, this led to the establishment of the GMP regulations, which started out being specific but have now become generic (21 CFR §110) except for infant formula (21 CFR §107), thermally processed low-acid foods (21 CFR §113), and acidified foods (21 CFR §114) as separate categories. The regulations in 21 CFR §110 include general guidelines on

- groundskeeping
- personnel cleanliness
- separation of ingoing streams from outgoing streams
- quality control
- procedures
- packaging
- storage

In general, if one follows these procedures, one should not be found in violation of the FFDCA under section 402(a)(4) for either functional foods or dietary supplements as long as all the prior parts are also adhered to and labeling is done correctly. Some current dietary supplement manufacturers follow these GMPs and in fact go further in that they adhere to drug-manufacturing GMPs (21 CFR § 210). When the DSHEA was passed on October 25, 1994, it included section 402(g), which gave the secretary of HHS discretion to establish specific GMPs for dietary supplements. The major players in the dietary supplement industry acted quickly and submitted a proposal to FDA on October 10, 1995, to ensure that FDA would promulgate it on the basis of current dietary supplement industry practice and 21 CFR §110. Five months later (February 6, 1997), FDA published in the *Federal Register* an Advanced Notice of Public Rulemaking (ANPR) for dietary supplement GMPs.[40] In this proposed rulemaking, FDA has gone a little further, borrowing some parts from the infant formula and drug industry GMPs. It has also listed a series of concerns related to dietary supplement manufacture:

1. Since dietary supplements are consumed in greater quantities than are spices, and potentially by more immunocompromised elderly people who may have health problems, there is a need to establish more stringent standards for filth and safety in the raw agricultural product with respect to both pathogens and metabolites present.
2. Since these products are marketed with structure-function claims, and since some of the new dietary ingredients may be harmful, manufacturers should be required to establish the safety of these ingredients. Obviously, they are required to do so for food in that ingredients must be approved as either GRAS or food additives. But for the new category of "new dietary supplement" in the DSHEA, this requirement for safety has been circumvented. Logically, most manufacturers would not want to use an incoming ingredient that is potentially harmful, but the

question of the level at which the ingredient is harmful should not be considered under GMPs. Also, firms should be required to report on any injuries from use of the product. This is similar to the requirement in 21 CFR §211 for prescription and over-the-counter drugs, so FDA is looking at new dietary substances as "pseudodrugs" because of their structure-function claims.

3. Since the basis for use of these products is some structure-function claim, purportedly due to the presence of an active ingredient "X," there is a need to

- *Establish an analytical method to easily identify the herb/botanical that is the source of "X."* There are many accounts that unknown or ineffective substitutes are being substituted for particular herbal/botanicals when the availability of high quality product is limited or when someone takes advantage of the lack of knowledge of the industry. Even with a known source there is a problem, since only a specific part of the plant is effective and that part must be labeled in the supplement fact panel on the package label. For example, for cloves to be used in food, FDA, under CPGM 7109.05, section 525.250, allows up to 5% of stems by weight in the finished product. Such guidelines do not exist for dietary supplements, so the chance for illegal substitution exists. On July 17, 1997, FDA asked the Bureau of Customs to blocklist a plantain weed being imported as a dietary ingredient on the basis that it was sub-stituted for in part with harmful levels of the leaves of the digitalis plant, a very toxic sub-stance. The plantain weed was being promoted for regularity. In addition, FDA also had a blocklist put on a product called "Sleeping Buddha" that illegally contained a drug "estaza-loma," presumably to enhance the effect of the botanical.

- *Establish the identity of "X" from a biochemical standpoint.* This is a big question that is not discussed in this chapter. It should be noted that the National Institutes of Health are con-ducting a major study on the efficacy of St Johns wort at Duke University. In addition, since herbs/botanicals are a significant fraction of prescribed medicines in Germany, the Germans have studied many of these products. Some of the data are available from the German Commission E Report, a translation of which was published by the American Botanical Council (ABC).[41] Blumenthal, director of ABC, has concerns over the fact that many prepa-rations being sold in the United States are not the same materials as those that have been studied in the clinical trials submitted for approval of herb/botanicals under the Commission E studies. Thus, they may be totally ineffective.[42]

- *Establish a standardized analytical procedure for "X."* R Lenoble,[43] of Hauser Inc, an ana-lytical testing lab specializing in herbs and botanicals, stated that there are many herbs/botanicals for which the peaks that they detect using high-performance liquid chro-matography (HPLC) are complete unknowns, yet many companies use the total area under the peaks as a quality control procedure, even though there is no evidence that these are the effective biochemical agents upon which the structure-function claim is based.

- *Establish a knowledge base of how variety, climate, and growing area affect the amount of "X" in an herb/botanical.* This concern has some basis, since in 1998, at a symposium in New York initiated by Consumer Reports, they[44] presented a study in which they examined products of six different manufacturers of St John's wort and found a zero level of hypericin in one product and a 17-fold variation in the other five products. The compounds of interest, hypericin and pseudohypericin, supposedly are the key metabolites on which the claims of efficacy against anxiety and depression are based. It is possible that the difference in levels were the result of processing and storage conditions. Dr Craig Hassel (personal communica-tion, 1999) heads an agricultural group that is linked with traditional Chinese medicine prac-titioners in Minneapolis. This group is looking for new cash crops that could help save farm-ing in the Midwest and could allow such practitioners to get their herbs/botanicals locally.

He has been told that the ginseng grown in the Midwest is essentially ineffective, since the farmers do not know the several steps taken by growers in China to prepare the product for traditional Chinese medicine practitioners.

- *Establish different types of GMPs for different categories of products.* In a way, this would be a reversion to 20 years ago, when FDA began writing GMPs for specific industries. This practice has gone by the wayside.
- *Establish a requirement that plants manufacturing dietary supplements document GMP procedures and that daily records be kept that would be available to FDA.* This is based on the fact that under 21 CFR § 210 for prescription and over-the-counter drugs, FDA has that authority. Under food regulations, FDA's authority in this area is limited to the manufacturing of infant formula, which is much more like a drug since it is a sole source of food in the diet, and of low-acid thermally processed foods, in which FDA also requires that the process be registered and in which FDA has authority to see retort temperature/time data.[45] FDA also requires processors of acidified foods to file their process.[46] Thus, there is some precedent for FDA to see records, especially since these products are more like prescription or over-the-counter drugs.

FDA also noted that there may be other ways to approach regulation besides GMPs—for example, through an HACCP plan. But use of HACCPs seems inappropriate, since for most products, microbial pathogens are not a problem. FDA also considered whether the manufacturer should be required to prove a supplement's safety before introducing it into the market, as is the case for all other food and drug materials, and suggested that, as with infant formula and over-the-counter and prescription drugs, perhaps firms should be required to report all adverse events that become known because of use of the product. It should be noted that in the DSHEA, Congress requires that FDA give a 10-day notice to a company before it can seize the product.

Because of all these concerns, several of the associations involved with dietary supplements have been preparing white papers for FDA or offering assistance. The United States Pharmacopoeia (USP), for example, through its senior vice president, JG Valentino, has stated emphatically in a letter to FDA on June 8, 1999, that the companies need to ensure to FDA the consistency, identity, purity, potency, and quality of any dietary supplement, as is already required with over-the-counter drugs. He stated that USP is willing to work with any manufacturer to achieve this and that perhaps FDA should adopt USP or National Formulary (NF) standards, as is done with over-the-counter drugs. He also said that FDA should begin to review current product labeling.[47]

MISBRANDING: PRODUCT QUALITY AND SHELF LIFE

Requirements and the Law

The FFDCA, under section 403(a), states that a product is misbranded and thus illegal if its label is false or misleading in any particular. Further, under section 201(n), it states that a product is misbranded (and possibly adulterated) if its label fails to reveal facts material in light of such representations or material facts with respect to the consequences that may result from the use of the article. The basic purpose of a functional food or dietary supplement is to attract the consumer to its use because it will provide a structure-function benefit. Presumably, this benefit is based on some ingredient that is or contains the active agent "X." Thus, it would seem both logical and legally required that the manufacturer know the ingredients used to make the product and the influences of the process used, the packaging, and the effect of distribution conditions on the amount of "X" in the finished consumer product.

It is thus logical that in the ANPR for the GMP for dietary supplements,[48] FDA suggested that dietary supplements carry an expiration date. Under 21 CFR §211.66 for prescription and over-the-counter drugs, there is such a requirement. For these products, the date has to be based on the time the concentration decreases to 90% of the label amount in a serving (tablet, capsule, teaspoon, etc), which is in turn based on the lower of the 95% confidence limits for loss of the active ingredient "X." In the ANPR, FDA says that the date must be supported by data, that the manufacturer can use accelerated shelf life testing procedures, and that the date can be adjusted after the product enters the market because of potential effects of different environmental conditions.

Experimental Design

Many factors can influence the rate of deterioration of food products and the active ingredient "X." In general, the rate of deterioration is a function of composition and environmental factors. Composition includes the concentration of the reactive component, catalysts, pH, and water activity (a_W). Environmental factors include temperature, relative humidity, light, and gas composition in the package. The permeability of the package film to oxygen and water vapor is also critical. The factors that influence the deterioration of the product need to be incorporated into the shelf life study. Assuming that the composition of the product is not variable (which is, of course, a question that has not been answered with respect to the presence of "X"), the above environmental factors are the ones that need to be incorporated to enable shelf life prediction. Because some experiments take an extended amount of time (ie, for products with inherently long shelf lives, such as dry tablets and powders), accelerated shelf life testing (ASLT) is often used. The primary application of ASLT for dietary supplements and perhaps functional foods (note that the ANPR for GMPs is for supplements only) is to determine if the functional ingredient "X" maintains an effective dose over its projected distribution time so that an expiration date can be set. ASLT uses higher temperatures or humidity to accelerate the rate of deterioration and then some basic physical chemical principles or a magic number to project what the shelf life would be at normal storage and distribution conditions. One concern in using too high a temperature in testing is that the physical structure of the system can be different at higher temperatures, as in the melting of lipids, the denaturing of proteins, and the glass-rubber transitions of polymeric carbohydrates.[49] In addition, reactions that led to deterioration at high temperatures may be different from those at normal storage conditions so that the projection is false. The different activation energies of the reactions above and below the glass transition could also result in erroneous prediction from high temperature.[50] Other problems of ASLT have been discussed by Labuza and Riboh,[51] Labuza,[52] Labuza and Schmidl,[53] and Taoukis et al.[54]

Kinetic Modeling

Experimental Design Factors

Before we describe the kinetic models, several factors important for the accumulation of kinetic data of good quality and reliability should be mentioned. First, the error in the reaction rate must be reduced by good analytical precision. Benson[55] calculated the theoretical relative maximum error in the rate on the basis of the precision of the method and the extent to which the degradation is carried out. As shown in Table 2–3, the reaction should be carried out beyond 50% change of the initial value so that the proper kinetic model can be selected.

Typically this is done with drugs, even though the shelf life is determined at the 10% change level. With foods, this is generally impossible; the shelf life predictions have greater error, especially when

Table 2–3 Estimate of Maximum Percent Error in Reaction Rate Constants

Analytical Precision (%)	Change in reactant species monitored							
	1%	5%	10%	15%	20%	30%	40%	50%
0.1	14	2.8	1.4	1.0	0.8	0.6	0.5	0.5
0.5	71	14	7	4.8	3.7	2.7	2.2	2
1.0	>100	28	14	9.7	7.4	5.3	4.4	4
2.0	>100	57	29	19	15	11	8.9	8
5.0	>100	>100	72	48	37	27	22	21
10.0	>100	>100	>100	97	74	53	44	41

based on sensory testing.[56] Importantly, with herbs and botanicals, even though the HPLC procedure for determination of "X" may have a precision error of $\pm 0.1\%$, the plant material must first be extracted to get "X" out, which can increase the precision error to greater than $\pm 5\%$. At that level and a 10% end point, the error in predicting shelf life can be as much as 75% to 100%, obviously very unacceptable unless the shelf life is so long that the product would be consumed much before the end point. On the basis of this, Labuza and Kamman[57] have shown that the data should consist of at least eight data points, including zero time, spaced over the 0% to 50% loss, which, when regressed, will yield reasonable confidence limits so that a good prediction of shelf life is possible.

Overall, the loss of "X" should follow the following equation:

$$-d[X]/dt = k_{app}[X]^n$$

where k_{app} is an apparent rate constant and n is the power factor termed the *reaction order*. It should be pointed out that in this expression, the reaction order is not based on true reaction molecularities, so it should be called *apparent* or *pseudo-order*. The value of n is generally set to 0 or 1 for most modes of loss of "X" deterioration, though for some—for example, a free-radical oxidation of a lipid material such as carotenoids—there may be a more complex expression.[58] In complex food systems, the order by which to model kinetic data is usually determined by a variation of the integration method, plotting attribute value X as a function of time (apparent zero order, $n = 0$) and the natural log of attribute value as a function of time (apparent first order, $n = 1$). If the factors for producing good kinetic data discussed previously are employed, then the highest coefficient of determination (r^2) of the linear regression is usually the best apparent order by which to model loss rate and set the shelf life. In 21 CFR §211.66, this is the method of choice for drug expiration dates, and in fact, the regulation allows for zero order to be used, since, as was shown by Labuza and Kamman,[57] the differences between zero and first order up to 50% loss are negligible if the overall precision is good.

Apparent Zero Order

Apparent zero-order kinetic equations in which $n = 0$ have the differential form of

$$-d[X]/dt = k_{obs}[X]_0 = k_{obs}$$

The integrated form of the apparent zero-order expression for a loss of "X" is

$$[X] = [X]_0 + k_{obs}\,t$$

where $[X]_0$ is the initial concentration and $[X]$ is the concentration at some time t. A plot of $[X]$ as a function of time yields a straight line whose slope equals the apparent zero-order rate constant, k_{obs}.

The units of the rate constant are amount of "X" per day or per any time unit used. Linear regression routines available in most spreadsheet programs and statistical packages can then be used to determine the best straight line from the data collected over time, the r^2, and the 95% confidence limits. Figure 2–1 shows a zero-order plot with the lower 95% confidence limit being used to determine expiration date. It must be noted that this is done at constant temperature and moisture content. As noted on the graph, the expiration date based on the 95% confidence limit is less than the value on the regression line for the data. To improve upon this—that is, to narrow the 95% confidence limit—one can take more points over time but the cost of that improvement is generally not worthwhile from a statistical standpoint.[57] Generally, it is better to improve on the analytical method, especially the extraction process used to analyze for "X."

Assuming that the setting of a standard for the expiration date will be less rigorous for dietary supplements and functional foods than it is for drugs, one can surmise that FDA might use the time for

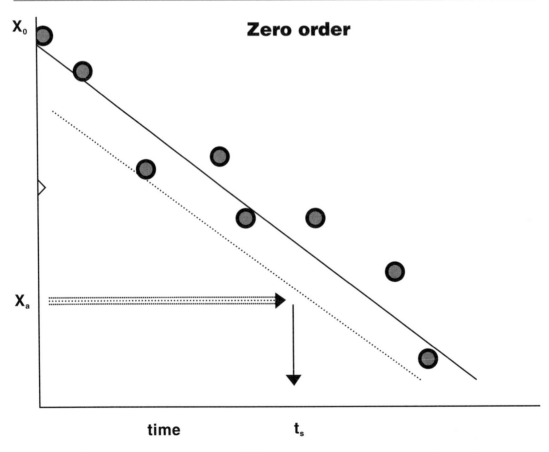

Figure 2–1 Plot of loss of the active ingredient "X" as a pseudo–zero-order reaction with expiration date at t_s. The dotted line represents the lower 95% confidence limit. The expiration date is based on reaching some level of "X" that makes the product ineffective on the basis of its structure-function claim. *Source:* Copyright © 1999, Theodore Labuza.

20% loss, as is done in nutritional labeling for labile vitamins such as Vitamin C. The regulations for label compliance appear in 21 CFR §101.9(g). The 80% retention time for zero-order $t_{0.8}$ is defined as

$$t_{0.8} = 0.8[X]_0/k_{obs}$$

and gives the amount of time for the concentration of $[X]_0$ to be reduced by 20%. Apparent zero-order kinetics have found ample use in modeling the Maillard reaction.[59–61]

Apparent First Order

Apparent first-order kinetics is probably the most widely used kinetic model for food and drug deterioration. The differential of the first-order kinetic expression has the form

$$- d[X]/dt = k_{obs}[X]$$

and the integrated equation for a loss of the quality factor "X" takes the form

$$\ln[X/X_0] = -k_{obs} t$$

Thus, a plot of either $\ln[X]$, $\log [X]$, $\ln[X/X_0]$, or $\log [X/X_0]$ as a function of time gives a straight line with a slope equal to the pseudo–first-order rate constant k_{obs}. Note that log denotes "log" to the base 10, while "ln" represents the natural log. The units of the rate constant are reciprocal time (eg, 1 per day). To make things easier, one can plot the fraction remaining, X/X_0 (or percentage remaining $= 100 \times X/X_0$), on semilog graph paper to get a straight line. In this case, the slope must be multiplied by 2.303 to convert to the true k_{obs}. Figure 2–2 shows a zero-order semilog plot, with the lower 95% confidence limit being used to determine the expiration date. It must be noted that the experiment is carried out at constant temperature and moisture content. From a kinetic standpoint, if the data show a straight line on this type of plot for more than two half-lives (ie, <25% remaining), then the order is most likely first order. In drug expiration testing, generally the experiment is carried out to three half-lives (12.5%) remaining with lots of points in order to reduce the 95% confidence limits.[62] As before, assuming that the setting of a standard for the expiration date will be less rigorous than for drugs, one can surmise that FDA might use the time for 20% loss, as is done in nutritional labeling for loss of natural vitamins under 21 CFR §101.9(g).

The time for 20% loss for a first-order model is defined by

$$t_{0.8} = 0.223/k_{obs}$$

Many reactions can be modeled as pseudo–first-order reactions, including vitamin C degradation,[50] aspartame degradation,[63–65] and amino acid or reducing sugar loss due to the nonenzymatic browning.[66]

Effect of Temperature on Shelf Life

The fact that most reaction rates increase with increasing temperature is well recognized. Increased reaction rates because of temperature abuse during distribution and storage are a critical factor in the design of the shelf life testing of both foods and drugs and, for the same reason, of dietary supplements. The effect of temperature is used in the design of ASLT experiments to speed determination of the expiration date. The use of ASLT was recognized in the GMP ANPR with respect to expiration dating of dietary supplements. The influence of temperature on the rate of deterioration has been shown to follow the Arrhenius relationship, which is given by

$$k_{obs} = k_A \exp(-E_a/RT)$$

where k_A is the pre-exponential constant, E_a is the Arrhenius activation energy in Kjoules/mole and is a measure of the temperature sensitivity of the reaction (ie, the higher the value, the faster the reaction

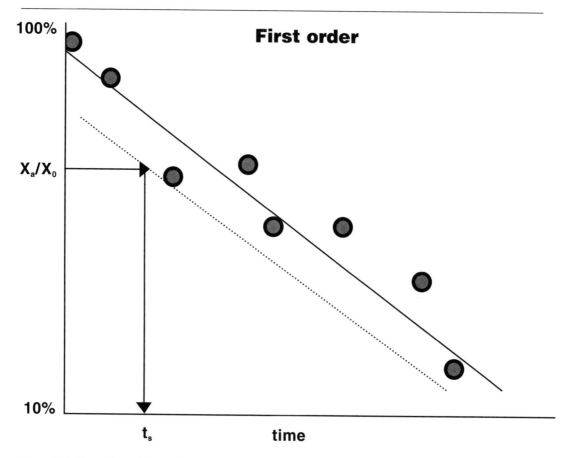

Figure 2–2 Plot of loss of the active ingredient "X" as a pseudo–first-order reaction with expiration date at t_s. The data are plotted on a semilog scale with the value of $X/X_0 \times 100$, which is equal to the percentage remaining. The dotted line represents the lower 95% confidence limit. The expiration date is based on reaching some level of X/X_0 that makes the product ineffective on the basis of its structure-function claim. *Source:* Copyright © 1999, Theodore Labuza.

as temperature increases), R is the ideal gas constant (1.986 Cal/mol K or ~8 Js/mol K), and T is the temperature in Kelvins. A plot of the natural log of the rate constant from either an apparent zero-order or an apparent first-order reaction as a function of the reciprocal of absolute temperature yields a straight line with a slope of $-E_a/R$, as shown in Figure 2–3. For confidence in the value of the slope for extrapolation to a lower temperature, either rate constants at a minimum of five different temperatures, rate constants at three temperatures analyzed by the point-by-point method, or nonlinear regression of all the data simultaneously from at least two temperatures must be used.[57,67] If only three k values are available (ie, only three test temperatures are used), the confidence range is usually wide since the Student's $t_{a/2}$ value is large. Because of the cost of shelf life testing of foods, especially by sensory analysis, only rarely are more than two temperatures used, but this may be possible for dietary supplements where only "X" is being measured. Figure 2–3 also shows the value of the Arrhenius plot from an accelerated shelf life testing standpoint. If the linear relationship holds, one can use

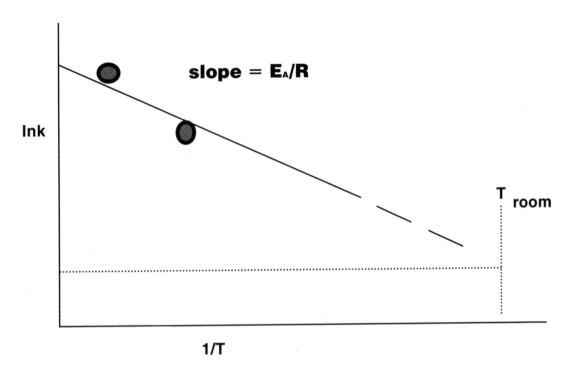

Figure 2–3 Typical Arrhenius plot with extrapolation to room temperature. *Source:* Copyright © 1999, Theodore Labuza.

high-temperature data to extrapolate to lower temperature, the "magic number" approach. Of course, the more temperatures you use, the better the prediction, but the minimum is two to be able to draw a straight line. The magic number, then, is the factor you multiply by the shelf life at the highest temperature of the test to get the shelf life at the presumed storage temperature to which the product will be exposed. Obviously, one needs a minimum of two test temperatures to do this and a reasonable guess as to the conditions of storage and distribution. The latter is not well known even for foods.[68]

Because use of Arrhenius activation energy plots is confusing to many people in terms of understanding natural logs of rate constants and inverse absolute temperature in Kelvin, another method of expressing the temperature dependence of reactions is the Q_{10} approach. The Q_{10} is defined as the ratio of the rate of reaction at one temperature compared to that at a temperature 10°C higher or lower. It can also be defined as the ratio of the shelf life at one temperature to that at a temperature 10°C higher or lower. For example, a Q_{10} of 2 means that if the shelf life at 30°C is 3 months, the shelf life at 40°C is 1 1/2 months, while at 20°C it is 6 months. Plotting the log of the shelf life (using any definition for end point, such as the 20% loss value) as a function of temperature (or shelf life vs temperature on a semilog plot) yields a straight line with a slope of b, as shown in Figure 2–4. As with the Arrhenius plot, one can extrapolate from high temperature to room temperature, assuming that all factors remain constant, which is not always the case, although this is the basis of ASLT and the magic number.

The Q_{10} is thus expressed by the equation

$$Q_{10} = \exp(10b) = \text{Shelf Life at } T \, / \, \text{Shelf Life at } (T + 10)$$

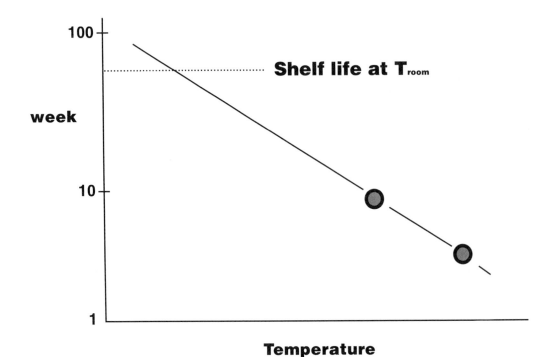

Figure 2–4 Typical shelf life plot with two temperatures, showing extrapolation to room temperature. *Source:* Copyright © 1999, Theodore Labuza.

The Q_{10} and activation energy are related as shown in the following equation, where T is in Kelvin, or degrees Centigrade plus 273.15:

$$\ln(Q_{10}) = 10E_a/RT(T + 10)$$

The Q_{10}, being temperature dependent, holds constant over a narrow range of temperatures (generally a range of 15–25°), while E_a holds over much larger range; thus, for extrapolating data over large temperature ranges, the Arrhenius equation yields more accurate results. In any case, given the caveats about extrapolation, from the Arrhenius activation energy or the Q_{10} plots, shelf life can be predicted at any temperature.

Figure 2–5 a and b show examples of an Arrhenius plot and shelf life plot for actual components "X" from the food field. These plots are the basis for determining the magic number. It should be made clear that there is no universal number because the activation energy and Q_{10} will vary for each reaction, so at least two temperatures are needed to determine a magic number. Even if you know the Q_{10} range, which is about 2 to 10 for many typical reactions, the extrapolation will give an enormously large range for prediction of potential shelf life, as shown in Figure 2–6. The shelf life of 3 weeks found at 43°C projects to a range of 12 weeks (Q_{10} = 2, magic number = 2 × 2 = 4) to almost 4 years (Q_{10} = 10, magic number = 10 × 10 = 100) at 23°C. Obviously, one needs a minimum of two temperatures to refine the prediction with the correct Q_{10}. In any case, in the ANPR for dietary supplements, FDA stated that one can adjust the ASLT-determined expiration date after gathering information on actual losses during real-time distribution.

Aspartame pH 7 Arrhenius plot

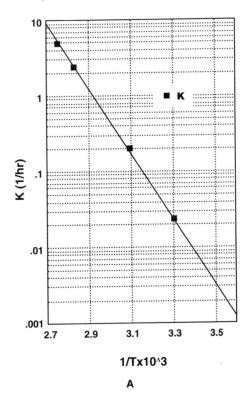

A

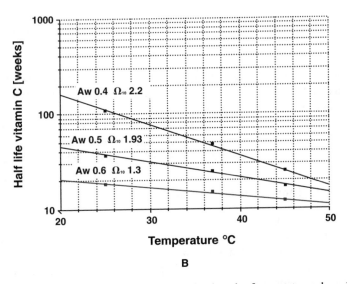

B

Figure 2–5 Influence of temperature on kinetics. **A,** Arrhenius plot for aspartame degradation in a dry model system at pH 7. **B,** Shelf life plot in time to 25% loss for ascorbic acid in a ready-to-eat cereal at three different moisture contents (water activities). *Source:* Copyright © 1999, Theodore Labuza.

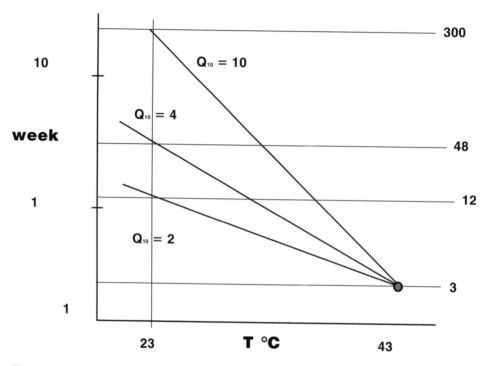

Figure 2–6 Extrapolation of shelf life found at 43°C (ie, using one temperature) to shelf life at room temperature (23°C). The magic number for extrapolation to 23°C is the Q_{10} squared. *Source:* Copyright © 1999, Theodore Labuza.

It also must be made clear that the effect of temperature is critical in processing. Most herb/botanicals are generally dried under mild conditions so that the functional factor "X" is not destroyed. If it is added to a drink that is thermally processed or a cereal dough that undergoes extrusion or baking to form a bar, there are serious concerns about the loss of "X." It would seem that these losses would be much greater than they would be in the formation of a caplet or tablet. No data exist concerning the losses that can occur in processing for the critical substance "X" in most herbs/botanicals.

Effect of Moisture Content and Water Activity on Shelf Life

In the mid-1970s, water activity came to the forefront as a major factor in understanding the control of the deterioration of reduced-moisture and dry foods, drugs, and biological systems.[69,70] It was found that the general modes of deterioration, namely physical and physicochemical modifications, microbiological growth, and both aqueous- and lipid-phase chemical reactions, were all influenced by the thermodynamic availability of water (termed a_w or water activity), as well as by the total moisture content of the system. It is the difference in the chemical potential of water (micrograms in joules per mole) between two systems that results in moisture exchange, and above a certain chemical potential, as related to the a_w of a system, there is enough water present to result in physical and chemical reactions.

Water activity, or the equilibrium relative humidity of a system, is defined as

$$a_w = p/p_o = \%ERH$$

where p is the vapor pressure of water in equilibrium with the system and p_o is the saturation vapor pressure of pure water at the same temperature.

Figure 2–7 shows a plot of the moisture content versus the water activity in equilibrium with the product. This sigmoid-shaped curve results from different factors related to the binding of water. To maximize shelf life of a dried material, it is imperative that an isotherm be determined. The methods for determining isotherms are reviewed by Labuza.[71]

The physical structure of a food or biological product, important from both functional and sensory standpoints, is often altered by changes in water activity due to moisture gain or loss. For example, the caking of powders, which many dietary supplements are, is attributed to the amorphous-crystalline state transfer of sugars and oligosaccharides that occurs as water activity increases above the glass transition point. This is close to an a_w of 0.3 to 0.4 at room temperature.[72,73] This caking inter-

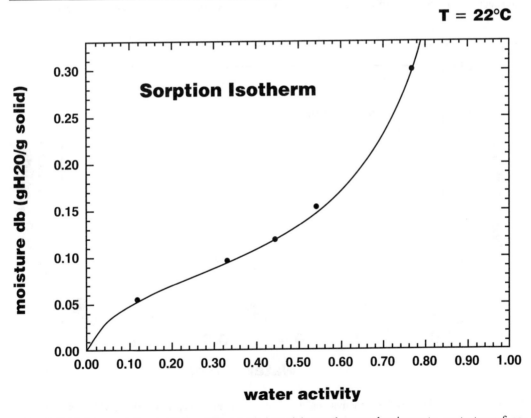

Figure 2–7 Moisture sorption isotherm of freeze-dried specialty mushrooms showing water content as a function of water activity at 22°C. *Source:* Copyright © 1999, Theodore Labuza.

feres with the powder's ability to dissolve or be free flowing, and phase transitions can lead to volatile loss or oxidation of encapsulated lipids. The desirable crispiness of crackers, dry snack products such as potato chips, and breakfast cereals is lost if a moisture gain results in a water activity above 0.40 to 0.45.[74] This is important to the stability of dietary supplement bars. Conversely, bars filled with higher moisture soft centers may harden due to the loss of water to the surrounding baked dough. The hardness is associated with a decreased water activity, usually below 0.55. Hyman and Labuza[75] have reviewed the potential solutions to prevent moisture transfer in multidomain systems. Finally, as a_w increases, the permeability of packaging films to oxygen and water vapor increases, due to swelling in the rubbery state. This allows greater transport of these gases across the film barrier, shortening the shelf life.

Water activity has been shown to influence the kinetics of many chemical reactions, such as the loss of vitamin C in the dry state. The rates of chemical reactions generally increase with increasing water activity up to a maximum and then decrease again as moisture or water activity is increased, as depicted in Figure 2–8.[70] Interestingly, if the active ingredient "X" is oxidized by oxygen through a free-radical mechanism, the reaction rate can also increase below the monolayer, showing the importance of water activity in designing long–shelf life dry products.

Generally, the minimum reaction rate for aqueous-phase reactions is found at the monolayer moisture content, which is the point where theoretically all polar groups have adsorbed one molecule of water vapor. It is at or just above this point that an aqueous phase just begins to form such that chemical species can dissolve, diffuse, and react. From moisture sorption isotherm data, the monolayer can

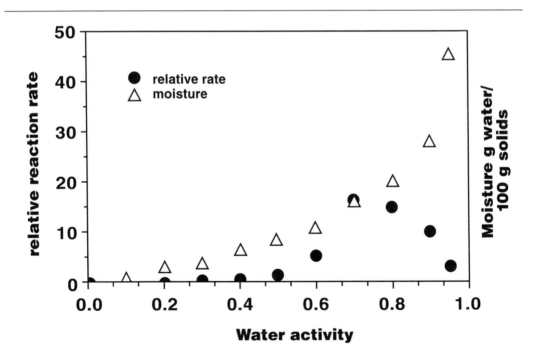

Figure 2–8 The influence of water activity on chemical reaction rates such as ascorbic acid loss ● in the aqueous phase overlaid on a sorption isotherm of moisture versus water activity ▲. *Source:* Copyright © 1999, Theodore Labuza.

be determined from either the Brunauer-Emmet-Teller (BET) equation, using data below a water activity of 0.5, or the Guggenheim-Anderson-deBohr (GAB) equation, which is based on the thermodynamics of multilayer moisture adsorption and uses all the data.[52] The GAB equation can be written as

$$m = m_o k_b C a_W / [(1 - k_b a_W)(1 - k_b a_W W + k_b C a_W)]$$

where m is the moisture content, a_W is the water activity, m_o is the monolayer, k_b is a multilayer factor, and C is a heat constant. Either a polynomial solution or a nonlinear regression is used to solve for the constants in this equation so as to determine the monolayer value. This is currently the most accepted equation to generate an isotherm. It can generate the whole curve from five data points at different water activities. It is thus important from a shelf life-testing standpoint to evaluate the moisture sorption isotherm of each system and then to calculate the monolayer water activity to determine the optimum point for stability from either the BET equation or the GAB equation. From this, a window for initial moisture content can be set to be used as a production goal to maximize shelf life.

There are many examples of the influence of water activity above the monolayer value on the rates of chemical reactions in the dry state that lead to loss of shelf life or biological activity.[76] For example, the rate of ascorbic acid degradation increases dramatically as water activity increases from the dry state.[77] Similarly, the rate of glucose utilization during nonenzymatic browning increases as water activity increases.[66] Recently, Davies et al[78] showed that the active ingredient in soy protein from a dietary supplement standpoint, genestein, reacts through the Maillard reaction and thus will be lost in storage unless water activity is controlled. Below the monolayer, the rate of lipid oxidation increases again; thus, for systems that also contain unsaturated lipids, such as the membranes of all biological systems, the shelf life is at a maximum near the monolayer.[58] Since many of the reactions in the aqueous phase are acid-base catalyzed, the amount of water and the availability of H^+ and OH^- ions in the reduced moisture state will also significantly influence reaction rates.[79–81]

One of the most studied reactions as influenced by both temperature and water activity is that of the acid-base–catalyzed degradation of aspartame in both liquid- and dry-state systems.[64,65,82] The influence of the system water activity state on aspartame degradation can be used to illustrate many shelf life–related chemical phenomena causing the degradation of "X" in reduced-moisture systems. Figure 2–9 shows the simple case where water activity is the only factor that is varied to determine the effect on aspartame degradation in the dry state. An increase in water activity from the dry state increases the aspartame degradation rate. It also should be noted that the reaction follows first order, since a plot of percentage remaining versus time follows a straight line through almost two half-lives (ie, <25% remaining).

As was shown in Figure 2–8, the rates of reaction increase above the monolayer; then, above about an a_W of 0.6 to 7, reaction rates fall again. In general, the pattern of a linear rise and then fall of reaction rate with water activity can be modeled on a semilog plot as shown in Figure 2–10. If the expiration date or shelf life (eg, time to 20% loss of "X") is plotted on a semilog graph paper, two straight lines also occur with the opposite slopes.[83] From this, the influence of water activity can be calculated, where Q_A is the shelf life time at one water activity divided by the shelf life at a water activity of 0.1 units higher. The Q_A for aspartame degradation ranges from 1.3 to 2 depending on temperature, which means that for an increase in water activity of 0.1, the rate of aspartame degradation increases by 30% to 100%. For most aqueous-phase chemical reactions in the dry state from $a_W = 0$ to 0.7, the Q_A is between 1.2 to 3. This emphasizes the need for maintaining the water activity as close as possible to the monolayer, since an increase in water activity of 0.1 units would decrease the shelf life by up to three times. This also suggests that any water activity–sensitive product must be packaged in a very moisture-impermeable film. Note that most drugs that are also moisture sensitive are packaged in plastic or glass bottles and that many have an added desiccant sachet to help maintain a low a_W in

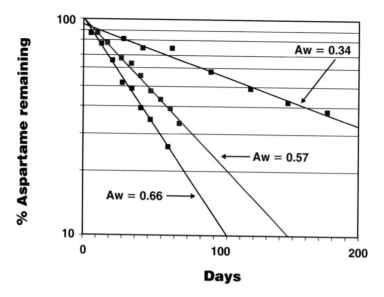

Figure 2–9 Degradation of aspartame in a dry model system as a function of water activity at pH 5 and 30°C. *Source:* Copyright © 1999, Theodore Labuza.

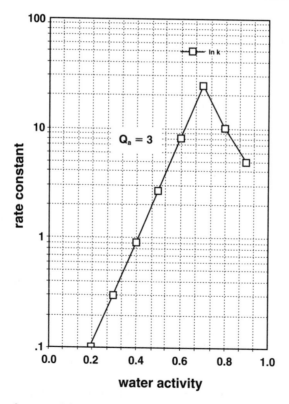

Figure 2–10 Influence of water activity on shelf life—Q_A plot. *Source:* Copyright © 1999, Theodore Labuza.

the package. Labuza et al,[84] Labuza and Contreras-Medellin,[85] and Taoukis et al[86] have reviewed the kinetics of reactions as a function of the moisture transfer rate through packaging materials.

Figure 2–11 shows the effect of the initial pH before drying on the stability of aspartame in a system during storage at a water activity of 0.34.[80] It is clear that even at low water activities, the initial pH before drying is another important factor that influences acid-base–catalyzed degradation rates in the dry state. This suggests that even in dry systems the influence of the isoelectric point of the reacting species is critical. This phenomenon should be important in controlling the stability of many dry food, drug, and biological systems, as well as dietary supplements. Bell and Labuza[81] have shown that the actual pH in the condensed water phase of a dry system can decrease by up to 1.5 pH units during drying.

As a last point, going back to Figure 2–5b, the temperature sensitivity (E_a or Q_{10}) of reactions is also a function of water activity, with the product being more temperature sensitive (higher E_a or Q_{10}) at a lower moisture content. Thus, using an ASLT condition with both high temperature and high humidity could lead to a serious underprediction of actual shelf life. The minimum number of conditions to be able to predict to low temperature/low water activity would be four; that is, there is no empirical magic number.

Oxygen Concentration

Many fresh foods, as well as dry and semimoist foods packaged in oxygen-permeable packaging, are sensitive to oxygen, losing shelf life faster as the amount of oxygen increases. An example of this is the oxidation of fat in potato chips or the lipids of fresh roasted ground (FRG) coffee reacting with

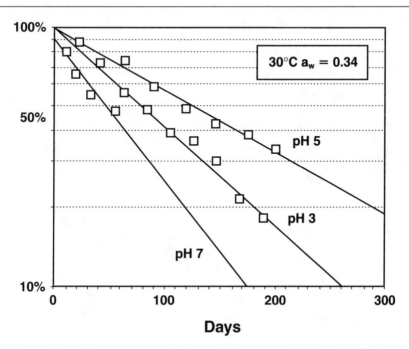

Figure 2–11 Degradation of aspartame as a function of pH at an a_w of 0.34 and 30°C. *Source:* Copyright © 1999, Theodore Labuza.

oxygen permeating through the package film. Above a certain critical oxygen level, an increase in oxygen does not change the rate of degradation, as shown for FRG coffee in Figure 2–12. A hyperbolic function, as derived from studies of lipid oxidation,[58] can be used to express the reaction rate as a function of oxygen concentration, where

$$r = -d[O_2]/dt = k_1\{([O_2]/k_2) + [O_2]\}$$

This can be rewritten as

$$1/r = (1/k_1)\{1 + (k_2/[O_2])\}$$

where k_1 and k_2 are rate constants. By plotting $1/r$ versus $1/[O_2]$, one obtains a linear region where the slope is k_2/k_1 and the intercept is $1/k_1$. When $[O_2]$ is very small, the rate is linear with oxygen concentration; when $[O_2]$ is large, the rate is somewhat independent of oxygen concentration, as shown in Figure 2–12, but the magnitude depends on the temperature. Both of these situations could occur in the same food product depending on temperature, surface-to-volume ratio, oxygen pressure, extent of reaction, and packaging material. Figure 2–12 indicates the combined effect of both oxygen pressure and temperature on the oxidation rate of FRG coffee. It is obvious that lowering the oxygen can increase shelf life significantly. For FRG coffee, the shelf life in air at room temperature is 10 days, while at 0.5% oxygen it is more than a year. It is also clear from Figure 2–12 that controlling temperature in distribution has very little effect if the packaging system can maintain a low oxygen level. Ragnarrson and Labuza[87] have reviewed the critical issues related to shelf life testing of oxidizable products.

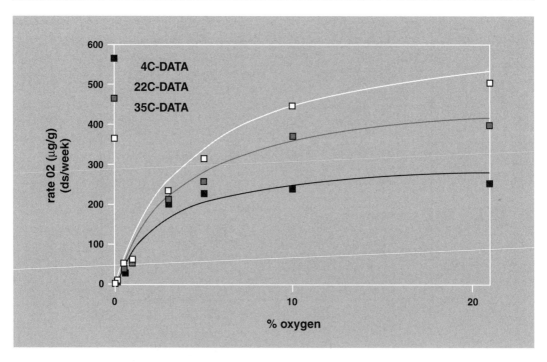

Figure 2–12 Rate of oxidation of fresh roasted ground coffee as a function of temperature and oxygen level. *Source:* Copyright © 1999, Theodore Labuza.

For highly oxidizable ingredients, one method to increase shelf life is to encapsulate them during spray drying. Figure 2–13 shows the loss of carotenoids from carrots prepared by encapsulating in a maltodextrin matrix and stored at several temperatures and two relative humidities. As shown, there is an initial fast rate in each system that then decreases dramatically for the rest of the storage time after some critical time.[88] During encapsulation of an oxygen-sensitive ingredient that is lipid based, some of the material is not completely embedded in the matrix and sits especially on the particle surface or in large pores accessible to oxygen. It is this surface material that oxidizes faster, so better methods of encapsulation are needed. It also should be noted in Figure 2–13 that there was no significant difference in the product at an a_w of 0.11 versus 0.33 (2% vs 5% moisture) at any temperature. This suggests that the minimum rate near the monolayer (0.24 in this case) is over a fairly wide range, giving the processor a generous window for moisture content in production.

Environmental Fluctuations

In real life, the distribution of a product through the marketing channel is not done at a constant temperature but rather occurs at variable temperatures and humidities except in a few well-controlled instances. If a good impermeable package is used, the major problem is temperature fluctuations. Kamman and Labuza[89] have reviewed and tested the mathematical models for prediction of shelf life

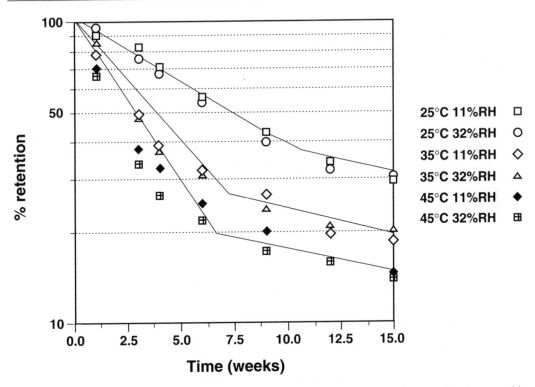

Figure 2–13 Effect of temperature and relative humidity on the degradation rate of encapsulated carotenoids. *Source:* Copyright © 1999, Theodore Labuza.

Figure 2–14 Change in appearance of the 3M Monitor Mark shelf life time-temperature indicator tag as a function of time. Courtesy of 3M Company.

under fluctuating temperature conditions. Time temperature indicators (TTIs) are small devices that can be placed on a product to indicate the expiration date of the product. TTIs provide a temperature-sensitive phenomenon that is accompanied by a visual change; the extent of change is theoretically correlated to a food or drug product's mode of deterioration, thereby integrating time-temperature exposure in the same manner as the theoretical models. To do this, the device must have the same activation energy as the food deterioration reaction and be able to be set with a visual color index at the end of shelf life, a run-out time.

The general types of TTIs currently available commercially are based on a diffusion or chemical reaction. The 3M Monitor Mark is based on a time-temperature–dependent diffusion of a polymer through a matrix. The measurable (or visual) response is the darkening of a white bar passing through a colored (gray) circle. When the bar is the same color as the circle, as shown in Figure 2–14, the end of shelf life has been reached. Currently, this is being used on hamburger by the CUB Foods chain in the Midwest. The Lifelines Freshness Monitor, a chemical reaction tag, is based on a solid-state polymerization of thinly coated colorless acetylenic monomer that polymerizes to a dark color. In this case, a smaller circle containing the reactant is in a colored circle, and when the inner circle matches the outer one in color, the end of shelf life has occurred. It is being employed on deli items in several food chains in Europe. Both these TTIs have great potential for use with dietary supplements. Experiments have been performed to establish the applicability for TTIs as shelf life monitors along with their kinetic parameters.[90–92]

CONCLUSIONS

Functional foods and dietary supplements take the manufacturer into the realms of drug regulations, food regulations, and new regulations such as the DSHEA. Following good manufacturing practices will be a necessity to ensure that if a structure-function claim is made on a product, several requirements are met, including

1. an assurance that the specific effective substance "X" that is the basis of the structure-function claim is present in the initial ingredient
2. an assurance that the ingredients and substances used to deliver "X" are safe
3. an assurance that that substance "X" will last through the processing procedures used (see Chapter 3)

4. an assurance that the product is labeled with an expiration date that must be based on the date at which the amount of "X" falls to a level where the structure-function claim can no longer be met

Adherence to these four critical factors will thus ensure that the product is not adulterated and is not misbranded, at least with respect to the claim.

REFERENCES

1. DeFelice S. Foundation for Innovation in Medicine. (http://www.fimdefelice.org)
2. Dietary Supplement Health and Education Act of 1994. FFDCA §201 (ff).
3. 62 *Federal Register* 49868 (1997).
4. Food, Drug, and Cosmetic Act 21 USC 301.
5. 21 CFR §110.
6. Nutritional Labeling and Education Act of 1990. Pub Law No. 101-535.
7. Food Quality Protection Act 7 of 1996. Pub Law No. 104-170.
8. FDA Modernization Act of 1997. Pub Law No. 105-115.
9. Administrative Procedures Act. Pub Law No. 79-404.
10. Compliance Policy Guides Manual. (http://vm.cfsan.fda.gov)
11. United States v Park, 421 US 658 (1975).
12. United States v Hata, 52 F2d 508 (1976).
13. United States v Starr, 535 F2d 512 (1976).
14. United States v 133 Cases of Tomato Paste, 22 F Supp 515 (1938).
15. United States v Lazere, 56 F Supp 30 (1944).
16. Olsen AR. Regulatory action criteria for filth and other extraneous matter III. Allergenic mites: an emerging food safety issue. *Reg Tox Pharm.* 1998;28:190–198.
17. Olsen AR. Regulatory action criteria for filth and other extraneous matter II. Review of flies and food borne enteric disease. *Reg Tox Pharm.* 1998;28:199–211.
18. Freedom of Information Act. 5 USC §552.
19. 21 CFR §110.110.
20. United States v Walsch, 331 US 432 (1937).
21. United States v Seabrook Int'l Foods, 501 F Supp 1086 (1980).
22. Burini G, Curini M, Marcotullio MC, Policichio. Determination of agaratine in cultivated mushrooms using high performance liquid chromatography with fluorometric determination. *Ital J Food Sci.*1999;11(1):39–46.
23. Sheppard SE, Gunz D, Schlatter C. Genotoxicity of agaratine in the lactyl mouse mutagen assay: evaluation of the health risk of mushroom consumption. *Food Chem Toxicol.* 1995;33:257.
24. Federal Insecticide, Fungicide and Rodenticide Act. 7 USC § 136 et seq.
25. Les v Reily, SCt 1361 (1993).
26. California v Browner, February 2, 1995, 9th DC CA.
27. EPA ban of methyl parathion. 64 *Federal Register* 57877-57881 (1999).
28. United States v Select Natural Herb Tea Civ #73-1370 RF; DC Cal; July 15, 1973.
29. Fmali Herb Inc v Heckler, 715 F2d 1385 (1982).
30. 53 *Federal Register* 16544 (1988).
31. 62 *Federal Register* 18938–18964 (1997).
32. Neff J. Benecol, Take Control cases crumble FDA regulatory walls. *Food Proc* Functional Foods Supp. July 1999 F-28–F-29.
33. See 62 *Federal Register* 39886–49882 (1997), which is the *Federal Register* notice and comment process for the new category.
34. 21 CFR §§109.6(b) (a) through 109.6(b) (c) & 209.6(b) (4).
35. §409(c) (3) (a).
36. §201 (s) (6).

37. DSHEA §402 (f) (1) (D).

38. 62 *Federal Register* 30678 (1997).

39. Pharmanex v Shalala, Case 2:97 CV 0262K DC Utah.

40. 62 *Federal Register* 5700–5709 (1997).

41. Blumenthal M, ed. *The Complete German Commission E Monographs: Therapeutic Guide to Herbal Medicines.* Boston, Mass: American Botanical Council; 1998.

42. Blumenthal M. Herbal products: current trends and future prospects. Presented at the annual meeting of the National IFT; June 25, 1999; Paper #4-1.

43. Lenoble R. Standardization of nutraceutical products. Presented at the annual meeting of the IFT; July 25, 1999; Paper #4-5.

44. New Good Housekeeping Institute study finds drastic discrepancy in potencies of popular herbal supplement [news release]. New York: Good Housekeeping Institute, Consumer Safety Symposium on Dietary Supplements and Herbs; March 3, 1998.

45. 21 CFR §113.

46. 21 CFR §114.

47. These comments are available on the Internet at http://www.usp.org.

48. 62 *Federal Register* 5704 (1997).

49. Roos Y, Karel M. Application of state diagrams to food processing and development. *Food Technol.* 1991;45(12):66–71.

50. Nelson K, Labuza TP. Arrhenius vs WLF kinetics in the rubbery and glassy state. *J Food Eng.* 1994;22:261–279.

51. Labuza TP, Riboh D. Theory and application of Arrhenius kinetics to the prediction of nutrient losses in foods. *Food Technol.* 1982;36(10):66–74.

52. Labuza TP. Applications of chemical kinetics to deterioration of foods. *J Chem Educ.* 1985;61:348–358.

53. Labuza TP, Schmidl MK. Accelerated shelf-life testing of foods. *Food Technol.* 1985;39:57–62, 64, 134.

54. Taoukis P, Labuza TP, Sagy S. Kinetics of food deterioration and shelf life prediction. In: Sagy S, Rotstein K, eds. *Food Engineering Handbook.* Denver, CO: CRC Press; 1997:363–405.

55. Benson SW. *The Foundations of Chemical Kinetics.* New York, NY: McGraw-Hill; 1960.

56. Labuza TP, Schmidl MK. Use of sensory data in the shelf life testing of foods: principles and graphical methods for evaluation. *Cereal Foods World.* 1988;33:193–206.

57. Labuza TP, Kamman JF. Reaction kinetics and accelerated tests simulation as a function of temperature. In: Saguy I, ed. *Computer-Aided Techniques in Food Technology.* New York, NY: Marcel Dekker; 1983:71–115.

58. Labuza TP. Kinetics of lipid oxidation in foods. *Crit Rev Food Technol.* 1971;2:355–405.

59. Labuza TP, Massaro S. Browning and amino acid loss in model total parenteral nutrition solutions. *J Food Sci.* 1990;55:821–826.

60. Mizrahi S, Labuza TP, Karel M. Computer-aided predictions of extent of browning in dehydrated cabbage. *J Food Sci.* 1970;35:799–803.

61. Warmbier HC, Schnickles RA, Labuza TP. Effect of glycerol on nonenzymatic browning in a solid intermediate moisture model food system. *J Food Sci.* 1976;41:528–531.

62. Connor KA, Amidon AL, Stella VJ. *Chemical Stability of Pharmaceuticals.* 2nd ed. New York, NY: John Wiley; 1986.

63. Prudel M, Davidkova E. Stability of α-L-aspartyl-L-phenylalanine methyl ester hydrochloride in aqueous solutions. *Die Nahrung.* 1981;25:193–199.

64. Bell LN, Labuza TP. Aspartame degradation kinetics as affected by pH in intermediate and low moisture food systems. *J Food Sci.* 1991;56:17–20.

65. Bell LN, Labuza TP. Aspartame degradation in limited water conditions. In: Levine H, Slade L, eds. *Water Relationships in Foods.* New York, NY: Plenum Press; 1991:337–349.

66. Kamman JF, Labuza TP. A comparison of the effect of oil versus plasticized vegetable shortening on rates of glucose utilization in nonenzymatic browning. *J Food Proc Preserv.* 1985;9:217–222.

67. Saguy I, Karel M. Modeling of quality deterioration during food processing and storage. *Food Technol.* 1980;34:78–85.

68. Labuza TP, Szybist L. Playing the open dating game. *Food Technol.* 1999;53(7):70–85.

69. Duckworth RB, ed. *Water Relations in Foods.* London, England: Academic Press; 1975.

70. Labuza TP. Sorption phenomena in foods. In: Rha C, ed. *Theory, Determination and Control of Physical Properties of Foods.* Dordrecht, the Netherlands: D. Reidel; 1975:197–219.

71. Labuza TP. *Sorption Isotherms: Practical Use and Measurement.* St. Paul, MN: AACC Press; 1984.

72. Saltmarch M, Labuza TP. Influence of relative humidity on the physicochemical state of lactose in spray-dried sweet whey powders. *J Food Sci.* 1980;45:1231–1236, 1242.

73. Chuy L, Labuza TP. Caking and stickiness of dairy based food powders related to glass transition. *J Food Sci.* 1994;59:43–46.

74. Katz EE, Labuza TP. The effect of water activity on the sensory crispness and mechanical deformation of snack food products. *J Food Sci.* 1981;46:403–409.

75. Hyman C, Labuza TP. Moisture migration in multidomain systems. *Trends Food Sci Technol.* 1998;9:47–55.

76. Labuza TP. The effect of water activity on reaction kinetics of food deterioration. *Food Technol.* 1980;34(1):36–41, 59.

77. Lee S, Labuza TP. Destruction of ascorbic acid as a function of water activity. *J Food Sci.* 1975;40:370–373.

78. Davies CGA, Netto F, Glassenap N, Gallaher CM, Labuza TP, Gallaher DD. Indication of the Maillard reaction products during storage of protein isolates. *J Agric Food Chem.* 1998;46:2485–2489.

79. Bell LN, Labuza TP. Potential pH implications in the dry state. *Cryo Lett.* 1991;13:335–344.

80. Bell LN, Labuza TP. Compositional influence on the pH of reduced moisture systems. *J Food Sci.* 1992;57:732–734.

81. Bell LN, Labuza TP. Evaluation and comparison of simple methods for pH measurement of reduced moisture solid systems. *J Food Proc Preserv.* 1992;16:289–297.

82. Bell L, Labuza TP. The influence of pH on reactions in the reduced moisture state. *J Food Eng.* 1994;22:281–302.

83. Loncin M, Bimbenet JJ, Lenges J. Influences of activity of water on the spoilage of foodstuffs. *J Food Technol.* 1968;3:131–142.

84. Labuza TP, Mizrahi S, Karel M. Mathematical models of flexible film packaging of foods for storage. *Trans ASAE.* 1972;15:150–155.

85. Labuza TP, Contreras-Medellin R. Prediction of moisture protection requirements for foods. *Cereal Foods World.* 1981;26:335–343.

86. Taoukis PS, ElMeskine A, Labuza TP. Moisture transfer and shelf life of packaged foods. In: Hotchkiss J, ed. *Food Packaging Interactions.* ACS Press; 1998:244–261.

87. Ragnarsson JO, Labuza TP. Accelerated shelf-life testing for oxidative rancidity in foods: a review. *Food Chem.* 1977;1:291–308.

88. DeSobry SA, Netto FM, Labuza TP. Comparison of spray drying, drum drying and freeze drying for encapsulation and stability of β-carotene. *J Food Sci.* 1997;62:1158–1162.

89. Kamman JF, Labuza TP. Kinetics of thiamin and riboflavin loss in pasta as a function of constant and variable storage conditions. *J Food Sci.* 1981;46:1457–1461.

90. Taoukis PS, Labuza TP. Applicability of time-temperature indicators as shelf life monitors of food products. *J Food Sci.* 1989;54:783–788.

91. Taoukis PS, Labuza TP. Reliability of time-temperature indicators as food quality monitors under nonisothermal conditions. *J Food Sci.* 1989;54:789–792.

92. Labuza TP, Fu B, Taoukis PS. Time temperature indicators. *Food Technol.* 1991;45(10):70–82.

CHAPTER 3

Thermal and Nonthermal Preservation Methods

L. Lotte Dock and John D. Floros

In the past, the only purpose of food preservation was to achieve safety and extend the shelf life of food products. In the last two decades, however, more attention has been given to maintaining the freshlike characteristics of certain foods; retaining important organoleptic attributes such as color, texture, taste, and appearance; and improving foods' functional properties. With the recent introduction of functional foods and nutraceuticals in the marketplace, the retention of nutrients such as vitamins, the bioavailability of functional food components such as antioxidants, and the preservation of other beneficial food compounds have become increasingly important.

To obtain high retention of heat-sensitive nutrients and functional components, reduced thermal processes or nonthermal processes are being sought out. Terms such as *minimally processed* and *freshlike* have become commonplace in food industry and consumer terminology in the 1990s.[1] Manvell[2] defined *minimal processing* as the least possible treatment to achieve a purpose, such as distributing the food safely under specified storage conditions. Under this definition, minimal processing does not specify the quality to be retained. Therefore, a more appropriate term that covers both safety and quality would be *optimal processing*. Optimal processing would thus be defined as the process needed to obtain safe products with the highest possible quality.

This chapter discusses thermal and nonthermal preservation methods that are appropriate for processing functional foods.

THERMAL PRESERVATION METHODS

Conventional Thermal Processing

Conventional thermal processes, such as cooking, blanching, pasteurization, and sterilization have different objectives. Cooking aims at producing a more palatable food, blanching is primarily aimed at destroying enzymes, pasteurization is aimed at destroying some but not all vegetative cells, and sterilization is aimed at destroying all microorganisms.[3] Sterilization used for food processing prevents growth of microorganisms at some expected storage conditions, so the term *commercial sterility* should be used.[3] For example, vegetables meant for frozen storage are only blanched, and conventionally pasteurized milk, yogurt, and other functional foods containing pre- and probiotics must be refrigerated. Thermal treatments needed to produce commercially sterile and shelf-stable foods, such as infant formulas and some weight control products (eg, Slim Fast and Nestle's Sweet Success), depend on product characteristics (eg, pH, viscosity, and water activity), product type, heat resistance and initial load of target microorganism(s), and heat transfer characteristics of the food.

Target Organisms

The amount of heat needed for commercial sterilization depends on the organisms that survive and grow in the product. Most sterile containers are hermetically sealed and contain low oxygen concentrations that inhibit the growth of aerobic microorganisms. Hence, the microorganisms of concern are the anaerobes or facultative anaerobes. Spore-forming organisms are of particular concern because spores are more heat resistant than vegetative cells. Consequently, anaerobic spore-forming pathogens that produce lethal toxins set the constraints needed for heat sterilization, and in many cases, the organism of concern is *Clostridium botulinum*. Products with a pH higher than 4.6 (low acid) must receive a treatment resulting in a 12-log reduction of *C. botulinum*. At pH 4.6 or lower, *C. botulinum* does not grow, and for these products microbial destruction is based on other target microorganisms. Many other organisms are more heat resistant than *C. botulinum* and are used to calculate the extent of thermal treatments. For example, another *Clostridium* species, *C. sporogenes* which is nontoxic, is used to determine safe thermal processes for low-acid foods.[3] Another anaerobic spore former, *Bacillus stearothermophilus*, is more heat resistant than *C. botulinum*, but its optimal growth temperature is about 49 to 55°C, and it does not often grow at temperatures below 37°C.[3] Therefore, a product containing spores of this organism should be considered commercially sterile when stored at temperatures below 43°C (110°F).

Processing Parameters

Thermal destruction of microorganisms, most nutrients, enzymes, and other quality factors such as flavor, color, and texture obeys first-order reaction kinetics.[3] This means that the reaction rate depends on the concentration of the substance of interest. Two parameters are important in thermal destruction kinetics: (1) the destruction rate at a given temperature and (2) the relationship of destruction rate to temperature. The first parameter is known as the *decimal reduction time* (D) and is defined as the time it takes to reduce the concentration by one log cycle, or, in other words, the time to produce a 90% reduction in concentration. In depictions of concentration versus time, D is determined by the negative reciprocal of the slope (Figure 3–1A).

The second parameter is known as the z value and describes the relationship of D to temperature. When D values are depicted versus temperature, z is the negative reciprocal of the slope (Figure 3–1B). In other

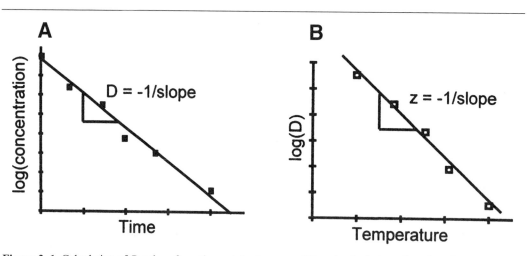

Figure 3–1 Calculation of D values from thermal death curves (**A**) and calculation of z values from D-versus-temperature plot (**B**).

words, z is the change in temperature needed to change D by a factor of 10. If z is small, the microbe or reaction is sensitive to temperature change, and only a small temperature increase is needed to decrease D by 10-fold. On the other hand, if z is big, the temperature change needed to change D 10-fold is big, and the microbe or reaction is less influenced by temperature. Sometimes, the total time to achieve microbial inactivation is given by the so-called F value or thermal death time. The most commonly used F value is the F_{250}, where 250 denotes the processing temperature in degrees Fahrenheit. For low-acid foods, $F = 12D$ corresponding to a 12-log reduction of the target microorganism(s), whereas for high-acid foods $F = 6D$. Another method to measure substance deterioration is the Arrhenius equation:

$$k = A \cdot e^{\frac{-E_a}{RT}}$$

where k is the reaction rate ($\min^{-1}$), A is a constant ($\min^{-1}$), E_a is the activation energy (cal/mol), R is the universal gas constant (1.987 cal/mol $\times$ K), and T is the absolute temperature (°K). E_a describes the energy needed to get 1 mol of substance into the active state. The Arrhenius equation is often used in reaction kinetics theory, while the thermal death time method is used for microbial death time studies. However, using D and z is the basis of thermal process calculations and is the most common method used in industry.[3]

Factors Influencing Thermal Resistance of Microorganisms

Factors affecting the thermal resistance of microorganisms can be divided into three major categories: (1) inherent resistance, (2) environmental factors during formation of cells or spores, and (3) environmental factors during thermal treatment.[3] The inherent heat resistance of different strains of the same bacteria can vary grossly. For example, the aciduric strain of *Escherichia coli*, *E. coli* O157:H7, has increased heat resistance in low-pH foods as compared to other nonaciduric strains of *E. coli*. Several factors influence the heat resistance of cells and spores during their growth. For example, spores produced at higher temperatures are more heat resistant than spores produced at lower temperatures.[3] Other factors that may affect heat resistance are ionic strength, organic compounds, lipids, and age of cells. For the latter, the most common belief is that heat resistance is lower during the logarithmic phase and then increases during the stationary phase. Environmental factors that affect heat resistance during thermal treatment include pH, buffers, ionic environment, and composition of the medium. An important factor to consider at all times is water activity, a_w. Generally, spores and cells cannot grow at a_w lower than 0.85, so in dry and semidry foods there is little concern for bacterial growth. However, molds and yeast grow at low a_w and become major spoilage and pathogenic (ie, toxins produced by molds) concerns in dry and semidry foods.

Thermal Degradation of Food Substances

Destruction of pathogenic microorganisms to achieve a high level of food safety is the main objective of thermal processing. However, retention of quality attributes such as flavor, color, texture, and nutritional value must also be considered. Thermal processing may adversely affect some food constituents. In cases where retention of specific nutrients and functional food components is required, their thermal destruction kinetics must be determined, and their dependence on temperature and other environmental factors must be calculated. In certain cases, thermal processing produces positive effects on functional ingredients. For example, lycopene, a carotenoid found primarily in tomatoes and suspected to lower the risk of several types of cancer, becomes more bioavailable when tomatoes are processed. Therefore, lycopene absorption increases 2.5 times in tomato paste as compared to the absorption from raw tomatoes.[4] Similarly, heating blueberries by a mild steam treatment increases the levels of available antioxidants such as anthocyanins.[5] Anthocyanins are flavonoids, functional substances found in many fruits, specific to blueberries. In another case, processed wheat bran showed an enhanced ability to reduce the development of cancer-related cells as compared to nonprocessed bran.[5] Last, phytochemicals present in rice were more active in rice bran oil rather than in rice bran, implying that processing improves the health-promoting action of functional ingredients such as phytochemicals, probably by increasing their bioavailability.[5]

Denaturation of Proteins. Food proteins are denatured by mild heat, 60 to 90°C, resulting in insolubility and decreased functional properties. However, protein denaturation also improves digestibility and biological availability of essential amino acids. For example, several raw plant proteins exhibit poor digestibility, but moderate heating improves their digestibility without toxin production.[6] In other cases, plant proteins contain antinutritional factors. For example, trypsin and chymotrypsin inhibitors that are present in plant protein impair efficient digestion. Similarly, lectins agglutinate red blood cells, impair protein digestion, and cause malabsorption of other nutrients.[6] Fortunately, both protease inhibitors and lectins are thermolabile. Moderate heat also inactivates proteinaceous toxins such as the *Clostridium botulinum* toxin, which is deactivated at 100°C.[6]

Enzyme Inactivation. Most oxidative and hydrolytic enzymes (ie, proteases, lipases, lipoxygenases, amylases, and polyphenoloxidase) are inactivated by moderate heat treatments. The treatment, termed *blanching*, consists of heating the product to about 70 to 105°C and holding it there for several minutes. D and z values can describe inactivation of enzymes, but often the activation energy method is used (described in more detail below in the section "Aseptic Processing"). The D and z values for enzyme inactivation are of the same order of magnitude as those for bacteria. However, some heat-resistant isozymes exist, which may cause off flavors and colors.[3] The target is usually the most heat-stable enzyme that causes quality deterioration during storage. In most cases, this means either peroxidase, lipoxygenase, or catalase. Peroxidase is the most frequently used because its activity is easy to measure and it is the most heat stable. However, peroxidase is not directly involved in quality deterioration of unblanched products. On the contrary, the quality of blanched and frozen products is better if some peroxidase activity remains.[7] Therefore, the best quality products may be obtained if some peroxidase activity remains, the amount of which depends on the product.[7] To determine the optimal point at which peroxidase activity should remain in the product, all possible factors (flavor, color, texture, etc) must be considered at once, and an optimization should be performed. Studies on green beans showed that methods detecting inactivation of catalase, lipoxygenase, and pectin methylesterase, rather than peroxidase, were better ways to prevent quality deterioration.[8] Similar results were found for green peas.[9]

Vitamin Losses. Vitamin losses are not caused solely by processing. Instead, loss of vitamins starts from the time of harvest for most foods,[10] and it is important to consider the loss of specific vitamins. For example, if a food is a major source of a specific vitamin, or a claim is made, then the food processor must maintain and prove the specific vitamin's bioavailability and concentration. Inactivation of enzymes often has a stabilizing effect on vitamins during storage. Heat is not the primary factor of vitamin loss. Instead, losses occur primarily by oxidation and leaching (aqueous extraction).[10] The extent of thermally induced vitamin loss depends on the chemical nature of the food (eg, pH, a_w, dissolved oxygen) and the likelihood of leaching.[10] High-temperature-short-time (HTST) processing is very effective in increasing retention of heat-labile vitamins and other nutrients. The reason is that the z values for nutrients and quality factors (eg, color, flavor) generally are larger than those for microorganisms and heat-labile enzymes.[10] Therefore, HTST has a greater effect on microorganisms than on most nutrients. For example, when the temperature is increased 10°C, the reduction in the number of microbial survivors is considerably larger compared to the degradative effect on colors[11] (Figure 3–2). Likewise, D values are much greater for nutrients than for microorganisms[11] (Figure 3–3). This allows thermal processes to be optimized for maximum quality retention while achieving the appropriate microbial reduction.

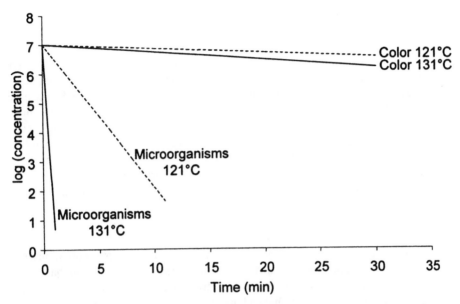

Figure 3–2 The effect of temperature on microbial survivors and color retention. *Source:* Based on D and Z values from P. Jelen, Review of Basic Technical Principles and Current Research in UHT Processing of Foods, *Canadian Institute of Food Science and Technology Journal*, Vol. 16, No. 3, pp. 159–166, © 1983.

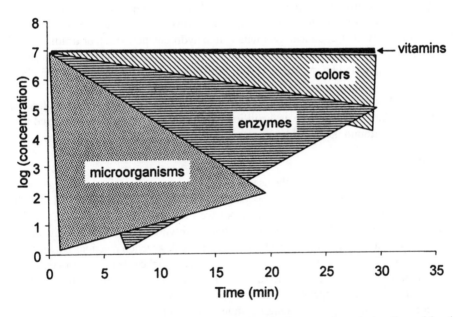

Figure 3–3 Destruction of vitamins, colors, enzymes, and microorganisms during thermal heating at 121°C. The shaded areas indicate the range of D values for each of the quality characteristics or inactivation of microorganisms. *Source:* Based on D and Z values from P. Jelen, Review of Basic Technical Principles and Current Research in UHT Processing of Foods, *Canadian Institute of Food Science and Technology Journal*, Vol. 16, No. 3, pp. 159–166, © 1983.

Minerals. Most research shows that heat, pH, oxidation, and other factors affecting organic nutrients do not affect minerals. These findings are usually related to ash values, which are largely unaffected by heating.[12] The only way to significantly decrease the mineral content of a food is by removing/losing it through leaching, milling, or other physical means.[13] The most extensive loss of minerals occurs during milling. For example, during milling of wheat, up to 86% of manganese, 76% of iron, 78% of zinc, and 68% of copper are lost.[14] Loss of calcium during draining of whey in low-pH cheeses has also been shown.[13] However, instead of measuring ash values, one study examined the bioavailability of minerals before and after processing and found some significant differences.[15] Some minerals may become unavailable or less available for digestion due to their interaction with conutrients or nonfood components during processing. Increased mineral bioavailability may also occur. For example, thermal destruction of binding ligands, such as phytates, increases mineral bioavailability.[12] Fortification after processing usually includes iron only,[13] but some cereal products are now fortified with several minerals and vitamins.

Aseptic Processing

Aseptic products are free of putrefying bacteria and packaged under aseptic conditions into a presterilized container.[16] The aseptic process is different from conventional retorting in that the product is continuously heated and then continuously cooled before packaging (Figure 3–4). This results in several advantages, including better nutritional and organoleptic quality retention.

Advantages of Aseptic Processing

In addition to better nutritional and quality retention, aseptic processing also has the advantage of using flexible, plastic/paper laminated containers because the product is cooled prior to packaging. On the contrary, during retorting, the container must be able to withstand sterilization temperatures because the product is filled prior to sterilization. In aseptic processing, the package is sterilized before filling by either heat (steam), chemical treatment (often hydrogen peroxide), or irradiation, while in other cases the heat of extrusion can be used to sterilize the packages as they are being formed. Even thin plastic films supported by a cardboard box, known as the bag-in-box system, can be used for aseptic products.[17]

Another advantage of aseptic processing technology is its ability to package products in large aseptic bulk storage tanks. Containers as large as 1 million gallons (about 4 million L) are used to store and ship aseptic products around the world.[18]

Quality Retention

An important part of aseptic processing is the use of HTST processing. Aseptic processing's effectiveness in retaining quality can be partly explained by the activation energy of important food ingredients. For example, oxidation and enzyme-catalyzed reactions have lower activation energies (E_a = 2 to 15 kcal/mol) as compared to other quality degradation reactions involving color, flavor, texture, and nutrients (E_a = 15 to 30 kcal/mol), while microbial and some enzymatic destruction reactions have the highest E_a at 50 to 100 kcal/mol.[19] The activation energy reflects the change in the rate of the reaction with temperature. High E_a indicates that reaction rates increase rapidly with temperature, while processes with low E_a are minimally influenced by temperature. Therefore, high-E_a reactions are favored during HTST processes, so microbial destruction proceeds rapidly, while degradation of quality factors is minimized.[19]

Vitamin Retention

Nutritional losses are minimized in aseptically processed food. In particular, vitamin retention is high in most HTST processes. For example, 91% of vitamin C was retained when it was aseptically

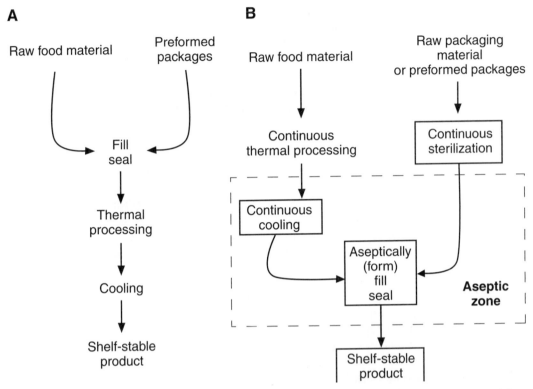

Figure 3–4 Comparing conventional (**A**) and aseptic (**B**) processing and packaging. *Source:* Reprinted from "Principles of Aseptic Processing & Packaging," J.V. Chambers & P.E. Nelson, editors, 1993, Figure 2, Chapter 6, page 118, with permission of the Food Processors Institute, Washington, DC, 202-393-0890.

processed, as compared to 59% retention achieved by retorting.[20] Likewise, 82% of thiamin was retained aseptically, compared to only 27% by retorting.[20] However, oxygen-sensitive vitamins, like C and A, may sometimes degrade faster in aseptically processed foods. Oxidative reactions are a problem in aseptic processing because in this type of processing, unlike retort processing, there is no vacuum in the container, and flexible aseptic containers allow some oxygen to permeate. The inclusion of a deaeration step in aseptic processing and the use of better oxygen barrier materials could further improve vitamin retention in aseptic products.

Another important factor is storage temperature. Most aseptic products are stored at ambient temperature, and some unwanted reactions may occur. The minimum reaction temperature for vitamin C degradation is about 23°C. Thus, to retain vitamins C and A, proper attention must be given to storage temperature and oxygen content of the container. Additionally, residual hydrogen peroxide may accelerate ascorbic acid degradation.[19]

Enzymatic Destruction

Enzyme denaturation or reversible denaturation sometimes occurs in aseptically processed foods. In some aseptic products, enzyme activity was undetectable immediately after processing, but slowly enzymes regained their activity throughout storage. For example, peroxidase (POD) activity was found to regenerate during storage of aseptically processed peas in starch.[19] Enzymes encapsulated in particulates, such as peas, may be somewhat protected within the tissue, resulting in the need for additional heating or preprocessing to properly inactivate them. An enzymatic problem associated with aseptically processed orange juice not from concentrate is settling and separation of pectin caused by pectinesterase. Therefore, aseptically processed orange juice is distributed and stored under refrigerated temperature and sold in nontransparent cartons, and the consumer is told to shake the carton before pouring. Orange juices from concentrate and pulp-free, not-from-concentrate juice (a newly marketed product from Tropicana, FL) do not exhibit this problem because little or no pulp is present. Amylase also exhibits high thermal resistance, and thinning of starch-based puddings is caused by bacterial α-amylase. Similarly, lipase and protease activities in ultrahigh-temperature (UHT) milk have been associated with growth of psychrophilic bacteria in raw milk prior to processing.[21] Therefore, it is important to ensure the quality of raw materials because although bacteria are killed during processing, the enzymes they produced may not be completely inactivated. In an aseptically processed enriched soymilk product marketed by Vitasoy USA, Inc (San Francisco, CA), unwanted beany flavor was minimized by using a UHT process. The process deactivated the enzymes, but it did not affect other important nutrients added to soymilk, such as calcium, riboflavin, and vitamins A, D, and B_{12}.[22]

Color Retention

Nonenzymatic browning, especially Maillard browning, is a problem in some aseptically processed foods stored at ambient temperatures. Maillard browning occurs only above 27°C, which is sometimes found in warm warehouses. Oxidative browning is also a problem in aseptic processing due to headspace oxygen and oxygen that permeates through the package.

Additionally, oxygen may cause lipid oxidation and changes in natural pigments. Although natural pigments like chlorophylls, anthocyanins, betalins, and carotenoids are retained better in HTST products than in retorted ones, they are susceptible to oxygenation. For example, carotenoids may lose color in oxygen-permeable containers due to oxidation.[23] Anthocyanins may react with residual hydrogen peroxide to form colorless end products.[24] To retain maximum amount of anthocyanins in cranberry juice, low temperature storage was recommended.[25] HTST treatment of spinach retained 68% of the chlorophyll content, but it quickly degraded during storage.[26] Finally, color loss in aseptically packaged papaya puree during the first month of storage was attributed to a decrease in carotenoids.[27]

Flavor Retention

The ultrahigh temperatures often used in aseptic processes cause some off flavors, such as the "cooked" flavor in UHT milk that arises from the release of free –SH groups cleaved from serum proteins.[28,29] Adsorption of D-limonene (a flavor component in orange juice) by polyolefin films used for packaging has also been demonstrated.[30,31] The problem may be eradicated by adding flavor or packaging in multilayer films, with a low-adsorbing material next to the juice. Some flavor components have been shown to bind proteins, resulting in suppression and/or release of flavor components.[19]

Texture Retention

Whole peeled tomatoes subjected to a HTST process exhibited firmer texture than tomatoes processed in a rotary pressure cooker.[32] However, the HTST product was not subjected to the abuse of a continuous-flow process.[19] Peas processed aseptically were significantly firmer than an identical product processed by conventional canning.[19] Further, additives such as Ca^{2+} salts can be added to aseptically processed products to improve their texture even more.[33]

Ohmic Heating

Ohmic heating is also called resistance or electroheating.[34] The process consists of passing an electric current through the product to be treated. The product acts like a resistor and heats up. The concept of this technology is not new, and several patents from the 19th century have described the use of electrical energy for heating flowable materials.[35] More recently, ohmic heating gained particular interest for processing particulate foods, especially in combination with aseptic processing. Ohmic heating is capable of fast, uniform heating and can heat particulates as fast or faster than the surrounding liquid medium.[35] It is this rapid and uniform heating of food throughout that makes ohmic heating different from traditional thermal processing, which depends on heat transfer from liquid to particulate surface and eventually into the particulate's center.[36,37] Furthermore, ohmic heating offers better uniformity of heating than microwave heating.[35]

The effect of ohmic heating on spores and other microorganisms is the same as that of regular heating. Therefore, the focus of research has been to validate the lethal kill of microorganisms in ohmic heating systems and to explore the factors that affect the heating rates of various constituents in a system.

Theory

The basic principle of ohmic heating is that an electrical current passing through an electrically conducting medium results in energy generation, leading to a rise in temperature similar to that from heating an electrical hotplate.[35] Most foods contain water with dissolved ionic salts or acids, so they possess relatively high electrical conductivity.[38] Furthermore, in most foods the conductivity of the liquid phase is higher than that of particulates.[37] The relationship between rate of energy generation (Q) and conductivity (σ) is

$$Q = \sigma E^2$$

where E is the electric field strength. Field strengths commonly used in ohmic heating do not cause microbial lethality, so only thermal inactivation of the microbial population takes place.[35] Conductivity increases with temperature, especially around 60°C, as a result of breakdown of cell wall material.[35]

Generation of Ohmic Heat

Ohmic heating of foods is obtained by placing the food between two electrodes equipped with an alternating current power supply (Figure 3–5). A typical continuous ohmic system consists of a col-

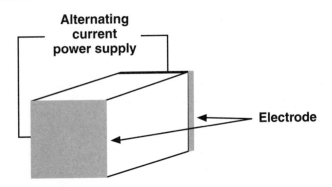

Figure 3–5 Principle of ohmic heating.

umn of heating units (electrode housings) that are connected by spacer tubes,[39] while the remainder of the system is similar to that of an aseptic processing system (Figure 3–6).

Factors Affecting Ohmic Heating of Particulate Foods

The effect of thermal treatment on particles and liquids in an ohmic heating system depends on several factors. Particle size, shape, orientation, concentration, density, conductivity, specific heat capacity, and liquid viscosity of the product affect the temperature increase.[35,36,38,40]

In conventional heating, heat transfer is mostly by conduction, which means the product is heated by the heated tube, resulting in rapid heating of the liquid, which in turn heats up the particles from the outside in. Due to slow heating of the particles' interior, overprocessing of the liquid phase is necessary to ensure that the center of all particles is sterilized.[39] Large particles pose a serious problem for conventional heating but not for ohmic heating. Large particles that block the flow of the liquid produce higher resistance, causing higher energy production, which in turn causes the particles to heat up faster than the liquid.[35,37,41] The same principle applies to the orientation and concentration of particles in the liquid. As the resistance increases (particles oriented perpendicular to electrodes, or high concentration of particles), particles are heated faster than the liquid.[40,42]

The size of particles is mostly restricted by filler capability and consumer response.[38] Particle density is mostly of concern when it differs extremely from the viscosity of the carrier medium. Very dense particles in a low-viscosity liquid tend to sink and cause excessive overprocessing, while light

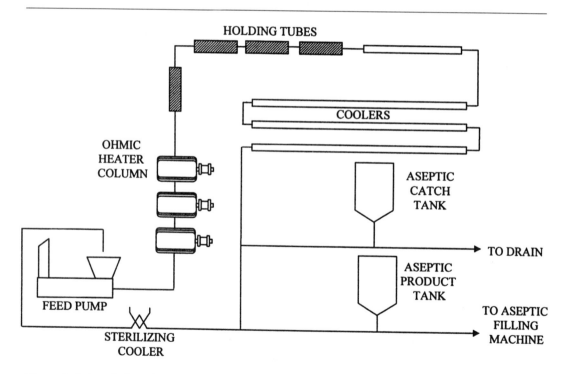

Figure 3–6 A typical continuous ohmic system. Heating units are connected by spacer tubes and produce a tower of ohmic heating units. *Source:* Reprinted with permission from P. Zoltai and P. Sweringen, Product Development Considerations for Ohmic Processing, *Food Technology*, Vol. 50, No. 5, pp. 263–266, © 1996, Institute of Food Technologists.

particles may float. Either sinking or floating makes residence time calculations virtually impossible. Changes of liquid viscosity during processing, due to thinning or starch gelation, or release of moisture from particles can cause similar problems.[38]

The conductivity of both particles and liquid is very important in ohmic heating and should be thoroughly investigated before an ohmic process is established. The difference between liquid and particle conductivity should be minimized to achieve even heating,[38] and presence of electrically nonconductive particles (eg, fats) should be studied. Such particles will tend to heat up mostly by conduction (heat applied from the liquid phase) and can result in cold spots in the center of particles.[40] Preprocess treatments of particulates and liquids can be done thermally, enzymatically, or chemically to obtain a certain conductivity—for example, by melting and expelling nonconductive fat from particles or soaking them in salt solutions.[38]

The actual temperature rise in particles and liquid depends on their specific heat capacities. Low-moisture content particles have a lower specific heat capacity than the high-moisture content liquid, which means that the particles will heat faster than the liquid,[38] especially if uniform conductivity is achieved through formulation.

Other critical processing variables are temperature and flow rate, which ultimately determine the residence time of product.[38] Calculation of residence time is similar to that for conventional heating and is based on the fastest-moving particle in the system.[40] Validation of the lethal effect of ohmic heating has been done with both chemical markers and bacterial spores.[36,37] Considerable efforts to model the flow of particulate foods in ohmic systems have been undertaken in order to identify any cold spots that may exist, especially in the processing of low-acid particulate foods.[43]

Applications in the Food Industry

Ohmic heating is currently being used to process low-acid particulate foods. A 300-kW commercial system produces shelf-stable low-acid foods such as beef stew with meat chunks and ravioli packaged in #10 cans.[44] This product is one of the first ohmically processed foods. Ohmic heating could also be used with success for tomato-based sauces, with a considerable increase in product quality.[39] Another potential application is heating of liquids that are difficult to heat by conventional technologies—for example, liquid eggs.[39]

In relation to functional foods, ohmic heating presents a good alternative to thermal heating because of the ability to heat particles as fast as or faster than liquid. Vitamins, proteins, and other functional ingredients that are sensitive to heat can be processed by ohmic heating in a more tender fashion, providing better retention of heat-sensitive ingredients than processing by traditional retorting.

Microwave Heating

Microwave energy has been used for heating food since the first microwave ovens were developed in 1947. Consumer microwave ovens were introduced by Tappan in 1955, and today they are present in more than 90% of American homes.[45] Microwaves kill microorganisms by generating heat. However, shorter time, up to fivefold less,[46] is needed to heat food by microwaves as compared to conventional heating. Therefore, microwaves yield foods with better texture, taste, and appearance.[46]

Theory and Method of Heating

Microwaves carry far too little energy to break any chemical bonds and are thus a form of nonionizing radiation.[47] Microwaves consist of a magnetic and an electrical field oriented perpendicular to each other. The electrical field is the primary cause of heating by promoting rotation of polar (dielectric) molecules and resulting in heat caused by molecular friction.[47]

Microwave heating kinetics is described by three constants, unique to different food systems. The dielectric constant (k') describes a material's capacity to store electrical energy. The dielectric loss (k'') describes a material's ability to dissipate electrical energy as heat. The loss tangent (δ) refers to the material's capacity to be penetrated by an electrical field and subsequently to dissipate that energy as heat and is expressed as the ratio of k''/k'.[47] The frequency used for commercial microwave ovens is 2450 MHz, but some industrial ovens use 915 MHz, producing wavelengths of 0.12 m and 0.33 m, respectively.[48] The frequency used influences the penetration ability of the waves, and the relationship between wavelength and penetration is given by

$$D = \frac{\lambda_o \sqrt{k'}}{2\pi k''}$$

where D is the penetration depth (in centimeters) at which 67% of the incident energy is absorbed and λ_o is the wavelength (in centimeters) in free space. It is this penetration of microwaves into food that results in shorter heating time as compared to conventional heating. In turn, shorter heating time is the main incentive for using microwave energy.

Microwave Equipment

A commercial microwave system (Figure 3–7) consists of a magnetron, a waveguiding system, and a treatment chamber. The magnetron generates microwaves by converting electrical power to microwave energy. The waveguide system focuses the energy directly on a small area for optimum efficiency, and as food is conveyed through the treatment chamber, it is exposed to microwaves. Magnetrons, which have a service life of 10,000 to 20,000 hours, are usually placed directly above and below the product in an alternating fashion. Due to the focused treatment, it is possible to give

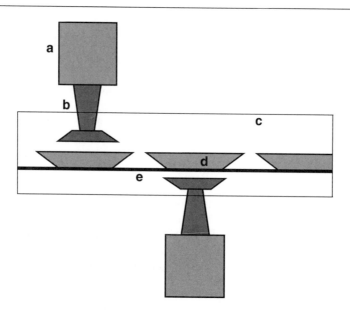

Figure 3–7 A schematic representation of a commercial microwave system: magnetron (microwave generator) (a), waveguiding system (b), treatment chamber (c), product tray (d), conveyor belt (e). *Source:* Reprinted with permission from W. Schlegel, Commercial Pasteurization and Sterilization of Food Products Using Microwave Technology, *Food Technology*, Vol. 46, No. 12, pp. 62–63, © 1992, Institute of Food Technologists.

different treatments to different compartments in a food tray. For example, in a compartmentalized tray, less heat can be applied to pasta or vegetables than to meat.[49]

Effect on Microorganisms

The method of microbial destruction by microwaves is generally believed to be entirely thermal. However, some research using microwave frequencies beyond the normal range of 30 to 30,000 MHz showed evidence of nonthermal effects.[47] Microbial destruction kinetics from microwaves is not as easily calculated as conventional thermal destruction because the temperature in the microwave is not easily controlled. For this reason, the general method used for conventional thermal destruction cannot be applied. In an attempt to quantify microbial destruction by microwaves, the Arrhenius equation was coupled with a model of nth order reaction kinetics to yield[47]

$$k_1 = k_2^{\left[\frac{Ea\,(T-T_r)}{RTT_r}\right]}$$

where k_1 is the thermal inactivation rate (min^{-1}) at product temperature (T), k_2 is the thermal inactivation rate (min-1) at a reference temperature (T_r), E_a is the activation energy (in calories per mole), and R is the universal gas constant (1.987 cal/mol °K). The time-temperature effects should be determined at the cold spots to guarantee acceptable microbial destruction.

Nonuniform heating is the most important issue regarding the safety of microwave energy. Cold spots and large temperature gradients may be present, leading to insufficient kill of pathogens in those areas. Therefore, much effort has been devoted to predicting and modeling temperature profiles in various food systems.[45,50] Relatively simple mathematical models can give direct insight into microwave heating and temperature distribution profiles in homogeneous products with basic geometries.[50]

The major parameters influencing the dielectric behavior of food systems are moisture, salt, and solids content.[47] In particular, the presence of NaCl increases the temperature gradient in microwaved foods. For example, destruction of *Salmonella* species in a liquid system showed that NaCl increased the survivability of microorganisms by inducing large temperature gradients in the system.[51] The amounts of other ingredients present in the system, like lipids, proteins, and carbohydrates, did not influence significantly the temperature achieved.[51] Other salts such as KCl, $CaCl_2$, and $MgCl_2$ did not produce temperature gradients as large as that for NaCl.[52] In general, higher water content generates greater absorption of microwaves, which decreases penetration depth, increases the dielectric loss factor, and causes the product to heat more efficiently.[53] However, some low-moisture products may also heat well because of greater penetration depth and lower specific heat capacity.[54] Density, mass, and food geometry also influence the absorption of microwave energy. Density has a nearly linear relationship with the dielectric constant.[47] Air pockets in the food produce an insulation effect to microwave heating and decrease k'.[55] For large pieces of food, more time is needed to conduct heat throughout and to equilibrate temperature gradients. Furthermore, localized heating may occur due to the multidirectional distribution of microwave energy.[47]

Effect on Food Components

In a study of microwave heating of breast milk, the effect on vitamins B_2 (thiamin) and E (α-tocopherol), polyunsaturated fatty acids (linoleic and linolenic), and specific immunoglobulins was evaluated.[56] The study showed that vitamins and fatty acids were stable over the temperature range tested (37–77°C) and that immunoglobulins were stable up to 57°C but were completely destroyed at 77°C. Loss of vitamin C during microwave and conventional heating for various vegetables was studied.[57] Steaming of vegetables in a microwave showed the lowest loss of vitamin C, which depended on the amount of water used. Similarly, a study on inactivation of pectin methyl esterase in orange juice by microwave and conventional heating showed that inactivation during microwave heating was significantly faster, indicating that some nonthermal effects may be present during microwave heating.[58] A study of α-tocopherol stability in virgin olive oil cooked in a frying pan or in a microwave oven for 8 minutes showed that the retentions of α-tocopherol were 38% and 51%, respectively.[59]

Commercial Applications

Microwave sterilization is used widely for pasta, bakery, and prepared meals.[46] Microwaves are also used for meat tempering; cooking of bacon, sausage, and chicken; and drying of pasta, snacks, and vegetables.[45] Most of the commercially available microwave systems are of European origin.[45]

NONTHERMAL PRESERVATION METHODS

Irradiation

Food irradiation is one of the most effective nonthermal methods of destroying microbes while retaining food quality. However, the public has not accepted irradiation as a safe technology. In the early 1990s, the US regulatory agencies classified irradiation as an additive.[60] Thus, scientists had to prove that irradiation did not "add" or produce any dangerous substances in the food. This obstacle was unique to irradiation, unlike other new processing techniques that were never considered to be or produce additives. All irradiated foods must be labeled with a radura (the international symbol of irradiation) and the words *treated by irradiation* or *treated with radiation*.[61] Irradiation can be used for various purposes depending on the level of exposure (Table 3–1).

Irradiation Mechanism and Generation

Irradiation is achieved by exposing the food to high-energy radiation, called ionizing radiation. Food is passed through an enclosed chamber (irradiator), where it is exposed to an ionizing energy source contained within stainless steel rods placed in racks.[61] In the United States, gamma (γ) rays from cobalt 60 are used exclusively as the energy source.[61] Other sources are also used around the world, including X-rays, electron beam generators, and caesium 137.[62] The γ-rays emitted are of very short wavelength and are not able to elicit neutrons. Therefore, meltdown or chain reactions do not occur, and irradiation does not make either the package or the food radioactive.[63] As γ-rays penetrate the food, most pass through the food, and only a small amount is retained within the product to cause a

Table 3–1 Irradiation Levels, Their Function, and Typical Product Uses

Dose (kGy)	Function	Examples of Products
0.05–0.15	Inhibition of sprouting	Potatoes, onions, garlic
0.15–0.5	Insect and parasite disinfestation	Cereals, fruits (fresh and dry), dry fish and pork
0.5–1.0	Delay of ripening	Fruit and vegetables
1.0–3.0	Extension of shelf life	Strawberries, fresh fish
1.0–7.0	Destruction of pathogens	Seafood, poultry, and meat
10–50	Decontamination	Spices, enzyme preparations
30–50 (+mild heat)	Industrial sterilization	Meat, poultry, seafood

Source: Data from World Health Organization, *Safety and Nutritional Adequacy of Irradiated Food,* © 1994, WHO, Geneva.

very small temperature rise.[61] Even if all the energy in a 10-kGy (kilo-gray) treatment were absorbed by a food with a heat capacity similar to that of water, the temperature would only rise by 2.4°C.[62]

Irradiation Effect and Radiolytic Compounds

Radiation destroys microorganisms and other contaminants such as insects and parasites by partial or total inactivation of their genetic material, or in other words, by directly attacking DNA or by producing radicals and ions that attack DNA.[62] The breaking of chemical bonds by irradiation, called radiolysis, produces unstable reactive products that are subsequently converted to stable final products.[62] However, these radiolytic products have been shown to be virtually identical to thermolytic products produced by conventional heating.[62]

The amount of radiolytic products generated increases linearly with the irradiation dose.[62] The duration of γ-ray exposure, the density of food, and the amount of energy emitted determine the dose of irradiation. Other factors like oxygen, pH, and temperature also affect the amount and nature of radiolytic products.[61] For example, a frozen product will have longer-lived radiolytic compounds because diffusion within the food is limited.[62] This has a protective effect on the substrate because, upon thawing, the radiolytic compounds will react primarily with each other rather than with the substrate.[62]

A study by the US army[64] showed that 40 of the 65 volatile components found in irradiated beef (50 kGy) were unique to the radiation process when compared to thermally treated beef. However, only six of these compounds were not present in any other nonirradiated food.[65] The structure of the remaining six compounds was similar to that of molecules occurring in other food volatile fractions.[61]

Effects on Microorganisms

Irradiation can destroy microorganisms at levels as low as 1 to 3 kGy. The resistance of microorganisms to irradiation is influenced by several factors, including pH, temperature, and chemical composition of the food. Because irradiation affects the cell's DNA, the cell's ability to repair this damage is important. For example, the vegetative cells of some bacteria have a resistance to irradiation similar to that of spores because they are able to repair large numbers of breaks in their DNA.[66] However, spores are generally more resistant than vegetative cells of bacteria, yeasts, and molds, viruses are even more resistant,[67] and toxins are not affected.[62]

The dose necessary to kill 90% of a population (D) of vegetative cells is less than about 0.9 kGy.[67] For example, the D value for *E. coli* on pork at 10°C in air was 0.34 kGy,[68] and for *Salmonella* species on ground beef at 4°C in air it was about 0.66 kGy.[69] *Salmonella* species are probably the most resistant gram-negative pathogens with respect to irradiation. Therefore, any treatment designed to kill *Salmonella* species will also eliminate other gram-negative bacteria, including *E. coli*, *Yersinia enterocolitica*, and *Campylobacter jejuni*. A pathogenic species of *E. coli* that has gained much attention lately, *E. coli* O157:H7, is very sensitive to irradiation. D values of about 0.27 kGy were found on deboned chicken at 0°C,[70] and in ground beef at −16°C the D value was 0.31 kGy.[69] Spores are more resistant to irradiation. For example, spores of *Bacillus cereus* at 0 to 4°C in air had D values of up to 4.0 kGy,[71] and *C. botulinum* spores in phosphate buffer (pH 7) at −30°C and vacuum had D values of up to 3.2 kGy.[72]

Effect on Food Components

All nonthermal preservation techniques have the potential of retaining superior quality and nutritional value of food as compared to conventional methods like retorting. Studies on the effect of irradiation on the three major food components—carbohydrates, lipids, and proteins—showed that doses as high as 50 kGy induce only very small changes.[62] Carbohydrates react with hydroxyl radicals to form ketones, aldehydes, or acids as end products, leading to a drop in pH.[61] Starch degrades to dextrins, maltose, and glucose, resulting in a decrease in the viscosity of polysaccharides in solution.[61] Glucose alone produces at least 34 radiolytic products.[73] In general, the nature and number of products formed depend on the amount of water present.[62] Furthermore, pure carbohydrates are more susceptible to irradiation effects than carbohydrates that are present in a food.[62]

Irradiation of fruits and vegetables often causes tissue softening due to breakdown of pectin.[74] The effect of irradiation on lipids is production of cation radicals and molecules in an excited state, which leads to deprotonation, followed by dimerization or disproportionation.[62] Excited triglycerides may undergo several reactions producing many different products, such as fatty acids, propanediol esters, aldehydes, ketones, diglycerides, diesters, alkanes, methyl esters, and shorter-chain triglycerides.[62] However, production of potential carcinogenic substances, such as aromatic or heterocyclic rings, has not been proven.[62]

Some antioxidative effects of protein-carbohydrate interaction products increase with irradiation dose and may mitigate oxidative changes in lipids.[62] Usually, irradiation of foods sensitive to oxidation is done under anaerobic conditions because in the presence of oxygen auto-oxidation may be rapid.[62] Proteins present in a food matrix are not greatly accessible and are therefore less susceptible to attack. The radicals formed are mostly immobile and are prone to recombination rather than to reaction with other components in the food matrix.[62] Splitting of carbon-nitrogen bonds and disulfide bridges may occur, degrading proteins into smaller polypeptides. On the other hand, globular proteins may aggregate, producing solutions with higher viscosity.[62] Enzymes are not affected by irradiation to any great extent. Therefore, enzyme activity must be prevented by other means. Similarly, irradiation of amino acids in proteins with up to 50 kGy showed limited changes in amino acid composition,[62] although some off flavors can be formed from breakdown products of sulfur containing amino acids.[74]

Most vitamins are not affected severely by irradiation up to 10 kGy, which is the limit for food irradiation (Table 3–2). In general, the most thermolabile vitamins are also the ones most sensitive to irradiation.[61] Vitamin C and thiamin (vitamin B$_1$) are the most sensitive of the water-soluble vitamins, and

Table 3–2 Sensitivity of Vitamins to Irradiation

Vitamin	Soluble to	Sensitivity to Irradiation
Thiamin (B$_1$)	Water	High
α-Tocopherol (E)	Oil	High
Retinol (A)	Oil	High
Vitamin C	Water	Medium
Beta carotene	Oil	Medium
Vitamin K (in meat)	Oil	Medium
Riboflavin (B$_2$)	Water	Low
Niacin (B$_5$)	Water	Low
Pyridoxine (B$_6$)	Water	Low
Biotin (B$_{10}$)	Water	Low
Choline	Water	Low
Cobalamin (B$_{12}$)	Water	Low
Folic acid	Water	Low
Pantothenic acid	Water	Low
Vitamin D	Oil	Low
Vitamin K (in vegetables)	Oil	Low

Source: Data from D. Kilcast, Effect of Irradiation on Vitamins, *Food Chemistry*, Vol. 49, No. 2, pp. 157–164, © 1994.

their losses usually increase with increased irradiation dose and temperature.[74] Nevertheless, thiamin is more sensitive to heat than to irradiation,[75] while a study on oranges showed no significant loss of vitamin C at irradiation levels up to 1 kGy through 6 weeks of storage.[76] Also, no loss of vitamin C was observed in onions and garlic at sprout inhibition doses (0.15 kGy). The remaining water-soluble vitamins (Table 3–2) show minimal or no loss of activity due to irradiation up to about 10 kGy.[74]

Of all the fat-soluble vitamins, α-tocopherol (vitamin E) is the most sensitive. However, foods containing significant amounts of vitamin E (ie, oils and dairy products) are not suitable for irradiation due to off-flavor production. A few other fat-soluble vitamins (Table 3–2) are somewhat susceptible to irradiation.[74]

In conclusion, irradiation causes some changes in foods, but in general, they are small and without significant health effects. The safety of irradiated foods has been studied for longer than 40 years in multispecies animal studies that showed no toxic effects.[77] In a human study, volunteers were fed 100% irradiated foods and no ill effects were observed.[63]

High Hydrostatic Pressure

High-pressure technology was originally developed for production of ceramics, carbide components, steels, and superalloys.[78,79] In these applications, the technology uses an inert gas as the pressure medium, pressures up to about 100 MPa, and temperatures of about 1000°C.[80] More recently, high-pressure processes have been developed for food materials. The first studies of pressure effects on foods and microflora were conducted in the late 19th century on milk and meat.[80] Recently, renewed interest in this technology has surfaced, particularly in Japan, where the first nonthermal high-pressure food products have been marketed.

Theory

The basis for applying pressure to foods is to compress the medium (usually water) surrounding the food (Figure 3–8). The compression of water under pressure is much smaller than that of gases. At ambient temperature (20°C), the volume of water decreases by 4% at 101.3 MPa (1000 atm), as compared to 15% at 607.8 MPa (6000 atm).[79] Due to the small change in volume, high-pressure machines using water do not present the same hazards as machines utilizing gases, which are used in high-pressure processing of ceramic and other nonfood products.[79]

The effects of high pressure on microorganisms are multidimensional, including biochemical, genetic, and morphological changes, as well as changes in cell wall and membrane.[80] At high pressures, microbial death is considered to be caused primarily by permeabilization of the cell membrane.[81] However, it is also believed that pressure affects a combination of processes and does not inhibit or destroy one specific function.[80] Pressure acts in two ways on the cell biosystem: the available molecular space is decreased, and interchain reactions are increased.[80] Reactions that produce a volume increase tend to be inhibited, while the opposite is true for volume-decreasing reactions.[79]

Applications of High Pressure

High pressure can be applied by several different methods (Table 3–3) in cold, warm, or hot isostatic systems.[82,83] Cold isostatic pressing (CIP) is the most promising technique for the food industry.[78] Recent studies showed pulsed high pressure to be more effective than traditional static pressure for inactivation of *Saccharomyces cerevisiae* in pineapple juice.[84] Thus, pulsed high-pressure techniques could be a promising new process.

The pressure used in the ceramic industry is usually between 100 to 300 MPa with a holding time of 10 seconds to 1 minute. For food processing, high-pressure processing requires pressures greater

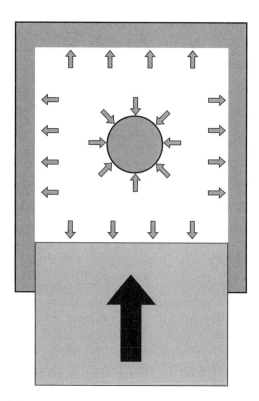

Figure 3–8 The principle of hydrostatic pressure. Courtesy of Flow Pressure Systems.

than 400 MPa and holding times between 5 and 20 minutes.[78] CIP machinery operating at pressures up to 390 MPa and volume capacities of 3000 L is routinely used in other industries.[79]

In commercial applications, the food material to be processed is packaged in high-barrier plastic films such as ethyl-vinyl-alcohol or poly-vinyl-alcohol.[85] The package is then placed in the pressure chamber, which is closed and filled with the pressurizing medium consisting of water mixed with a

Table 3–3 Pressure Application Techniques and Their Uses

Method	Description	Uses
Direct compression	Generated by pressurizing the food with a piston that has a smaller diameter at the end	Restricted to laboratory or pilot plant systems
Indirect compression	Pressure medium is pumped into a closed high-pressure vessel until the desired pressure is obtained	Used in most industrial systems
Heating of pressure medium	Expansion of pressure medium is achieved by heating to the desired pressure	Restricted to uses where high pressure is applied in combination with high temperature

Source: Compiled with data from U.R. Pothakamury et al., The Pressure Builds for Better Food-Processing, *Chemical Engineering Progress*, Vol. 91, No. 3, pp. 45–53, © 1995; and G. Deplace and B. Mertens, The Commercial Application of High Pressure Technology in the Food Processing Industry, Vol. 224, in *High Pressure and Biotechnology*, C. Balny et al., eds., Colloque, INSERM, © 1992.

small amount of soluble oil for anticorrosion and lubricating purposes.[78] The size and geometry of the package do not affect the effectiveness of the pressure process because, according to the Pascal principle, high hydrostatic pressure is applied instantaneously and uniformly throughout the system.[86] Once the desired pressure is obtained, it can be maintained without any further energy input.[79]

Effects on Vegetative Cells

Most microorganisms are baroduric, which means that they can endure high pressures but grow best at ambient pressure. Thus, most bacteria can grow at pressures up to 20 to 30 MPa, while some barophiles can grow at pressures higher than 40 to 50 MPa.[78] Irreversible inactivation of certain vegetative cells does not occur until much higher pressure (ie, about 400 MPa). The sensitivity of microorganisms to pressure increases in the order of gram-positive, to yeast, to gram-negative bacteria. Inactivation of *Listeria monocytogenes* suspended in phosphate-buffered saline showed no appreciable inhibition at pressures up to 240 MPa for 40 minutes at 23°C. When the pressure was increased to 310 MPa, a three-log reduction of *Listeria monocytogenes* was achieved in 20 minutes.[87]

Variables that influence the effect of pressure on microorganisms include the magnitude and duration of compression, the stage of microbial growth, the composition of the growth medium, and environmental parameters such as temperature, pH, and water activity (a_w) during processing.[80,86] Normally, an increase in duration and magnitude of the pressure treatment increases vegetative cell death.[80] Cells from the log phase were more sensitive to pressure than stationary, dormant, or death-phase cells.[88]

The growth media may sometimes protect microorganisms against pressure. For example, milk protected bacteria against pressure, although it was possible to obtain a six-log reduction of *L. monocytogenes* in UHT and raw milk subjected to 345 MPa at 23°C for 80 and 60 minutes, respectively.[87] Application of high pressure decreased the pH of a medium and narrowed the pH range for growth.[89] A study on the effect of pressure treatment (400 MPa, 15 minutes) on the yeast *Rodotorula rubra* at various water activities showed a seven-log reduction at $a_w = 0.96$, less than a two-log reduction at $a_w = 0.94$, and no reduction at $a_w = 0.91$.[90]

Effect on Spores

Bacterial spores are resistant to several treatments, such as temperature, homogenization, irradiation, and pressure.[91] Bacterial spores may survive pressures higher than 1200 MPa, in part due to the thickness and structure of their coat.[92] Certain ranges of pressure cause germination of spores. Fairly low pressure (25 MPa for 30 minutes) treatments at 50°C caused 50% and 64% germination of *Bacillus coagulans* and *B. cereus* respectively.[93] Investigating pressures between 100 and 800 MPa, Gould and Sales[93] found that the optimal inactivation of spores occurred at 200 MPa at 20°C for 1 hour. The survival rate was greater at high pressure (400–800 MPa), probably because fewer spores germinated, enabling more spores to be recovered as viable cells after the pressure treatment.[91] Combining heat and pressure to treat heat-resistant spores of *B. stearothermophilus* showed limited destruction at ambient pressure (1 atm) and 90°C, and at 400 MPa and 20°C, but more than a seven-log reduction occurred at 200 MPa and 90°C.[94] The largest spore inactivation was achieved by a cycle of treatments where spores were first treated with 100 MPa at 50°C, then held at ambient pressure at 20°C for 45 minutes, and then treated with 200 to 400 MPa at 50°C.[94]

Effect on Enzymes

High-pressure treatments may cause either reversible or irreversible changes to different enzymes and may result in complete or partial inactivation or activation. The effect depends on the level and duration of the treatment, the chemical nature of the enzyme, the substrate, temperature, and other environmental parameters.[95] At pressures between 98 and 390 MPa, most effects on proteins can be reversed,[96] probably due to conformational changes and subunit dissociation/association processes.[79]

Enzymatic browning is usually accelerated in pressure-treated foods. For example, enzymatic activity remained in jams treated with 600 MPa, and refrigerated storage was required to reduce color and flavor deterioration.[97] On the contrary, high-pressure treatment of meat showed that proteolytic activity was modified to improve meat quality.[98] Amino- and carboxy-peptidases were completely inactivated by 500 and 400 MPa, respectively, and acid, neutral, and alkaline proteases were only slightly inactivated by 400 MPa or higher.[98] Studies on polyphenol oxidase (PPO) and POD from apples and carrots showed that pressures of 100 to 500 MPa caused enzyme activation, especially at neutral pH.[99] Higher pressures of 800 to 900 MPa caused complete inactivation of both PPO and POD. A study with pectin methyl esterase (PME) in orange juice showed a dual effect of pressure treatments: (1) an instant pressure inactivation, dependent on pressure level, and (2) a second inactivation, dependent on holding time, which was described by first-order reaction kinetics.[100] PME inactivation also depended on pH and total soluble solids in the juice.

Effect on Textural and Rheological Changes

Contrary to thermal processing, high-pressure processing does not destroy covalent bonds but attacks only noncovalent bonds such as hydrophobic and ion pair.[79] Thus, conformational changes in proteins happen because of deprotonation of charged groups, disruption of salt bridges, and the formation of hydrophobic bonds.[78]

Good retention of quality characteristics of foods can be achieved by high-pressure treatments because covalent bonds are not destroyed.[86] For example, texture and taste of rice and soybeans did not change, while texture of potatoes, taproot vegetables, and sweet potatoes changed to pliable (flexible) when subjected to 500 MPa at 20°C for 15 minutes.[101] The firmness of carrots subjected to 100 MPa was not significantly different from that of raw carrots, while carrots cooked for 3 minutes were considerably less firm.[102] Furthermore, a jam treated by ultrahigh pressure retained 95% of its ascorbic acid, and it was preferred by a taste panel over a heat-processed one.[97]

High-pressure treatment of egg yolk at 390 MPa for 30 minutes at 25°C produced a softer gel that kept more of its native color and flavor than did heat-induced gels.[103] Pressure-induced gels of egg white possessed a natural flavor, contained all original vitamins and amino acids, and were digested more easily than heat-induced gels.[78] Pressure-induced gels of rabbit meat and soy protein were glossy, smooth, soft gels with more elasticity than heat-induced gels.[98] These superior qualities of pressure-induced gels could be due to the nondestructive effect of pressure on covalent bonds.[79]

Experiments with high-pressure treatment of tomatoes resulted in a strong rancid taste. The taste was caused by very high concentrations of specific chemical compounds (eg, n-hexanal) produced by oxidation of free fatty acids, particularly linoleic and linolenic acid.[104] In small concentrations, n-hexanal imparts a fresh flavor, but at higher concentrations the flavor becomes rancid.

The effect of pressure treatment on polyunsaturated fatty acids was studied in fish products. Extracted sardine oil treated with 600 MPa for 60 minutes did not show any changes in peroxide values (POV) or thiobarbituric acid values.[105] However, in cod muscle, POV increased with increasing pressure and process time.[106] This indicated that extracted oils may be more resistant to oxidation than oils existing in the muscle during high-pressure treatments.[105]

Applications in the Food Industry

High pressure can be used to extend the shelf life of foods by destroying microorganisms and inactivating undesirable enzymes, while retaining desirable quality characteristics. Several applications of high pressure are already used in the food industry. Examples include application to prolong the shelf life of milk, guacamole, and fruit juices and to reduce microbial loads of meats, eggs, fish, pickles, cheese, and spices.[78] High pressure in combination with subzero temperatures has also been suggested as a means of storing food products without formation of ice and thus avoiding freezing-related damage.[79] Furthermore, thawing of foods is much faster under pressure. For example, 2 kg of beef thawed in 80 minutes at 200 MPa, while it took 7 hours at atmospheric conditions.[78] The juiciness and flavor of the beef were the same at both conditions, but the pressure-thawed beef was slightly discolored.

High-Intensity Pulsed Light

High-intensity pulsed light (HIPL) is a novel technique for sterilization, and little research has been done to confirm its effectiveness. Most reports stem from the company that recently developed this technique (Purepulse Technologies, Inc, San Francisco, CA) and show significant reduction of both microorganisms and spores.[107]

Basic Principles of Pulsed Light

High-intensity light inactivates microorganisms through a combination of photochemical and photothermal reactions, whereas ultraviolet (UV) light commonly used to sterilize packaging material acts primarily in a photochemical way.[2] High-intensity light, as produced by Purepulse Technologies, consists of about 25% UV (300–380 nm), 45% visible (380–780 nm), and 30% infrared (780–1,100 nm) light.[108] The wavelengths used are nonionizing because they are too long to cause ionization of small molecules.[107] High-intensity light may be used to treat UV-sensitive foods. In that case, the UV portion of the bandwidth is filtered out, and the inactivation of microorganisms is thermal.[2] Thus, when UV-sensitive material is treated, high-intensity light is simply a way to quickly transfer large amounts of energy to the surface of the material in order to heat up a very thin layer of the surface.[1]

Generation of Pulsed Light

Pulsed light can be generated by electrically ionizing a xenon gas lamp, causing it to emit a broadband "white" light.[108] Power is magnified by accumulating electrical energy in a capacitor over a relatively long time (fractions of a second) and then releasing the energy in very short periods (thousands of a second).[107] The spectral distribution and peak of HIPL are similar to those of sunlight, but the intensity of one flash of HIPL is about 20,000 times the intensity of sunlight at the surface of the earth.[108] Up to 20 pulses per second can be applied, and the duration of each light pulse is about 200 to 300 μs.[108]

Effect on Microorganisms and Spores

The antimicrobial effects of HIPL are greater than those of nonpulsed or continuous wave (CW) UV light.[107] *Aspergillus niger* spores on packaging material were reduced by more than seven log cycles with a few pulsed light flashes, whereas UV light (using a CW UV-C mercury vapor lamp) yielded a three-log reduction after 2 seconds and only one additional log reduction after 16 seconds.[107]

HIPL affects all types of bacteria. For example, reductions of seven to nine log cycles were obtained for *A. niger* spores, *Bacillus pumilus*, *E. coli* O157:H7, and *Listeria monocytogenes* when subjected to a few flashes with a fluence of 1 J/cm^2 per flash.[107] HIPL is most effective when it hits the bacteria directly on simple surfaces, such as packaging material. However, on complex surfaces, such as most food materials, HIPL is not as effective because surface recesses, fissures, and folds protect microorganisms from direct exposure.[107] Reductions of about two to three log cycles of *Salmonella*, *Listeria*, and *Pseudomonas* species have been demonstrated on meats.[107] On shelled eggs, reductions of up to eight log cycles of *Salmonella enteritidis* were achieved after treatment with eight flashes of 0.5 J/cm^2 per flash.[108]

Effect on Texture and Enzymes

Frankfurters treated with 30 J/cm^2, a very severe treatment for microorganisms, did not show any difference in nutrient quality when compared to untreated controls.[107] Some of the components measured were protein, riboflavin, nitrosamine, benzpyrene, and vitamin C. However, it is well known that many nutrients, especially vitamins, unsaturated fatty acids, certain amino acids, and pigments, are sensitive to light.[109]

Applications

HIPL has been tested on a variety of baked products, including breadsticks, chocolate cupcakes, pizza, tortillas, and bagels,[107] and has resulted in longer shelf life by retarding mold growth. Fish fillets and several meats, including chicken wings and frankfurters, had significantly longer shelf life when treated with HIPL.[107] Similarly, preliminary tests on potable water showed high inactivation of *Klepsiella terrigena* and *Cryptosporidium parvum* cysts.[107] Furthermore, it has been suggested that HIPL could be used to treat food packaged in transparent containers, thus providing a "clean" sterilization method without chemical residues or recontamination.[108] However, the most promising HIPL application is the sterilization of packaging materials, which may replace chemical methods such as hydrogen peroxide.[110]

High-Intensity Pulsed Electric Fields

High-intensity pulsed electric field (PEF) pasteurization involves application of a short burst of high voltage to foods placed between two electrodes.[111] The use of electric fields to inactivate microorganisms and enzymes in foods was first studied in the late 1920s.[112] A process called Electropure for pasteurizing milk involved heating milk to about 70°C and then passing it through carbon electrodes. In the 1950s an electrohydraulic treatment was developed to inactivate microorganisms in liquid foods.[112] Inactivation was attributed to a shock wave generated by an electric arc, which in turn produced reactive radicals from chemical species in the food.[112] Application of high-intensity PEF has been studied extensively in the past few years, and prototype equipment is now available. Therefore, it is expected that in the near future some fruit juices and liquid eggs will be processed using PEF.[112]

Theory

Application of an external electric field to a cell induces a transmembrane potential ($\Delta\varphi_g$), as indicated in Figure 3–9. When the transmembrane potential becomes greater than 1 V (which is the natural potential of the cell membrane), permeabilization of the cell membrane is induced, due to accumulation of ions at the surface of the membrane bilayer.[113]

The relation between electric field strength (E) and transmembrane potential is given by[114]

$$\Delta\varphi_g = \left(\frac{l}{l - 0.67a}\right) \cdot a \cdot E \cdot \cos \alpha$$

where l is cell length, a is cell radius, and α is the orientation of the electric field (always 0° or 180°; hence, $\cos \alpha$ is either 1 or −1, and only the sign of the membrane potential is affected). When a cell is exposed to a critical electric field (E_c), electroporation is induced (Figure 3–10).

The mechanism of pore formation in the membrane (ie, membrane permeabilization) is not entirely understood. However, Zimmerman[115] proposed that opposite charges induced on the inner and outer cell membrane surfaces result in compression of and decrease in thickness of the cell membrane. When critical electric field strength is applied, pores are formed in the membrane. These pores may be reversible or irreversible depending on field strength, number, and duration of pulses.[112,116]

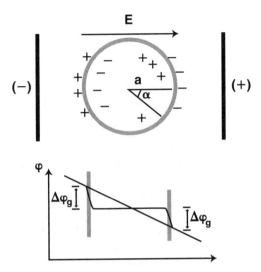

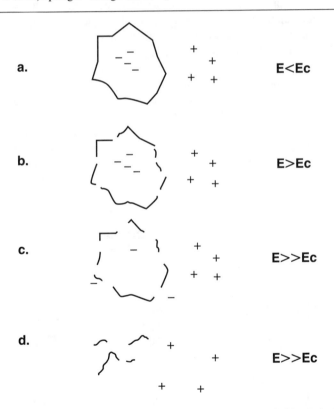

Figure 3–9 Induction of transmembrane potential ($\Delta\varphi_g$). *Source:* Reprinted with permission from T. Grahl and H. Märkl, Killing Microorganisms by Pulsed Electric Fields, *Applied Microbiology & Biotechnology*, Vol. 45, No. 1-2, pp. 148–157, © 1996, Springer-Verlag New York, Inc.

Figure 3–10 Electroporation of the cell membrane when exposed to electric fields. E_c is the critical electric field intensity. *Source:* Adapted from Qin et al., *Critical Reviews in Food Science and Nutrition*, Vol. 36, No. 6, pp. 603–627, © 1996, CRC Press, Inc.

Generation of Pulses

In liquid foods, large amounts of ions are available as electric carriers. To generate a high-strength electric field in a food, a large flux of electrical current must flow through a treatment chamber for a very short period of time, namely microseconds.[117] The pulse duration is much shorter than the time between pulses. A capacitor is used to store energy and produce slow charging and fast discharging of energy. An ignitron can be used as a discharge switch. A typical layout of a PEF circuit is given in Figure 3–11.

Effect on Vegetative Cells

The effect of PEF on vegetative microorganisms has been proven in several studies. Zhang et al[118] showed a nine-log reduction of *E. coli* in simulated milk ultrafiltrate (SMUF) using an electric field strength of 70 kV/cm and 80 pulses at 2 μs. Several variables in the PEF process affect the extent of microbial death, including electric field strength, duration, number, and shape of pulses applied. Grahl and Märkl[114] used cultures of *S. cerevisiae* in orange juice and *E. coli* in UHT milk to demonstrate the effect of electric field strength. Cell concentrations were reduced by four logs when the electric field strength was increased from 4.5 kV/cm to 6.5 kV/cm for *S. cerevisiae* and from 12 kV/cm to 22 kV/cm for *E. coli*. The relatively greater sensitivity of the yeast is due to its greater cell diameter (a smaller electric field is needed to create a given transmembrane potential, as shown in Equation 1). Martín et al[119] showed that the inactivation of *E. coli* in skim milk at 15°C was greater with increased duration and number of pulses at constant field strength and temperature. In earlier studies showing similar effects, the temperature was allowed to rise to 45°C due to the PEF process, so inactivation may have been partly due to thermal treatment.[120]

Electric fields can be applied in the form of exponentially decaying, square-wave, oscillatory, or bipolar pulses.[111] Square-wave pulses were more lethal than exponentially decaying pulses to *E. coli* suspended in SMUF,[121] probably because the energy efficiency of square-wave pulses is greater.[111] Bipolar pulses cause stress to the cell membrane because they continuously reverse the orientation of the electric field, thus enhancing electric breakdown of the cell membrane.[111] Oscillating pulses were least efficient in inactivating *S. cerevisiae* when tested at 40 kV/cm and a frequency of 100 Hz.[111]

Other variables that affect cell death are pH and ionic concentration, temperature, fat content of food, cell concentration, microbial growth state, and type of microorganism.[112,114,121–123] For example, the inactivation of *E. coli* in SMUF was about 1.5 logs greater at pH 5.7 as compared to pH 6.8, using 55 kV/cm electric fields. A similar reduction was observed when ionic strength of SMUF was reduced from 168 to 28 mM.[122]

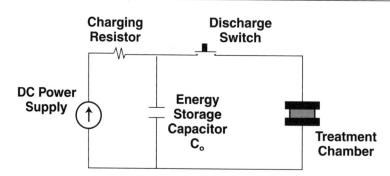

Figure 3–11 A typical layout of a PEF circuit. *Source:* Reprinted from *Journal of Food Engineering,* Vol. 25, Q. Zhang, G.V. Barbosa-Cánovas, and B.G. Swanson, Engineering Aspects of Pulsed Electric Field Pasteurization, pp. 261–281, Copyright 1995, with permission from Elsevier Science.

Coster and Zimmerman[124] suggested the existence of synergistic effects between temperature and PEF treatments on the inactivation of microorganisms. It was shown that *E. coli* inactivation increased from two to three log cycles when the temperature increased from 7 to 40°C.[121]

PEF treated suspensions of *E. coli* in alginate and UHT milk with 1.5% fat content showed that the survival rate was much lower in the alginate solution. This suggests that the fat particles in the milk protected the bacteria against electric pulses.[114] Other researchers suggested that microorganisms in skim milk (no fat) were protected by the complex solution—that is, high protein content.[125] Proteins diminished the effect of PEF by absorbing free radicals and ions that are active in the cell breakdown.[126]

The effect of microbial growth state and initial concentration of *E. coli* cells suspended in SMUF was studied. Cells in the logarithmic growth state cells were more sensitive to PEF treatments than stationary and lag-phase cells.[121] Initial cell concentrations ranging between approximately 10^3 and 10^8 did not influence the effect of PEF treatment (70 kV/cm, 16 pulses of 2 μs).[118] However, inactivation of *S. cerevisiae* was significantly lower at initial cell concentration of 10^6 as compared to 10^4 using one pulse of 25 kV/cm and 25 μs.[123] This could be related to cluster formation of the yeast in low electric fields.[111]

Inactivation of *E. coli* by PEF follows first-order reduction kinetics.[119] However, reduction kinetics in semisolid products may be different from that in fluid foods because *E. coli* is fixed in a gel matrix that increases the uniformity of inactivation.[119,127]

Effect on Microbial Spores

The effect of PEF on spores has been addressed in only a few studies. In UHT milk (1.5% fat), the survival rate of *Bacillus cereus*, *Clostridium tyrobutyricum*, and *Byssochlamus nivea* spores was not affected by PEF treatments.[114] The study concluded that it is impossible to sterilize spore-containing media with PEF. In another study,[128] a 3.4-log reduction was observed when *B. cereus* spores were subjected to 30 pulses of 50 kV/cm, and a five-log reduction was seen when *B. subtilus* spores were treated with 50 pulses of 50 kV/cm. Spore viability did not change after 1 month, indicating that the damage was irreversible. Fewer spores survived when the process temperature and the time between pulses increased.[128]

Effect on Enzymes

Enzymes are stabilized by weak noncovalent forces, such as hydrogen bonds and hydrophobic interactions, and they must maintain their three-dimensional structure to remain active. The effect of PEF on enzymes varies. Some enzymes are activated by PEF, while others are partly or completely inactivated.[112,129,130] For example, the activity of lipase, glucose oxidase, and heat-stable α-amylase showed 70% to 80% reduction in activity when subjected to 87 kV/cm and 30 pulses of 2 μs duration at 20°C.[129] The activity of peroxidase and polyphenol oxidase was reduced by 30% to 40%, and that of alkaline phosphatase by only 5% when subjected to the same PEF treatment.[129] On the other hand, pepsin showed an increased activity (up to 240% of the control), while lysozyme showed a wavelike activity correlation with electric field strength (high reduction at low field strength, low reduction at higher field strength).[129] This stimulating effect on enzyme activity could be due to creation of more reactive sites or an increase in the size of the existing sites.[129] A 90% inactivation of plasmin suspended in SMUF was achieved when the enzyme was subjected to 50 pulses of 30 kV/cm at 15°C.[130]

Effect on Organoleptic Quality

Sensory evaluation of milk and orange juice subjected to PEF treatments did not indicate any deterioration of taste.[114] Similarly, a sensory panel found no significant differences between juice treated with PEF and the nontreated control.[131] However, about 90% of ascorbic acid in milk was destroyed when subjected to 20 pulses of 21 kV/cm.[114]

Ultrasonics

Most applications for ultrasonics (sonication) are nonmicrobial. To mention a few, ultrasound has been used to enhance drying and dewatering,[132,133] accelerate extraction processes,[134] assist heat transfer,[135] and facilitate diffusion through membranes and biomaterials.[136] Ultrasonic equipment has been developed for cleaning, emulsification, and cell disruption.[137] Development of ultrasonics for microbial destruction has been studied with some success.

Effect on Microorganisms

Ultrasound can cause disruption of biological structures.[2] An array of acoustic effects occurs—for example, cavitation, which is the formation, growth, and collapse of small voids in liquids; localized heating, especially at membrane interfaces; and free-radical formation.[136] Of these, cavitation is recognized as the mechanical effect responsible for the destruction of bacterial cells.[138] The effect of ultrasound is amplified when used in combination with prolonged treatment time, heat, pH, or chlorine.[139–141] For example, *Salmonealla typhimorium* was destroyed in a time-dependent manner, with the greatest decrease after 30 minutes of ultrasonic treatment.[141] Treatments with heat and ultrasound showed that *Staphylococcus aureus* was more sensitive to the combined treatment than to each treatment alone.[142,143] Likewise, the sensitivity of *Bacillus* spores to heat increased when they were subjected to ultrasound.[144] However, other researchers who conducted ultrasonic treatments on broiler drumsticks submerged in deionized water at 25 or 40°C for up to 30 minutes found no effect on aerobic plate counts.[145] This observation may be explained by the irregular skin surface of the drumsticks, which may protect bacteria against cavitation.[145] Combination of ultrasonic treatments with chlorinated water showed a 2.5 to 4-log reduction of salmonellae in peptone, while sonication or chlorination alone yielded only about 1-log reduction.[140] The medium in which the treatment was carried out had a significant effect. A 4-log reduction was obtained in peptone but only a 0.8 log reduction in chocolate milk. Research on the effect of ultrasonic treatment on bacterial endospores and enzymes is scarce.[146]

In conclusion, appropriate combination of ultrasonics with other methods may make ultrasonics a useful nonthermal method in the future. A combination method, manothermosonication, where ultrasound treatment is coupled with pressure and a mild thermal treatment, has been tested for inactivation of enzymes.[147,148]

Oscillating Magnetic Fields

Magnetic fields have long been known to influence living organisms. For example, magnetotactic bacteria are capable of synthesizing magnetosomes composed of magnetite, which help the bacteria migrate in certain directions following the earth's magnetic field.[149] Growth of bacteria may also be inhibited if subjected to a certain magnetic field. The use of oscillating magnetic fields (OMF) as a pasteurization technique is a new and rather unexplored area. However, efforts are being made to develop this method and especially to study the effects of magnetic fields on microorganisms.

Theory

Magnetic fields may be static or oscillating. In a static field, the intensity is constant with time and the direction remains the same. On the other hand, OMF is applied in the form of pulses, it reverses the charge from pulse to pulse, and the intensity of each pulse decreases with time by about 10% of the initial intensity.[149] The field can be homogeneous or heterogeneous, depending on whether the intensity is uniform or not in the area enclosed by the magnetic coil. Certain magnetic fields may either stimulate or inhibit microbial growth.[149]

Generation of OMF

OMF with flux densities of 5 to 50 Tesla (T) can be produced by superconducting coils or coils energized by discharge of energy stored in a capacitor.[150] The latter option is used in the OMF equipment shown in Figure 3–12. The capacitor is charged, and when the circuit is completed, an oscillating current is generated between the plates of the capacitor, which in turn generates an OMF.[149] As the current changes direction, the magnetic field changes polarity and deteriorates rapidly as the intensity is reduced drastically after about 10 oscillations.[149]

Effect on Microorganisms and Food Components

Approximately a two-log reduction was observed when microorganisms were subjected to a single pulse of 5 to 50 T and 5 to 500 kHz.[151] For the successful application of OMF to foods, the electrical resistivity of the food should be high (10–25 Ohms-cm). As a point of reference, orange juice has an electrical resistivity of 30 Ohms-cm.[149] The thickness of the food is also important, and for thicker food pieces, stronger magnetic field intensities must be used.[149] The magnetic field exists only within and immediately around the coil and is reduced quickly at any distance from it. Studies on three enzyme substrate systems showed no effect on the rate of enzyme-catalyzed reactions using magnetic fields.[152–154]

In conclusion, both stimulatory and inhibitory effects of OMF on microorganisms have been reported, and reductions of only about two log cycles have been noted. Furthermore, reports on the effect on spores and important microbial pathogens are not available at this point. Therefore, more research is required before OMF is used as a food-processing method.

Chemical Treatments

Early attempts to preserve food consisted of treatments of heat, cold, drying, and fermentation. Chemical preservation methods were limited to salting of fish and meat and pickling of fruit and vegetables, although the mechanism of preservation was poorly understood. It is now known that manipulation

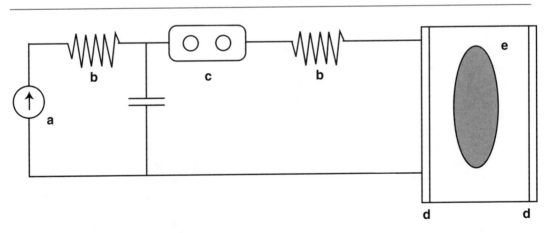

Figure 3–12 Circuit for oscillating magnetic field processing. High-voltage DC power supply (a), resistor (b), switch (c), magnet coil (d), food sealed in a plastic bag and (e), Magneform 7000 coil. *Source:* Reprinted with permission from U.R. Pothakamury, G.V. Barbosa-Cánovas, and B.G. Swanson, Magnetic Field Inactivation of Microorganisms and Generation of Biological Changes, *Food Technology*, Vol. 47, No. 12, pp. 85–93, © 1993, Institute of Food Technologists.

of the chemical characteristics of food, such as pH, water activity, and redox potential, prevents growth of spoilage organisms and/or inhibits activity of harmful enzymes.

Over time, many chemicals, organic and inorganic, have been examined for their antimicrobial and antienzymatic properties. The most important of these chemicals are mentioned in the following sections, with an emphasis on new and inventive treatments.

Benzoate and Sorbate

Sodium benzoate and potassium sorbate are the two most commonly used preservatives. Sodium benzoate, the first chemical preservative approved by the Food and Drug Administration (FDA) for use in foods, is characterized by low cost, lack of color, and relatively low toxicity.[155] Sorbic acid and its salts, especially potassium sorbate, became well known as preservatives in the 1950s. Both sorbate and benzoate were first recognized for their inhibition of yeasts and molds.[155] However, benzoate also inhibits or delays growth of many bacteria. Sorbate affects bacteria to a lesser degree, and some bacteria may even metabolize it.[156]

Organic acids affect microbial growth mainly through the undissociated molecule of the acid, which is responsible for the antimicrobial activity.[155,157] However, some studies showed that for sorbate the dissociated acid also exhibited antimicrobial activity, particularly in environments of pH greater than 6.[158]

The specific effect of benzoic and sorbic acids on microorganisms is not fully understood. Perhaps they interfere with the permeability of the microbial cell membrane.[159] Salmond et al[160] found that benzoate decreased the intracellular pH of *E. coli*. It is widely believed that the undissociated form of the acid diffuses freely through the cell membrane. It then ionizes in the cell, causing uncoupling of both substrate transport and oxidative phosphorylation from the electron transport system, yielding protons that acidify the basic interior of the cell.[155,159]

The salts of benzoic and sorbic acid are more soluble than the acids. Of the salts, sodium benzoate and potassium sorbate are the most soluble, and they are preferred for use in foods. The dissociation constants (the pH at which 50% of the acid is dissociated) of benzoate and sorbate are 4.19 and 4.75, respectively. As pH increases, less acid is undissociated, so the antimicrobial activity diminishes. The pKa may explain the effectiveness of sorbate at higher pH as compared to benzoate and propionate. Sorbates are effective at pH up to 6.5, whereas the limits for benzoate and propionate are 4.5 and 5.5, respectively.[157] The effect of pH on the dissociation of benzoic and sorbic acid is given in Figure 3–13.

Sorbate and benzoate are both approved as GRAS (generally recognized as safe) substances. Sodium benzoate can be used in concentrations of up to 0.1% in the United States, whereas other countries allow concentrations as high as 0.25%.[155] There is no legal limit for potassium sorbate in foods, but it is usually added in concentrations below 0.2%. Sorbates exhibit mild organoleptic properties and neutral taste, and they are often used in combination with benzoate.[155]

Benzoate and sorbate are used mostly as antifungal agents. Most yeasts and molds are inhibited by 0.05% to 0.1% of undissociated benzoic acid.[155] Effective antimicrobial concentrations of sorbate in foods are in the range from 0.05% to 0.3%.[161] Benzoate concentrations of 0.1% to 0.3% had strong bacteriostatic but only modest bactericidal effect on *Listeria monocytogenes* when suspended in a liquid medium.[162] Inhibition or inactivation of *Bacillus cereus*, *E. coli*, *Lactobacillus* species, and other bacteria has also been shown.[155,163] Inhibition of bacteria by sorbate appears to cause an extension of the lag phase, with smaller influence on rate and extent of growth.[161] Overall, sorbate can inhibit many bacteria, including gram-positive, gram-negative, aerobic, and anaerobic bacteria.[161] Sorbate's effect on spore-forming bacteria may be due to competitive and reversible inhibition of amino acid–induced germination.[164]

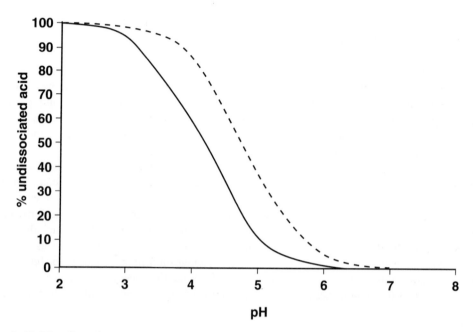

Figure 3–13 The effect of pH on the dissociation of benzoic (—) and sorbic (- - -) acid.

Individual and synergistic effects of pH; combination of benzoate, sorbate, and other acids; and their concentration have been reported.[165–167] Other factors that influence the effect of acids are water activity, food-processing treatments, temperature of storage, and gas atmosphere.[161] For example, the synergism between benzoic and sorbic acid was pH dependent in the growth rate of the yeast *Zygosaccharomyces bailii*.[167] Synergistic effects on yeast and molds by combining benzoic acid and heat were also reported.[168,169] A combination of benzoic acid and low temperatures was used effectively to reduce yeast in grape juice.[170] A recent study by Splittstoesser et al[171] investigated the effect of adding either sorbate or benzoate to apple juice incubated with *E. coli* O157:H7. They reported D_{50} values of 5.2 minutes and 0.64 minutes in apple cider with 0.1% sorbic acid and 0.1% benzoic acid, respectively.

Organic Acids

Adjustment of pH is a powerful way to inhibit or delay growth of most microorganisms, especially bacteria. It is also a way to reduce the sterilization time needed because increased acidity reduces the heat resistance of microorganisms.[172] Adjustment of pH can also be coupled with other treatments such as salt, sugar, gas concentration, and water activity modification to prevent outgrowth of bacteria. This method, using several inhibitors at once rather than just one, is often referred to as *hurdle technology*.

The effectiveness of an acidulant depends on its dissociation constant. The dissociation constant of most acids is between pH 3 and 5,[172] so acids are mostly used in foods with pH close to or below 5. As with the preservatives, it is the undissociated portion of the acid that is believed to be antimicrobial. However, the direct effect of pH manipulation also causes inactivation of bacteria and fungi.[172]

Acids contribute to the taste and tartness of a product and may also control gel formation, aid inversion of sucrose, prevent browning, and protect color.[172] Most organic acids can be metabolized by the tricarboxylic acid cycle, and many occur naturally in different foods. Therefore, most acids appear to be safe for use in food products.

Edible Coatings

Over the past decade, more than 90 patents and scientific papers on edible coatings have been granted/published.[173] Edible coatings have the ability to work as barriers to gas, oil, and water or as adhesives for seasoning or glaze on baked goods to enhance their appearance.[174] Edible films and coatings cannot replace traditional packaging, but they could reduce packaging needs and limit the exposure of foods or food components to moisture, aroma, and lipid migration.[175]

Edible film formulations must contain at least one component capable of forming a structural matrix with a sufficient cohesiveness.[173] Most edible films contain several substances complementing each other to minimize the disadvantages of each component. For example, hydrophilic edible polymers such as cellulose require a hydrophobic material, such as fatty acids or waxes, to form a moisture barrier.[175] Several materials are being tested for film and coating formulations, including carbohydrates, proteins, lipids, and mixtures.[173] Incorporation of antimicrobial agents can be used to retard spoilage and microbial growth. For example, chitosan coatings on fruit may induce production of chitonase (a natural antifungic agent). Ghaouth et al[176] found that decay by *Botrytis cinerea* and *Rhizopus stolonifer* on strawberries coated with chitosan (10 mg/mL) was significantly reduced after 14 days.

Trisodium Phosphate

A patented process was developed consisting of immersing or spraying poultry meat with a solution of trisodium phosphate (TSP) to reduce the incidence of *Salmonella* during poultry processing.[177] Excess solution is allowed to drip from the bird, leaving a minimal amount of phosphate on the meat. The mode of action is believed to be removal of fat coating, allowing bacteria to be washed off the surface more effectively. TSP has been approved by the US Department of Agriculture Food Safety and Inspection Service as a hog scald agent,[178] while approval as antimicrobial treatment is pending further studies.[179]

Inoculated carcasses of chicken were dipped in 10% TSP for 15 minutes and rinsed, resulting in a two-log reduction of *S. typhimurium*, whereas only minimal reduction of aerobic plate count was detected.[180] When TSP was not rinsed off the carcass, an 8.6-log reduction was found. The study also revealed that this reduction was most likely an artifact of the high pH (11–12) of the TSP solution, which prevented recovery of *Salmonella*. A two-log reduction was found on pork carcasses dipped in 8% TSP for 5 to 15 seconds.[178] This study also detected little or no effect on aerobic plate count.

The effect of TSP on the organoleptic quality of chicken was studied.[181] The scores for flavor, texture, and visual appearance were not significantly different for treated and untreated broilers. On pork loin, Morris et al[178] reported some significant effects on discoloration, overall color, texture, and off odor, depending on the method of TSP application (dip, spray, or dip with scalding).

Bacteriocins

Bacteriocins are peptides or proteins with antimicrobial properties produced by bacteria.[182] Colicins produced by strains of *Escherichia coli* were the first bacteriocins to be discovered, but lactic acid bacteria now account for a substantial portion of the research efforts on bacteriocins.[182] This is partly because only nicin, produced by *Lactococcus lactis* subsp. *lactis*, is the only bacteriocin that has met the criterion of an effective suitable antimicrobial compound approved by FDA in 1988 for use as a preservative in pasteurized processed cheese spreads.[182] Furthermore, nicin inhibits *Clostridium botulinum* spores and together with other bacteriocins produced by *Pediococcus* and *Lactobacillus* is quite effective against *Listeria monocytognes*.

Bacteriocins are natural preservatives because they are produced by lactic cultures, which have been consumed for thousands of years without ill effects. This should shorten the regulatory process and increase consumer acceptance. The interest in bacteriocins as effective preservatives is still growing. However, one of the criteria for use of naturally occurring antimicrobial substances is that the substance should be effective at relatively low concentrations. One of the concerns with bacteriocins is that they are effective only at high concentrations, far exceeding those normally ingested in fermented foods containing bacteriocin-producing strains.[182]

Ozone

Ozone (O_3) is a naturally occurring gas with strong oxidizing properties. It is readily soluble in water, and its solubility increases with decreasing temperature. It may be produced by several chemical methods, but commercially it is produced by passing dry air or dry oxygen through an electric arc.[183] During autodecomposition, ozone produces numerous free-radical species, among which the hydroxyl free radical (OH^-) is the most prominent.[184] The main reaction by-product of ozone decomposition is oxygen, so no harmful products are produced. The rate of decomposition of ozone increases with temperature and pH.[184] The half-life of gaseous ozone is about 12 hours, while in water its half-life is a matter of seconds and depends on the amount of ozone-demanding material present.[184] Chemically, ozone is a powerful oxidizer that reacts up to 300 times faster than chlorine with organic material.

Ozone has been used as a disinfectant of potable water in Europe for almost a century, but the first potable treatment with ozone in the United States did not occur until 1940.[184] Recently, several studies[185–188] investigated the effect of aqueous ozone on bacteria and spores. Ozone destroys bacteria by rupturing their cellular membrane[183] and dispersing their cytoplasm in the water. The effects on microbial destruction from gaseous ozone have also been investigated.[189–191] For example, *Botrytis cinerea* and *Sclerotinia sclerotiorum* growth on carrots in chambers flushed with 7.5 to 60 μL/L of ozone showed a 50% reduction in daily growth rates.[189] Ozone is GRAS for treatment of bottled water[192] when used in accordance with good manufacturing practices (GMPs). It was also declared GRAS by an expert panel for use in food processing. The panel concluded that available information supports GRAS classification of ozone as a disinfectant or sanitizer for foods when used at levels and by methods of application consistent with GMPs.[184] This may allow US processors to use ozone in processes such as fish harvest and processing, control of fruit ripening, minimal processing of fruits and vegetables, sanitation of bottles, and disinfection of poultry, beef, and cereal products.

Chlorine Dioxide

In the past, chlorine compounds have been used extensively in the food industry as sanitizing agents on equipment and surfaces due to their strong antimicrobial effects. More recently, chlorine dioxide (ClO_2), both as gas and as liquid, has been recognized as an effective antimicrobial substitute for chlorine. Compared to chlorine, ClO_2 has 2.5 times the oxidation capacity,[193] and is five times as soluble in water.[194] It also maintains its bactericidal activity longer because it is not affected by alkaline pH or high concentration of organic matter.[194] Furthermore, unlike chlorine, ClO_2 does not react with ammonia or nitrogenous compounds, so it does not form chloramine compounds, which may be carcinogenic.[195,196] Destruction of bacteria in poultry-processing chill water showed that ClO_2 was seven times more effective than chlorine.[197]

The antimicrobial effect of ClO_2 has been demonstrated against several bacteria, including *Escherichia coli*, *Listeria monocytogenes*, *Salmonella* species, spores of *Bacillus cereus* and *Clostridium perfringens*,[198–201] fungi,[202] and viruses.[203]

In Europe, ClO_2 has been used to treat water supplies since the late 19th century and is still used widely as an alternative to chlorine.[204–206] Today, aqueous ClO_2 is used in over 400 drinking-water treatment plants in the United States.[207] Applications of ClO_2 in the food industry have been presented in several studies. For example, Reina et al[208] found that ClO_2 effectively controlled microbial buildup in cucumber hydrocooling water but that it did not affect microorganisms on the fruit. Lin et al[198] found ClO_2 to be more effective than chlorine in killing *L. monocytogenes* on fish cubes. On poultry carcasses, ClO_2 treatments eliminated all *Salmonella* species, as compared to a 65% reduction by an in-plant chlorination system.[199] However, the reduction of fecal matter on beef using a ClO_2 spray was the same with that of pure water spray.[209] Aseptic tanks used for fruit juice storage were treated with ClO_2 gas, and a complete inactivation of selected spoilage organisms (eg, *Lactobacillus*, *Leuconostoc*, and *Penicillium* species) was achieved.[210] The researchers recommended a treatment of 10 mg/L of ClO_2 gas for 30 minutes coupled with high relative humidity and temperatures of 9 to 28°C.

In 1995, FDA amended the food additive regulations to specify 3 ppm residual ClO2 for controlling microbial populations in poultry chill water.[211] More recently, ClO_2 (amounts not exceeding 3 ppm) was also allowed as an antimicrobial agent in wash water for fruit and vegetables that are not raw agricultural commodities.[212]

REFERENCES

1. Mertens B, Knorr D. Developments of nonthermal processes for food preservation. *Food Technol.* 1992;46(5):124–133.

2. Manvell C. Minimal processing of food. *Food Sci Technol Today.* 1997;11(2):107–111.

3. Lund D. Heat transfer in foods. In: Fennema OR, ed. *Physical Principles of Food Preservation.* New York, NY: Marcel Dekker, Inc; 1975: chap 3.

4. Broihier K. 1999. A tomato a day keeps cancer away. *Food Proc.* 1999;4:58.

5. Zind T. The functional foods frontier. *Food Proc.* 1999;4:45–50.

6. Damodaran S. Amino acids, peptides, and proteins. In: Fennema OR, ed. *Food Chemistry.* New York, NY: Marcel Dekker, Inc; 1996.

7. Böttcher H. Enzyme activity and quality of frozen vegetables. I. Remaining residual activity of peroxidase. *Nahrung.* 1975;19:173–179.

8. Adams JB. *Effect of Blanching on Enzyme Activities and Their Relationship to the Quality of Frozen Vegetables. Part 4C: Green Beans.* Chipping Campden, Gloucester, England: Campden Food Preservation Research Association; 1983. Campden Food Preservation Research Association, Technical Memorandum, No. 322.

9. Williams DC, Lim MH, Chen AO, et al. Blanching of vegetables for freezing: which indicator enzyme to choose. *Food Technol.* 1986;40(6):130–140.

10. Gregory JF. Enzymes. In: Fennema OR, ed. *Food Chemistry.* New York, NY: Marcel Dekker, Inc; 1996.

11. Jelen P. Review of basic technical principles and current research in UHT processing of foods. *Can Inst Food Sci Technol J.* 1983;16(3):159–166.

12. Watzke HJ. Impact of processing on bioavailability examples of minerals in foods. *Trends Food Sci Technol.* 1998;9:320–327.

13. Miller DD. Minerals. In: Fennema OR, ed. *Food Chemistry.* New York, NY: Marcel Dekker, Inc; 1996: chap 9.

14. Rotruck JT. Effect of processing on nutritive value of food: trace elements. In: Rechcigl M, ed. *Handbook of Nutritive Value of Processed Food.* Boca Raton, FL: CRC Press; 1982:521–528.

15. Halzell T, Johnson IT. Influence of food processing on iron availability *in vitro* from extruded maize-based snack foods. *J Food Agric.* 1989;46:365–374.

16. Holdsworth SD. *Aseptic Processing and Packaging of Food Products.* New York, NY: Elsevier Applied Science Co: 1992:1–10.

17. Floros JD. Aseptic packaging technology. In: Chambers JV, Nelson PE, eds. *Principles of Aseptic Processing and Packaging.* Washington, DC: Food Processors Institute; 1993:115–148.

18. Floros JD, Ozdemir M, Nelson PE. Trends in aseptic packaging and bulk storage. *Food Cosmetics Drug Packaging.* 1998;21:236–239.

19. Schwartz JS. Quality considerations during aseptic processing of foods. In: Singh RK, Nelson PE, eds. *Advances in Aseptic Processing Technologies.* New York, NY: Elsevier Applied Science Co; 1992:245–259.

20. Wilhelmi F. *Soups and Sauces: The Aseptically Packed Feast for the Table.* Holzminden, Germany: DRAGOCO Gerberding & Co.AG; 1988. DRAGOCO report 3:63–77.

21. Anderson JE, Adams DM, Walter WM. Conditions under which bacterial amylases survive ultrahigh temperature sterilization. *J Food Sci.* 1983;48:1622–1625.

22. Ohr LM. Functional foods put flavor up front. *Prepared Foods.* 1999;2:59–64.

23. Saguy I, Goldman M, Karel M. Prediction of beta-carotene decolorization in model system under static and dynamic conditions of reduced oxygen environment. *J Food Sci.* 1985;50:526–530.

24. Hrazdina G. Oxidation of anthocyanidin-3-5-diglucosides with H_2O_2: structure of malvone. *Phytochemistry.* 1970;9: 1647–1652.

25. Toledo RT. Postprocessing changes in aseptically packed beverages. *J Agric Food Chem.* 1986;34:405–408.

26. Schwartz SJ, Lorenzo TV. Chlorophyll stability during continuous aseptic processing and storage. *J Food Sci.* 1991;56:1059–1062.

27. Chan HT, Calvaletto CG. Aseptically packaged papaya and guava puree: changes in chemical and sensory quality during processing and storage. *J Food Sci.* 1982;47:1164–1169.

28. Hansen AP, Swarzel KR. Taste panel testing of UHT fluid dairy products. *J Food Qual.* 1982;4:203–216.

29. Shipe WF, Bassette R, Deane D, et al. Off flavors of milk: nomenclature, standards, and bibliography. *J Dairy Sci.* 1978;61:855–869.

30. Manheim CH, Miltz J, Letzter A. Interaction between polyethylene laminated cartons and aseptically packed citrus juices. *J Food Sci.* 1987;52:737–740.

31. Durr P, Schobinger U, Waldvogel R. Aroma quality of orange juice after filling and storage. *Lebensmittel Verpackung.* 1981;20:91–93.

32. Leonardo SJ, Merson RL, Marsh GL, et al. Estimating thermal degradation in processing of foods. *J Agric Food Chem.* 1986;34:392–396.

33. Floros JD, Ekanayake A, Abide GP, et al. Optimization of a diced tomato calcification process. *J Food Sci.* 1992;57: 1144–1148.

34. Reznick D. Ohmic heating of fluid foods. *Food Technol.* 1996;50(5):250–251.

35. Sastry SK, Palaniappan S. Ohmic heating of liquid-particle mixtures. *Food Technol.* 1992;46(12):64–67.

36. Kim HJ, Choi YM, Yang APP, et al. Microbiological and chemical investigation of ohmic heating of particulate foods using a 5kW ohmic system. *J Food Proc Preserv.* 1996;20:41–58.

37. Kim HJ, Choi YM, Yang TCS, et al. Validation of ohmic heating for quality enhancement of food products. *Food Technol.* 1996;50(5):253–261.

38. Zoltai P, Swearingen P. Product development considerations for ohmic processing. *Food Technol.* 1996;50(5):263–266.

39. Parrott DL. Use of ohmic heating for aseptic processing of food particulates. *Food Technol.* 1992;46(12):68–72.

40. Larki JW, Spinak SH. Safety consideration for ohmically heated, aseptically processed, multiphase low-acid food products. *Food Technol.* 1996;50(5):242–245.

41. deAlwis AAP, Fryer PJ. A finite element analysis of heat generation and transfer during ohmic heating of food. *Chem Eng Sci.* 1990;45:1547–1559.

42. deAlwis AAP, Halden K, Fryer PJ. Shape and conductivity effects in the ohmic heating of foods. *Chem Eng Res Design.* 1989;67(3):159–168.

43. Sastry SK, Qiong L. Modeling the ohmic heating of foods. *Food Technol.* 1996;50(5):246–248.

44. Rice J. Ohmic adventures. *Food Proc.* 1995;56(3):87–91.

45. Schiffmann RF. Microwave processing in the US food industry. *Food Technol.* 1992;46(12):50–57.

46. Harlfinger L. Microwave sterilization. *Food Technol.* Dec. 1992;46:57–61.

47. Heddelson RA, Doores S. Factors affecting microwave heating of foods and microwave induced destruction of foodborne pathogens: a review. *J Food Protect.* 1994;57:1025–1037.

48. Annis PJ. Design and use of domestic microwave ovens. *J Food Protect.* 1980;43:629–632.

49. Schlegel W. Commercial pasteurization and sterilization of food products using microwave technology. *Food Technol.* 1992;46(12):62–63.

50. Remmen HHJ Van, Ponne CT, Nijhuis HH, et al. Microwave heating distribution in slabs, spheres and cylinders with relation to food processing. *J Food Sci.* 1996;61:1105–1117.

51. Heddelson RA, Doores S, Anantheswaran RC, et al. Survival of *Salmonella* species heated by microwave energy in a liquid menstruum containing food components. *J Food Protect.* 1991;54:637–642.

52. Heddelson RA, Doores S, Anantheswaran RC, et al. Destruction of *Salmonella* species heated in aqueous salt solutions by microwave energy. *J Food Protect.* 1993;56:763–768.

53. Schiffmann RF. Food product development for microwave processing. *Food Technol.* 1986;40(6):94–98.

54. Mudgett RE, Smith AC, Wang DIC, et al. Electrical properties of foods in microwave processing. *Food Technol.* Feb. 1982;36:109–115.

55. Schiffmann RF. Microwave foods: basic design considerations. *TAPPI J.* 1990;73:209–212.

56. Lassen A, Ovesen L. Nutritional effects of microwave-cooking. *Nutr Food Sci.* 1995;No.4:8–10.

57. Schnepf M, Driskell J. Sensory attributes and nutrient retention in selected vegetables prepared by conventional and microwave methods. *J Food Qual.* 1994;17(2):87–99.

58. Tajchakavit S, Ramaswamy HS. Thermal vs microwave inactivation kinetics of pectin methylesterase in orange juice under batch mode heating conditions. *Lebensmittel-Wissenschaft Technologie.* 1996;30(1):85–93.

59. Ruiz-Lopêz MD, Artacho R, Pineda MAF, et al. Stability of α-tocopherol in virgin olive oil during microwave heating. *Lebensmittel-Wissenschaft Technologie.* 1995;28:644–646.

60. Pauli GH. Food irradiation in the United States. In: Thorne S, ed. *Food Irradiation.* New York, NY: Elsevier Science Publishers; 1991:235–259.

61. American Dietetic Association. Position of the American Dietetic Association: food irradiation. *J Am Diet Assoc.* 1996; 96(1):69–72.

62. World Health Organization. *Safety and Nutritional Adequacy of Irradiated Food.* Geneva, Switzerland: World Health Organization; 1994.

63. Diehl JF. *Safety of Irradiated Foods.* New York, NY: Marcel Dekker, Inc; 1995.

64. Merritt C Jr. Qualitative and quantitative aspects of trace volatile components in irradiated foods and food substances. *Radiat Res Rev.* 1972;3:353–368.

65. Food and Drug Administration. *Recommendations for Evaluating the Safety of Irradiated Foods: Final Report.* Washington, DC: Food and Drug Administration; July 1980.

66. Moseley BEB. Radiation damage and its repair in non-sporulating bacteria. In: Andrew MEH, Russell AD, eds. *The Revival of Injured Microbes.* London, England: Academic Press; 1984:147–174.

67. Monk JD, Beuchat LR, Doyle MP. Irradiation inactivation of food-borne microorganisms. *J Food Protect.* 1995;58:197–208.

68. Grant IJ, Nixon CR, Patterson MF. Effect of low-dose irradiation on growth and toxin production by *Staphylococcus aureus* and *Bacillus cereus* in roast beef and gravy. *Int J Food Microbiol.* 1993;18(1):25–36.

69. Clavero MRS, Monk JD, Beuchat LR, et al. Inactivation of *E. coli* O157:H7, salmonellae and *Campylobacter jejuni* in raw ground beef by gamma irradiation. *Appl Environ Microbiol.* 1994;60:2069–2075.

70. Borsa J, Chelack WS, Marquardt RR, et al. Comparison of irradiation and chemical fumigation used in grain disinfection on production of ochratoxin A by *Aspergillus alutaceus* in treated barley. *J Food Protect.* 1992;55:990–994.

71. Kamat AS, Nerkar DP, Lewis NF. *Bacillus cereus* in some Indian foods: incidence and antibiotic, heat and radiation resistance. *J Food Safety.* 1989;10(1):31–41.

72. Anellis A, Berkowitz D, Kemper D, et al. Production of type A and B spores of *Clostridium botulinum* by the biphasic method: effect on spore production, irradiation resistance and toxigenicity. *Appl Microbiol.* 1972;23:734–739.

73. Sonntag C von. Free-radical reactions of carbohydrates as studied by irradiation techniques. *Adv Carbohydr Chem Biochem.* 1980;37(1):7–77.

74. Kilcast D. Effect of irradiation on vitamins. *Food Chem.* 1994;49:157–164.

75. Josephson ES, Thomas MH, Calhoun WK. Nutritional aspects of food irradiation: an overview. *J Food Proc Preserv.* 1978;2:299–313.

76. Nagai NY, Moy JH. Quality of gamma irradiated California Valencia oranges. *J Food Sci.* 1985;50:215–219.

77. Thayer DW. Wholesomeness of irradiated foods. *Food Technol.* 1994;48(5):132–135.

78. Pothakamury UR, Barbosa-Canovas GV, Swanson BG, et al. The pressure builds for better food-processing. *Chem Eng Prog.* 1995;91(3):45–53.

79. Farr D. High pressure technology in the food industry. *Trends Food Sci Technol.* 1990;1(1):14–16.

80. Hoover DG, Metrick C, Papineau AM, et al. Biological effects of high hydrostatic pressure on food microorganisms. *Food Technol.* 1989;43(3):99–107.

81. Morita RY. Psychotrophic bacteria. *Bacteriol Rev.* 1975;39:144–167.

82. Mertens B, Deplace G. Engineering aspects of high pressure technology in the food industry. *Food Technol.* 1993;47(6):164–169.

83. Deplace G, Mertens B. The commercial application of high pressure technology in the food processing industry. In: Balny C, Hayashi R, Heremans K, Masson P, eds. *High Pressure and Biotechnology.* Vol. 224. Montrouge, France: John Libbey Eurotext, Ltd, 1992:469.

84. Aleman GD, Ting EY, Mordre SC, et al. Pulsed ultra high pressure treatments for pasteurization of pineapple juice. *J Food Sci.* 1996;61:388–390.

85. Masuda M, Satio Y, Iwanami T, et al. Effects of hydrostatic pressure on packaging materials for food. In: Balny C, Hyashi R, Hermans K, Masson P, eds. *High Pressure and Biotechnology.* Vol. 224. Montrouge, France: John Libby Eurotext Ltd, 1992:545.

86. Knorr D. Effects of high-hydrostatic-pressure processes on food safety and quality. *Food Technol.* 1993;47(6):156–161.

87. Styles MF, Hoover DG, Farkas DF. Response of *Listeria monocytogenes* and *Vibrio parahaemolyticus* to high hydrostatic pressure. *J Food Sci.* 1991;56:1404–1407.

88. ZoBell CE. Pressure effects on morphology and life processes of bacteria. In: Zimmerman AM, ed. *High Pressure Effects on Cellular Processes.* New York, NY: Academic Press; 1970.

89. Marquis RE. High-pressure microbial physiology. *Adv Microb Physiol.* 1976;14:159–241.

90. Oxen P, Knorr D. Baroprotective effects of high solute concentrations against inactivation of *Rhodotorula rubra. Lebensmittel-Wissenschaft Technologie.* 1993;26:220–223.

91. Lechowich RV. Food safety implications in high hydrostatic pressure as a food processing method. *Food Technol.* 1993;47(6):170–172.

92. Larson WP, Hartzell TB, Diehl HS. The effects of high pressure on bacteria. *J Infect Dis.* 1918;22:271–279.

93. Gould GW, Sales AHJ. Initiation of germination of bacterial spores by hydrostatic pressure. *J Gen Microbiol.*1970;60:335–346.

94. Seyderhelm I, Knorr D. Reduction of *Bacillus stearothermophilus* spores by combined high pressure and temperature treatments. *Z Lebensmitteltechnologie.* 1992;43(4):17–20.

95. Cheftel JC. Effect of high hydrostatic pressure on food constituents and overview. In: Balny C, Hayashi R, Heremans K, Mason P, eds. *High Pressure and Biotechnology.* Vol. 224. Montrouge, France: John Libby Eurotext Ltd, 1992:195–209.

96. Morild E. The theory of pressure effects on enzymes. *Adv Protein Chem.* 1981;34:93–166.

97. Horie Y, Kimura K, Ida M, et al. Jam preparation by pressurization. *Nippon Nogeikagaku Kaushi.* 1991;65:975–980.

98. Ohmori T, Shigehisa T, Taji S, et al. Effect of high pressure on the protease activities in meat. *Agric Biol Chem.* 1991;55:357–361.

99. Anese M, Nicoli MC, Dall'Aglio G, et al. Effect of high pressure treatments on peroxidase and polyphenoloxidase activities. *J Food Biochem.* 1995;18:285–293.

100. Basak S, Ramaswamy HS. Ultra high pressure treatment of orange juice: a kinetic study on activation of pectin methyl esterase. *Food Res Int.* 1996;29:601–607.

101. Shimada A, Kasai M, Yamamoto A, et al. Effect of hydrostatic pressurization on the palatability of foods. In: Hayashi R, ed. *High Pressure Science of Food: Research and Development.* Kyoto, Japan: San-ei Publication Co; 1990:249–261.

102. Kato N, Teramoto A, Fuchigami M. Pectic substance degradation and texture of carrots as affected by pressurization. *J Food Sci.* 1997;62:359–362.

103. Okamoto M, Kawamura Y, Hayashi R. Application of high pressure to food processing: textural comparison of pressure- and heat-induced gels of food proteins. *Agric Biol Chem.* 1990;54:183–189.

104. Porretta S, Birzi A, Ghizzoni C, et al. Effects of ultra-high hydrostatic pressure treatments on the quality of tomato juice. *Food Chem.* 1995;52(1):35–41.

105. Ohshima T, Ushio H, Koizumi C. High-pressure processing of fish products. *Trends Food Sci Technol.* 1993;4:370–375.

106. Ohshima T, Nakagawa T, Koizumi C. Effect of high hydrostatic pressure on the enzymatic degradation of sockeye salmon on the changes in lipid composition during frozen storage. In: Bligh EG, ed. *Seafood Science and Technology.* Cambridge, MA: Fishing News Books, Blackwell Scientific Publications; 1992:64–75.

107. Dunn J, Ott T, Clark W. Pulsed-light treatment of food and packaging. *Food Technol.* 1995;49(9):95–98.

108. Dunn J. Pulsed light and pulsed electric field for foods and eggs. *Poultry Sci.* 1996;75:1133–1136.

109. Gravani RB. Food deterioration and spoilage caused by light. *Dairy Food Sanitation.* 1985;5:386–387.

110. Rice J. Sterilizing with light and electric impulses. *Food Proc.* 1994;55(7):66.

111. Qin BL, Pothakamury UR, Barbosa-Cánovas GV, et al. Nonthermal pasteurization of liquid foods using high-intensity pulsed electric fields. *Crit Rev Food Sci Nutr.* 1996;36:603–627.

112. Vega-Mercado H, Martín-Belloso O, Qin BL, et al. Non-thermal food preservation: pulsed electric fields. *Trends Food Sci Technol.* 1997;8(5):151–157.

113. Sales AJH, Hamilton WA. Effect of high electric fields on microorganisms. III. Lysis of erythrocytes and protoplasts. *Biophys Acta.* 1968;148:781–788.

114. Grahl T, Märkl H. Killing of microorganisms by pulsed electric fields. *Appl Microbiol Biotechnol.* 1996;45:148–157.

115. Zimmerman U. Electrical breakdown, electropermeabilization and electrofusion. *Rev Physiol Biochem Pharmacol.* 1986;105:175–256.

116. Knorr D, Geulen M, Grahl T, et al. Food application of high electric field pulses. *Trends Food Sci Technol.* 1994;5(3):71–75.

117. Zhang Q, Barbosa-Cánovas GV, Swanson BG. Engineering aspects of pulsed electric field pasteurization. *J Food Eng.* 1995;25:261–281.

118. Zhang Q, Qin BL, Barbosa-Cánovas GV, et al. Inactivation of *E. coli* for food pasteurization by high-strength pulsed electric fields. *J Food Proc Preserv.* 1995;19(2):103–118.

119. Martín O, Qin BL, Chang FJ, et al. Inactivation of *Escherichia coli* in skim milk by high intensity pulsed electric fields. *J Food Proc Eng.* 1997;20:317–336.

120. Dunn JE, Pearlman JS. Methods and apparatus for extending the shelf life of fluid food products. US Patent 4,695,472 (1987).

121. Pothakamury UR, Vega H, Zhang Q, et al. Effect of growth stage and processing temperature on the inactivation of *E. coli* by pulsed electric fields. *J Food Protect.* 1996;59:1167–1171.

122. Vega-Mercado H, Pothakamury UR, Chang FJ, et al. Inactivation of *Escherichia coli* by combining pH, ionic strength and pulsed electric fields hurdles. *Food Res Int.* 1996;29(2):117–121.

123. Zhang Q, Monsalve-González A, Qin BL, et al. Inactivation of *Saccharomyces cerevisiae* in apple juice by square wave and exponential-decay pulsed electric fields. *J Food Proc Eng.* 1994;17:469–478.

124. Coster HGL, Zimmerman U. The mechanisms of electrical breakdown in the membrane of *Valonia utricularis*. *J Membr Biol.* 1975;22(1):73–90.

125. Goff HD, Hill AR. Chemistry and physics. In: Hui YH, ed. *Dairy Science and Technology Handbook.* Vol 1. New York, NY: VCH Publications; 1993:1–82.

126. Allen M, Soike K. Sterilization by electrohydraulic treatment. *Science.* 1966;154:155–157.

127. Zhang Q, Chang FJ, Barbosa-Cánovas GV, et al. Inactivation of microorganisms in a semisolid model food using high voltage pulsed electric fields. *Lebensmittel-Wissenschaft Technologie.* 1994;27:538–543.

128. Marquez VO, Mittal GS, Griffiths MW. Destruction and inhibition of bacterial spores by high voltage pulsed electric field. *J Food Sci.* 1997;62:399–401.

129. Ho SY, Mittal GS, Cross JD. Effects of high field electric pulses on the activity of selected enzymes. *J Food Eng.* 1997;31(1):69–84.

130. Vega-Mercado H, Powers JR, Barbosa-Cánovas GV, et al. Plasmin inactivation with pulsed electric fields. *J Food Sci.* 1995;60:1143–1146.

131. Qin BL, Pothakamury UR, Vega H, et al. Food pasteurization using high-intensity pulsed electric fields. *Food Technol.* 1995;49(12):55–60.

132. Fairbanks HV. Ultrasonically assisted drying of fine particles. *Ultrasonics.* 1974;12:260–262.

133. Ensminger D. Acoustic and electroacoustic methods of dewatering and drying. *Drying Technol.* 1988;6:473–499.

134. Kim SM, Zayas JF. Processing parameters of chymosin extraction by ultrasound. *J Food Sci.* 1989;54:700–703.

135. Lima M, Sastry SK. Influence of fluid rheological properties and particle location on ultrasound-assisted heat transfer between liquid and particles. *J Food Sci.* 1990;55:1112–1115.

136. Floros JD, Liang H. Acoustically assisted diffusion through membranes and biomaterials. *Food Technol.* 1994;48(12):79–84.

137. Gunasekaran S, Chyung A. Evaluating milk coagulation with ultrasonics. *Food Technol.* 1994;48(12):74–78.

138. Shukla TP. Microwave ultrasonics in food processing. *Cereal Foods World.* 1992;37:332–333.

139. Earnshaw RG, Appleyard J, Hurst RM. Understanding physical inactivation processes: combined preservation opportunities using heat, ultrasound and pressure. *Int J Food Microbiol.* 1995;28:197–219.

140. Lillard HS. Bactericidal effect of chlorine on attached salmonellae with and without sonification. *J Food Protect.* 1993;56:716–717.

141. Wrigley DM, Llorca N. Decrease of *Salmonella typhimurium* in skim milk and egg by heat and ultrasonication wave treatment. *J Food Protect.* 1992;55:678–680.

142. Ordonez JA, Aguilera MA, Garcia ML, et al. Effect of combined ultrasonic and heat treatment (thermal-ultrasonication) on the survival of a strain of *Staphylococcus aureus. J Dairy Res.* 1987;54:61–67.

143. Ordonez JA, Sanz B, Hernandez PE, et al. A note on the effect of combined ultrasonic and heat treatments on the survival of thermoduric streptococci. *J Appl Bacteriol.* 1984:56:175–177.

144. Sanz B, Paklacios P, Lopêz P, et al. Effect of ultrasonic waves on the heat resistance of *Bacillus stearothermophilus* spores. *Fed Eur Microbiol Soc.* 1985;18:251–259.

145. Sams AR, Feria R. Microbial effects of ultrasonication of broiler drumstick skin. *J Food Sci.* 1991;56:247–248.

146. Hoover DG. Minimally processed fruits and vegetables: reducing microbial load by non-thermal physical treatments. *Food Technol.* 1997;51(6):66–71.

147. Vercet A, Lopêz P, Burgos J. Inactivation of heat-resistant lipase and protease from *Pseudomonas fluorescens* by manothermosonication. *J Dairy Sci.* 1997;80(1):29–36.

148. Lopêz P, Sala F, Fuente J, et al. Inactivation of peroxidase, lipoxygenase, and polyphenol oxidase by manothermosonication. *J Agric Food Chem.* 1994;42:252–256.

149. Pothakamury UR, Barbosa-Cánovas GV, Swanson BG. Magnetic field inactivation of microorganisms and generation of biological changes. *Food Technol.* 1993;47(12):85–93.

150. Gersdorf R, de Boer FR, Wolfrat JC, et al. The high magnetic facility of the University of Amsterdam. In: Date M, ed. *High Field Magnetism: Proceedings of the International Symposium on High Field Magnetism.* Amsterdam, Netherlands: North Holland Publications; 1983.

151. Hoffman GA. Deactivation of microorganisms by an oscillating magnetic field. US Patent 4,524,079 (1985).

152. Rabinovitch B, Maling JE, Weissbluth M. Enzyme substrate reactions in very high magnetic fields, I. *Biophys J.* 1967;7:187–204.

153. Rabinovitch B, Maling JE, Weissbluth M. Enzyme substrate reactions in very high magnetic fields, II. *Biophys J.* 1967;7:319–327.

154. Maling JE, Weissbluth M, Jacobs EE. Enzyme substrate reactions in high magnetic fields. *Biophys J.* 1965;5:767–776.

155. Chipley JR. Sodium benzoate and benzoic acid. In: Davidson PM, Braner AL, eds. *Antimicrobials in Foods.* New York, NY: Marcel Dekker, Inc; 1993.

156. Sofos JN. *Sorbate Food Preservatives.* Boca Raton, FL: CRC Press; 1989.

157. Sofos JN, Busta FF. Antimicrobial activity of sorbate. *J Food Protect.* 1981;44:614–622.

158. Eklund T. The antimicrobial effect of dissociated and undissociated sorbic acid at different pH levels. *J Appl Bacteriol.* 1983;54:383–389.

159. Freese E, Sheu CW, Galliers E. Function of lipophilic acids as antimicrobial food additives. *Nature.* 1973;241:321–325.

160. Salmond CV, Kroll RG, Booth IR. The effect of food preservatives on pH homeostasis in *Escherichia coli. J Gen Microbiol.* 1984;130:2845–2850.

161. Sofos JN, Busta FF. Sorbic acid and sorbates. In: Davidson PM, Branen AL, eds. *Antimicrobials in Foods.* New York, NY: Marcel Dekker, Inc; 1993.

162. El-Shenawy MA, Marth EH. Sodium benzoate inhibits growth of or inactivates *Listeria monocytogenes. J Food Protect.* 1988;51:525–530.

163. Davidson PM, Juneja VK. Antimicrobial agents. In: Banen AL, Davidson PM, Salminen S, eds. *Food Additives.* New York, NY: Marcel Dekker, Inc; 1990.

164. Smoot LA, Pierson MD. Mechanisms of sorbate inhibition of *Bacillus cereus* T and *Clostridium botulinum* 62A spore germination. *Appl Environ Microbiol.* 1981;42:477–483.

165. Banks JG, Morgan S, Stringer MF. *The Influence of Combinations of Chemical Preservatives on the Growth of Food Poisoning and Spoilage Microorganisms.* Chipping Campden, Gloucester, England: Campden Food Preservation Research Association; 1988. Camden Food Preservation Research Association, Technical Memorandum, No. 471.

166. Parish ME, Caroll DE. Effects of combined antimicrobial agents on fermentation initiation by *Saccharomyces cerevisiae* in a model broth system. *J Food Sci.* 1988;53:240–242.

167. Cole MB, Keenan MHJ. Synergistic effects of weak-acid preservatives and pH on the growth of *Zygosaccharomyces bailii. Yeast.* 1986;2:93–100.

168. Beuchat LR. Synergistic effects of potassium sorbate and sodium benzoate on thermal inactivation of yeasts. *J Food Sci.* 1981;46:771–777.

169. Beuchat LR, Jones WK. Effects of food preservatives and antioxidants on colony formation by heated conidia of *Aspergillus flavus. Acta Alimentaria.* 1979;7:373–384.

170. Pederson CS, Albury NN, Christensen MD. The growth of yeasts in grape juice at low temperature. IV. Fungistatic effect of organics acids. *Appl Microbiol.* 1961;9:162–167.

171. Splittstoesser DF, McLellan MR, Churey JJ. Heat resistance of *Escherichia coli* O157:H7 in apple juice. *J Food Protect.* 1996;59:226–229.

172. Doores S. Organic acids. In: Davidson PM, Branen AL, eds. *Antimicrobials in Foods.* New York, NY: Marcel Dekker, Inc; 1993.

173. Debeaufort F, Quezada-Gallo JA, Voilley A. Edible films and coatings: tomorrow's packagings: a review. *Crit Rev Food Sci.* 1998;38:299–313.

174. Edible films solve problems. *Food Technol.* 1997;51(2):60.

175. Krochta JM, Mulder-Johnston C De. Edible and biodegradable polymer films: challenges and opportunities. *Food Technol.* 1997;51(2):61–74.

176. Ghaouth A El, Arul J, Grenier J, et al. Antifungal activity of chitosan on two postharvest pathogens of strawberry fruits. *Phytopathology.* 1992;82:398–402.

177. Giese J. Experimental process reduces *Salmonella* on poultry. *Food Technol.* 1992;46(4):112.

178. Morris CA, Lucia LM, Savell JW, et al. Trisodium phosphate treatment of pork carcasses. *J Food Sci.* 1997;62:402–405.

179. Cohen PM. *Petition for Rule Making: Decision Memorandum for Associate Administrator, USDA-FSIS.* Washington, DC: 1995.

180. Lillard HS. Effect of trisodium phosphate on salmonellae attached to chicken skin. *J Food Protect.* 1994;57:465–469.

181. Hollender R. Research note: consumer evaluation of chicken treated with a trisodium phosphate application during processing. *Poultry Sci.* 1993;72:755–759.

182. Hoover DG. Bacteriocins with potential for use in foods. In: Davidson PM, Branen AL, eds. *Antimicrobials in Foods.* New York, NY: Marcel Dekker, Inc; 1993.

183. Electrical Power Research Institute. Ozone: GRAS affirmation for use in food. *Food Ind Curr.* 1997;1(1):1–6.

184. Graham DM. Use of ozone for food processing. *Food Technol.* 1997;51(6):72–75.

185. Liyange LRJ, Finch GR, Belosevic M. Sequential disinfection of *Cryptosporidium parvum* by ozone and chlorine dioxide. *Ozone Sci Eng.* 1997;19:409–423.

186. Restaino L, Frampton EW, Hemphill JB, et al. Efficacy of ozonated water against food related microorganisms. *Appl Environ Microbiol.* 1995;61:3471–3475.

187. Finch GR, Black EK, Labatiuk CW, et al. Ozone inactivation of *Cryptosporidium parvum* in demandfree phosphate buffer determined by *in vitro* excystation and animal infectivity. *Appl Environ Microbiol.* 1993;59:4203–4210.

188. Rickloff JR. An evaluation of the sporicidal activity of ozone. *Appl Environ Microbiol.* 1987;53:683–686.

189. Liew CV, Prange RK. Effect of ozone and storage temperature on postharvest diseases and physiology of carrots. *J Am Soc Hortic Sci.* 1994;119:563–567.

190. Naitoh S. Studies of the application of ozone in food preservation - ozone inhibition of yeast. *J Antibacterial Antifungal Agents—Jpn.* 1992;21:341–346.

191. Isizaki K, Shinriki N, Matsuyama H. Inactivation of *Bacillus* spores by gaseous ozone. *J Appl Bacteriol.* 1986; 60(1):67–72.

192. Food and Drug Administration. Beverages: bottled water, final rule. *Fed Reg.* 1995;60:57075–57130.

193. Benarde MA, Israel BM, Olivieri VP, Granstrom ML. Efficiency of chlorine dioxide as a bactericide. *Appl Microbiol.* 1965;13:776–780.

194. White GC. *Handbook of Chlorination.* New York, NY: Van Nostrand Reinhold Co; 1972:596.

195. Aieta EM, Roberts PV, Hernandez M. Determination of chlorine dioxide, chlorine, and chlorate in water. *Am Water Works Assoc J.* 1984;76:64–70.

196. Alliger H. *An Overall View of Chlorine Dioxide.* Farmingdale, NY: Alcide Corp; 1980.

197. Lillard HS. Levels of chlorine and chlorine dioxide of equivalent bactericidal effect in poultry processing water. *J Food Sci.* 1979;44:1594–1597.

198. Lin WF, Huang TS, Cornell JA, et al. Bactericidal activity aqueous chlorine and chlorine dioxide solutions in a fish model system. *J Food Sci.* 1996;61:1030–1034.

199. Villarreal ME, Baker RC, Regenstein JM. The incidence of *Salmonella* on poultry carcasses following the use of slow release chlorine dioxide (Alcide). *J Food Protect.* 1990;53:465–467.

200. Foegeding PM, Hemstapat V, Giesbrecht FG. Chlorine dioxide inactivation of *Bacillus* and *Clostridium* spores. *J Food Sci.* 1986;51:197–201.

201. Thiessen GP, Usborne WR, Orr HL. The efficacy of chlorine dioxide in controlling *Salmonella* contamination and its effect on product quality of chicken broiler carcasses. *Poultry Sci.* 1984;63:647–653.

202. Roberts RG, Reymond ST. Chlorine dioxide for reduction of postharvest pathogen inoculum during handling of tree fruits. *Appl Environ Microbiol.* 1994;60:2864–2868.

203. Chen YS, Vaughn JM. Inactivation of human simian rotaviruses by chlorine dioxide. *Appl Environ Microbiol.* 1990;56:1363–1366.

204. Katz J. *Ozone and Chlorine Dioxide Technology for Disinfection of Drinking Water.* Park Ridge, NJ: Noyes Data Corp; 1980.

205. Symons JM, Carswell JK, Clark RM, et al. Ozone, chlorine dioxide, and chloramines as alternatives to chlorine for disinfection of drinking water. In: *Water Chlorination: Environmental Impact and Health Effects,* vol. 2. Ann Arbor, MI: Ann Arbor Science Publishers, Inc; 1978.

206. McCarthy JA. Bromine and chlorine as water disinfectants. *J New Engl Water Works Assoc.* 1944;58(1):55–68.

207. Richardsen SD, Thruston AD, Collette TW, et al. Future use of chlorine, chlorine dioxide, and other chlorine alternatives. *Institute of Food Technology Annual Meeting: Book of Abstracts.* New Orleans, LA, 22–26 June 1996. IFT: 140.

208. Reina LD, Fleming HP, Humphries EG. Microbiological control of cucumber hydrocooling water with chlorine dioxide. *J Food Protect.* 1995;58:541–546.

209. Cutter N, Dorsa WJ. Chlorine dioxide spray washes for reducing fecal contamination on beef. *J Food Protect.* 1995;58:1294–1296.

210. Han Y, Guentert AM, Smith RS, et al. Efficacy of chlorine dioxide gas as a sanitizer for tanks used for aseptic juice storage. *Food Microbiol.* 1999;16:53–61.

211. Food and Drug Administration. Secondary direct food additives permitted in food for human consumption. Department of Health and Human Services. *Fed Reg.* 1995;60:11899–11900.

212. 21 CFR pt 173.300.

CHAPTER 4

Flavoring Systems
for Functional Foods

Gary A. Reineccius

It is worthwhile to begin this chapter by defining its scope and some of the terms that will be used. I will be discussing flavor issues in functional foods that include dietary supplements and medical foods. Isolated active ingredients, supplements, and complete foods (both mainstream foods and complete diets prescribed or recommended by a physician as a result of illness) will be considered. An overview of strategies for flavoring functional foods will be provided rather than a detailed "how to" guide.

Since medical foods are prescribed and the consumer has little choice in his or her purchase or consumption, one might assume that there will be little concern for the flavoring of these products. However, the flavoring of these foods also has some benefit as will be discussed later in this chapter. Dietary supplements are purchased voluntarily so flavor acceptability is a significant issue. Many of these products have inherent off aromas and tastes and are quite objectionable (Table 4–1).[1] However, one must recognize that the individual taking these items may be very committed to their benefit and be willing to suffer some unpleasant flavor. Also, many of these products are sold as gelatin capsules that dissolve in the gut and thus do not offer any objectionable aroma or taste to the consumer. Although capsules offer taste and aroma management, these products often are difficult for the elderly to swallow and may be shunned by this market segment.

The taste and aroma of functional foods are critical if there is a desire to make these products appeal to the mainstream customer. One can envision that major food companies may choose to fortify an existing product line with vitamins, minerals, or botanicals. In this case, extreme care must be exercised in considering the flavor quality of the product. It has been repeatedly shown that the consumer will not purchase foods in the long term if they do not taste good. This has been demonstrated in the 1970s and 1980s for microwave foods, earlier "healthy" trends for meat analogues, and more recent "healthy" trends for low- or no-fat foods. The mainstream consumer initially embraced microwave foods and low- or no-fat foods but quickly refused to continue purchasing them. Food is one of the few means of obtaining immediate gratification that is not illegal, immoral, unsafe, or expensive. Additionally, eating does not require a group or any organization but can be done alone at any time. The average consumer is not interested in long-term deprivation, and food products that do not offer elements of pleasure tend to stay minor players in the industry. The functional food industry faces a similar challenge if it is to grow into a significant market and become an integral part of the food sector in the long term.

CHALLENGES IN FLAVORING FUNCTIONAL FOODS

The challenges in flavoring functional foods (in all of the contexts noted earlier), include problems with off flavors inherent to the functional ingredient or associated ingredients, possible alteration of flavors in the manufacture of the functional food, negative interactions between flavorings and the

Table 4–1 Flavor Character of Selected Dietary Supplements

Supplement	Potential Health Benefit*	Dosage	Sensory Properties
Trace minerals (14 minerals in aqueous solution)	Prevent deficiencies	1 tsp 3×/day in water or juice	Clean—no off notes
Ginseng (extract of four ginsengs)	Aphrodisiac; increases resistance to stress and disease	15–20 drops/day	Gingerlike; vegetable; bitter caffeine notes
Kava kava (extract)	Aids relaxation, sleeplessness	1 mL 2×/day	Cassis; hay; berry
Echinacea (extract)	General immune enhancer	1 cc 2×/day in 1/2 glass water	Fruity; mild; gelatin
Echina Gold (extract) (blend of echinacea and golden seal)	Immune enhancer; used for gastric and genitourinary disorders	1 cc 2×/day in 1/2 glass water	Green vegetable; green peas; mild
St John's wort (extract)	Mild antidepressant, diuretic	1 cc 2×/day	Resinous; chemical
Golden seal (extract)	Remedy for various gastric genitourinary disorders	5–10 drops 3×/day in water or juice	Bitter; acrid; lingering off taste
Aged garlic (extract)	Anticancer, anticoronary and vascular disease, antihypertension	30–60 drops 2×/day in juice or as garnish	Heated garlic oil
Saw palmetto (infusion)	Aids in prostate cancer prevention	Hot tea	Dry—haylike; as tea—green; soapy; lavender; herbal
Kava kava (infusion)	Stress relief	Hot tea	Dry—haylike; tea—mild; savory
Golden seal herb (infusion)	Remedy for gastric and genitourinary disorders	Hot tea	Dry—haylike; as tea—bitter, acrid, herbal lingering off taste
Isolated soy powder (vanilla flavored)	Estrogen replacement, cardiovascular disease	3 heaping tbsp in 1 cup of 2% milk	Slimy; bitter; astringent; beany
Vitamin C (crystalline)	Cold prevention, antioxidant	1/2 tsp/day	Slightly tart but clean
Multivitamin/multimineral (liquid)	General health (deficiency-related diseases)	1 tb/day	Meaty; brothy; lingering off taste (potent "thiamin" aroma)
Vitamin E (liquid)	Antioxidant, cardiovascular	1 mL/day	No notes of significance

*Potential benefits are variable and depend upon source.

functional ingredient, and any storage issues that may be directly or indirectly relate to the functional ingredient. While functional foods as a category are relatively new in the United States, or perhaps more correctly, have been recently rediscovered by the food industry, the flavoring problems associated with them are not new. Working with off flavors in a food system is a common task and an ongoing battle. For example, the food industry has considerable experience in flavoring soy-based foods in this context (covering beany off flavor).[2] The flavor industry also has considerable experience in flavoring pharmaceutical products that often have very pronounced off flavors and require intense flavor systems to make them palatable. Fortunately for pharmaceutical products, the consumer purchases and consumes them over a long time period not by choice but by need. Although the flavoring problems may be similar, the flavoring needs for functional foods are much more stringent. Manufacturers want the consumers to purchase their products not for the immediate benefit of reducing pain (symptoms of illness) but for some more remote possible benefit of better health. This will greatly decrease consumer tolerance for poor sensory quality. Nevertheless, there is a body of knowledge in companies that have traditionally provided flavorings for pharmaceutical products that can be directly applied to functional foods.

Similarly, other challenges mentioned above (processing, interaction, and storage issues) are not unique to functional foods. It is unlikely that the processes used to recover or further process a functional food will differ significantly from those currently used in the food industry for an existing product. Some of these processes, retorting, for example, are extremely harsh on flavor stability. The interactions one might expect are also well known through either existing food products or pharmaceuticals. While there may be differences in the degree to which each of these problems occurs in a functional food that make flavoring more difficult, the solution to the problem will typically follow a traditional and similar strategy.

STRATEGIES FOR FLAVORING FUNCTIONAL FOODS

Ingredient Quality

The types of problems one faces in flavoring functional foods are well known. For example, masking the beany note of soy is a general concern. The solution to this problem begins with the soy protein manufacturing process itself. One must process the soybeans in a manner that minimizes this off note (early enzyme inactivation). This strategy carries over to any functional ingredient—one must produce functional ingredients of the best possible sensory quality. In manufacturing one cannot simply focus on the functional quality of the ingredient but also its sensory quality. The range in quality of a given ingredient in the industry drives the point home that large variations exist and more focus needs to be placed on sensory quality. Psyllium is a product that can be very objectionable in taste, texture, and odor or pleasantly bland, depending upon the manufacturing process, and the temperature and humidity conditions during storage and transport. The point is to demand the best quality ingredient one can obtain. One does not make a good flavored product starting with an off-flavored ingredient.

Flavor Selection: Compatible Flavorings

If the functional food carries off notes and they cannot be adequately (or economically) removed, then one has to work with the off notes, either by choosing flavors such that the objectionable notes are positive contributors to the flavor or by adding other ingredients that will compete for sensory binding sites and thus minimize the perception of the objectionable notes (masking). The first approach is most widely used—choose the flavor such that the objectionable sensory notes are positive contributors to the flavoring. For bitter functional ingredients, for example, one might choose a cheese, coffee, dark chocolate, or tea flavoring. All of these products have bitter notes as a desirable part of the flavoring, and the bitterness inherent to the ingredient may be used to provide this aspect of the flavoring. It would be more difficult to flavor a bitter ingredient strawberry, for example, since strawberry does not have any characteristic bitter notes. Fischer[3] has suggested flavors that are compatible with some of the common taste defects associated with functional foods (Table 4–2).

Cooked protein off notes may be flavored with caramel, butterscotch, or chocolate flavorings, since the cooked notes would be desirable. Functional foods containing thiamin can be delivered in a meat soup or broth form since the thiamin degradation products would be actually desirable in these products. Milk chocolate is often used to hide oxidized notes. Hershey's milk chocolate derives its characteristic flavor from the use of carefully oxidized milk. Thus, the American consumer who has grown up with Hershey's chocolate finds an oxidized note to be very acceptable as part of the chocolate flavor profile. Another example of a complementary flavor was suggested by Manley,[4] who recommended the use of citrus and fruit flavors to cover fish oil off notes. The consumer is very tolerant of oxidized citrus oils, and some companies intentionally "age" orange oil to produce a characteristic oxidized character that the consumer finds quite acceptable. The use of berry flavors to cover oxidized oil off notes is less obvious except for in the case of raspberry. Raspberry has a sensory character that can cover oxidized sensory notes quite well.

Masking

There is a limited amount one can do to truly mask an off note—that is, to block the sensory receptors so that the objectionable notes are not perceived. While there are examples of masking in the sensory literature, those that are well documented are few. One of the best proven examples is the masking of bitterness by sweetness. If one adds sugar to a bitter product, the bitter sensation will be decreased since the same taste receptor serves to detect bitter and sweet.[5,6] Thus, if there is a great deal of sugar in a product, the taste receptors are occupied with it simply due to mass action, and the bitter sensation is decreased. Anecdotal information in the literature suggests that salt masks a given note or that MSG enhances or masks a given note, but there is little agreement in these cases, and the documentation is often weak.

Flavor/Ingredient Interactions

The literature on food/flavor interactions has grown tremendously in recent years.[7,8] This knowledge base has been greatly enhanced by concerted efforts such as the European COST 96 Action (European Co-operation in the Field of Scientific and Technical Research) on flavor interactions with

---

Table 4–2 Flavors Most Compatible with Off Tastes in Functional Foods

Taste Sensation	Recommended Flavor Selection
Sweet	Mixed fruits; berries; specific fruits and berries; vanilla; maple and honey
Sour	Citrus; root beer; anise; licorice; raspberry; cherry; strawberry
Salty	Melon; raspberry; mixed citrus and fruits; maple; butterscotch; nut
Alkaline	Mint combinations; chocolate; cream; vanilla
Metallic	Grape; burgundy; lemon-lime

Courtesy of L. Fisher, Flavors of North America, Carol Stream, Illinois.

food matrices.[9] Thus, we are gaining a greater appreciation for how a given food ingredient or component will interact with a flavor component and thereby influence flavor perception. We have come to appreciate that proteins and certain carbohydrate polymers are good binders of flavorings. We understand that the physical properties of a food (eg, viscosity or friability) will influence how the flavor is released and subsequently sensed. The influence of fat content on food flavor has been well documented and modeled. Thus, as we create functional foods that have known ingredients and physical properties, we are better able to manage the interactions that occur.

In terms of interactions between flavors and specific ingredients, the flavoring of pharmaceutical products offers substantial knowledge. For example, the flavor industry has learned how to make a cherry flavor without benzaldehyde when the pharmaceutical has an active ingredient with an amine function, or grape without methylanthranilate when the active ingredient is a carbonyl. This is required to eliminate the amine/carbonyl reactions that result in the loss of both the flavor and the active ingredient (Maillard reaction). The problem with functional foods is that often the active ingredient of the food may be unknown (eg, in herbal extracts). Thus, the creative flavorist may have to experiment with the flavoring system and intuitively make such changes in flavor formulation.

Flavor Stability

Flavor stability may or may not be a problem in functional foods. Stability will depend upon whether the base product itself is stable or will develop off notes during storage. This will depend upon many factors, such as the functional component being delivered, its concentration, the food matrix itself, the presence of stabilizing ingredients such as antioxidants or chelating agents, packaging, and storage conditions (see review on lycopene stability[10]). The stability of flavorings added to the product is dependent on many of the same factors. However, one must again recognize that the food and flavor industries have dealt with delivering many of the same components already. Foods containing carotenoids (oxidation), thiamin (degradation to meaty or vitamin notes), or fish oils (oxidation) are very unstable during storage, and various techniques have been developed to minimize oxidized off flavors in these products. The most common approach is to use antioxidants when the products are subject to oxidative mechanisms.[11] The antioxidants may be natural products such as vitamin E (or derivatives thereof), ascorbic acid (when effective), or rosemary extracts. Vitamin E is expensive and has limited effectiveness, while rosemary extracts are effective but carry a residual rosemary odor. Natural antioxidants fit the healthy image better than synthetic antioxidants (eg, BHA or TBHQ) and are therefore preferred in functional foods. Teas and some herbal extracts have a significant antioxidant function as well and thus may serve as a natural antioxidant and flavoring.

Encapsulation may also be used to slow the deterioration of oxidizable components. Some work has been published on the encapsulation of carotenoids and fish oils. Through the use of antioxidants and the proper choice of encapsulation matrix and process, substantial improvements in oxidative stability can be achieved.[12] Some encapsulation matrices, such as those containing corn syrup solids (DEs ranging from 20–36) have been found to greatly increase the oxidative stability of spray dried citrus oils. One would expect similar protection for the carotenoids and fish oils. One should be reminded that encapsulation is a flavoring alternative only for functional foods that are delivered in the dry form (or may be reconstituted by the final consumer).

ENCAPSULATION TO MINIMIZE FLAVOR DEFECTS

Dietary supplements can be delivered to the consumer in several forms: as liquids, powders, infusions (teas), or encapsulates (tablets, soft or hard gels). The flavoring strategies discussed thus far have assumed that the functional food is to be delivered to the consumer in a form that can contribute an off flavor, that is, a nonencapsulated form. However, many functional foods are delivered to the consumer as an encapsulate, since the off notes associated with the product are not thereby perceived.

Encapsulating a dry herbal extract in a hard gelatin capsule (essentially filling a preformed two-part capsule with the dry ground herbal) has two purposes: (1) it acts to control serving size, and (2) it hopefully results in no perceptible flavor during swallowing. This delivery approach does not require any flavoring or flavor protection if the glassy gel is impermeable to odors. The technology used to form these capsules is quite old.[13] It can be as simple as dipping form pins (molds) into a gelatin solution, passing the coated pins through a drying chamber, and then removing the parts from the molds for filling and assembly.

Soft gelatin capsules offer the advantage of delivering a liquid functional ingredient (eg, fish oil or carotenoid). These capsules are simply swallowed without rupture, so there is no issue of off flavors or necessity for flavoring. The gelatin wall protects the liquid core from moisture, light, and oxygen. Gelatin can be a very effective oxygen barrier. Soft capsules can be formed by several techniques.[13] One approach is to use two opposing soft gelatin sheets that have preformed chambers (the core area). The chambers can be filled, sealed, and then cut cleanly from the sheet to form capsules.

Tablets are an additional form for the delivery of functional ingredients. Tableting is done by dry blending the desired formulation usually with an expedient if the added ingredient is in low concentrations, and then compressing it into tablets. The formulation can be made of various materials that vary in binding ability to form tablets that easily decompose in the mouth to those that are bound very tightly and do not decompose in the mouth. Additionally, the tablets can be coated to make them less soluble (not dissolving in the mouth) and to facilitate swallowing. As mentioned earlier in this chapter, encapsulation via these techniques solves the flavoring issue but creates problems for the elderly.

FLAVORINGS FOR FUNCTIONAL FOODS: NATURAL VERSUS ARTIFICIAL

There are pros and cons to both natural and artificial flavorings. Natural flavorings offer the advantages of clean labeling, a "healthy" perception, and marketability. Natural flavorings dominate the US market. Problems with natural flavorings include their higher cost, limited shelf life (potential), and possible incompatibility with the functional food (nonflavoring components). Artificial flavorings offer strength, stability, and good character but suffer from an unhealthy image with the consumer that is very contrary to the desired image of functional foods. Thus, the vast majority of functional foods are flavored with natural flavorings.

FLAVORINGS AS FUNCTIONAL FOODS

Both Caragay[14] and Manley[4] have given several examples of flavorings that have been shown to have positive health benefits. (Table 4–3 shows those with a potential for cancer prevention.) The stability of the flavoring becomes doubly important in this case, since the flavoring serves two purposes: active ingredient and palatability enhancer. Manley[4] noted that chili, mustard, allium, citrus, and some spice flavorings have health benefits. The chili flavorings contain high levels of capsaicins (the hot burning notes associated with hot peppers) that have been shown to have anticancer activity.[15,16] The capsaicins are relatively stable in most food environments, but the problem may be one of improving their palatability. They may be easily delivered in capsule forms that do not dissolve in the mouth but dissolve later in the stomach or digestive system.

Mustard-type flavorings contain isothiocyanates, which give mustard, horseradish, and radish their characteristic "bite." The isothiocyanates are found in much lower concentrations in Brassica vegetables (eg, cabbage, broccoli, and Brussels sprouts). Unfortunately, isothiocyanates are unstable in

Table 4–3 Food Flavorings That Are Believed To Have Cancer Preventative Benefits

	Sulfides	Mono-terpenes	Tri-terpenes	Phenolic Acids	Indoles	Isothio-cynates	Phthalides
Garlic	X	X	X	X			
Green tea				X			
Soybeans			X	X			
Cereal grains			X	X			
Cruciferous	X	X	X	X	X	X	
Umbelliferous		X	X	X			X
Citrus		X	X	X			
Solanaceous		X	X	X			
Cucurbitaceous		X	X	X			
Licorice root			X	X			
Flax seed				X			

Source: Reprinted with permission from A.B. Caragay, Cancer-Preventative Foods and Ingredients, *Food Technology*, Vol. 46, No. 4, p. 85, © 1992, Institute of Food Technologists.

aqueous systems and slowly degrade into nonflavorful and nonbioactive sulfur and nitrogen-containing components. Thus, they must be made in dry products. As with capsaicins, delivery in high doses is problematic due to their pungent odor, so delivery in an insoluble capsule is possibly the best choice.[2,4,17,18]

Onions, chives, shallots, garlic, and scallions make up the allium family. This group (particularly garlic) is considered to have anticancer activity as well as possible benefits related to cardiovascular diseases.[19,20] The bioactive components responsible for this health benefit are assumed to be allylic sulfur compounds. These compounds are offensive in sufficiently high dosages but are much less offensive than either the capsaicins or isothiocyanates. They are reasonably stable in most food systems and can be delivered more easily than the capsaicins or isothiocyanates.

The last group includes citrus and spice oils (caraway, dill, and spearmint),[4,19,20] and vanilla.[21] The "healthy" components of these oils are considered to be the monoterpenes d-limonene and d-carvone. These components are prone to oxidation (as are other components of the oils) and must be protected through the use of antioxidants or be encapsulated. The literature offers substantial information on the use of antioxidants for this purpose as well as methods for inhibiting oxidation through encapsulation.[22]

MEDICAL FOODS

Medical foods,[23] like other functional foods, may encompass many different products, each offering its unique flavoring challenge. Although the consumer generally has little choice in the consumption of these products and does so under a doctor's recommendation, this does not necessarily mean that there is no need for sensory quality in these products. Manley[4] has pointed out that flavor has several benefits other than simply making the food taste good. He has noted that food flavor stimulates the release of saliva and associated enzymes, enhances the release of gastric acid in the stomach, increases nutrient absorption in the intestines, and influences glycogen storage in the liver. He also notes that flavor influences the physiological well-being of a person. These benefits are arguably more critical in the case of a patient taking medical foods than the healthy consumer. Thus, it is important that medical foods offer the same sensory stimuli as conventional foods. Even if the food is fed by nasogastric or intestinal tube, the pleasant odor of the food can be beneficial to the patient.

While there are numerous medical foods on the market, potential flavoring issues will be considered for only three commercial products as examples. The first product to be considered is a "high nitrogen elemental" diet and is intended for patients suffering from inflammatory bowel disease, short-gut syndrome, transition from total parenteral to enteral therapy, cystic fibrosis, or nonspecific malabsorptive/maldigestive states. This product is a liquid and has at least 100% of the daily requirements for vitamins, minerals, and protein in a 2000-calorie diet. The protein is in the form of free amino acids (50%) and small peptides (50%), which bring in considerable aroma and taste. Protein hydrolysates are traditionally produced by acid treatment, which gives the product a characteristic hydrolyzed vegetable protein character. While this character has become almost synonymous with process flavors (reaction products used to simulate meats), these sensory notes are undesirable in other food products. The hydrolysates may also bring in bitter tastes as well. Additionally, the vitamins and minerals are problematic. The vitamins, most notably thiamine and potentially vitamin A, will also contribute off odors on degradation. This problem is well known to the breakfast cereal industry, which receives a considerable number of consumer complaints due to off notes caused by the vitamins used in fortification.

The off notes from ingredients and need for a sterile process make flavoring this type of product extremely difficult. The most successful approach may create a meaty flavor. Meat flavors are compatible with the character of the protein hydrolysate and quite tolerant to aseptic processing. However, meat flavors are associated with heated foods, which then would require that the product be heated before consumption. This reduces the convenience of the product. The manufacturer may also investigate the use of the newer, enzymatically produced protein hydrolysates. These products are much blander than the corresponding acid hydrolysates. Another alternative is to highly flavor (fruit) and sweeten (high potency sweeteners) the product. The consumer expects fruity-flavored products to be served cold, so the flavor can meet the consumer's expectations in this respect. The sweetness may mask some of the bitterness, and the flavor "cover" some of the off notes. However, the sensory acceptability of this type of product will be marginal at best.

A second example is a liquid product and is intended for use by individuals with diabetes, glucose intolerance, or stress-induced hyperglycemia. It is high in protein (unhydrolyzed milk protein) and balanced in fat and carbohydrate. It contains a full complement of vitamins and minerals. This product may be used in hospital settings or by individuals as a meal replacement or supplement. This product would be expected to carry some typical vitamin off notes but otherwise would not present unique challenges in flavoring. The vitamin notes may be quite compatible with berry flavors (particularly raspberry or citrus).

The third example is a very low-fat dry product (about 0.2% on a reconstituted basis) that is recommended for individuals with cancer, pressure ulcers, cerebrovascular accident, oral surgery, geriatrics, protein-calorie malnutrition, pre- and postoperative diet, or anorexia. However, it is also purchased by individuals who desire protein supplements. This product would not bring in particular off flavors other than those associated with the vitamins or their deterioration products during storage. One of the limiting aspects of flavoring this product would be to deal with the very low fat content. The flavor industry has not dealt particularly well with low fat/no-fat foods in the general marketplace. The lack of fat results in an imbalanced flavor that often appears unnatural. An alternative strategy for flavoring is to use beverage-type flavorings, since beverages typically do not have any fat.

In conclusion, a wide variety of food products falls into the category of functional foods. Success in giving these products acceptable sensory properties will depend strongly on the flavor quality of the products themselves. It is important that the best quality ingredients be used to minimize off flavor. It is difficult or impossible to make a severely off-flavored product taste good through flavoring. Additionally, the flavor stability of these products depends upon the food itself, the required processing, and storage. Each of these factors can pose problems for sensory acceptability. Fortunately, while some of these products are very poorly defined and their flavoring will take the flavorist into "uncharted waters," the vast majority will not present new challenges. Previous experience in the flavoring of pharmaceutical or food products will provide guidance in flavoring. While functional foods are a relatively new and undefined category of products for the food industry, they are sufficiently similar to existing products on the market that the flavor industry should be able to deal with them knowledgeably.[24]

REFERENCES

1. Reineccius GA. *Food Flavor Workshop: Flavor Applications*. Offered by the Dept. of Food Science and Nutrition, University of Minnesota, St Paul, MN, 1999.
2. Milo OL. Functional foods put the flavor up front. *Prepared Foods*. 1999; 168(2):59–60, 62, 64.
3. Fischer L. Flavoring nutraceuticals and functional foods. Presented at the IFT workshop on Flavor/Food Interactions, New Jersey, 1998.
4. Manley C. The flavoring of nutraceuticals, dietary supplements, functional foods and medical foods. Presented at Nutracon 97, Las Vegas, NV, 1997.
5. Heijden A. van der. Sweet and bitter tastes. In: *Flavor Science: Sensible Principles and Techniques*. TE. Acree and R. Teranishi, eds. ACS Publ: Washington, DC, 1993;67–116.
6. Walters DE, Roy G. Taste interactions of sweet and bitter compounds. *Institute of Food Technology Annual Meeting Abstracts*. 1995:171.
7. Gaonkar A. *Ingredient Interactions: Effects on Food Quality*. New York, NY: Marcel Dekker, Inc; 1995.
8. Contis E, Ho CT, Mussinan C, Parliament T, Shahidi F, Spanier A, eds. *Food Flavors: Formation, Analysis and Packaging Influences*. Amsterdam, The Netherlands: Elsevier Science Publications; 1998.
9. Schieberle P. COST (European Co-operation in the Field of Scientific and Technical Research) 96. Interaction of Food Matrix with Small Ligands Influencing Flavour and Texture. Brussels, Belgium: European Commission; 1998.
10. Nguyen ML, Schwartz SJ. Lycopene: chemical and biological properties. *Food Technol*. 1999; 53(2):38.
11. Elliott JG. Application of antioxidants vitamins in foods and beverages. *Food Technol*. 1999; 53(2):46.
12. Desobry SA, Netto FM, Labuza TP. Comparison of spray-drying, drum-drying and freeze-drying for beta-carotene encapsulation and preservation. *J. Food Sci*. 1997;62(6):1158–1162.
13. Gutcho A. *Capsul Technology and Microencapsulation*. Park Ridge, NJ: Noyes Data Corp; 1972.
14. Caragay AB. Cancer-preventative foods and ingredients. *Food Technol*. 1992;46(4):85.
15. Walker M. Piperine in black pepper: one of the newly recognized class of thermonutrients. *Health Foods Business*. 1997;43(4):40.
16. Surh YJ, Lee SS. Capsaicin in hot chili peppers: carcinogen, co-carcinogen or anticarcinogen. *Food Chem Toxicol*. 1996;34(3):313–316.
17. Rhodes MJC. Physiologically-active compounds in plant foods. *Proceed Nutrit Soc*. 1996;55(1B):371–384.
18. Ahmad JI. Garlic—a panacea for health and good taste? *Nutrition Food Sci*. 1996;1:32–35.
19. Matsukura T. Antimutagenic and anticarcinogenic substances obtained from flavor materials. I. *Koryo*. 1994;183:61–68.
20. Matsukura T. Antimutagenic and anticarcinogenic substances obtained from flavor materials. II. *Koryo*. 1994;184: 113–119.
21. Kometani T, Tanimoto H, Nishimura T, Okada S. Glucosylation of vanillin by cultured plant cells. *Bioscience Biotechnol Biochem*. 1993;57(8):1290–1293.
22. Reineccius GA. Flavor encapsulation. *Food Reviews Internat*. 1989;5(2):147–176.
23. Mermelstein N. Medical foods. *Food Technol*. 1992;46(4):87.
24. Berge P van, Evenhuis B. Flavors into the 21st century. *Perfumer Flavor*. 1998;23(3):1–2, 4–6, 8–10, 12–13.

CHAPTER 5

Measurements of Nutrients and Chemical Components and Their Bioavailability

Jonathan W. DeVries and Karlene R. Silvera

With functional foods, as with conventional foods, measurement of key characteristics and parameters is essential to ensure the quality of the food—that is, to ensure its safety and its claimed functionality as it is being consumed. Obviously, measurements that are not essential are sometimes made on a food or food product, but such occurrences are rare and are usually the result of a poor understanding of the real purpose for carrying out any measurements at all. It behooves all analysts to determine why analyses are being carried out and not just to make a measurement "because someone told me we had to do it." On the other hand, discontinuing a measurement protocol and data collection after a superficial fruitless survey of the reason for the assay can be a dangerous decision. Some digging in the archives and discussions with staff experienced with the situation may reveal that in fact some quality aspect of the product has been maintained through the years by the assay. The assay data may be used for controlling the quality of the product so well that the reason has not been passed along with the method.

Measurements are the building blocks of quality. Defining quality in its broadest sense requires ensuring that everything about the product is up to standard. This encompasses the safety, authenticity, performance, and desirable characteristics of the product. Developing, producing, and distributing a product are integral parts of a whole system, and quality is one of the wheels that keeps the parts interconnected and running smoothly. If one cog of the quality wheel is missing or breaks off, the whole machine grinds to a halt. Measurements further serve to develop an understanding of the products and processes they relate to. This understanding leads to adequate process control, which in turn leads to consistent quality.

PURPOSES OF MEASUREMENT

Detection of Adulteration or Misbranding

A primary purpose of making measurements is to ensure that a functional food is not adulterated or misbranded—that is, to ensure that what is represented as being sold for consumption is indeed what is being sold. This assurance goes beyond the mere identification of the main constituent(s) of the product. It extends to ensuring the presence of all the components that should have been added as well as ensuring that constituents not expected to be present are indeed absent. For example, a fortified product that lists a relevant level of vitamin C on the label must have vitamin C present at or above that la-

The authors wish to thank Karen Shrake for all of her assistance in preparing this chapter.

beled level in the product to not be considered misbranded or adulterated. Likewise, if that product does not list peanuts as an ingredient but there are peanut proteins present, the product label will be out of compliance because, without an adequate warning, an allergenic danger exists for peanut-sensitive individuals. Of course, over and above the existing legal requirements, all ethical and reputable manufacturers want to produce safe products, maintain company credibility, and credibility in the products they produce. Various tools can be applied to aid the analyst in ensuring authenticity. For whole plant species or discernible fragments of plants, microscopy is an effective tool. For ground powders, especially where plants may be similar but different varieties have different characteristics, as in the case of the various varieties of ginseng, molecular fingerprinting technology may be necessary to identify the material and ensure against misbranding and adulteration.

Safety

A number of measurements apply directly to ensuring the safety of a food, the most evident being microbiological assays (and related assays, such as pH and water activity, which aid in controlling/preventing the growth of pathogens) and sterility testing. An equally obvious safety issue is product tampering, although measuring the occurrence of or measuring to prevent the occurrence of product tampering is not necessarily straightforward. Measurements are also made to ensure against the presence of or excess levels of naturally occurring undesirable toxins such as mycotoxins, the species-specific by-products of mold growth. Heavy metal residues or heavy metal–derivative residues, whether incurred due to their natural presence in the soil where the plant is grown or incurred due to industrial contamination, are always of concern. Pesticides (whether fungicides, herbicides, insecticides, or rodenticides), highly desired natural or synthetic compounds to ensure adequate supplies of a plant or commodity, become less than desirable if still present in excess quantities in the plant product at the point of consumption. Further, there is a need to ensure against the unlabeled presence of allergens, a situation that might put sensitive individuals at risk. At this time, the US Food and Drug Administration (FDA) recognizes peanuts, tree nuts, fish, shellfish, wheat, soy, eggs, and dairy foods as true allergens. Many compounds that have a positive functional effect at one level are toxic at a higher level, so measurements are necessary to ensure against excessively high quantities of these biologically active materials. One example of this is vitamin A; another is ephedra, a component of ma juang. These will be discussed further in the Toxicology section later in this chapter.

Economics

Economics is a tremendous driving force for accurate analytical methods. A simple example is the analysis of moisture in a product or ingredient. Obviously, water is the least expensive component of most ingredients or products. There is no question, then, that none of us want to pay the expensive price of an ingredient, only to find that we are receiving excess water for part of that expenditure. Therefore, we want to control the water content of the ingredients we buy. We also often want to add water or remove water as part of our formulations to obtain key textural and safety end points. Nonetheless, from an economics standpoint, we do not want to dry a product excessively, since this practice results in increased ingredient costs and energy expenditures.

Major commodities are typically traded on the basis of a key, measurable characteristic. Many of the cereal grains are traded at a price adjusted on the basis of moisture and protein content. Similarly, the oil content of oilseeds can be a determining factor in the trading price. Other measurements related to the economics of the commodity being tested include dockage for chaff and debris in the material, insect damage, and functional component content. Examples of functional component content would include plumpness in grains as an indication of edible portion of the kernel; protein content of wheat as a predictor of the quality of the bread ultimately baked from the flour made from the wheat; ginsenoside content of ginseng as an indication of medicinal potency; and beta-glucan content of oatmeal, oat bran, and oat flour and their products as an indication of heart health impact.

Quality Assurance

Traditional quality assurance is perhaps the most obvious reason for conducting analytical measurements. Quality assurance (besides having readily apparent safety ramifications) works to ensure that the desirable characteristics that have been designed into a product are present when the product is at the point of consumption. Quality control testing is applicable to incoming ingredients, processing and in-process product, finished product, and stored product to the point of use. The most obvious test measurements applicable are those of sight, taste, touch (texture), and smell. Obviously, no one wants to purchase ingredients or products that are spoiled or rancid or will become so in short order. On the other hand, these obvious test measurements are perhaps the most difficult to quantitate accurately and precisely because everyone's perceptive capabilities are slightly different. In addition to producing highly desirable products, reasons for carrying out quality assurance measurements include ensuring that the products produced are in compliance with all applicable laws and regulations, ensuring that the products are consistent from one batch to the next, and ensuring continuous safety of the product to the consumer.

GENERAL CONSIDERATIONS OF ANALYSIS

Sampling

While this chapter cannot fully discuss the key factors in adequate sampling of the material to be characterized by analysis, some of the key factors should be pointed out. The most obvious requirement, easily stipulated but difficult to fulfill, regarding sampling is that the sample being analyzed should be representative of the entire quantity of material to be characterized. Ascertaining the representative character of the sample can be approached in a number of ways. For instance, if the entire lot of material is already homogeneous or can be homogenized before sampling, almost any portion of the material can be taken as a sample for analysis. An excellent example of this is stirred, pasteurized, homogenized milk. It is difficult to envision any subsample that would not be representative of the entire lot. Likewise, a lot of juice beverage produced from juice concentrate and mixed thoroughly in a large-scale mixer would be expected to be homogeneous, and any subsample would be representative. On the other hand, dealing with the sampling of particulates presents a major challenge. In fact, one of the most challenging situations facing the analyst is the sampling of naturally contaminated particulate commodities. Natural contamination often is not evenly dispersed on the commodity in which it occurs. For example, *Aspergillus* mold growing on peanuts or corn typically attacks relatively few of the kernels, but with such a high concentration on those particular kernels that it contaminates the entire quantity of crop in the vicinity of the contaminated kernels to the level of tens of parts per billion. The challenge to the analyst, then, is to ensure that the contaminated kernel is represented in the analytical portion taken but not so overrepresented as to be a "hot spot" that indicates an artifically high level of the contaminant present. Therefore, the analyst needs to aim for the maximum practical size reduction combined with thoroughness of mixing to achieve distribution of contaminated portions. One contaminated peanut weighing about 0.5 g can contain enough aflatoxin to result in a significant aflatoxin level when mixed with 5 kg of peanuts. If the analyst is taking a 50-g subsample, the single bad nut has to be broken into a minimum of 100 pieces, and those pieces need to be uniformly blended through the entire 5-kg mass. To achieve the required size reduction, the ground sample must be able to pass through a 20-mesh sieve. The sample must then be thoroughly mixed.[1] Whitaker[2] provides an excellent discussion of the effect of sampling on the probability of correctly accepting or rejecting a lot of material when analyzing low-level contaminants on particulates. When a single, sufficiently large subsample of the material to be characterized cannot be obtained, compositing of a random selection of a number of small samples randomly selected from the lot to be analyzed followed by thorough homogenization can be an effective means of achieving representative results. This is the approach taken by FDA when sampling food products. Twelve random samples are selected from a particular lot of product. The samples are then composited to achieve a subsample considered representative of the lot before analyses are conducted. Care must be taken when using the composite approach. If one is looking for variability within a lot of material, compositing too many subsamples will average the variability to the extent of obscuring high or low levels of analyte that could be significant.

Measurement Accuracy and Precision Concepts

It is the objective of all analysts to provide the most accurate and precise data possible, suitable for the purpose of the analysis being run. Despite this desire, analytical results can rarely be repeated exactly from one run to the next. If one assumes that the results of the method are accurate—that is, in close agreement with the amount of analyte truly present—and the analyst carries out the measurement more than one time, then the data points from the individual measurements will tend to cluster around the true value. The dispersion around the true value can be calculated mathematically as the standard deviation of the results, with approximately 65% of the results falling within one standard deviation, 95% within two standard deviations, and 99% within three standard deviations. The standard deviation is a useful tool for estimating the likely interval (confidence interval) that a single measurement will fall in: that is, a 65% chance of falling within one standard deviation, a 95% chance within two standard deviations, and a 99% chance of falling within three standard deviations. The variability expected from a single measurement is sometimes referred to as the *uncertainty* of the result. In some cases, attempts are made to estimate the uncertainty of a method by summing the uncertainties of the various steps—that is, $U_t^2 = U_w^2 + U_x^2 + U_y^2 + U_z^2$. Although this may be realistic for physical measurements such as weighing procedures, applicability to chemical and microbiological procedures is limited. In most cases variability of one or more steps is unknown, and in some cases the source of variability is unidentifiable. Therefore, it is necessary that variability of methods be determined in the laboratory, preferably using multilaboratory protocols, as discussed later. The term *uncertainty* has been applied to both the results of an analytical test and the assigned value for reference materials and standards (see the sections "Measurement Standards and Traceability" and "Reference Materials" below). Application of the term to reference materials is logical: the producer of the reference material is uncertain about the exact value to assign for the analyte in the material. An assigned value with an uncertainty interval indicates to the user that caution is advised in relying entirely on the assigned value. On the other hand, users of analytical results desire to have confidence in the data provided by the analytical measurement. The analyst, on the basis of method validation and experience, is confident that the result obtained lies within a defined interval of the true value. Therefore, it is preferable, when reporting analytical results, to refer to the *confidence interval* of the reported data, applying the appropriate terminology and reducing the confusion regarding the meaning of results.

When replicate measurements are made, if the average result of the replicate measurements is calculated, the calculated result should tend to move closer to the true value as more individual results are averaged. Thus, the mean of the averaged measurements will still have some variability, but the variability (ie, the standard deviation of the average) will be reduced by the square root of the number of measurements taken. This variability in the mean result, termed the *standard error of estimate of the mean*, is distinguished from the standard deviation in that it indicates the variability of the analyst's estimate of the true value as compared to the variability to be expected when a single measurement is made. Because the standard error of estimate is reduced by the square root of the number of replicates, running duplicates reduces the variability of the estimate of the mean by nearly 30%, running quadruplicates cuts the variability in half, and running nine measurements cuts the variability to one third.

Two factors affecting analytical precision should always be kept in mind when reviewing analytical data. First, the analytical standard deviation will remain fairly constant over a broad range of analyte concentrations if there are a significant number of manual handling steps as part of the method involved. For example, the analysis of dietary fiber involves numerous manual manipulation steps. Each step adds a small increment of variability to the result—that is, it adds slightly to the standard deviation. The standard deviation is fairly constant at approximately 1% dietary fiber across the range of dietary fiber from 0% to 100%. Of course, the relative standard deviation shifts from roughly 1% at the 100% fiber level to 100% at the 1% fiber level. Second, in instrumental assays with relatively few manual steps, the standard deviation of the result decreases with decreasing analyte level, and the relative standard deviation stays relatively constant, increasing slowly (as a logarithmic function) as the analyte level decreases. Horwitz[3] has delineated this change by developing the following equation:

$$RSD_R = 2^{(1 - 0.5 \log C)}$$

where

RSD_R is the average relative standard deviation to be expected (SD_R/concentration), R is the between-laboratory designation, and C is the concentration of analyte in decimal terms (100% = 1, 1% = .01, etc). This equation was derived from close scrutiny and study of hundreds of multilaboratory collaborative study reports published through the years and represents the average variability one can expect for a particular level of analyte. When it is applied to various method studies conducted in multilaboratory settings, one can rapidly develop an appreciation for the difficulties in obtaining precise results with a particular method or for a particular analyte and for the degree of replication necessary to build confidence in the average value obtained for the analyte.[4] Table 5–1 illustrates the expected increase in analytical result variability with decreasing analyte concentration level.

Earlier in our discussion, we stated that chances are 95% that the true value of a data point lies within two standard deviations of a given value obtained by measurement. If we look closely at the table, we find that we can be only 95% confident that the result we obtain is indeed not zero when the

Table 5–1 Expected Variability of Analytical Results Relative to Analyte Concentration

Concentration (Decimal)	Common Units	Expected %RSD$_R$*
1.00	100%	2
0.10	10%	2.83
0.01	1%	4
0.001	0.1%	5.66
0.0001	0.01%	8
10^{-5}	10 ppm	11.31
10^{-6}	1 ppm	16
10^{-7}	100 ppb	22.63
10^{-8}	10 ppb	32
10^{-9}	1 ppb	45.25
10^{-10}	100 ppt	64

*Note that these are the average expected relative standard deviations for a given analyte level, based on data obtained from published collaborative study reports. The actual RSD_Rs will range up to twice the average.

analyte level is at or above 5.2×10^{-10} (ie, 520 ppt, at 50% relative standard deviation). At or below this level, we absolutely need to run replicate assays to reduce the estimated variability of our result to give us confidence we have a real number value. Alternatively, we need to do sufficient method validation to show that the variability of the method we are using is substantially lower than typically found.

Measurement Standards and Traceability

To have meaning, the results of any measurement must be useful to the parties using the data generated. In many cases, this usefulness can be established contractually between two parties working together, and the interaction regarding the commodity in question is limited to those two parties. On the other hand, for many measurements, a means of tracing the measurement back to a single reference standard is desirable. Let us take a simple example. Can you imagine what it would be like if everyone made length measurements with rulers having different scales? Chaos would result. If one desires to measure length in metric units, one typically uses a meter stick or a shorter rule calibrated or sectioned in centimeters or millimeters. Although it is not explicitly stated, one expects that the meter stick being used is somehow related through a series of serial comparisons to the "original meter" located in Sevres, France, and maintained as the ultimate unit against which lengths are measured. In theory, then, the measurement we just made can be traced back to the well-defined unit of measure, the meter. Similarly, other basic units of measure can be traced back to their respective standards.

The traceability system described above works well for simple units of measure, such as length. However, when more complex measurements are to be made, say, a measurement by a complex analytical chemistry or microbiological procedure requiring numerous steps, obtaining agreement and widespread acceptability of results is certainly more involved. Rarely do two laboratories desiring to measure the same analyte in a particular matrix approach the analysis in exactly the same way. As a result, two different laboratories obtain somewhat different results. Even an individual analyst introduces minor changes (that he or she is often not aware of) into a procedure when it is repeated, resulting in some variability. To enhance agreement of results, scientific and other professional societies have developed programs to promote agreement among measurements, primarily by standardizing the methods used and secondarily by providing reference materials to ensure the proper performance of those methods.

Standardized Methodology

AOAC International (formerly the Association of Official Analytical Chemists), through its cadre of scientific volunteers, adopts analytical methods as Official Methods of Analysis[5] for use in making measurements related to regulatory compliance and action and for commerce and trade, primarily in the areas of foods, feeds, and fertilizers. To qualify as an Official Method of Analysis, the method must have demonstrated adequate performance when used in multiple laboratories. This performance is assessed by means of a collaborative study, the design of which was developed by the participants at the International Union of Pure and Applied Chemistry (IUPAC) workshop on Harmonization of Collaborative Analytical Studies in Geneva, Switzerland, in 1987.[6,7] For quantitative methods, this requires data on a minimum of five materials analyzed in a minimum of eight laboratories. For qualitative methods, data on at least 15 samples analyzed in eight laboratories are required. For qualitative methods of analysis, these samples include high, low, and blank levels of the analyte in question.

The entire collaborative study process is handled by a designated associate referee, a person chosen for the task by virtue of his or her expertise in the assay and assay subject area being studied. Prior to the collaborative study, the method to be studied is typically peer reviewed by a "general referee" who is expert in the analytical subject area and the appropriate methods committee, which is also knowledgeable on broad methodology needs in the subject area. After completion of the collaborative study, the method and method performance data collected are peer reviewed to ensure that methods suitable for their intended purpose are adopted as Official Methods of Analysis. Finally, all methods are reviewed by the Official Methods Board to ensure consistency in method quality across diverse subject areas and to ensure that adequate peer review has been completed before adoption. If accepted as an Official Method of Analysis, the method is placed in Official First Action Status. Following Official First Action Status adoption, a method must be published and available for general use without modification and without significant problems reported for at least 2 years before it is eligible for consideration as an Official Final Action Method, adoption of which occurs by vote of the AOAC membership.

AOAC International has taken steps to stay up to date with methods related to the functional foods area. Before the term *functional foods* became commonplace, methods of a "functional food" nature were addressed by AOAC. Before total dietary fiber was considered a standard nutrient factor, AOAC led the way in method validation for dietary fiber and related methods. Similarly, in anticipation of methods needs in the functional foods arena, AOAC has now set up, under its Methods Committee on Food Nutrition, two general referee assignments on the subjects "Phytonutrients" and "Botanicals and Other Supplements." An associate referee exists in the phytonutrients area, focusing on isoflavone in soy and soy-based products. Specific topics handled by the "Botanicals and Other Supplements" general refereeship include associate referee positions in botanical microscopy and the analysis of skullcap and germander, ephedra, ephedra alkaloids, hydroxycitric acid in garcinia cambogia, and proamphocyanines, all of which are expected to be of relevance to functional foods at some point in the future.

The United States Pharmacopoeia (USP) emphasizes the optimization and standardization of methods of pharmacological significance. Like AOAC, USP depends on expert volunteers for peer review of methods and standards. However, unlike AOAC International, USP does not require collaborative study in multiple laboratories as a prerequisite for adoption of the methods that are included in its compendium. Pharmacological preparations are typically well-characterized matrices requiring the measurement of a single entity or two. Method adoption approaches that do not require collaborative study may not prove adequate when complex matrices such as functional foods are involved.

The American Society for Testing and Materials (ASTM) is primarily known for adopting standards with an emphasis on physical characteristics and physical measurements. Examples are production of standards for packaging materials and standards for packaging, handling, and shipping practices. The ASTM standards adoption procedure incorporates collaborative studies, where applicable, followed by scientific and professional agreement on all standards before adoption. Agreement on standards is achieved by an extensive balloting system that includes all members of working subcommittees, committees, and the society as a whole. All negative ballots and respective comments are resolved before standards are adopted.

Two organizations that are more focused with regard to the scope of their interests, namely the American Association of Cereal Chemists (AACC) and the American Oil Chemist's Society (AOCS), have developed memoranda of understanding with AOAC International regarding development, acceptability, and publication of standardized methods. Like AOAC, AACC and AOCS follow the International Harmonized Protocol of Guidelines for Collaborative Study Procedure To Validate Characteristics of a Method of Analysis for method evaluation and adoption.

Whenever available, methods validated in multiple laboratories and adopted through stringent scientific peer review systems should be used for making measurements. In spite of this, individual laboratories will always carry out procedures slightly differently from other laboratories. This leads to minor variability in results between laboratories, even when the same method is used. When laboratories use different methods (many analysts would rather use another analyst's toothbrush than that analyst's method), the results can be dramatically divergent. The use of standardized methods can obviously reduce this divergence.

Reference Materials

An additional tool available to analysts that helps to reduce between-lab variability is the use of reference materials. An ideal reference material contains the analyte or property of measurement interest at an accurately and precisely defined level in a matrix identical to the matrix the analyst is testing. Unfortunately, this ideal is rarely achieved. Providing and using such an array of reference materials would be prohibitively expensive. Preparing reference materials requires that a representative matrix be selected on the basis of its expected analytical behavior. The sample must then be homogenized, or at least an adequate level of homogeneity must be ensured. The effort involved in establishing the analyte level in the reference material will depend upon the difficulty of the assay and the accuracy and precision desired and necessary for the analyte under consideration. Typically, the analyte is determined in replicate, using more than one method in a number of different laboratories over a period of time. The results are expressed as a mean value along with an uncertainty interval around that mean. As discussed previously, *uncertainty* is the correct terminology to use with reference materials, since the reference material producer cannot be absolutely certain of the concentration or where the value of the concentration lies in the uncertainty range. The analyst using the reference material should be sure to determine the nature by which the uncertainty interval provided with the analyte value is expressed. The uncertainty interval may be expressed as that expected for a single determination in the future—that is, as some function of the standard deviation of the results for the analyte. Alternatively, the uncertainty interval may be the standard error of estimate of the quantitative value of the analyte (ie, some function of the standard error of estimate of the mean). Means and uncertainty intervals reported in the latter format will typically be unachievable by a laboratory in a single or a few assays of the reference material. In this case, the analyst will need to multiply the standard error of estimate uncertainty interval by the square root of the number of assays used to establish that interval to arrive at the uncertainty interval expected for a single assay. And, of course, the reference material producer needs to establish the stability of the analyte or, if it is not completely stable, to understand the kinetics of any degradation going on. In the United States, the National Institute of Standards and Technology (NIST) is the best known provider of reference materials. Other providers of potential interest for analysts of functional foods are USP, AACC, and AOCS. Ensuring that one's assay results fall within the acceptable range of values of a reference material ensures that the method is in control, at least for the matrix of the reference material. The individual laboratory needs to validate its methodology for applicability to other matrices.

To reduce operating costs, laboratories can develop their own in-house reference materials. A matrix equivalent to or similar to the matrices expected to be analyzed is selected with an analyte level similar to that likely to be encountered. After sample preparation and homogenization, the material is divided into subsamples and placed in storage. Different laboratories use different routines for establishing the reference value of the in-house reference material. Some laboratories analyze the sample in quadruplicate each day for four different days; some analyze in triplicate each day for five days. In general, a database of 20 or more quantitative analytical data points is desirable before the reference material is used to ensure performance of a method. Collecting the data over as many days as possible is preferred to running many replicates on the same day so that the influences of as many uncontrolled variables as possible have been taken into account. An excellent overview on establishing in-house reference materials was published in *The Referee* in 1993.[8]

Proximate Analysis

In the world of foods analysis, as in most aspects of life, everyone desires to have clear-cut, black-and-white answers to qualitative and quantitative questions. Unfortunately, there are a number of food assays where the analyte being quantitated is not clearly defined—that is, the analyte is neither a single chemical entity nor the sum of specific chemical entities. In such cases, the analyst must use a *proxi-*

mate analysis. *Proximate* analyses are those analyses where the final analytical result (and in essence the analyte) is defined by the method used. An example is dietary fiber. Dietary fiber has a functional physiological definition that is supported by a defining method. According to the current working definition, "Dietary fiber consists of the remnants of plant cells, polysaccharides, lignin, and associated substances that are resistant to hydrolysis (digestion) by the alimentary enzymes of humans."[9] This is a "gold standard" definition, based on the physiological function of the diet component, as opposed to a chemical definition. The chemistry, then, from an analytical perspective, is adjusted to match this definition. As the scientific understanding of dietary fiber changes, the chemistry related to its analysis may need to change as well. No analyst can perfectly mimic the human digestion system at the laboratory bench. Therefore, methodology for dietary fiber is aimed at simulating, on the laboratory bench, the digestion that goes on in the stomach and small intestine of the body. Thus, the method defines the outcome of the assay. The specifics of the dietary fiber assay will be discussed later on. Moisture, perhaps the most commonly and frequently run assay for foods after product weight, is generally run as a proximate assay. Absolute methods, such as the Karl Fischer titration method and the gas chromatographic method (water is extracted into dimethylsulfoxide before injection onto a gas chromatograph equipped with a thermal conductivity detector), are available that quantitate all the water present in a sample. However, the majority of all moisture analyses are carried out by proximate procedures. Of this majority, the most common moisture methods are based on measuring a change in sample weight while drying under a specific set of conditions. Other approaches include measuring changes in the conductivity or the dielectric constant of the sample. Until recently, the analysis for the fat content of food was clearly a proximate analysis: fat was defined as the lipidous material that was extracted from the food samples by a defined set of lipid-solubilizing solvents, with the extraction being carried out under a clearly defined set of conditions. With the recent adoption of regulations[10] for compliance with the Nutrition Labeling and Education Act, fat is more clearly defined as the sum of all the fatty acids present in the food, expressed as their triglyceride equivalents. This means that fat is now recognized as a definitive chemical entity (sum of chemical entities) and that methods of analysis need to quantitate all the fatty acids present in the sample to the exclusion of other lipid solvent–soluble materials. Protein, determined by measuring the nitrogen present in the sample by a defined procedure (essentially all food nitrogen methods are expected to give results equivalent to the Kjeldahl digestion method using mercury salt catalysis) is clearly being determined by a proximate procedure. Further, except for certain matrices such as wheat and wheat products (factor = 5.7) and dairy products (factor = 6.38), a factor of 6.25 for converting nitrogen to protein is used,[11] although a wide range of factors for a variety of matrices could potentially or theoretically be applied.[12] Other analytes determined by proximate methods include betaglucans, ash, polyfructans (ie, inulin and oligofructosaccharides), soluble and insoluble components of dietary fiber, and carbohydrate content by calculated difference.

ANALYTICAL CHEMISTRY

One of the most powerful tools available for measurement of food parameters is analytical chemistry. The use of analytical chemistry applies not only directly to the analysis of the food itself but also to the monitoring of the effects of the food after it is consumed. Analytical chemistry is used to determine the bioavailability of essential constituents. It is also used as part of the assessment of the toxicity of food components. Even in the field of microbiology, analytical chemistry is being directly applied. GC coupled with mass spectrometry can be used to profile unique fatty acid patterns so that microbes can be positively identified. Riboprinting, the technique currently achieving wide use for the identification of microbes, is also heavily dependent on analytical chemistry techniques for successful qualitative results.

Specific Measurements

pH

One of the very important measurements made on food products from a safety and quality perspective is the measurement of the hydrogen ion concentration present in the aqueous phase of the product. In common terms, this concentration is referred to as pH. The importance of this measurement will be discussed further in the microbiological analysis section. Normally, the hydrogen ion concentration of solutions varies from $>10^{-1}$ (very acid) to 10^{-7} (neutral—water) to 10^{-14} (very alkaline). For convenience, this hydrogen ion concentration is expressed as the negative logarithm of the concentration. Therefore, a very acid solution has a pH of <1, a neutral solution a pH of 7, and a very alkaline solution a pH of 14. The fact that pH is actually the negative logarithm of the hydrogen ion concentration is important to keep in mind when setting specifications and making measurements. A shift in pH of 1 unit represents an order of magnitude ($10x$) change in hydrogen ion concentration; even a shift in pH of only 0.1 unit represents a 25% change. Therefore, it is essential, especially when tight tolerances are involved, that pH equipment be properly calibrated at two points of the pH scale (ie, one point on either side of the pH being measured) on a frequent basis and that appropriate temperature compensation be in place.

Titratable Acidity

A measurement that is often associated with pH is titratable acidity. Although the pH and titratable acidity of a given food may correlate with one another as long as the constituent makeup of the food is constant, there is no direct absolute relationship between titratable acidity and pH. Titratable acidity measures the total quantity of acidic material present in a sample. pH is dependent on the total quantity of acids as well as the strength of the acids present. Titratable acidity is a useful measurement to make to monitor the progress of acid-producing fermentations. It is also useful for monitoring the breakdown during production and storage of fats to fatty acids in fat-containing products. Methods useful for this purpose include AOAC Official Method of Analysis 939.05.[13]

Water Activity

Another chemical measurement closely related to microbiological safety is water activity (A_w). For microorganisms to survive and grow in or on an edible item, a sufficient quantity of water has to be present. Not only must water be present in sufficient quantity, it must be available for the microorganism to use. One might think of this water as being "free" to be used. If one pictures "free" water as quantitatively related in some way to those molecules available to equilibrate with the atmosphere around the sample, one has a means of applying a quantitative measure to the usable water. This "free" water is related to the "activity" level of the water. Water activity is defined as the relative vapor pressure of water over a sample compared to the vapor pressure of pure water under the same conditions. This may be more simply pictured as the relative humidity of the air over a sample compared to 100% relative humidity, with 100% relative humidity being the saturation quantity of water that the air can hold at a given temperature. Water activity measurements (as measured by the water vapor pressure over the sample in a closed vessel) depend on measuring a change in some other measurable physical property that changes with a change in humidity. Examples are measuring changes in the length of a humidity-sensitive fiber such as human hair, measuring the change in capacitance of polymer films, measuring the weight gain of desiccating solids, measuring a change of conductivity of salt solutions, or measuring the dew point of the air above the sample. This dew point is most easily measured by condensing the moisture of the air on a mirror when the system is at equilibrium, identifying the corresponding temperature, and using the observed dew point to calculate the water activity of the sample. From a chemistry perspective, a high water activity, generally above 0.5 A_w, leads to rapid enzymatic degradation of a food. An A_w above 0.3 leads to rapid nonenzymatic browning if the product is susceptible to such reactions, while a very low water activity, generally below 0.15, leads to rapid lipid oxidation and consequent rancidity of fat in the sample.[14,15]

Antioxidants

Enhancement of human health and prolongation of human life (of adequate quality to be enjoyed) are the primary or underlying goals of all human nutrition research. One of the factors emerging from the intense nutrition research that has occurred in recent years is the importance of the role of antioxidants in the diet. Obviously, the role of antioxidants in preserving foods is important. Reducing fat oxidation, with its chemically active peroxides and potentially damaging free-radical reactions and reaction byproducts, is important not only from a sensory perspective but also from a health perspective. Relationships between increased health risk and oxidation products such as cholesterol oxides[16] continue to emerge, as do relationships between increased antioxidant consumption and improved health states. More than a dozen cholesterol oxidation derivatives in various states of oxidation have been identified. An excellent review of the state of understanding of the role of cholesterol oxides in health and the state of analytical testing was published by Addis et al in 1996.[17]

It is crucial that antioxidant functionality be preserved to the maximum extent possible during processing for maximum product effectiveness. Obviously, inherent antioxidants in foods, such as beta carotene, anthocyanins, and ascorbic acid, are quite sensitive to mishandling. To be effective, they need to be ingested in their unoxidized forms. Processing steps, particularly exposure to heat and air, result in the degradation of the sensitive antioxidants, rendering them ineffective. Processors interested in preserving the antioxidant properties of a functional food need to design process systems that minimize exposure to oxygen, particularly in the presence of heat. An analytical program to ensure that degradation is not occurring is also essential. Where the antioxidant is well characterized, as in the case of beta carotene, analysis can be straightforward, if somewhat difficult due to the antioxidant's sensitivity to handling. Not so for the less well-characterized materials.

Measurement of a food's content of synthetic antioxidants, namely butylated hydroxy anisole (BHA), butylated hydroxy toluene (BHT), and tetrabutylhydroquinone (TBHQ), is relatively straightforward. Because these compounds are synthetic, their chemical characteristics are well defined, and methods have historically been in place to control their addition to food products, such as AOAC Official Method 983.15.[18]

What is more difficult to measure than the quantity of synthetic antioxidants in foods is the antioxidant effect of compounds or combinations of compounds that occur naturally in foods. Two of the procedures that have been found to provide consistent data are the AOCS Methods Cd 12-57[19] and Cd 12b-92.[20]

Vitamins

As a result of historical scientific research, all compounds officially designated as vitamins are functional. The scientific information base regarding their functionality is extensive enough that compounds in the vitamin class are now considered as "traditional nutrients." Although the term *vitamin* is derived from the conjugation of the words "vital" and "amine," from the fact that the first vitamins discovered were amines vital to some function of the human body, the term came to be more broadly applied to a number of compounds of numerous functionalities. One thing these compounds have in common is that they are vital (essential) for the body to function properly in some aspect, but the compound needs to be present only in small or minute quantities. Considering the diversity of compounds included in this category, a better term than vitamins might be *essential micronutrients*. Currently, for labeling purposes, the regulations for the Nutritional Labeling and Education Act have defined the compounds listed in Table 5–2 as vitamins. Also included in the table are methods applicable to the analysis of these compounds.

Table 5–2 Vitamins Specified for Nutritional Labeling

Vitamin Designation	Chemical Name	Common Synonyms	Method(s) for Assay	Therapeutic Function(s)
Vitamin A	3,7-Dimethyl-9-(2,6,6-trimethyl-l-cyclohexen-1-yl)-2,4,6,8-nonatetraen-1-ol	Retinol; anti-infective vitamin; lard factor; vitamin A alcohol; opthalamin	AOAC 974.29	Antixerophthalmic vitamin; nutritional factor
Vitamin B_1	3-[(4-Amino-2-methyl-5-pyrimidinyl)methyl]-5-(2-hydroxyethyl)-4-methylthiazolium chloride	Thiamine chloride hydrochloride; thiamine mononitrate; aneurine hydrochloride; Thiavit; Vitaneuron	AOAC 938.12 AOAC 942.23	Energy metabolism
Vitamin B_2	7,8-Dimethyl-10-(D-ribo-2,3,4,5-tetrahydroxypentyl)isoalloxazine; 7,8-dimethyl-10-ribitylisoalloxazine	Riboflavine; lactoflavine; vitamin G; Flavaxin; Beflavine	AOAC 970.65 AOAC 940.33 AOAC 981.15	Neuromuscular function
Vitamin B_6	5-Hydroxy-6-methyl-3,4-pyridinedimethanol hydrochloride	Pyridoxine hydrochloride; pyridoxal; pyridoxamine dihydrochloride; Pyridox	AOAC 961.15	Amino acid and protein metabolism
Vitamin B_{12}	Cobinamide cyanide phosphate 3′-ester with 5,6-dimethyl-1-α-D-ribofuranosyl-benzimidazole inner salt	Cyancobalamine; antipernicious anemia principle; extrinsic factor; Antipernicin; B-Twelve	AOAC 952.20	Cell division
Vitamin C	3-Oxo-L-gulofuranolactone (enol form)	Ascorbic acid; anti-ascorbutic vitamin; cevitamic acid; Ascorvit; Scorbu-C; Lemascorb	AOAC 985.33 AOAC 967.21 AOAC 984.26	Immune, muscular, and antioxidant functions
Vitamin D	(D2) 9,10-secoergosta-5,7,10(19),22-tetraen-3-ol; (D3) 9,10-secocholesta-5,7,10(19)-trien-3-ol	(D2) calciferol; ergocalciferol: activated ergosterol; (D3) cholecalciferol; cholecalciferol	AOAC 979.24 AOAC 981.17	Antirachitic vitamin; calcium mediation in intestinal absorption, bone metabolism, and muscle activity

	Chemical name	Synonyms	AOAC method	Function
Vitamin E	3,4-Dihydro-2,5,7,8-tetramethyl-2-(4,8,12-trimethyltridecyl)-2H-1-benzopyran-6-ol; 2,5,7,8- tetramethyl-2-(4',8',12'-trimethyltridecyl)-6-chromanol	α-tocopherol; antisterility vitamin; Etavit	AOAC 971.30	Antioxidant, promotes fertility, interactive with selenium
Vitamin K	2-Methyl-3-(3,7,11,15-tetramethyl-2-hexadecenyl)-1,4-naphthalenedione; 2-methyl-3-phytyl-1,4-naphthoquinone	phytonadione; phylloquinone; antihemorrhagic vitamin	AOAC 992.27	Antihemorrhagic vitamin; prothombogenic vitamin
Pantothenate	(R)-N-(2,4-dihyrdroxy-3,3-dimethyl-1-oxobutyl)-β-alanine	Pantothenic acid; chick antidermatitis factor	AOAC 945.74	Energy metabolism
Biotin	Hexahydro-2-oxo-1H-thienol[3,4-d]imid-azole-4-pentanoic acid	Biotin; vitamin H, coenzyme R; Bioepiderm		Muscular function; energy metabolism
Niacin	3-pyridinecarboxylic acid	Nicotinic acid; vitamin B_3; nicotinamide; vitamin PP; Amide PP	AOAC 944.13 AOAC 961.14 AOAC 975.41 AOAC 981.16	Neuromuscular function; antipellegral
Folate	N-[4-[[(2-amino-1,4-dihydro-4-oxo-6-pteridinyl)methyl]amino]benzoyl]-L-glutamic acid	Folic acid; folacin; liver *Lactobacillus casei* factor; vitamin Bc; vitamin M; folsäure; Folvit	AOAC 944.12	Cell division

From a functional foods perspective, the antioxidant vitamin content of the food is often of particular interest to the user. Most well known for their antioxidant properties are the fat-soluble vitamins A, E, and beta carotene, and the water-soluble vitamin C. In addition, the water-soluble folic acid (or, more properly, the folates, a series of homologues and derivatives of pteroylglutamic acid) exhibits antioxidant properties along with the ability to transfer single carbon units in several oxidation states. Folate is also important for the effective biosynthesis of DNA and RNA. Folate deficiency results in reduced cell division as evidenced by anemia, decreased growth, depression, dermatologic disorders, and neural tube defects. Neural tube defects are severe congenital malformations that include spina bifida, encephalocele, and anencephaly. Incidences of neural tube defects may be reduced by folate supplements.

Allergens

Focused attention to allergens is relatively recent with regard to food safety from a regulatory perspective. At present, FDA officially recognizes only eight foods as truly allergenic. These eight foods are peanuts, tree nuts, fish, shellfish, wheat, soy, eggs, and dairy foods. Once ingested, food allergens can cause a number of reactions, ranging in severity from hives and itching to anaphylaxis. Anaphylaxis is a severe reaction involving vomiting; diarrhea; difficulty breathing; swelling of the mouth, tongue, and throat; and a rapid drop in blood pressure.[21] Without quick medical attention, death can occur from a condition known as anaphylactic shock. Dr Steve Taylor, head of the Department of Food Science and Technology at the University of Nebraska, has identified some 160 foods that have been found to cause allergic reactions[19] (personal communication, 1996). Of that number, the eight foods listed above account for 90% of all severe reactions in the United States. Peanuts are the most common allergenic food. Although significant effort has gone into developing rapid, rugged, and sensitive methodology for the unintentional or unlabeled presence of these foods in other foods, the obstacles to effective testing are substantial. It is easy to appreciate the difficulty of determining a characteristic protein that is a natural component part of a particulate, the particulate in turn being typically mixed (but not homogeneously) into a particulated product. Before the analysis actually proceeds, the sampling steps can be formidable (see the previous section "sampling"). At this time, the best approach to allergenicity issues is prevention of exposure. Although food allergens affect a small portion of the population, the potential life-threatening reaction for some susceptible individuals warrants food companies' making allergen risk reduction a high food safety priority. An effective allergen risk reduction program starts at the supplier level and continues through production and distribution. It focuses on label control, prevention of cross-contact/contamination, development of "allergen-clean" sanitation practices, and establishment of practical employee training programs. Taking these steps can enhance the confidence that food products are safe for all consumers.

In addition to the eight allergenic foods listed above, sulfiting agents are considered allergenic in their behavior, especially to asthmatics. Those sulfiting agents legally allowed for uses in foods are listed in Table 5–3.[22–24]

Sulfites interact with foods in a variety of ways. As a result, sulfites are present in a particular sample in a variety of forms—free, reversibly bound, and irreversibly bound. Analytical methods developed for sulfites do not necessarily agree in the quantitative level of sulfite determined, due to their varying ability to measure the various forms of sulfite present. Thus, a reference method is necessary. Treating a sample with acid and heat for an extended period of time is effective in releasing the free and reversibly bound sulfites.[25] Federal regulations specify a reference method for sulfites for labeling purposes. AOAC Official Method of Analysis 990.28,[26] the Optimized Monier-Williams Method for Sulfites in Foods, is the reference method that is the basis for judging the level of sulfites in foods.

Table 5–3 Sulfiting Agents Accepted as Legal for Food Use

Sulfur dioxide	SO_2
Sodium metabisulfite	$Na_2S_2O_5$
Sodium bisulfite	$NaHSO_3$
Sodium sulfite	Na_2SO_3
Potassium metabisulfite	$K_2S_2O_5$
Potassium bisulfite	$KHSO_3$

Source: Reprinted from The Potential Role of Analytical Chemistry in a HACCP Program-Foodborne Sulfites, in *Food Safety from a Chemistry Perspective: Is There a Role for HACCP?* J.W. DeVries, J.A. Dudek, M.T. Morrissey, and C.S. Keenan, eds., p. 60, © Analytical Progress Press.

Mycotoxins

Unfortunately, molds and mold by-products are potentially present regardless of the plant species being grown for food or as a food ingredient. Often the conditions that most effectively promote plant growth most effectively produce molds as well. Many of the molds at some stage of their growth cycle produce toxic by-products, referred to categorically as *mycotoxins* (*myco*-being derived from the Greek word *mykes,* meaning "fungus," or, more basically, "muck, slippery"). Over the past 30 to 40 years, significant gains have been made in identifying many of the molds and their by-product toxins. These will be discussed further below. On the other hand, as food researchers seek out new ingredients, particularly those related to functional foods, it is likely that additional, plant-specific species of molds and toxins will be identified. An example of this is the toxin fumonisin, a by-product of the fusarium molds that was virtually unknown prior to 1990 but in recent years has surfaced as a major concern in corn production. The toxin (or perhaps the toxin in conjunction with other mold metabolites present when the mold produces the toxin) has effects that are species dependent. The most significant toxic effect appears to be the induction of leukoencephalomalacia in horses, the result being death by brain decay when the horses consume corn containing greater than 5–10 ppm of fumonisins. Pulmonary edema and kidney damage are effects observed in other animals at higher levels of intake. The effects on humans are unknown, and fumonisins are currently under intensive study. It would seem likely that as more plant species gathered from wider regions of the globe are scrutinized, additional molds and their respective toxins will be identified. Some of the toxins of current concern, the respective foods of concern, and the respective analytical methods are listed in Table 5–4.

In addition to the Official Methods of Analysis for mycotoxins, a number of analytical kits have been certified as to their performance claims by the AOAC International Research Institute. The Performance Tested certification system, while not as stringent a test as the 8– or 15–laboratory test regimen for Official Methods, is very thorough and ensures that the kit's performance meets all the claims the manufacturer makes on the kit label and/or accompanying literature. Table 5–5 lists the Performance Tested, certified kits for mycotoxins that were available at the beginning of 1999.

Sugars

The most straightforward procedure for the analysis of the sugars most commonly found in foods, namely fructose, glucose, sucrose, maltose, and lactose, is high-pressure liquid chromatography (HPLC). AOAC Official Method of Analysis 977.20[27] (Separation of Sugars in Honey, Liquid

Table 5–4 Mold Toxins of Concern for Foods

Mold Toxin	Mold Species	Food(s) of Concern	Assay Method
Aflatoxins B, G	Aspergillus flavus	General	AOAC 970.43, 970.44, 970.47, 970.48, 971.22, 975.35, 975.36, 975.37, 977.16, 985.17
		Almonds	AOAC 994.08
		Brazil nuts	AOAC 994.08
		Cocoa beans	AOAC 971.23
		Coconut	AOAC 971.24
		Copra	AOAC 971.24
		Corn	AOAC 972.26, 979.18, 990.32, 990.33, 990.34, 991.31, 993.16, 993.17, 994.08
		Cottonseed	AOAC 980.20, 989.06, 990.34
		Eggs	AOAC 978.15
		Feed, mixed	AOAC 989.06
		Green coffee	AOAC 970.46
		Liver	AOAC 982.24, 982.25
		Peanuts	AOAC 968.22, 970.45, 979.18, 990.32, 990.33, 990.34, 991.31, 991.45, 993.17, 994.08
		Pistachio nuts	AOAC 974.16, 994.08
		Soybeans	AOAC 972.27
Aflatoxin M	Aspergillus flavus (produced by animals ingesting plants containing aflatoxins)	Dairy products	AOAC 974.17, 980.21, 986.16
Deoxynivalenol	Fusarium species	Liver	AOAC 982.24, 982.25, 982.26
Fumonisins	Fusarium moniliforme	Wheat	AOAC 986.17 (TLC), 986.18 (GC)
Ochratoxins	Aspergillus and Penicillium species	Corn	AOAC 995.15
		General	AOAC 970.43
		Barley	AOAC 973.37, 991.44
		Coffee	AOAC 975.38
		Corn	AOAC 991.44
Patulin	Aspergillus and Penicillium species	Apples, apple juice	AOAC 974.18, 995.10
Zearalenone, zearalenol	Fusarium species	Corn	AOAC 976.22, 985.18, 994.01
		Feed	AOAC 994.01
		Wheat	AOAC 994.01

Table 5–5 "Performance Tested" Certified Test Kits for Mycotoxins

Test Kit Name	Manufacturer	Mycotoxin	Matrices
Veratox AST	Neogen Corporation	Aflatoxin residues	Grain and grain products, cottonseed, cottonseed meal, raw peanuts and roasted peanuts
Afla Test P	Vicam L.P.	Aflatoxin residues	Grain and grain products
Veratox for Vomitoxin	Neogen Corporation	Deoxynivalenol (DON or vomitoxin)	Corn, corn meal, corn screenings, wheat, wheat mids, wheat flour, wheat bran, and barley

Source: Data from Anon, © 1999, AOAC International Research Institute, Gaithersburg, Maryland.

Chromatographic Method) extracts the sugars from the sample matrix with water. Acetonitrile is added to the extract to clarify it by precipitating the proteins and starches. After filtering, the extract is injected into a stream of acetonitrile/water in a HPLC system equipped with an amine-modified silica column. Sugars are separated on the basis of their ability to hydrogen-bond with the amine functionality of the column's stationary phase. Small polyhydroxyl compounds elute first, namely water, then glycerol, followed by monosaccharides such as fructose and glucose, then disaccharides such as sucrose, maltose, and lactose, then trisaccharides such as maltotriose and raffinose, then tetrasaccharides such as maltotetrose and stachyose, and so on. The separated sugars are detected using a refractive index detector as they elute from the column. For sugars other than the common five and for sugar alcohols, alternative methods are necessary. For example, AOAC Official Method 971.18[28] (Carbohydrates in Fruit Juices, Gas Chromatographic Method) quantitates sorbitol along with fructose, alpha and beta glucose, and sucrose; AOAC Official Method 973.28[29] (Sorbitol in Food, Gas Chromatographic Method) quantitates sorbitol in a variety of foods; and AOAC Official Method 997.08[30] (Fructans in Food Products, Ion Exchange Chromatographic Method) quantitates glucose, fructose, sucrose, lactose, galactose, and maltitol.

Dietary Fiber

As mentioned earlier, dietary fiber has the following working definition: "Dietary fiber consists of the remnants of plant materials, polysaccharides, lignin, and associated substances that are resistant to hydrolysis (digestion) by the alimentary enzymes of humans." The dietary fiber methods adopted as Official Methods of Analysis are proximate methods of analysis, with a goal of simulating, as closely as possible on the laboratory bench, the digestive action of the human alimentary system. The typical approach is to digest the nonfiber portions of the sample away from the fiber component. This is usually done using an amylase enzyme, a protease enzyme, and an amyloglucosidase enzyme for stepwise digestion of the sample. The enzymes used must pass rigorous purity and activity tests to meet the specifications of the assay.[31] AOAC Official Method of Analysis 985.29[31] was the first Official Method adopted for the quantitation of total dietary fiber using the principle of simulating the human digestive system. After the starch in the sample is gelantinized in boiling water using a high-temperature amylase, protease and amyloglucosidase are used to digest the protein and starch in the sample and remove them from the dietary fiber. Four parts ethyl alcohol are added to precipitate the soluble dietary fiber. The combined mass of insoluble fiber and precipitated soluble fiber is filtered, defatted, dried, and weighed. The residue is analyzed for protein and ash, and appropriate corrections are applied.

Insoluble dietary fiber is important for its effect in improving laxation when consumed. Increased consumption of soluble dietary fiber, on the other hand, results in improvements in coronary health, as indicated by the recent acceptance of two health claims allowed on the labeling of food, particularly the whole-oat health claim[32] and the psyllium husk health claim.[32] The distinction between insoluble and soluble fiber is not necessarily clear cut. Again, as with the case of total dietary fiber, the analytical methodology attempts to simulate the human digestive system. Insoluble dietary fiber is insoluble in a buffered aqueous solution of pH optimized for enzymatic digestion. Soluble dietary fiber is the digestion-resistant residue that is soluble in the buffered aqueous solution but that precipitates when four parts alcohol are added to the solution, a technique long used by analysts to separate molecules of some complexity from simple molecules such as sugars. To measure insoluble and soluble dietary fiber as well as the total, the principles of AOAC Official Method 985.29[31] are incorporated into Official Methods 991.42[33] and 993.19,[34] which quantitate the insoluble and soluble dietary fiber components, respectively. The methods were adopted in 1985, 1991, and 1993, respectively. The reason for the time sequence of adoption relates to the progression of knowledge regarding dietary fiber and its effects and improvements in methodology. These three methods are based on the use of phosphate buffering systems for the enzymatic digestions. Official Method 991.43 incorporates the same principles as the three methods outlined above except that organic buffers are used in place of the phosphate buffer system. Organic buffers allow a slightly shorter analysis time, reduced alcohol use, and easier filtration of the fiber residue. Other Official Methods with specific applicability have been developed that give results equivalent to Official Methods 985.29, 991.42, and 993.19. Official Method 992.16 determines total dietary fiber as the sum of a neutral detergent fiber determination and an alcohol precipitant of an autoclaved, enzymatically digested second portion of the sample.[35] For foods with less than 2% starch, Official Method 993.21 can be used.[36] Official Method 994.13 goes beyond the enzymatic digestion followed by precipitation and gravimetric steps of the previous methods and analyzes the carbohydrate makeup of the fiber by a series of digestion, derivatization, and GC quantitation steps. This provides information on the individual sugars making up the dietary fiber as well as the uronic acid and lignin content.[37]

Beta Glucans

As discussed above, whole-oat products with a whole-oat content that exceeds a regulated minimum quantity per serving are allowed to make a health claim on the product label. This claim relates increased oat consumption and reduction in the risk of coronary heart disease. Beta glucans are a significant component of the soluble dietary fiber of oats. Beta glucans are mixed 1,3 and 1,4 beta-linkage glucose polymers resistant to digestion because of the beta linkages. Unlike cellulose, a homogeneously 1,4 beta-linked glucose polymer that tends to be rigid because of the ability of adjacent polymers to align themselves in parallel with strong intermolecular hydrogen bonding, beta glucans tend to be soft in texture and quite viscous in solution. Official Methods 992.28[38] and 995.16[39] are available for beta-glucan assay. Both are based on the principle of digesting the sample with a lichenase enzyme that specifically digests beta glucans to oligosaccharide segments; these segments in turn are digested with a beta-glucosidase enzyme to free up glucose. The glucose is measured colorimetrically after reaction with glucose oxidase and 4-aminoantipyrine.

Polyfructans (Inulin) Methods

Polyfructans (carbohydrate polymers of various lengths typically made up of a glucose molecule attached to a linear chain of fructose molecules) in the form of inulin or oligofructans have been shown to provide a probiotic function in the intestinal tract of humans.[40] In particular, the polyfructans not only promote the overall growth of bacteria in the tract but shift the relative populations of the bacteria species present to the more beneficial *Bifido* species. The Official Method for polyfructans, AOAC 997.08,[30] is an enzymatic HPLC method. Samples are analyzed for sugar makeup and content by HPLC, then digested enzymatically to release the fructose and glucose of the polyfructan. Analysis of the resulting digestate is done by HPLC. The results of the first assay are used to correct for the quantities of sugars present in the original profile. The calculation is necessarily complex because many of the disaccharides present in the food sample, such as sucrose and lactose, are hydrolyzed during the enzymatic step, and their contribution to the fructose and glucose levels from the oligofructan after enzymatic digestion must be taken into account.

SECONDARY METHODS

Once a primary method for an analyte has been established, a secondary, typically high-speed, low-cost method can often be applied for routine analysis of that particular analyte. By *secondary method*, we are referring to a method that is not calibrated against primary standards or on the basis of primary principles. Rather, it measures a signal that can be correlated to the results of a primary method, including many cases where the primary method used is a proximate method. An example of an effective secondary method is the use of near-infrared reflectance (NIR) spectroscopy to measure protein. Protein is typically measured by the traditional Kjeldahl method (a proximate method), and the results are used to calibrate the NIR instrument. Because the NIR signal is made up of weak and overlapping signals from the sample, an extensive number of samples from the commodity being analyzed need to be analyzed by both methods to establish a reliable calibration curve. Samples selected to be used for generating the calibration curve need to cover the entire range of all parameters likely to be present. In order to have a rugged, reliable calibration, these parameters need to vary in the calibration samples as they are likely to vary in the future. For example, in establishing the protein calibration for wheat, samples need to be selected that cover the full range of wheat varieties, moisture, starch, fiber, and fat, preferably across a number of crop years, to ensure that all variants have been accounted for in the calibration. Multifactorial analysis of the secondary method measurements versus the primary method measurements results in a calibration curve. It is not unusual that 50 or more samples need to be analyzed by both methods for a satisfactory calibration to ensure that all factors are adequately included. Further, a secondary method may often be used on the basis of a preliminary calibration with a large calibration set; then the calibration may be upgraded on an ongoing basis by including samples taken from different growing seasons, different geographic areas, or different varieties. Because of this need for continual updating, a word of caution is always in order for applying secondary methods to formulated products. Changes in formulation typically affect the calibration factors sufficiently to induce inaccuracy in the measurement results. Therefore, every time a formulation change is made, recalibration should be done, or at the very least, the calibration should be checked. Typical of the principles used for secondary methods are near-infrared spectroscopy (reflectance and transmittance), dielectric properties, conductivity, refractive index, and specific gravity.

PHYSICAL TESTING

Weight Measurements

One of the most fundamental measurements to be made for any saleable products is the net weight determination. This usually consists of weighing the empty package prior to filling (tare weight), weighing the full package post fill (gross weight), and subtracting to obtain the difference (net weight). Determining net weight on filled packaging is the reverse procedure, namely weighing the full package (gross weight), emptying the package, weighing the empty package (tare weight), and calculating the difference. It seems obvious that the results by the two procedures should be the same, but difficulties in obtaining agreement between the two results are sometimes encountered. Those result when the product clings to the packaging, or when there is interaction between the product and the packaging—for example, when water evaporates from the product through the package during shipment and storage or when product components are adsorbed onto or absorbed into the packaging material.

Bulk Density and/or Apparent Volume

During processing, the bulk density of many products can be adjusted up or down with process control. Users of products (and the law) justifiably expect that the container of product they purchase, even if the product is sold on a weight basis, will essentially be filled with that product. Otherwise, the product quantity appears to be misrepresented. In some cases, where a product truly settles during shipping and handling, the less-than-full container received by the consumer can be explained by the statement "This product is sold by weight and not by volume. Some settling of contents may have occurred during shipping and handling." The use of this statement is reserved for products where settling does occur. Perhaps one of the easiest means of ensuring proper bulk density is to build easy-to-use measuring devices that simulate the container the product is shipped in. Visual inspection and graduations indicate the volume, while weight is readily monitored with an electronic balance.

Tensile Strength

Tensile strength measurements apply primarily to the packaging component of a product but can be applied directly to the product itself if the product's strength and texture are important. They are particularly useful for determining the resistance of packaging materials to tearing. Tensile strength measurements are also useful for determining the ease or difficulty of opening the seal on a package. Tensile tests are typically carried out by placing a portion of the material to be tested between two clamps (in the case of testing seals for "openability," the sample is clamped on each side of the seal) and clamping tightly. One clamp is held stationary; the other is a moveable clamp interconnected with a pressure-sensitive device to measure the force necessary to pull the sample apart. Typically, properties measured are yield and "burst" strength as well as elongation of the sample. Since the parameters being measured are sensitive to the test conditions, the samples must be preconditioned to a specific temperature and moisture. Samples are typically conditioned to TAPPI (Technical Assistance to the Pulp and Paper Industries) test conditions, namely 23°C (73°F) and 50% relative humidity. To obtain agreement between laboratories, the sample must be cut to a uniform size and shape and stretched using a uniform velocity protocol, defined in the test method.

Shear Strength

Measurements of shear strength, the resistance to tearing or breaking in a direction perpendicular to the longitude of the product, have a variety of useful purposes for food products. Viscosity measurements are technically shear measurements and will be discussed below. Resistance to bite can often be related to a shear measurement. This can be particularly useful for characterizing chewy materials or materials that harden with age. Burst strength and impact strength of packaging are modified shear tests.[41,42]

Viscosity

For products produced from mixed components that result in liquids, gels, or slurries and for products meant to be mixed with liquid, viscosity measurements can be used to ensure that the desired appearance and texture of the product are consistently obtained. The simplest type of viscometer is the "hole"- or cup-type viscometer. A cup of fixed volume has a precisely sized hole drilled in the bottom. The cup is filled with liquid, and a stopwatch is used to measure the time necessary for the cup to empty. The longer the time necessary to empty the cup, the higher the viscosity. A similar principle applies to allowing the liquid to be analyzed to drain through a capillary tube of precise size and length.[43]

For samples that have higher viscosities than traditional liquids, such as spreads, frostings, and glazes, a spindle viscometer is very useful. A rotating spindle attached to a tension-sensing device (strain gauge) is placed in the sample and rotated at a fixed shear rate (revolutions per minute). The torque necessary to rotate the spindle is converted to viscosity units and recorded.[44] Variations on this technique include using a screw device to "spiral" the viscometer tip through the sample, continually exposing it to fresh sample as the test progresses. This can be particularly useful for materials that "bridge" around the viscometer spindle or for materials that are thixotropic (become less viscous as they are moved, shaken, or stirred). Spindle designs can be varied from straight round shafts to shafts with side arms to shafts with "propeller" blades attached to the end. Obviously, strict test condition criteria must be followed to ensure within-laboratory and between-laboratory precision with this type of testing. Examples of this type of measurement are elucidated in the AACC Approved Methods,[45,46] the ASTM standards,[47] and the AOCS Methods.[48]

Additional means of measuring viscosity have been designed for specific purposes. Examples include rotating sleeve in cylinder designs and vibrations-dampening designs.

Thermal Analysis

Thermal analytical techniques are designed to quantitate some aspect of the sample that changes with either a change in temperature or a change that occurs when heat is applied or removed. The most easily recognized of the thermal analytical techniques is the measurement of melting points and boiling points, neither determination of which requires further explanation. Additional process-related determinations of importance are specific heat, thermal conductivity, heat of fusion (melting), and heat of vaporization. These are usually measured using simple apparatuses such as vacuum-insulated flasks in combination with sensitive temperature-measuring devices. For more complex heat, temperature, and matrix interactive effects, more sophisticated techniques such as differential scanning calorimetry (DSC) can be used. During a DSC assay, a sample in a holding vessel is placed in a thermal chamber along with a blank (usually an empty, equivalent vessel is used). Heat (or cooling) is applied to the chamber, and the heat absorption or desorption characteristics of the sample and vessel are compared to that of the blank. Characteristics measured by this technique include crystallization point, melting point, heat of fusion, glass transition temperatures (onset, inflection, completion), and specific heat (both below and above the glass transition point). A single pass through a heating or cooling cycle elucidates characteristics of the sample related to its inherent physical properties and its recent thermal history. Cycling a sample through repetitive controlled heating and cooling cycles elucidates characteristics of the sample related to its inherent physical properties as affected by the controlled heating and/or cooling of the test sample. Pressurizable thermal chambers are available options for setting alternative conditions under which to study samples.

MICROWAVE PACKAGING SAFETY

Functional foods seek to provide a functionality beyond the action of traditional nutrition. If this functionality is fragile, rapid preparation, particularly with regard to the heating steps, may be desirable. In addition, speed of preparation is a convenience for the consumer. Microwave susceptor packaging is one means of achieving rapid product heating. The metallic layer of the microwave susceptor interacts with the microwave field in the oven to provide a cooking surface hot enough to pop kernels of popcorn, for example. Microwave susceptor packaging must be properly manufactured to ensure the safety of the product with which it is used. The high temperatures reached by the susceptor increase the ability of compounds and/or polymers used in the package construction to migrate into the food product. ASTM has adopted a series of standards for assessing microwave susceptor packaging. Standard F1479[49] lists and defines the standard terminology uniquely related to the testing of microwave food packaging. To measure the temperature profiles generated by microwave susceptors during use, apply Standard F874.[50] ASTM Standard F1317[51] is used for calibrating microwave ovens to be used for analytical testing of microwave packaging. ASTM Standard F1308[52] provides a method for quantitating volatile extractables from microwave susceptors for food product use. To establish the identity of the extractables, use Standard F1519[53] for the qualitative analysis of volatile extractables in microwave susceptors destined for use in heating food products. ASTM F1349[54] is used for quantitating nonvolatile, ultraviolet-absorbing extractables from microwave susceptors. For quantitating nonvolatile, nonultraviolet-absorbing extractables for microwave susceptor packaging, ASTM Standard F1500[55] applies. Solvents are used as food simulants for this test. Application of these standard methods is useful for ensuring the safety of the microwave package/product combinations.

Although there is understandable concern regarding the potential of a hot microwave susceptor to promote the transfer of packaging materials, components, additives, and/or by-products to the food product, all other forms of packaging need to be properly assessed as well. The Code of Federal Regulations[56] indicates that packaging and additives thereof must not adulterate food or cause taste and odor in the food. A number of standard tests are available to assist the product and packaging developer to ensure against problems.[57–59]

MICROBIOLOGICAL TESTING

From an acute toxicity standpoint, contamination of foods by pathogenic microorganisms typically represents a far greater food safety risk than do chemical factors. Pathogenic microbes can induce illness and even death very quickly. Further, since microbes are living, growing organisms, capable of multiplying rapidly, typically the only way to ensure against risk is to ensure the absence of pathogenic organisms or to ensure control of the food product's environment such that they cannot survive, grow, and reproduce. The analytical technology available to the microbiologist is extensive for ensuring the safety of food product's environment but must be applied with diligence and skill. Granted, there are many applications where "friendly" microbes are used and need to be controlled, such as fermentation of milk to yogurt, fermentation of fruits and grains to beverages, and leavening of baked goods, but their consequences pale in comparison to an outbreak of pathogenic bacteria in a food product. The current emphasis on microbiological concerns by food safety regulators and other experts is no coincidence. In the recent past, pathogens such as *E. coli* O157-H7, *Salmonella enteritidis, Listeria monocytogenes, Cyclospora, Campylobacter jejuni*, and *Salmonella typhimurium* DT104 have grabbed the headlines as outbreaks have caused illness and/or death. Considering the massive extent of some of the food distribution systems in place today and the consequent risk to large numbers of consumers, it is no wonder that the public is highly concerned when an outbreak occurs.

Microbes, like all other living organisms, are fierce competitors when it comes to existing and fighting for resources. As a result, they present particularly difficult challenges when they transform to become resistant to traditional food safety-related processes and treatments. The newly discovered pathogens, or pathogens that are discovered to be transformed, are typically labeled *emerging pathogens*. Some of the factors relating to pathogens that have recently emerged will be discussed below. Emerging pathogens present a major analytical challenge to develop rugged and reliable test methodology and a regulatory challenge to set regulations on a relatively unknown and uncharacterized microbe. Obviously, the regulatory challenge is lessened significantly when rugged, reliable, and reproducible methods are available for data gathering and routine assay.

With pathogenic bacteria, it is always important to keep in mind that often the very conditions that are optimal for producing a food or a food crop are also optimal for growth of microorganisms. Sprout foods such as alfalfa, radish, and bean sprouts are delicious delicacies that are grown under conditions ideal for microbial growth. Fruits and vegetables also are often grown under conditions that support microbial growth. In many cases, however, the microbial growth is beneficial rather than pathogenic to the crop or to the consumer. Obviously, the answer is not to sterilize everything, but to control the pathogens.

Emerging-pathogen issues arise for a variety of reasons. It is generally believed that *E. coli* O157:H7 is derived from a species of *E. coli* that acquired the Shiga toxin gene. It has been shown to be carried by and transmitted to humans by cattle, sheep, and deer and is believed also to have been transmitted to humans via produce and sprouts. The microorganism shows exceptional heat, drying, and acid tolerance. Not only has *Listeria monocytogenes* been responsible for illness and death, but infections appear to correlate with increased risk of spontaneous abortion. *Cyclospora* species are showing up on imported produce and also presenting problems as modifications are made to water treatment systems and the organism escapes removal from the water supply by filtration. Infections caused by *Campylobacter jejuni*, the leading cause of acute bacterial diarrhea, also show indications of preceding cases of Guillain-Barré syndrome, a neurologic disorder resulting in paralysis. Those who enjoy eating eggs are aware that *Salmonella enteritidis* can now be found in raw eggs as well as in chicken meat. And *Salmonella typhimurium* DT104 is now resistant to antibiotics used in humans, a likely result of the use of sublethal doses of these antibiotics in cattle.

Testing for the presence of and quantitation of undesirable microbes in foods requires a good deal of skill and a variety of methods. Some typical applicable methods[5] are listed in Table 5–6.

Often, one of the Official Methods for analyzing for microbes must be purposefully and carefully modified for a particular matrix. For the analysis of certain seeds to be grown into fresh sprouts destined for human consumption, a modification has been reported to improve isolation of *Salmonella* from the seeds. The seeds are first allowed to sprout[60]; then the standard *E. coli* O157:H7 methodology is modified by adjusting the antibiotic level in the enrichment broth, and the enrichment time is adjusted.

One of the major issues encountered in conducting microbiological assays is turnaround time, the elapsed time to complete the assay, as opposed to the analyst time necessary for the test. To determine the presence of a particular microbe requires substantial time to allow the microbe isolated from a food sample to grow and then reproduce and multiply to a discernible level on appropriate media. Verification then requires additional time for enrichment of the isolated organism prior to the actual verification procedure.

An efficient and accurate means of verifying the identity of an organism is the use of riboprinting. Riboprinting is based on the principle of restriction digestion and hybridization with ribosomal probes. Automated systems are available for generating reproducible fingerprints that are based on the structure of the genome of the microbe. The actual analytical output of the process is a fingerprint

Table 5–6 Typical Methods for the Determination and Quantitation of Undesirable Microbes

Microbe	Method
General	AOAC 966.23 Microbial Methods
Escherichia coli	AOAC 983.25 Total Coliforms, Fecal Coliforms, and *Escherichia coli* in Foods, Hydrophobic Grid Membrane Filter Method.*
Escherichia coli	AOAC 988.19 *Escherichia coli* in Chilled or Frozen Foods, Fluorogenic Assay for Glucuronidase
Escherichia coli Salmonella	AOAC 991.13 *Salmonella, Escherichia coli* and Other *Enterobacteriaceae* in Foods, Biochemical System Identification (Vitek GNI+) Screening Method
Escherichia coli	AOAC 991.14 Coliform and *Escherichia coli* Counts in Foods, Dry Rehydratable Films (Petrifilm *E. coli* Count Plate and Petrifilm Coliform Count Plate Methods)
Escherichia coli	AOAC 991.15 Total Coliforms and *Escherichia coli* in Water, Defined Substrate Technology (Colilert) Method
Escherichia coli	AOAC 992.30 Confirmed Total Coliform and *E. coli* in All Foods, Substrate Supporting Disc Method (ColiComplete)
Escherichia coli	AOAC 966.24 Coliform Group and *Escherichia coli* in Tree Nut Meats, Microbiological Method
Escherichia coli	AOAC 996.09 *Escherichia coli* O157:H7 in Selected Foods, Visual Immunoprecipitate Method
Escherichia coli	AOAC 996.10 *Escherichia coli* O157:H7 in Selected Foods, Assurance Polyclonal Enzyme Immunoassay
Escherichia coli	AOAC 984.34 Detection of *Escherichia coli* Producing Heat-Labile Enterotoxin, DNA Colony Hybridization Method
Escherichia coli	AOAC 986.34 Enterotoxigenic *Escherichia coli* DNA Colony Hybridization Method Using Synthetic Oligodeoxyribonucleotides and Paper Filters
Listeria	AOAC 993.12 *Listeria monocytogenes* in Milk and Dairy Products, Selected Enrichment and Isolation Method
Listeria	AOAC 992.18 *Listeria* Species, Biochemical Identification Method (Micro ID *Listeria*)
Listeria	AOAC 992.19 *Listeria* Species, Biochemical Identification Method (Vitek GNI and GPI)
Listeria	AOAC 993.09 *Listeria* in Dairy Products, Seafood, and Meats, Colorimetric Deoxyribonucleic Acid Hybridization Method (GENE-TRAK *Listeria* Assay)
Listeria	AOAC 994.03 *Listeria monocytogenes* in Dairy Products, Seafoods, and Meats, Colorimetric Monoclonal Enzyme Linked-Immunosorbent Assay Method (*Listeria* Tek)
Listeria	AOAC 996.14 *Listeria monocytogenes* and Related *Listeria* Species in Selected Foods, Assurance Polyclonal Enzyme Immunoassay
Listeria	AOAC 997.03 *Listeria monocytogenes* and Related *Listeria* Species in Selected Foods, Visual Immunoprecipitate Assay
Listeria	AOAC 995.22 *Listeria* in Foods, Colorimetric Polyclonal Enzyme Immunoassay Screening Method (TECRA *Listeria* Visual Immunoassay)
Salmonella	AOAC 967.25 *Salmonella* in Foods, Preparation of Culture Media and Reagents
Salmonella	AOAC 967.27 *Salmonella* in Foods, Identification
Salmonella	AOAC 967.28 *Salmonella* in Foods, Serological Tests
Salmonella	AOAC 978.24 *Salmonella* spp. in Foods, Biochemical Identification Kit Method
Salmonella	AOAC 975.54 *Salmonella* in Foods, Fluorescent Antibody (FA) Screening Method

continues

Table 5–6 continued

Microbe	Method
Salmonella Escherichia coli	AOAC 991.13 *Salmonella, Escherichia coli* and Other *Enterobacteriaceae* in Foods, Biochemical System identification (Vitek GNI+) Screening Method
Salmonella	AOAC 989.13 Motile *Salmonella* in All Foods, Immunodiffusion (1-2 Test) Method
Salmonella	AOAC 991.38 *Salmonella* in Foods, Automated Conductance Method
Salmonella	AOAC 994.04 *Salmonella* in Dry Foods, Refrigerated Preenrichment and Selective Enrichment Broth Culture Methods
Salmonella	AOAC 989.14 *Salmonella* in Foods, Colorimetric Polyclonal Enzyme Immunoassay Screening Method (TECRA *Salmonella* Visual Immunoassay)
Salmonella	AOAC 993.08 *Salmonella* in Foods, Colorimetric Monoclonal Enzyme Immunoassay (*Salmonella*-TEK), Utilizing Elevated Temperature for Selective and Post Enrichments
Salmonella	AOAC 996.08 *Salmonella* in Foods, Enzyme Linked Immunofluorescent Assay, Screening Method (VIDAS *Salmonella* [SLM] Assay)

*Coliform counts are often used as indicators of effectiveness of sanitation practices.

pattern, a particular fingerprint pattern being unique to a particular microorganism. To aid microbiologists worldwide in accurately identifying microorganisms, the National Center for Food Safety and Technology at Summit-Argo, Illinois, has established an electronic database of riboprint fingerprints and is working to include characterizing data for all known strains and species of microbes. The fingerprint data can be assessed rapidly through electronic telemedia.

A number of analytical kits for the analysis of microbes have been certified as to their performance claims by the AOAC International Research Institute. The Performance Tested certification system, while not as stringent a test as the 8– or 15–laboratory test regimen for Official Methods, is very thorough and ensures that the kit's performance meets all the claims the manufacturer makes on the kit label and/or accompanying literature. Table 5–7 lists the Performance Tested certified kits that were available at the beginning of 1999.

Assays Relevant to the Control of Microbes

In addition to the standard microbiological tests that identify and quantify the microbes of concern in the sample, a number of related assays are very important for controlling or eliminating the growth of undesirable microbes. These are primarily chemical assays, but it is important to know the relationship between microbial survival and growth and these measurable factors, namely pH, titratable acidity, and water activity. Some of the emerging pathogens do not necessarily yield to the same control mechanisms—a particular water activity level, or pH adjusted below a certain point—as did pathogens in the past. Knowledge of the means of controlling these emerging pathogens is essential for food safety.

pH Measurement

Even though measurement of pH is a chemical measurement, its importance from a microbiological food safety perspective cannot be emphasized enough (see discussion above regarding measurement of pH). In food systems having adequate water activity to support microbial growth, inhibition of that growth by lowering the pH is a very effective tool. Accurate and precise control of pH is important.

Table 5–7 Performance Tested Certified Test Kits

Test Kit Name	Manufacturer	Microbes	Matrices
Listertest Lift Test	Vicam L.P.	*Listeria* spp., including *Listeria monocytogenes*	Environmental surfaces
PATH-STIK	Lumac B.V.	*Salmonella* spp.	Food
Oxoid Rapid Test for Listeria	Oxoid Ltd	*Listeria* spp.	Food
Reveal	Neogen	*Salmonella* spp.	Food
Bioline ELISA	Bioline and Diffchamb	*Salmonella* spp.	Food and animal feed
Salmonella Screen/SE Verify	Vicam L.P.	*Salmonella Enteritidis*	Food, stainless steel, concrete
Salmonella Screen/*Salmonella* Verify	Vicam L.P.	*Salmonella* spp.	Food, stainless steel, concrete
Oxoid Rapid Test for *Salmonella*	Oxoid Ltd	*Salmonella* spp.	Food, animal feed, environmental samples
Salmonella DLP Assay	GENE-TRAK Systems Corporation	*Salmonella* spp.	Food, animal feed, environmental samples
SimPlate for TPC	IDEXX Laboratories Inc.	Food-borne bacteria	Food with some specific limitations
Dynabeads Anti-*Salmonella* kit	DYNAL A.S.	*Salmonella* spp.	Food, animal feed
BIND *Salmonella* Rapid Assay Kit	IDEXX Laboratories Inc.	*Salmonella* spp.	Food, dry animal feed
BAX for Screening *Salmonella*	Qualicon, a Dupont Subsidiary	*Salmonella* spp.	Fluid and dried milk, chicken, turkey, beef, and pork
TECRA Unique *Salmonella*	TECRA Diagnostic	*Salmonella* spp.	Food, Environmental samples
ElaFoss *Salmonella*	Foss Electric A/S	*Salmonella* spp.	Food
Listeria DLP Assay	GENE-TRAK Systems	*Listeria* spp.	Dairy products, meats, seafood, and environmental samples
VIDAS *Listeria* Assay	bioMerieux, Inc.	*Listeria* spp.	Raw meats and poultry, cooked and processed meats and poultry, seafood, vegetables, dairy products, and environmental samples
BAX for screening *E. coli* O157:H7	Qualicon	*E. coli* O157:H7	Ground beef

Source: Data from Anon, © 1999, AOAC International Research Institute, Gaithersburg, Maryland.

Titratable Acidity

Titratable acidity per se does not provide a measurably direct correlation to microbial control. However, titratable acidity can be useful for determining the relative buffering capacity of a product—that is, the product's ability to hold the desired pH for microbial control.

Water Activity (a_w) Measurements

Like pH, water activity is a chemical measurement (see discussion above regarding its measurement). However, the primary purpose of establishing and measuring the water activity of a particular product is likely to be microbial growth control. Without adequate water, microbes cease to reproduce. As a general rule, all microbes will survive and grow when the a_w is greater than 0.95; when the a_w ranges from 0.90 to 0.95, all molds, yeasts, and some bacteria will grow; in the a_w range of 0.80 to 0.90, all molds and some osmophilic yeasts will grow; an a_w range of 0.70 to 0.80 will support several mold species; in the range of 0.65 to 0.70 a_w some xerophilic molds can survive and grow; and a product with an a_w of less than 0.65 is relatively impervious to microbial growth. In the laboratory, some pathogenic bacteria have been shown to survive at an a_w of 0.86, but this does not appear to be an issue in practice with food products.

ANIMAL STUDIES

Ideally, human studies examining various health effects resulting from the consumption of functional foods and their components should be conducted in a representative sample population. Since this is not always possible, the behavior of components in human beings often has to be projected from the results of animal studies. Animal studies may range from testing of sites within the body for bioaccumulation of a particular food component to more thorough investigations of absorption, distribution, metabolization and excretion of functional food components. Such animal studies play an important role in research, but extrapolation to humans must be done with great care, as results may not truly represent biochemical processes in human beings.

Animal Study Design

One of the most critical elements in designing animal model studies is the establishment of goals that are suitable for the type of functional food or component being evaluated. The specific approach for studying functional foods and components is dictated by the parameters required to substantiate the health, structure-function, and/or advertising claims to be made by the manufacturer. It is important to note that some substantiated claims may result in a particular functional food being regulated much like pharmaceuticals.

Different categories of functional foods require slightly different animal study designs. Functional foods that are considered similar to traditional foods require, among other things, the evaluation of metabolic utilization as it relates to efficacy and bioavailability. In addition, the identification and safety of undesirable substances simultaneously present with efficacious components must be determined. In the case of functional foods that differ from traditional foods in specifically designed ways, safety assessment must be done on the basis of the differences in composition and properties and may require both in vitro characterization and in vivo testing. With novel functional foods, there are much higher expectations regarding which tests are warranted to investigate efficacy, bioavailability, and toxicology, since less information is currently available on such foods and their components.

Several parameters, falling within four main categories,[61] are suggested for assessment via animal studies. The specific parameters to be assessed are dictated by the experimental design: for example, the parameters for single-intake studies (see Table 5–8) are somewhat different from those for repeat intake studies (see Table 5–9).

Table 5–8 Assessment Parameters for Animal, Bioavailability, and Toxicology Studies

Single-Intake Studies

Category	Parameter
Metabolism	^{14}C label distribution
	Induction of metabolism
	Metabolite identification
Physiology	Gastrointestinal function
	Transit time
	Fecal/urine composition
Nutrition	Interaction with dietary nutrients

Source: Adapted with permission from D. Bechtel, *Institute of Food Technologists Short Course: A Global Perspective on Regulatory Approval for Food Ingredients, Nutraceuticals and Dietary Supplements,* Copyright © Institute of Food Technologists.

Table 5–9 Assessment Parameters for Animal, Bioavailability, and Toxicology Studies

Repeat Intake Studies

Category	Parameter
Metabolism	^{14}C label distribution
	Induction of metabolism
Physiology	Gastrointestinal function
	Organ function
Nutrition	Nutrient balance studies
Animal health	Food intake/weight gain
	Organ weight
	Serum chemistry
	Histopathology
	Hematology

Source: Adapted with permission from D. Bechtel, *Institute of Food Technologists Short Course: A Global Perspective on Regulatory Approval for Food Ingredients, Nutraceuticals and Dietary Supplements,* Copyright © Institute of Food Technologists.

BIOAVAILABILITY

Understanding Bioavailability

FDA defines bioavailability as the quantity of a single test dose of a food component that can be measured after absorption in the target tissue. As it relates to functional foods, *bioavailability* refers to the sum of impacts that may suppress or ameliorate utilization of a component.

Methods for Studying Bioavailability

Although bioavailability is subject to a complex set of influences, regulations require methods for quantifying bioavailability of food ingredients. Existing methods for food ingredients measure changes in serum or plasma levels of the component of interest after ingestion of food, as outlined by Abad and Gregory,[62] Graham et al,[63] and Kivisto et al[64]; urinary excretion of the component after ingestion of food following body saturation, as outlined by Colman et al[65] and Jusko and Levy[66]; and the absorption of labeled components from the gastrointestinal tract, as outlined by Ink et al,[67] Nguyen and Gregory,[68] Fairweather-Tait and Minski,[69] and Pecoud et al.[70]

Bioassays incorporating isotopic labeling are generally considered to be the most accurate methods for facilitating our understanding of the bioavailability of components. With animal bioassays, the accumulation of the isotopically labeled component is measured via special whole-body isotope counters. Considering the limited equipment available in most laboratories and the high cost/benefit ratio associated with whole-body counters, it is no small surprise that there is an ongoing search for simpler, more cost-effective in vitro methods of analysis. In vitro methods must be preceded by appropriate extraction procedures. In most cases, however, extraction procedures do not adequately model mammalian digestion and intestinal absorption processes. Obviously an in vitro assay can replace an animal bioassay only if a significant statistical correlation can be demonstrated between the results of the two methods for the components of interest.

To date, the most widely used in vitro analytical techniques for assessing bioavailability are microbiological assays, immunoassays, protein binding assays, HPLC, GLC, and flow-injection analysis (FIA). With functional foods and ingredients, as with other types of food analyses, standards are needed for chemical analyses. Standardized analytical methods are also needed to minimize interlaboratory variation.

Factors Influencing Bioavailability

It is important to reiterate that with animal and human studies the bioavailability of functional food components is affected by numerous factors. Physiological and pathophysiological factors are governed to a large extent by the animal's species and genotype; hence, results obtained in animal studies are not directly transferable or extrapolatable to human beings. The state of maturation is also important—infants tend to have immature intestinal function, while the elderly may suffer from loss of efficiency of component absorption. The sex, metabolic conditions, health status, hormonal status, and intestinal microflora of the study participants also influence bioavailability. Appropriately designed bioavailability experimental procedures should therefore include animal and human studies of both sexes and defined age groups. As physiological regulation is different for different food components, its impact on absorption, distribution, metabolism, and excretion needs careful and individual consideration to accurately quantify bioavailability.

The food containing the component of interest and the other foods eaten with it affect bioavailability, due to complex interactions. Component interactions may result in the inhibition of absorption of some components—for example, fiber content decreases absorption of some micronutrients, as indicated by Nguyen et al[71] and Lindberg et al.[72] On the other hand, synergistic interactions may lead to increased absorption. Kusin et al[73] show that the absorption of vitamin A increases when vitamins A and E are ingested together in high enough quantities. Fat content may also affect the absorption of fat-soluble vitamins. Interactions with medications, alcohol, and effects from unknown ingredients should also be taken into account.

Chemical and physical properties of a food or food component may be modified by processing and handling procedures, thereby altering the bioavailability of components in the final product. Coating or encapsulation procedures may render components less available or even totally unavailable. Processes to preserve or extend shelf life, such as pasteurization and irradiation, may also have an effect on bioavailability. In like manner, storage conditions such as temperature, moisture content, pH, and the presence or absence of oxygen are known to affect the properties and stability of certain components (see the section "Antioxidants" earlier in this chapter). Transportation, preparation, and service of functional foods cannot be overlooked in a consideration of the impact on bioavailability.

Quite often, there is a substantial loss of the component of interest due to instability during storage, particularly during the cycles of freezing and thawing. Product packaging is also very important. Component content may be reduced due to adsorption onto packaging material during storage; photodegradation may occur if the composition of the storage containers is such that substantial light is transmitted through to the food.

TOXICOLOGY

The term *toxicology* usually conjures up images of nasty materials, usually chemicals, invented and produced for the sole function of doing bodily harm to the consuming organism. It is certainly true that some products are produced solely for their toxic effects; insecticides are developed for their toxicity to insects, rodenticides for eliminating rodent problems, and herbicides for their toxicity to undesirable plants. However, the majority of toxic compounds we are exposed to are products of nature. Often, these naturally occurring substances are compounds that are beneficial at one dose level and hazardous or toxic at some higher dose level. Perhaps the best example of this is vitamin A. Vitamin A is essential to the body for the proper functioning of visual pigments, epithelial cell differentiation, the immune system, and fetal development. In addition, it serves as an antioxidant. Vitamin A in the food supply comes in two forms; the plant-based form, namely the carotenes, and the animal-based form, retinol. The recommended intake of vitamin A as retinol (or retinol equivalent amounts of retinol esters or carotenes) is 5000 IUs (international units) or 1500 µg daily for a 2000-calorie diet. At a level less than 1000 times as high—that is, at a dose of less than 1.5 g—retinol is toxic. Arctic explorers reported the poisoning of human beings within hours after a meal of polar bear liver.[74] Rodahl and Moore[75] estimated the vitamin A content in livers of three polar bears at 13,000 to 18,000 IU/g. Ephedra, a component of ma juang, an herb often promoted for consumption in weight loss, energy enhancement, and body building, is sufficiently toxic to have generated hundreds of adverse-event reports to FDA. This has resulted in an FDA proposal to limit ephedra usage to 8 mg per serving in a 6-hour period, 24 mg/d, and a maximum usage duration of 7 days for that ephedra dosage level.[76] Common table salt, sodium chloride, essential for life when consumed at low levels, shows long-term negative health effects when consumed at higher levels. At high enough levels, it shows acute toxicity (LD_{50} Oral in rats, 3000 mg/kg; LD_{50} Oral in mice, 4000 mg/kg). Even water, ingested at high enough levels, is toxic.[77]

As a result of their carcinogenic potency, food additives such as therapeutic drugs used in livestock, preservatives, stabilizers, naturally occurring toxicants, and mycotoxins are cause for concern. Food preparation, processing, and storage, excess biologically active compounds, heavy metal and heavy metal–derivative residues, industrial by-products, allergens, and unwanted residues are also cause for concern.

Several factors have the ability to affect an individual's sensitivity to the toxicity of a chemical component in foods. Factors that readily come to mind include, but are not limited to, species, gender, age, genetic predisposition, nutritional status, disease status, and exposure level, which is a product of daily intake and concentration of the component in question. The variability within these factors must be taken into consideration in designing experiments.

Toxicology Studies

There exists a long history of assessing the safety of and risks associated with synthetic chemicals and drug residues in foods. As with other types of food products, it is required that chemicals designed for use in functional foods be evaluated to ensure their safety and the safety of their components, especially since the components will be consumed in higher than usual amounts or in concentrated forms.

To date, limited information exists regarding the safety and toxicity of many functional foods and components. Often it seems that the study of toxicity of many functional foods and components and resulting regulation are related more to the available analytical methodology than to other factors. Appropriate methods and procedures for assessing the effects of long-term exposure to functional food ingredients must therefore be developed and optimized. Animal-based bioassay is one of several procedures being used currently to assess the risk associated with specified levels of exposure to chemicals present in food. While some scientists look favorably toward the replacement of animal bioassays in safety evaluations with in vitro methods, we cannot deny the wealth of information ascertained from in vivo animal studies. Although the in vitro methods may not be reflective of the digestive process in the gastrointestinal tract, they do serve as prominent prescreening processes for animal bioassays, thereby reducing the number of animals used in testing.

Experimental Design

With toxicological experiments, one of the most critical factors in the experimental design is the establishment of appropriate end points. Such end points are critical in order to quantify metabolic utilization. Selection of appropriate end points is largely dependent on the availability of suitable biomarkers for early and intermediary changes in function. Other factors to be considered in the experimental design include, but are not limited to, components in the background diet of the test animals, levels of exposure to the test components, suitability of material used as the control, measures taken to control bias, and effects across age and gender. There is one other challenge with experimental design: it is difficult to determine, through the use of biomarkers, whether needs are being met for presumably healthy individuals consuming functional foods in an effort to remain healthy.

Biomarkers

New methods in molecular toxicology that incorporate the use of biomarkers provide effective tools in risk assessment. The term *biomarker* refers to a wide range of alterations occurring at the biochemical, cellular, or molecular level on the continuum between exposure and disease, which can be measured by assays performed on body fluids, cells, or tissues, as stated by Perera et al.[78] DeVries[79] defines three broad categories of biomarkers: markers of exposure, markers of effect, and markers of susceptibility. The distinction, however, is not always strict. With a thorough understanding of the biological mechanism that connects exposure levels to effects, biomarkers may be used as a model to examine, in an organism, the ingestion, absorption, distribution, biotransformation, and excretion of functional food components.

Suitable biomarkers do not exist for all food components and even fewer exist for new food components. As expressed by DeVries,[79] the blood concentration of vitamin E, for example, is a relatively good indicator of dietary vitamin intake, but it is known that serum cholesterol concentration is a very poor marker of dietary cholesterol intake. It is therefore evident that the quest must continue for additional biomarkers, especially those needed for assessment of functional food components. Biomarkers must be relevant and possess the ability to measure changes that occur before the onset of disease. Procedures for detection and measurement should be inexpensive and easy to perform.

PRACTICAL ASPECTS OF MEASUREMENTS

For a measurement or series of measurements to be meaningful, some relationship to an end product being consumed must be present. This is very obvious when one is building a wooden structure, for example. How carefully one measures and cuts the wood is reflected in how tight fitting and well constructed the finished product looks. The impact on the end result is not so obvious for "invisible" measurements, such as the amount of vitamin C in a food, which is then related to the prevention of scurvy. Unfortunately, no matter how carefully one measures something, there is always some variability inherent in that measurement (see the previous section "Measurement Accuracy and Precision Concepts"). When the desired characteristic cannot be measured directly, a correlation with a measurable factor must be established.

Specifications

From a practical perspective, regarding a measurement and its related specification, the difference between the high and low ends of a specification should probably not be greater than 10 times or less than 4 times the standard deviation of the measurement. Of course, the inverse of this holds for the selection of an analytical method. If a specification has already been set and is meaningful, then the method's standard deviation should be no greater than one fourth the specification range. Secondary methods—those methods such as near-infrared spectroscopy, that are calibrated, not against primary standards or on the basis of primary principles, but by correlation with the results of a primary method—present a unique situation. The calibration data should extend several standard deviations (of the primary calibration method) beyond either end of the specification. This might seem paradoxical because often, after a secondary method is properly calibrated, its precision is better than that of the primary method from which it was derived. This improved precision results from a high replication level of the primary method during the calibration sequence to reduce the standard error of estimate of the calibration values to a low level.

Point-of-Application Measurements

With regard to producing safe functional foods with consistent quality characteristics, perhaps the most useful of all analyses are those whose results can be used immediately by a control system to make changes in a process automatically in response to changes in product characteristics. When truly effective, such feedback loops are a production engineer's delight. Almost as useful are analyses that can be carried out directly by, and have results immediately available to, the user of the data. Pocket-sized pH meters, compact conductivity and dielectric-based moisture meters, and rugged, well-calibrated near-infrared systems located near the production line are examples. Often it is cost-effective to install small, controlled-atmosphere analytical facilities near the production line. Although the individual performing the measurement may have limited skills in analytical techniques, the analytical chemist or microbiologist can serve as the expert for the assay, ensuring that it is properly validated and that adequate quality control procedures are in place. Immediate and accurate interpretation of results can assist in ensuring product quality and safety with minimal cost incurred.

THE FUTURE OF MEASUREMENTS IN FUNCTIONAL FOODS

As the overall food supply becomes even safer than it is today due to increased safety surveillance and public awareness, and as the knowledge regarding the functionality of foods in general and of functional foods in particular grows, the demand for testing and measurement will increase. The sophistication of the applicable tests will increase as well. Tests designed to prove and/or quantitate a particular functionality will be in demand, particularly to support claims regarding the functional food being assessed. Concurrently, new discoveries in microbiology will mean that additional safety concerns will need to be addressed.

Microbial Contamination

With the recent appearance and/or recognition of microbes that are having a serious impact on human health, improved test methods will be needed for identification, isolation, and quantitation of these organisms. These organisms are showing resistance to traditional methods of control and are proving very potent in inducing illness. A very low number of microbes per gram of food is infectious, so methods to detect such low levels are necessary to ensure product safety. Among the organisms of growing concern are *E. coli* O157:H7, *Listeria monocytogenes, Cyclospora* species, *Campylobacter jejuni, Salmonella enteritidis*, and *Salmonella typhimurium* DT104. In addition to low-level detection of these microorganisms, continuous, or monitoring, methods or instrumentation would improve food safety and reduce the costs incurred due to food recalls after a microbial contamination has been discovered.

Proof of Functionality

Proving the functionality of a food product will range from a routine task to a very involved endeavor. For some structure-function claims, no significant changes or improvements in analytical methodology will be necessary. For example, research has shown that increased calcium intake relates to a reduction in osteoporosis problems. Therefore, a claim that a product is a significant source of calcium will be supported by the traditional atomic absorption or inductively coupled argon plasma analyses currently in use. On the other hand, a claim for antioxidant properties of a functional product may require not only tests for the currently well-characterized antioxidant vitamins such as A, E, C, or beta carotene, or the synthetic antioxidants, but a test of the antioxidant properties of the product itself. Further, standardized testing may be in order to determine if the antioxidant properties under consideration survive the digestion process and provide antioxidant protection to targeted sites in the body. After functional antioxidant compounds have been identified and the antioxidant effect has been established, quantitative tests for the identified compounds will be necessary for quality control monitoring of the product.

REFERENCES

1. AOAC 977.16 (1977). Sampling of aflatoxins. preparation of sample. In: *Official Methods of Analysis of AOAC International.* 16th ed. Gaithersburg, MD: AOAC International; 1998.
2. Whitaker TB. Sampling granular foodstuffs for aflatoxins. *Pure Appl Chem.* 1977;49:1709–1717.
3. Horwitz W. Evaluation of analytical methods used for regulation of foods and drugs. *Anal Chem.* 1982;54:67A–76A.
4. Horwitz W, Albert R, Deutsch, M. Precision parameters of methods of analysis required for nutrition labeling. Part I. Major nutrients. *J Assoc Off Anal Chem.* 1990;73:661–680.
5. *Official Methods of Analysis of AOAC International.* 16th ed. 5th rev. Gaithersburg, MD: AOAC International.
6. Horwitz W. Protocol for the design, conduct and interpretation of collaborative studies. *Pure Appl Chem.* 1988;60:855–864.
7. Guidelines for collaborative study procedure to validate characteristics of a method of analysis. *J Assoc Off Anal Chem.* 1989;72:694–704.
8. Craft N, Boyer K. Guidelines for preparation of in-house quality assurance control materials. *Referee.* 1993;17(5):6.
9. Cho SC, DeVries JW, Prosky L. *Dietary Fiber Analysis and Applications.* Gaithersburg, MD: AOAC International;1997:2–9.
10. Food Labeling. 21 CFR § 101, (1999).
11. Protein. 21 CFR §101.9(c)(7), (1998).
12. Merrill AL, Watt BK *Energy Value of Foods: Basis and Derivation.* US Government Printing Office, US Dept of Agriculture Handbook 74. Washington, DC: 1955:4.

13. AOAC 939.05 (1939). Fat acidity-grains, titrimetric method. In: *Official Methods of Analysis of AOAC International.* 16th ed. Gaithersburg, MD: AOAC International; 1998.

14. Fritsch CW. Lipid oxidation-the other dimensions. *Inform.* 1994;5:423–436.

15. Labuza TP, McNally L, Gallagher D, Hawkes J, Hurtado F. Stability of Intermediate Moisture Foods. 1. Lipid Oxidation. *J Food Sci.* 1972;37:154–159.

16. Peng SK, Morin RJ. *Biological Effects of Cholesterol Oxides.* Boca Raton, FL: CRC Press; 1992.

17. Addis PB, Park PW, Guardiola F, Codony R. Analysis and health effects of cholesterol oxides. In: McDonald RE, Min DB, eds. *Food Lipids and Health.* New York, NY: Marcel Dekker Inc; 1996:199–240.

18. AOAC 983.15 (1998). Phenolic antioxidants in oils, fats, and butter oil. In: *Official Methods of Analysis of AOAC International.* 16th ed. Gaithersburg, MD: AOAC International; 1998.

19. AOCS Method Cd 12–57 (1957). Fat stability, active oxygen method. In: *Official Methods and Recommended Practices of the AOCS.* 5th ed. Champaign, IL: American Oil Chemists Society; 1997.

20. AOCS Method Cd 126–92 (1992). Oil stability index. In: *Official Methods and Recommended Practices of the AOCS.* 5th ed. Champaign, IL: American Oil Chemists Society; 1997.

21. Taylor SL. Food allergies. *Food Technol.* 1985;39(2):98–105.

22. Food labeling, declaration of sulfiting agents. *Federal Register.* Final rule July 9, 1986a;51:25012–25020.

23. Warner CR, DeVries JW. The potential role of analytical chemistry in a HACCP program: foodborne sulfites. In: DeVries JW, Dudek JA, Morrissey MT, Keenan ST, eds. Minneapolis, MN: Analytical Progress Press; 1996:59–75.

24. Food exemptions from labeling. 21 CFR §101.100(a)(4), (1998).

25. Wedzicha BL. *Chemistry of Sulphur Dioxide in Foods.* London, England: Elsevier Applied Science Publishers. 1984:86–87.

26. AOAC 990.28 (1990). Sulfites in foods, optimized Monier-Williams method. In: *Official Methods of Analysis of AOAC International.* 16th ed. Gaithersburg, MD: AOAC International; 1998.

27. AOAC 977.20 (1977). Separation of sugars in honey, liquid chromatographic method. In: *Official Methods of Analysis of AOAC International.* 16th ed. Gaithersburg, MD: AOAC International; 1998.

28. AOAC 971.18 (1971). Carbohydrates in fruit juices, gas chromatographic method. In: *Official Methods of Analysis of AOAC International.* 16th ed. Gaithersburg, MD: AOAC International; 1998.

29. AOAC 973.28 (1973). Sorbitol in food, gas chromatographic method. In: *Official Methods of Analysis of AOAC International.* 16th ed. Gaithersburg, MD: AOAC International; 1998.

30. AOAC 997.03 (1997). Fructans in food products, ion exchange chromatographic method. In: *Official Methods of Analysis of AOAC International.* 16th ed. Gaithersburg, MD: AOAC International; 1998.

31. AOAC 985.29 (1985). Total dietary fiber in foods, enzymatic-gravimetric method. In: *Official Methods of Analysis of AOAC International.* 16th ed. Gaithersburg, MD: AOAC International; 1998.

32. Health claims: soluble fiber from certain foods and risk of coronary heart disease (CHD). 21 CFR §101.81.

33. AOAC 991.42 (1998). Insoluble dietary fiber in food and food products, enzymatic-gravimetric method, phosphate buffer. In: *Official Methods of Analysis of AOAC International.* 16th ed. Gaithersburg, MD: AOAC International; 1998.

34. AOAC 993.19 (1998). Soluble dietary fiber in food and food products, enzymatic-gravimetric method, phosphate buffer. In: *Official Methods of Analysis of AOAC International.* 16th ed. Gaithersburg, MD: AOAC International; 1998.

35. AOAC 992.16 (1992). Total dietary fiber, enzymatic-gravimetric method. In: *Official Methods of Analysis of AOAC International.* 16th ed. Gaithersburg, MD: AOAC International; 1998.

36. AOAC 993.21 (1993). Total dietary fiber in foods and food products with ≤2% starch, non-enzymatic gravimetric method. In: *Official Methods of Analysis of AOAC International.* 16th ed. Gaithersburg, MD: AOAC International; 1998.

37. AOAC 994.13 (1994). Total dietary fiber (determined as neutral sugar residues, uronic acid residues, and klason lignin), gas chromatographic-colorimetric-gravimetric method. In: *Official Methods of Analysis of AOAC International.* 16th ed. Gaithersburg, MD: AOAC International; 1998.

38. AOAC 992.28 (1992). (1,3)(1,4)-*beta*-D-glucans in oat and barley fractions and ready-to-eat cereals, enzymatic-spectrophotometric method. In: *Official Methods of Analysis of AOAC International.* 16th ed. Gaithersburg, MD: AOAC International; 1998.

39. AOAC 995.16 (1995). *Beta*-D-glucans in barley and oats, streamlined enzymatic method. In: *Official Methods of Analysis of AOAC International.* 16th ed. Gaithersburg, MD: AOAC International; 1998.

40. Roberfroid M. Dietary fiber, inulin, and oligofructose: a review comparing their physiological effects. *Crit Rev Food Sci Nutr.* 1993;33:103–148.

41. ASTM D774/D774M-96a (1996). Standard test method for bursting strength of paper. In: *Annual Book of ASTM Standards*. West Conshohocken, PA: American Society for Testing and Materials; 1998.

42. ASTM D3420-95 (1995). Standard test method for pendulum impact resistance of plastic film. In: *Annual Book of ASTM Standards*. West Conshohocken, PA: American Society for Testing and Materials.

43. Corn starch analysis B-61 (1991). Inherent viscosity (one point). In: *Standard Analytical Methods of the Member Companies*. Washington, DC: Corn Refiner's Association, Inc.

44. ASTM D2196-86 (1991). Standard test method for rheological properties of non-Newtonian materials by rotational (Brookfield) viscometer. In: *Annual Book of ASTM Standards*. West Conshohocken, PA: American Society for Testing and Materials; 1991.

45. AACC Approved Method 56–79 (1994). Apparent viscosity of acidulated flour-water suspensions-spindle method. In: *American Association of Cereal Chemists, Approved Methods*. 9th ed. St Paul, MN: American Association of Cereal Chemists; 1998.

46. AACC Approved Method 56–80 (1994). Apparent viscosity of acidulated flour-water suspensions-wire method. In: *American Association of Cereal Chemists, Approved Methods*. 9th ed. St Paul, MN: American Association of Cereal Chemists; 1998.

47. ASTM D5400-93 (1993). Standard test methods for hydroxypropylcellulose. In: *Annual Book of ASTM Standards*. West Conshohocken, PA: American Society for Testing and Materials; 1993.

48. AOCS Ja 10–87 (1987). Brookfield viscosity. In: *Official Methods and Recommended Practices of the AOCS*. 5th ed. Champaign, IL: American Oil Chemists Society; 1997.

49. ASTM F1479–94 (1994). Standard terminology relating to microwave food packaging. In: *Annual Book of ASTM Standards*. West Conshohocken, PA: American Society for Testing and Materials; 1994.

50. ASTM F874–91 (1991). Standard test method for temperature measurement and profiling for microwave susceptors. In: *Annual Book of ASTM Standards*. West Conshohocken, PA: American Society for Testing and Materials; 1991.

51. ASTM F1317–90 (1990). Standard test method for calibration of microwave ovens. In: *Annual Book of ASTM Standards*. West Conshohocken, PA: American Society for Testing and Materials; 1990.

52. ASTM F1308–94 (1994). Standard test method for quantitating volatile extractables in microwave susceptors used for food products. In: *Annual Book of ASTM Standards*. West Conshohocken, PA: American Society for Testing and Materials; 1994.

53. ASTM F1519–94 (1994). Standard test method for qualitative analysis of volatile extractables in microwave susceptors used to heat food products. In: *Annual Book of ASTM Standards*. West Conshohocken, PA: American Society for Testing and Materials; 1994.

54. ASTM F1349–94 (1994). Standard test method for nonvolatile ultraviolet (UV) absorbing extractables from microwave susceptors. In: *Annual Book of ASTM Standards*. West Conshohocken, PA: American Society for Testing and Materials; 1994.

55. ASTM F1500–94 (1994). Standard test method for quantitating non-UV-absorbing nonvolatile extractables from microwave susceptors utilizing solvents as food stimulants. In: *Annual Book of ASTM Standards*. West Conshohocken, PA: American Society for Testing and Materials; 1994.

56. General provisions applicable to indirect food additives. 21 CFR §174.5 (1998).

57. ASTM E460–88 (1988). Standard practice for determining effect of packaging on food and beverage products during storage. In: *Annual Book of ASTM Standards*. West Conshohocken, PA: American Society for Testing and Materials; 1988.

58. ASTM E462–84 (1989). Standard test method for odor and taste transfer from packaging film. In: *Annual Book of ASTM Standards*. West Conshohocken, PA: American Society for Testing and Materials; 1989.

59. ASTM E679–91 (1991). Standard practice for determination of odor and taste thresholds by a forced-choice ascending concentration series method of limits. In: *Annual Book of ASTM Standards*. West Conshohocken, PA: American Society for Testing and Materials; 1991.

60. Gordenker A. Better detection methods needed for pathogens on fresh sprouts. *Inside Lab Manage.* 1999;3(1):21–23.

61. Bechtel D. A global perspective on regulatory approval for food ingredients, nutraceuticals and dietary supplements. Oral presentation at Institute of Food Technologists Short Course, Atlanta, GA: 1998.

62. Abad AR, Gregory JF. Determination of folate bioavailability with a rat bioassay. *J Nutr.* 1987;117:866–873.

63. Graham DC, Roe DA, Ostertag SG. Radiometric determination and chick bioassay of folacin in fortified and unfortified frozen foods. *J Food Sci.* 1980;45:47–51.

64. Kivisto B, Andersson H, Cederblad G, Sandberg A, Sandstrom B. Extrusion cooking of a high-fiber cereal product. 2. Effects on apparent absorption of zinc, iron calcium, magnesium and phosphorous in humans. *Brit J Nutr.* 1986;55:255–260.

65. Colman N, Green R, Metz J. Prevention of folate deficiency by food fortification. II. Absorption of folic acid from fortified staple foods. *Am J Clin Nutr.* 1975;28:459–464.

66. Jusko WJ, Levy G. Absorption, protein binding and elimination of riboflavin. In: Rivlin RS, ed. *Riboflavin.* New York, NY: Plenum Press; 1975.

67. Ink SL, Gregory JF, Sartain DB. Determination of pyridoxine beta-glucoside bioavailability using intrinsic and extrinsic labeling in the rat. *J Agric Food Chem.* 1986;30:801–806.

68. Nguyen LB, Gregory JF, III. Effects of food composition on the bioavailability of vitamin B6 in the rat. *J Nutr.* 1983;113:1550–1560.

69. Fairweather-Tait SJ, Minski MJ. Studies on iron availability in man using stable isotope techniques. *Brit J Nutr.* 1986;55:279–285.

70. Pecoud A, Donzel P, Schelling JL. The effect of foodstuffs on the absorption of zinc sulfate. *Clin Pharmacol Ther.* 1975;17:469.

71. Nguyen LB, Gregory JF, Damron BL. Effect of selected polysaccharides on the bioavailability in rats and chicks. *J Nutr.* 1981;111:1403–1410.

72. Lindberg AS, Leklem JE, Miller LT. The effect of wheat bran on the bioavailability of vitamin B6 in young men. *J Nutr.* 1983;113:2578–2586.

73. Kusin JA, Reddy V, Sivakimar B. Vitamin E supplements and the absorption of a massive dose of vitamin A. *Am J Clin Nutr.* 1974;27:774.

74. Richardson J. *Polar Regions.* Edinburgh, Scotland: Edinburgh Press; 1961:71.

75. Rodahl K, Moore T. The vitamin A content and toxicity of bear and seal liver. *Biochem J.* 1943;37:166.

76. FDA defends using illness, death reports to propose usage limits on ephedra. *Food Labeling Nutr News.* 1999;7(15):9–11.

77. Leon CG. How 8 glasses a day keep fat away. *J Nutr Educ.* 1986;18:149.

78. Perera F, Mayer J, Santella RM. Biologic markers in risk assessment for environmental carcinogens. *Environ Health Perspec.* 1991;90:247–254.

79. DeVries JW. *Food Safety and Toxicity.* Boca Raton, FL: CRC Press; 1997:208–209.

PART III

Product Groups

CHAPTER 6

Infant Formulas and Medical Foods

Alan S. Ryan, John D. Benson, and Ann Marie Flammang

There is an emerging body of scientific research that establishes the health benefits of certain functional foods that contain novel food components. These foods have varying amounts of synthetic and natural ingredients, vitamins and mineral supplements, pre- and probiotics, fish oils, and plant sterols. Whether these foods have a lasting effect on physical development, immunological mechanisms, and disease prevention is debatable. However, the challenge for those interested in functional foods is to develop valid research designs and programs that produce meaningful results that demonstrate both safety and efficacy.

The pediatric and adult nutritional product industry has focused its efforts on producing nutritional products that help meet both the routine and special dietary needs of people of all ages, from infancy through adulthood. The pediatric products include formulas for healthy, full-term infants; special formulas for infants with unique nutritional needs (eg, premature infants); and nutritional supplements for older children (eg, follow-on formulas). Adult nutritional products meet the dietary needs of ambulatory adults and address the nutritional requirements related to conditions such as diabetes, pulmonary disease, kidney disease, human immunodeficiency virus/acquired immune deficiency syndrome (HIV/AIDS), and acute respiratory distress syndrome (ARDS).

This industry is committed to using its resources and talent to identify novel ingredients that can improve the lives of all people. Our current understanding of the efficacy of nutrients and other components used in pediatric and adult nutritional products, the processes by which changes in formulas are considered and implemented, the composition of different nutritional supplements in the marketplace, and the future direction of nutritional research are described in this chapter.

INFANT FORMULAS

Over the last 50 years, the importance of infant formula has evolved from meeting the nutritional needs of infants during the first year of life to optimizing health throughout life. Breastfeeding is widely recognized as the best way to feed infants; nevertheless, infant formula is an excellent source of nutrition either as an alternative for mothers who choose not to breastfeed or as a supplement for breastfeeding mothers. For mothers who elect to wean their babies from the breast, infant formula also represents the obvious complement to weaning. Future changes in infant formula are likely to be designed to have a positive, lasting effect on physical development, immunological outcomes, and behavioral parameters.

The human brain is undergoing rapid growth throughout the first 2 years of life.[1] Nutrient deficits at the time of the brain growth spurt may lead to permanent alteration in learning abilities.[1,2] Lucas[3] has coined the term *nutritional programming* to describe the process by which the adequacy or deficiency of a nutrient at a critical period of development results in a long-term or permanent change in the infant or child. Two examples illustrate this concept. In the first example, Lucas et al[4,5] considered the long-term developmental outcomes of early feeding in premature infants who were fed either a standard formula for term infants or a nutrient-enriched formula for preterm infants. Infants assigned to the standard formula, for less than 6 weeks after birth, had major deficiencies in developmental and intelligence scores at 18 months and at 8 years of age.

The second example is with term infants. Infants who were iron-deficient anemic as infants[6,7] and given iron therapy to correct the anemia had significantly lower scores than nonanemic controls in tests that evaluated intellectual, linguistic, motor, psychoeducational, and visual-motor integrative abilities when they were 5 to 6 years old. These results suggest that optimal nutrition during the critical period of brain development during infancy may be linked to later developmental outcomes.

There is emerging evidence that diseases such as hypertension, cardiovascular disease, respiratory disease, and diabetes are related to poor health and nutrition of the infant and mother. Barker[8] and his colleagues argue that undernutrition during infancy permanently changes the body's structure, physiology, and metabolism and leads to coronary heart disease and stroke later in life. These studies provide convincing evidence that nutritional programming during sensitive periods in early life has long-lasting effects on traits that really matter. How we feed our infants appears to influence their long-term development and health, thus heightening the importance of improving infant formula.

The process involved in translating new knowledge into infant formulas has been discussed in detail previously.[9–12] Determining when to modify an infant formula can be challenging, and each proposed change carries a risk. Such changes must be intensively studied and weighed against the knowledge that the existing formula has been safely fed to millions of infants. Since our knowledge of infant nutrition continues to grow at an incredibly rapid pace, levels of nutrients and other ingredients in infant formulas will undoubtedly continue to be modified. However, manufacturers should institute only formulation changes that are based on new scientific knowledge of clear demonstrable benefit to the infant.

This section reviews the prevalence of infant formula use in the United States and its importance in meeting the nutritional needs of the growing infant. We also describe two models that serve as a reference for improving formula: the human milk composition model and the breastfed-infant–performance model. These models help identify not only the essential nutrients for optimal infant growth and development but also the preferred clinical outcomes that may have significant lasting effects. We discuss the processes involved in changing infant formulas and compare the composition of different infant formulas in the marketplace. Last, we speculate on the future direction of infant formula research.

Infant Feeding in the United States

Information on infant feeding has been surveyed by the Ross Products Division of Abbott Laboratories since 1955.[13–20] The Ross Laboratories Mothers' Survey (RLMS) was designed to determine patterns of infant feeding during the first 12 months of life. The RLMS is a large national survey that is periodically mailed to new mothers. Mothers are asked to recall the type of milk their baby was fed in the hospital and during each month of age. Mothers choose from a list that includes all commercial infant formulas, human milk, cow's milk, and milk from other sources (eg, goat and evaporated). Since 1997, questionnaires have been mailed to approximately 117,000 new mothers each month.

As indicated by the RLMS data, commercial infant formula is one of the most important sources of infant nutrition throughout the first year of life. As shown in Figure 6–1, in 1997, the prevalence of the initiation of exclusive breastfeeding was 46%. An additional 16% of mothers elected to supplement their breastfed infants with infant formula. By 5 to 6 months of age, the prevalence of exclusive breastfeeding or breastfeeding with a formula supplement declined to 28%. The remainder of the milk feeding at 5 to 6 months of age was almost exclusively commercial infant formula. By 12 months of age, approximately 15% of infants were fed human milk exclusively or with formula supplementation, and 60% of infants were fed formula exclusively. The use of cow's milk made its first appearance by 5 to 6 months, and by 12 months of age, approximately 20% of infants were fed cow's milk exclusively.

It is noteworthy that the use of cow's milk for infant feeding has declined markedly during the last 10 years (Figures 6–1 and 6–2). This finding probably coincides with the recommendation made in 1992 by the American Academy of Pediatrics' Committee on Nutrition (AAP/CON) that whole cow's milk and low-fat cow's milk should not be used during the first year of life.[21] This recommendation is based on the evidence that cow's milk is a poor source of iron and that the composition of cow's milk (ie, high calcium, high phosphorus, and low vitamin C) may decrease the bioavailability of iron in foods that are often fed to infants, such as iron-fortified infant cereals.[22] Optimal nutrition during infancy involves choosing the appropriate milk source and eventually introducing a judicious selection of solid foods. According to AAP/CON recommendations, the only acceptable alternative to human milk during the first year of life is iron-fortified infant formula.[21]

An infant's diet is a reasonable predictor of iron status in early childhood.[23,24] Approximately 20% to 40% of infants fed whole cow's milk or non–iron-fortified formula and 15% to 25% of infants fed human milk are likely to be at risk for developing iron deficiency by 9 to 12 months of age.[23,24] In contrast, infants fed iron-fortified formula are not likely to develop iron deficiency by 9 months of age.[24]

The reported decline in the prevalence of anemia in the United States from the 1960s through the 1990s has been linked to the generalized improvement of iron nutrition in infancy and childhood.[25–30] During this period, there were substantial increases in the use of iron-fortified infant formulas and human milk[31,32] (Figure 6–2); these increases have been primarily at the expense of whole cow's milk and low-fat cow's milk.

How To Improve Infant Formula

Human Milk Composition Model

Nutrient Concentration. The history of infant formula development has been characterized by producing formulations that attempt to mimic the composition of human milk. This has often been referred to as the human milk composition model. Following this model, levels of protein, fat, and carbohydrate—the fundamental ingredients of infant formula—have been selected to match the average levels as found in human milk. Nutritionists are in general agreement as to the nutrients required by infants for optimal growth and development; these nutrients are currently included in infant formula. However, several difficulties present themselves with implementation of this model. Concentrations of essential nutrients in infant formulas cannot match those of human milk because of existing federal laws. Formula composition must meet specifications set forth by the Infant Formula Act of 1985.[33] This law requires the declaration of a minimum and maximum level for 9 nutrients and a declaration of a minimum level for an additional 20 nutrients. Most of the nutrient levels of human milk are below the minimum level prescribed by the Infant Formula Act. This is because a safety factor is applied to account for the potential differences in bioavailability between nutrients present in human milk and those added to infant formula.

Another problem with the human milk model is the variation of nutrients in human milk. Nutrients vary from the beginning to the end of each breastfeeding event, between stages of lactation, and as the result of the maternal diet.[34] For example, in humans, as in other mammals, fat from the mammary gland is lower in concentration in the initial secretion than in the final secretion.[35] Human milk fat is

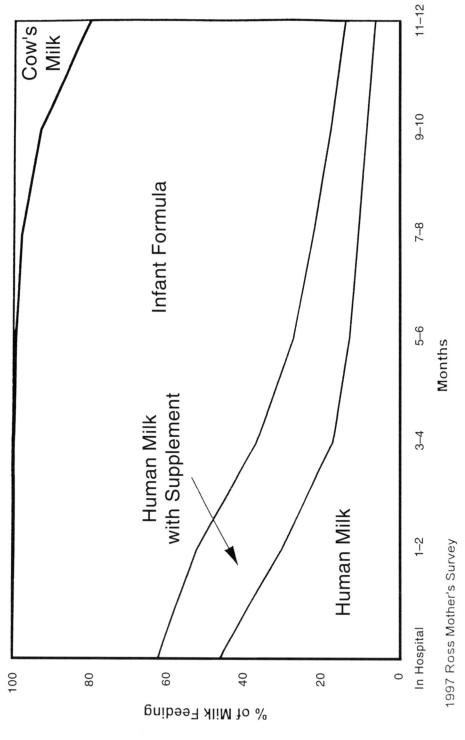

Figure 6–1 Infant feeding practices during the first year of life in the United States, 1997. Courtesy of Ross Products Division, Abbott Laboratories.

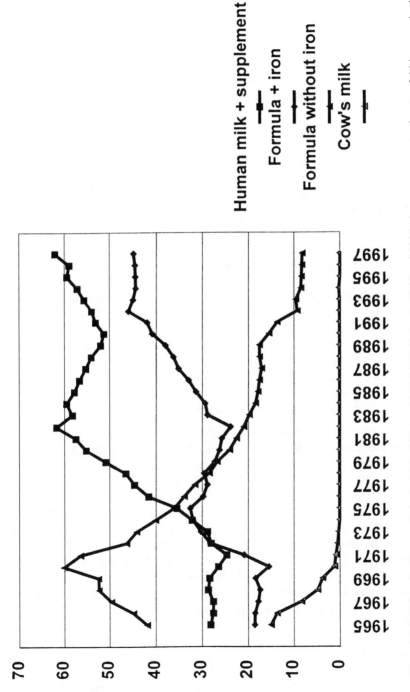

Figure 6–2 Infant feeding trends in the United States, 1965–1997. Between 1965 and 1997, there was a greater than twofold increase in the prevalence of breastfeeding (human milk + supplement) in the hospital. Concomitant with the long-term increase in breastfeeding was an increase in the use of iron-fortified infant formula at the expense of cow's milk and formula without iron. *Source:* Data from A.S. Ryan, *Yearbook of Physical Anthropology,* Vol. 40, p. 41, © 1997, Wiley-Liss (1965–1993 data); and Courtesy of Ross Products Division, Abbott Laboratories (1994 to 1997 data).

also highly variable between different mothers.[36] Further, the fatty acid composition of human milk differs from one geographical region to another and reflects differences in maternal eating patterns. Water-soluble vitamins are also highly variable and depend largely on the maternal diet.[37]

Protein concentration varies significantly with stage of lactation; it is higher in the first month of lactation and gradually decreases as lactation advances.[38] This change is probably beneficial to the infants, as their requirement for protein decreases with age.[39–41]

Unique Constituents.   As shown in Table 6–1, the number of known compounds in human milk is long and continues to grow with the development of more sophisticated analytical techniques. With this new knowledge, it becomes increasingly apparent that infant formula can never duplicate human milk, which contains living cells, hormones, active enzymes, immunoglobulins, a variety of bioactive compounds, and components with unique molecular structures.

Some of the components in human milk shown in Table 6–1 likely play an important nonnutritive role in an infant's diet. Some of these compounds are merely excretory products of the mammary gland and may serve no beneficial role to the infant. Some environmental contaminants found in human milk could actually be deleterious to the infant.[34] The latter possibility is illustrated by *trans*-fatty acids found in milk of mothers consuming a diet rich in *trans*-fatty acids.[42,43]

Many constituents in human milk are biologically active and most are likely to exist in balance with each other. The effect of adding single active ingredients to formula rather than considering their effects within a matrix of human milk may not be efficacious or beneficial.[44]

Stability.   Nutrients in human milk are intended for immediate consumption. Therefore, it is not surprising that some nutrients in human milk are not chemically stable over time. For example, vitamin B_6 in human milk, primarily as pyridoxal, is unstable. Infant formulas contain pyridoxine, a stable form of the vitamin necessary to ensure stability over the entire shelf life of the product (up to 2 years for liquid formulas and up to 3 years for powdered formulas).

One of the great challenges for formula manufacturers is to fully understand the effect of heat processing on the structure and function of compounds that are found in infant formula. Unlike human milk, infant formula is subject to periods of high temperature during processing to produce sterility. Heat treatment affects the structure of many proteins and significantly alters the biological activity of many compounds that are designed to simulate the properties of those in human milk.

Long-Chain Polyunsaturated (LCP) Fatty Acids Example.   One of the best examples of the shortcomings involved in using the compositional model is the proposed essentiality of certain LCP fatty acids for preterm and term infants. Many infant formulas derive their fat from vegetable oils and consequently do not contain the LCP fatty acids that are found in human milk. Excellent reasons exist for formula manufacturers to study the incorporation of two LCP fatty acids, docosahexaenoic (DHA) and arachidonic acid (AA), into infant formula. Of particular interest is DHA, which is one of the most abundant LCP fatty acids in the structural membranes of the central nervous system[45] and in the retina of the eye.[46,47] DHA is transferred across the placenta, particularly during the last trimester of pregnancy.[48,49]

Lower red blood cell and plasma phospholipid levels of DHA and AA have been reported for term infants who were fed formula not supplemented with DHA and AA compared with those fed human milk.[50–54] Visual acuity results, however, have been equivocal.[55–59] Recently, Auestad et al[59] considered 134 formula-fed and 63 breastfed term infants to evaluate whether providing a source of DHA and AA or DHA alone in formula would increase blood cell phospholipid levels, enhance visual acuity, or affect growth during the first year of life. There were no differences in growth or in visual acuity.

Language was also evaluated at 14 months of age using the MacArthur Communicative Development Inventory.[60] Differences were found among feeding groups for the two subscales of the MacArthur test: vocabulary comprehension and vocabulary production. Infants fed the DHA con-

Table 6–1 Unique Constituents of Human Milk

Antimicrobial Factors	Cytokines and Anti-inflammatory Factors	Hormones	Growth Factors	Digestive Enzymes	Transporters	Others
Secretory IgA, IgM, IgG	Tumor necrosis factor	Feedback inhibitor of lactation (FIL)	Epidermal (EGF)	Amylase	Lactoferrin (Fe)	Casomorphins
						Sleep peptides
Lactoferrin	Interleukins	Insulin	Nerve (NGF)	Bile acid–stimulated esterase	Folate binder	DNA, RNA
Lysozyme	Protaglandins	Prolactin	Insulinlike (IGF)		Cobalamin binder	
Component C3	α_1-antichymotrypsin	Thyroid hormones	Transforming (TGF)	Bile stimulated lipases	IgF binder	
Leucocytes	α-antitrypsin	Corticosteroids, ACTH	Polyamines		Thyroxine binder	
Bifidus factor	Platelet-activating factor: acetyl hydrolase	Oxytocin		Lipoprotein lipase	Corticosteroid binder	
Antiviral mucins, GAGs		Calcitonin		Ribonuclease		
Oligosaccharides		Parathyroid hormone				
		Erythropoietin				

Source: Adapted with permission from A. Prentice, Food and Nutrition Bulletin, Vol. 17, No. 4, table 1, © 1996, Food and Nutrition Bulletin, United Nations University.

taining formula had significantly lower scores for vocabulary comprehension than did infants fed human milk. They also had lower scores for vocabulary production than did infants in the unsupplemented control formula group. Post hoc analyses indicated that levels of plasma or red blood cell DHA at 4 months of age were negatively correlated with vocabulary production and comprehension in both formula-fed and breastfed groups. Possible explanations for the lower scores in the DHA group include the absence of a dietary source of AA, the reduction in AA levels in plasma and red blood cell phospholipids, the higher blood levels of DHA in the DHA group, or other lipid components included in the marine oil.

Using the compositional model, some scientists would argue that because the LCP fatty acids are in human milk they should be added to infant formula. However, in addition to the potential negative effects of DHA on vocabulary comprehension and production, as noted above, DHA and AA, like most nutrients, are not available as pure ingredients. The sources of these fatty acids include marine oil, fungal and algae extracts, and egg yolk phospholipids. These commercial ingredients are structurally different from the triglycerides or phospholipids in human milk; they also may contain other lipids not found in human milk.

Using the performance model (see below), researchers select fats in infant formula that provide important clinical benefits. At present, the premise that adding DHA and AA would be beneficial remains unproven. Thus, additional research is needed to shed further light on this important issue.

Breastfed-Infant–Performance Model

The goal of the performance model is to match the functional outcomes of the breastfed infant. Acceptance of this model recognizes that the chemical composition of breast milk cannot be replicated and that any attempts to do so with available ingredients could just as easily produce deleterious outcomes as improved outcomes. Those who use this model use nutrient levels in human milk as a starting point but recognize that the safety and efficacy of compounds in infant formula have to be demonstrated through carefully conducted clinical research. Levels and ratios of nutrients required in formulas may differ from those observed in human milk; the functional outcomes of infants receiving formula are targeted to be the same as those of breastfed infants.[11] The decision to change the composition of an infant formula requires an extensive, multifaceted, coordinated research effort. The chief characteristic of this research is that it is based on an evaluation of clinical benefits and risks. If a proposed change does not demonstrate a clinical benefit but is instead associated with a potential risk, it is not made. Users of this model are well aware that all changes are not necessarily beneficial to the infant and that millions of infants have been safely fed modern infant formulas.

It should be kept in mind that the breastfed infant performance model suffers from some of the same problems as those of the human milk composition model. Exclusively breastfed infants vary widely in growth rates, certain serum biochemical parameters, and some clinical outcomes. This variation cannot be controlled for in clinical studies by randomization because it would be considered unethical to assign an infant to a formula group if the mother chooses to breastfeed. The performance model will improve when it becomes clear what outcomes are preferred. At present, the outcomes that are commonly associated with breastfeeding that seem reasonable to achieve include similar gastrointestinal tolerance and bioavailability of nutrients, increased developmental scores, enhanced visual acuity, enhancement of the immune system, and reduced rates of infection.

Infant Formulas: Are They Different?

Clearly, the Infant Formula Act requires that infant formulas in the marketplace meet specific nutrient requirements. For most of the major nutrients, different infant formulas have very similar concentrations, and many nutrient levels are similar to but generally higher than those found in human milk. However, there are some noticeable differences that affect important clinical outcomes. Differences in clinical outcomes may be the result of variation in ingredient selection, types of processing used, and slight alterations in nutrient composition. As described below for cow's milk–based formulas, there are several differences in clinical outcomes when commercial infant formulas are compared. Modern infant formulas differ in tolerance, the amount of whey protein added, the types of fat added, and the amount of nucleotides added to the formulation.

Tolerance

Tolerance is defined as an infant's reaction to infant formula, as manifested by rates of emesis, stool consistency and frequency, fussiness, sleeping patterns, and so forth. It can be significantly affected by differing processing techniques and modification of certain ingredients. In one study, 82 breastfed infants were randomly assigned to be weaned to one of two commercial formulas when mothers chose to discontinue breastfeeding.[61] Parents kept dietary records for 3 consecutive days immediately preceding each office visit. Visits were scheduled before and after infants were weaned from the breast. Formula intake, the frequency of spit-up and vomiting, and the frequency, color, and consistency of stools were recorded. Notably, infants who were weaned to one commercial product maintained a stool consistency (mostly loose and mushy) that was similar to the stool consistency observed during exclusive breastfeeding (Figure 6–3). In contrast, infants weaned to the other commercial product developed firmer stools than when they were exclusively breastfed (Figure 6–3).

Cow's Milk Protein

Casein and whey proteins are the two major classes of protein in milk. The simplistic approach to selecting proteins for infant formula has been to duplicate the whey-to-casein ratio of human milk, which some formula manufacturers have defined as 60:40. However, this approach is flawed because in human milk, over the course of human lactation, the ratio of whey to casein proteins varies from 90:10 in early lactation to 60:40 in mature milk and 50:50 in late lactation.[62] Further, specific proteins in the casein and whey fractions differ markedly. For example, in cow's milk, whey β-lactoglobulin is the predominant protein; this protein does not exist in human milk. The differences in the casein fraction of cow's milk to that found in human milk are less distinct, but the caseins differ both antigenically and chemically.[41]

Another approach would be to match as closely as possible the pattern of plasma amino acids levels and ratios of formula-fed infants to the levels found in infants fed human milk. This approach is more meaningful than simply trying to match the whey-to-casein ratio of human milk because it considers formula performance: how well the formula matches human milk in delivering amino acids to the infant's plasma.

Paule and his colleagues[63] have developed a mathematical equation, the *plasma loss function*, that measures how closely the plasma essential amino acid pattern from a mixture of proteins (as used in infant formula) matches that from a breastfed infant (Figure 6–4). The equation is the sum of the absolute normalized deviations between the ratio of each essential amino acid to the total essential amino acids in the plasma of infants fed a particular mix of proteins in formula or in human milk. The equation not only controls for low and high concentrations of essential plasma amino acids but also takes into account both the absolute concentration and amino acid ratios. A value of zero would indicate no difference between plasma amino acid patterns—a perfect match. Figure 6–5 shows the essential amino acid plasma loss function for three milk-based infant formulas and human milk. These formulas include a partially hydrolyzed, exclusively whey formula (Good Start with Iron), a whey-dominant formula (Enfamil with Iron), and a casein-dominant formula (Similac with Iron). The for-

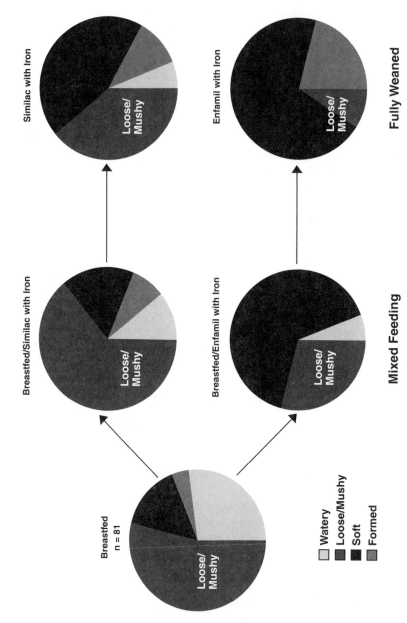

Figure 6–3 Stool consistency of infants who were initially breastfed and then fed one of two commercial formulas. Infants who were weaned to Similac with Iron maintained a stool consistency that was similar to the stool consistency observed during exclusive breastfeeding. In contrast, infants weaned to the other commercial product developed firmer stools than when they were breastfed exclusively. *Source:* Adapted with permission from B. Blennemann et al., *The American Pediatric Society Abstracts,* Vol. 39, Abstract 1814A, © 1996, American Pediatric Society.

$$\text{Loss Function} = \sum \frac{\left| \dfrac{EAA_M}{TEAA_M} - \dfrac{EAA_{HM}}{TEAA_{HM}} \right|}{\dfrac{EAA_{HM}}{TEAA_{HM}}}$$

Where   EAA=11 Essential Amino Acids
TEAA=Total Essential Amino Acids
M=Mixture (Formula)
HM=Human Milk

Figure 6–4 The plasma loss function used to estimate how closely a particular mixture of proteins in an infant formula matches that from a breastfed infant. Courtesy of Ross Products Division, Abbott Laboratories.

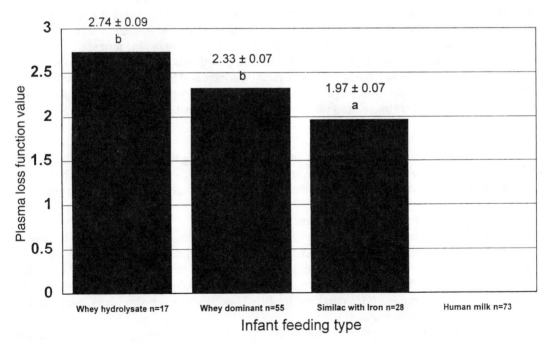

Figure 6–5 The plasma loss function values for human milk and a variety of commercial infant formulas. Means with unlike superscripts indicate significant differences (*P*<0.05). Courtesy of Ross Products Division, Abbott Laboratories.

mulas differed markedly in their plasma loss function values. Increasing the percentage of whey in the protein mixture did not produce plasma amino acid concentrations in the formula-fed infants that better matched those in the breastfed infants. Of significance is that the exclusively whey-containing and the whey-dominant formulas had plasma loss function values that were significantly higher than those of the formula that contained a nearly equal distribution of whey to casein (48:52).

Sources of Fat

The choice of the oil blend may have a major impact on the bioavailability of fat and calcium from infant formulas. In an effort to duplicate the composition of human milk, some manufacturers follow the compositional model and use palm olein oil to approximate the concentration of palmitic and oleic acids in human milk. However, the inclusion of palm olein oil reduces the bioavailability of fat and calcium from milk-based infant formulas.[64] The results of a crossover study with infants fed a formula containing a blend of palm olein (53%) and soy oil (47%) (Formula PO/S) or a formula containing a blend of soy oil (60%) and coconut oil (Formula S/C) indicated that mean absorption of fat was 91% of intake when the Formula PO/S was fed and 95% of intake when the Formula S/C was fed ($P<0.001$).[64] The difference in calcium absorption was even greater with infants fed the Formula S/C, achieving 48% absorption compared to 39% for those fed the Formula PO/S ($P<0.01$). In a second study[65] comparing commercial formulas, calcium and fat bioavailability were more significantly available from a formula without palm olein oil than with it (98.5% vs 90.0% for fat and 57.4% vs 37.5% for calcium). The greater bioavailability of palmitic acid from human milk compared with palm olein oil from formula is related to their differences in chemical structure. In human milk, palmitic acid is found largely on the 2-position of the triglyceride molecule. In palm olein oil, the palmitic acid is found on the 1- and 3-positions. When cleaved by lipases during digestion, the free palmitic acids in the 1- and 3-positions form soaps with calcium from the formula. This result diminishes the bioavailability of both fat and calcium from the palm olein–containing formula.

Preliminary results of a 4-month study[66] indicate reduced growth in boys fed a soy-based formula and in girls fed a cow's milk-based formula when palm olein oil is increased from 0% to 60% of the fat blend. An intermediate level of palm olein oil (45%) did not affect growth. These findings illustrate the importance of considering nutrient interactions when evaluating the nutritional quality of infant formula.

Carbohydrates

Carbohydrates are an important source of energy for the growing infant; they account for 35% to 42% of the energy intake of the breast-fed or formula-fed infant.[67] Although the carbohydrate in human milk consists primarily of lactose, glucose and a variety of oligosaccharides are also present.[67] Glucose represents 3% to 4% that of lactose.[68] The variation in lactose concentration in human milk is large, but 74 g/dL is considered the representative mean.[67]

Most cow's milk-based formulas provide lactose as the sole source of carbohydrates. Lactose is neither added to cow's milk-based lactose-free formulas nor to specialized non-milk protein formulas or protein hydrolysates fed to infants with milk protein sensitivity. Soy-based formulas are also free of lactose. Formulas without lactose use alternative sources of carbohydrate such as sucrose and glucose (found in corn syrup solids), natural and modified starches, starch hydrolysates, monosaccharides, and indigestible carbohydrates.

Lactose enhances the absorption of several minerals including calcium, magnesium, zinc, and iron.[69] Other sugars such as glucose, galactose, xylose, fructose, and sucrose also promote the absorption of calcium and other minerals but to a lesser degree than calcium.[67] Compared with a soy-based formula without lactose but containing a blend of sucrose and corn starch hydrolysate, a soy-based formula with lactose increased the absorption of calcium.[69] However, similar bone mineral content was observed in term infants fed soy-based formula without lactose or cow's milk-based formula or human milk with lactose.[70] Sucrose and corn syrup solids both promote calcium absorption from soy formulas.

During early infancy when lactase activity is at its highest, a high percentage of infants do not completely absorb the lactose. Roggero et al[71] reported that among 40 term and 10 preterm infants, 36% of the breastfed, 42% of the formula-fed, and 64% of infants fed human milk with formula did not completely absorb the carbohydrate. The unabsorbed lactose reaching the colon may provide a substrate for the growth of bifidus bacteria.[67]

Nucleotides

Nucleotides are compounds found in human milk that are involved in key metabolic processes, including energy metabolism and enzymatic reactions. They are also the building blocks of deoxyribonucleic acid (DNA) and ribonucleic acid (RNA). Nucleotides contribute 2% to 5% of the nonprotein nitrogen in human milk,[72] which is approximately 25% to 30% of the total nitrogen in human milk.[72,73] Human milk contains relatively high concentrations of cytidine, uridine, adenosine, and guanosine and little or no inosine. Cow's milk contains only low levels of uridine, inosine, and cytidine and small amounts of adenosine and guanosine.[74,75]

Because human milk contains nucleotides, several formula companies have added nucleotides to their infant formulas. Nucleotide levels in infant formula vary from 0 to 72 mg/L. However, only recently have the levels and ratios of nucleotides in human milk been investigated thoroughly. Studies have shown that the majority of dietary nucleotides are ingested in the forms of nucleoproteins and nucleic acids.[76] During digestion, these are degraded by pancreatic and intestinal proteases and nucleases to nucleotides, which are further broken down by phosphatases and nucleotidases to nucleosides that are absorbed into the gastrointestinal tract.[77] The nucleotide levels previously added to formulas were based on reports of the monomeric forms of nucleotides[78,79] and on certain nucleotide-containing adducts.[78]

A more accurate determination of the physiological availability of nucleotides in human milk was developed by Leach and his colleagues.[80] This required the precise and accurate measurement of nucleotides derived from all sources in human milk that could be absorbed as nucleosides during digestion—that is, the total potentially available nucleosides (TPANs). This study supported the addition of nucleotides to formula in quantities that most closely match the TPAN content in human milk (Figure 6–6)—that is, 72 mg of nucleotides per liter of infant formula.

This concentration of nucleotides was considered in a 12-month, controlled, randomized, blinded, multisite study.[81] In addition to two formula groups with (72 mg/L) and without (control group) nucleotides, a third group of infants fed human milk was also considered. Individual immune responses were evaluated by the antibody response to standard vaccines—*Haemophilus influenzae* type b polysaccharide (Hib), diphtheria and tetanus toxoids, and oral polio virus (OPV)—at 6, 7, and 12 months of age. Compared to the control group (no added nucleotides), 1 month after the third immunization (at 7 months of age), infants fed the nucleotide-containing formula had significantly higher Hib and diphtheria antibody concentrations. The significantly higher Hib antibody response in the nucleotide formula group persisted at 12 months of age. The antibody responses to tetanus and OPV were not enhanced by nucleotide fortification. Infants who were breastfed had significantly higher antibody titers to polio virus than either formula-fed group at 6 months of age.

Future Trends in Infant Formula Research

Future changes in infant formula composition will be based on incorporating compounds that have positive and possibly long-lasting effects. Hundreds of compounds found in human milk are not in-

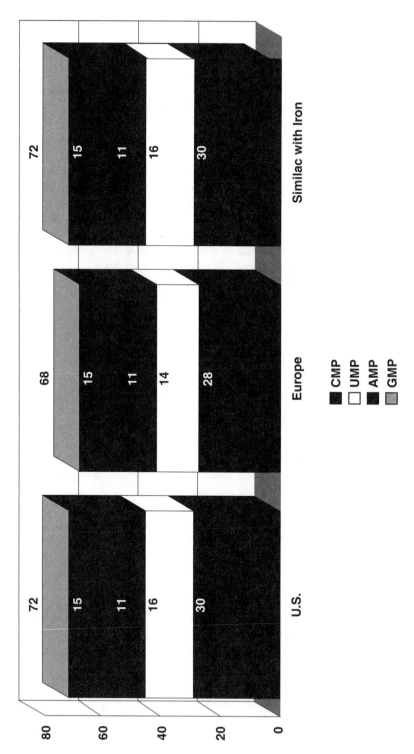

Figure 6–6 Total potentially available nucleosides (TPAN) content of human milk and formula (mg/L). CMP = cytidine monophosphate; UMP = uridine monophosphate; AMP = adenosine monophosphate; GMP = guanosine monophosphate; IMP = inosine monophosphate; US = mean content of 11 samples of human milk from the United States; Europe = mean content of 16 pooled samples of human milk from Europe. *Source:* Adapted with permission from J.L. Leach et al., Total Potentially Available Nucleosides of Human Milk by Stage of Lactation, *The American Journal of Clinical Nutrition,* Vol. 61, No. 6, © 1995, Federation of American Societies for Experimental Biology.

cluded in infant formula. The potential risks and benefits of incorporating new compounds into formula must be determined before changes can be made. Advances in molecular biology will undoubtedly make it possible to develop bioactive proteins, hormones, and immunologlobins. In the future, infant formula may be designed not only to enhance infant growth and development but also to prevent the development of diseases in adult life.

MEDICAL FOODS

Historical Background

Medical foods offer advantages similar to those of infant formulas in that they are foods that provide complete nutrition and/or enhance dietary management of a disease and are in either liquid enteral form or food form. Schmidl et al[82] and Schmidl and Labuza[83] presented two of the first reviews of the history and status of medical foods. Enteral nutrition can be traced back to ancient Egypt and Greece, where nutrient liquids were used both orally and as enemas.[84] Although rectal alimentation persisted until the early part of the 20th century, enteral feeding via a conduit was reported as early as the 12th century.[85] Periodic references from the 15th through the 18th centuries refer to enteral feeding by tube, but not until the end of the 19th century did tube feeding via the nose into the esophagus or stomach become well established.[84] Common tube-feeding products of the time included milk, eggs, meat extracts, jellies, sugar, and wine.

Feeding devices and their material composition have improved throughout the 20th century, and surgical access to the gastrointestinal tract has developed as an extension to nasal and orogastric feeding techniques.[84] Because surgical access bypasses a portion of the alimentary tract, it became necessary to develop nutritional formulas that were "partially digested." The first commercial enteral formula was introduced in 1942—a pediatric product containing partially hydrolyzed casein as a protein source. The late 1950s through 1970s saw exploration of low-residue defined formula diets for space travel by the National Aeronautics and Space Administration (NASA).[86–88] These chemically defined formulas, or "elemental" diets, consisted of crystalline amino acids, essential fatty acids, simple sugars, and vitamins and minerals.

As the role of nutrition in treating disease states and medical conditions became more widely recognized throughout the latter part of the 20th century, various methods of tube feeding and "disease-specific" enteral formulas were developed and refined. Modular products, single-nutrient additions to formulas, were also created as supplements to the general purpose, or specialized formulas. Understanding the limitations of parenteral nutrition, along with advances in access devices for administering enteral formulas, also helped increase the use of enteral nutrition products. However, as the complexity of enteral nutrition products for medical use evolved, issues arose regarding terminology, application, labeling, and regulation of these products.

Classification

"Foods for special dietary use" originally referred to foods that provide better nutrition than that supplied by the ordinary diet. This category was incorporated into the Code of Federal Regulations in 1941.[89] The differentiation of such foods from drugs occurred in 1972 when the Food and Drug Administration (FDA) classified a formula for infants with phenylketonuria, an inborn error on metabolism, as a "food for special dietary use" rather than as a drug.[90] Thus began the concept of medical foods. The following year, FDA promulgated that foods for use solely under medical supervision to meet nutritional requirements in specific medical conditions should be labeled similar to foods for special dietary use.[91] Originally, the Code of Federal Regulations (CFR) categorized labeling regulations for foods for special dietary uses into five distinct categories; however, recent updates now include only three categories: (1) hypoallergenic foods, (2) foods for reducing or maintaining body weight, and (3) infant foods.[92] The categories do not specifically include foods to be used solely under the supervision of a physician, so these foods essentially eluded labeling regulation. While FDA did not explicitly define medical foods, the concept of medical foods was established by the recognition of a category of foods to be used solely under the supervision of a physician.[93]

In 1982, a task force was commissioned by FDA to define medical foods. It recommended that policy to regulate medical foods primarily as foods should be adopted. Medical foods were defined as "specifically formulated or processed products represented for the dietary management of a specific disease, disorder, or medical condition, which products are consumed or administered enterally and are represented for use under the supervision of a physician."[94] FDA drafted regulations for medical foods 2 years later, but they were not published.[95] Periodic attempts were made to define medical foods more clearly, but no changes were incorporated into the CFR.

The Infant Formula Act of 1980 was passed by Congress in response to failure-to-thrive cases associated with metabolic acidosis linked to formulas low in chloride. Subsequent regulations[96] placed certain formulas for infants in a special category called "exempt." Infant formulas designed for specific diseases, such as inborn errors of metabolism, were classified as "exempt" because the nutritional requirements of infants with these diseases were different from those of healthy infants. FDA's usage of the term *exempt* refers to the exception to the requirement of infant formula manufacturers to include the full range of nutrients needed for growth that are normally present in formulas.

The Orphan Drug Act of 1973 was amended in 1988 to include medical foods.[97] The medical foods definition included in the amendment reads:

> The term "medical food" means a food which is formulated to be consumed or administered enterally under the supervision of a physician and which is intended for the specific dietary management of a disease or condition for which distinctive nutritional requirements based on recognized scientific principles are established by medical evaluation.[98]

While this definition was included to introduce a new subcategory of medical foods, orphan medical foods, it is currently the most authoritative definition available for medical foods. Orphan medical foods are products to be used in the management of "any disease or condition that occurs so infrequently in the United States that there is no reasonable expectation that a medical food for such disease or condition will be developed without assistance."[99] The FDA categorization thus includes medical foods as a subcategory of foods for special dietary use and includes orphan medical foods as a subcategory of medical foods. Orphan medical foods are similar to exempt infant formulas in that both have defined compositions and both are required for special and infrequently occurring medical conditions.

The FDA Compliance Program was initiated in 1989 to evaluate how manufacturers could provide appropriate formulations of medical foods, suitable microbiological standards, and reasonable therapeutic claims for medical foods. The program classifies medical foods into four categories based on their use recognized by FDA:

1. nutritionally complete formulas (eg, formulas that can be used as sole-source nourishment)
2. nutritionally incomplete formulas (eg, modular products of fat, protein, and carbohydrate)
3. formulas for metabolic disorders in patients over 12 years of age (eg, phenylketonuria formula, other than exempt infant formulas)
4. oral rehydration products (eg, electrolyte/water formulas for dehydration)[100]

In 1990, the Nutrition Labeling and Education Act (NLEA)[101] was passed by Congress. It included the definition of medical foods listed in the Orphan Drug Act Amendment. The NLEA provides regulations relating to nutrient content and health claims, among others. However, medical foods as defined in the Orphan Drug Act as amended are notably exempt from the new NLEA regulations because medical foods are to be used under the supervision of a physician.[102] Medical foods may employ only labeling text and format that are truthful and not misleading to provide useful information about the nutritional content of the product, providing they do not constitute a drug claim or violate other regulations. In November 1996, FDA issued an Advance Notice of Proposed Rulemaking[103] for the regulation of medical foods. At this time, a proposed rule has not been published.

Processing Requirements

Medical foods are remarkably unrestricted in their regulation compared to infant formulas, which are among the most tightly regulated of processed foods in quality control, labeling, nutrient requirements, formula recall, and new product notification. Medical foods do fall under FDA's guidelines for Good Manufacturing Practices (GMPs) for conventional foods and must also follow the FDA regulations to ensure the sterility of low-acid, thermally processed foods[104] if they are thermally processed. Most manufacturers meet or exceed the required quality regulations for their enteral products.[105] Premarket approval with necessary extensive qualitative and quantitative studies is required for infant formula. Medical foods do not currently require that level of regulation for new product clearance, with the result being the ability to market new products more quickly. The interests of reputable manufacturers and FDA are the same, however: to ensure the quality and safety of medical foods. Although it is not required, ethical manufacturers engage in extensive testing of their products before launch to ensure that they are safe and suitable for their intended purpose.

Enteral Nutrition Support

Medical foods are formulated to supply nutrition support to those individuals unable to ingest adequate quantities of conventional foods by mouth.[82] They may be formulated for oral or tube-feeding use, for supplemental or sole-source nutrition. Oral nutritional beverages can be formulated as ready-to-drink liquid or as a powder to be reconstituted with water, milk, or juice. Depending on the target population, a nutritional beverage can be a complete meal replacement with a full nutritional complement of vitamins and minerals and be considered "sole-source" nutrition when ingested at appropriate levels. Nutritional supplement beverages can also be formulated to provide adjunct nutrition to boost nutritional intake with only a portion of the required nutritional needs met.

Although selection of an oral supplement is based on fortification of nutrients and other factors such as age and health status of the consumer, commercial beverage success with the target population depends on taste. Creation of liquid nutritionals is the daunting task of simultaneously creating a good-tasting product and providing complete nutrition. Nutrient levels and various ingredients can interact, which can create stability, color, and organoleptic challenges. Oxidative degradation of vitamins and minerals must be accounted for to provide appropriate product labeling. Product pH and processing method used can also affect stability and taste of the final product. Enteral nutrition products formulated for tube feeding do not face the same organoleptic challenges faced by oral products. Hydrolyzed protein and fish oils, examples of strong-tasting ingredients, can be used in tube-feeding formulations for maximum impact in nutrition support without the taste issues encountered in oral products.

Numerous formulations exist for the dietary management of people with specific diseases and medical conditions. Enteral formulations have proliferated as tube-feeding delivery methods have improved and as the benefits and cost advantages derived from intraluminal nutrient delivery have been researched.[106,107] Patients in acute care, long-term care, or home care all have needs specific to not only their disease states but also their care level. Enteral formula selection should be based on the specific needs and disease state of each individual.

Among other considerations when choosing a tube-fed enteral formulation is the delivery method. Typically, enteral feedings are given via nasoenteric (nasogastric or nasojejunal) tube when the feeding is expected to be less than 3 weeks. If the patient is expected to receive the feeding longer than 3 weeks, enterostomal (gastrostomy or jejunostomy) tube feeding is appropriate. Enteric tubes offer a wide assortment of sizes, types, materials, accessories, and placement methods. The enteral formula feeding itself may be given via pump, gravity drip, or syringe. Feedings may be delivered continuously, intermittently, nocturnally, or by bolus. All of these factors are to be considered when designing a nutrition support regimen.

Specialized Nutritional Formulas and Ingredients

Standard nutrition products provide basic nutritional support, while specialized nutrition products are scientifically designed to meet specific patient needs. Standard nutritional formulas combine appropriate levels of carbohydrate, protein, and fat and recommended levels of vitamins and minerals. Specialized formulas contain unique ingredients and differing levels of both macro- and micronutrients appropriate for the disease state. Patient conditions such as ARDS, diabetes, HIV/AIDS, impaired gastrointestinal function, malabsorption, and renal failure are examples of disease states for which specialized products are available to meet the specific enteral nutrition needs.

Nutrition plays a key role in patient care. Hospitalized patients at risk for malnutrition have more complications, higher mortality rates, longer lengths of stay, and higher hospitalization costs.[108,109] Enteral feeding is adjunct therapy that can affect the outcome of critically ill patients because the physiological benefits of enteral nutrition support go beyond merely providing nutrients.[110] Like all health care interventions, enteral nutrition support requires careful identification of goals, selection of formula, and monitoring of therapeutic outcome. Unique and special nutrients in products can help in the implementation of effective enteral feeding therapy.

Protein

Protein is an important dietary constituent for all human beings; however, the optimal level of protein intake varies in persons with differing disease states. Appropriate protein intake depends on many factors, including energy intake, prior nutrition status, degree of stress, and stage of illness or recovery. Good-quality protein is an important component of most medical enteral formulas. Hydrolyzed protein in the form of peptides and free amino acids is used in defined formula diets, often termed *elemental* formulations. The existence of separate and noncompeting, carrier-mediated transport systems for free amino acids and small peptides in the gastrointestinal tract makes these elemental formulas popular for malabsorptive conditions such as Crohn's disease, short-bowel syndrome, metabolic stress, and acute trauma. Free amino acids, specifically glutamine and arginine, are also used in specialized enteral formulas.

Glutamine, a nonessential amino acid, is considered conditionally essential during critical illness.[111] Glutamine is involved in a wide variety of metabolic reactions and plays a significant role in the transfer of nitrogen between tissues.[112] It is used as an oxidative fuel for the intestine or is converted to other substrates for use by the liver. The need for glutamine appears to increase as the severity of injury increases.[113] Clinical research has shown the benefit of glutamine addition in an enteral formula preparation for critically ill patients.[114,115] Arginine is another conditionally essential amino acid and plays a key role as a precursor in the nitric oxide pathway.[116,117] Arginine supplementation is associated with accelerated wound healing[116,118] and support of immune function.[119,120] Supplementation at 1% to 3% of total calories has been shown to be beneficial in wound healing and nitrogen retention after injury.[116,121] Clinical studies with arginine-supplemented formula have shown a decreased incidence of infection and increased improvements in nitrogen balance and other indices of stress in major-trauma patients.[122]

Carbohydrates

Carbohydrates are an important fuel for the malnourished or severely ill patient.[123] Most people can efficiently digest and absorb all carbohydrate sources, with the exception of lactose. Adequate carbohydrate intake is particularly important for patients who require restriction of dietary proteins, such as people with renal or hepatic disease. Carbohydrates included in several medical formulas but not used for fuel sources include the components of dietary fiber: cellulose, hemicellulose, pectin, gum, and mucilage. Lignin, a noncarbohydrate substance of plant cell walls, is also a component of dietary fiber.[124] Sources of dietary fiber can be classified as providing soluble or insoluble fiber, based on the dispersability of the polysaccharide in water, rather than on true chemical solubility. Physiological responses associated with soluble dietary fiber include the lowering of plasma cholesterol[125] and modification of the glycemic response.[126] Ingestion of both insoluble and soluble dietary fiber is frequently associated with bowel-tract health.

The presence of fiber in the diet influences bowel function by reducing transit time, increasing stool weight and frequency, diluting large-intestinal contents, and providing a fermentable substrate for gut microflora. Ingested fiber passes unchanged by alimentary enzymes into the lower bowel, where constituents of dietary fiber are metabolized by fermentation by the microflora present in the colon. The fermentation results in the formation of ammonia and short-chain fatty acids (SCFAs) such as acetic, butyric, and proprionic acids, which can be reabsorbed and can contribute some calories to the diet.[127,128] Gases (carbon dioxide, hydrogen, and methane) are also formed as a result of the fermentation process.[125,126] Bacterial fermentation of the nondigested carbohydrates of dietary fiber is thus partially responsible for the physiological effects on bowel function.

Fructo-oligosaccharides (FOSs) are another carbohydrate ingredient that can be found in certain medical foods. FOSs are indigestible but highly fermentable by gut microbes. They occur naturally in foods such as onions, bananas, garlic, and tomatoes. FOSs, like other fermentable dietary fibers, serve as a substrate for colonic bacteria to produce SCFAs. FOSs help maintain and restore the balance of healthy gut flora[129] by creating an environment that supports the beneficial bifidobacteria but does not support the growth of some pathogenic bacteria.[130] FOSs play an important role in the maintenance of normal intestinal flora during antibiotic therapy.[131,132] Inclusion of FOSs in medical nutritional products is an efficient mechanism by which a desirable modification of the gut microflora may be achieved to induce beneficial effects on the digestive tract.

Fat

Dietary fats are necessary to provide energy, supply essential fatty acids (linoleic [n-6] and linolenic [n-3]), act as a fat-soluble vitamin carrier, and enhance palatability.[133] Fat consists of a complex mixture of triacylglercerol compounds that can differ greatly from one another in their physical and chemical properties. Differing roles and properties of the triacylglercerols and their constituent fatty acids must be taken into account for their effectiveness as ingredients in medical nutritional products. Recent recommendations for diet composition for the general public focus not only on the amount of fat in the diet in terms of total calories but also on the composition of fat with respect to type and amount of fatty acids. The American Heart Association has recommended that saturated-fat intake be less than 10% of total calories and that polyunsaturated-fat intake not exceed 10% of total calories.[134] This recommendation is consistent with the American Diabetes Association dietary guidelines for people with diabetes mellitus, which state that 10% to 20% of total calories should come from protein, less than 10% from polyunsaturated fat, less than 10% from saturated fat, and the remainder from carbohydrate and monounsaturated fats.[135]

Fat is a very useful energy source for glucose-intolerant patients. Absorption of carbohydrate and fat is different when these nutrients are ingested in liquid form rather than as solid food.[136] High-carbohydrate liquids are rapidly absorbed and result in a more rapid and higher rise in blood glucose levels than liquids with part of the carbohydrate replaced by fat. Liquid nutritionals for enteral or tube feeding formulated for persons with diabetes therefore need to be high in fat to achieve appropriate blood glucose control. Fat is also important for those individuals who suffer from pulmonary disease. The therapeutic goal in patients with chronic or acute retention of carbon dioxide is to decrease carbon dioxide production. High carbon dioxide production can precipitate acute respiratory failure in patients with chronic disease and can complicate weaning of ventilator-dependent patients.[137] Because the combustion of fat yields less carbon dioxide than combustion of either protein or carbohydrate, a high-fat diet may be preferable for patients with pulmonary disease.[137,138]

A medical nutritional product formulated for acute lung injury and those at risk for ARDS utilizes a novel blend of fats.[139] ARDS is characterized by pulmonary inflammation, poor lung compliance, acute gas exchange disturbances, and hypoxemia. It has traditionally been treated with chemotherapeutics and mechanical ventilation, with only limited success. Research has suggested that the usual inflammatory response in ARDS can be modulated by providing specific dietary nutrients, including gamma-linolenic acid found in oil from seeds of the borage plant; eicosapentaenoic acid (EPA), from fish oil; and specific dietary antioxidants.[140,141] These components have been shown to have important metabolic effects by altering the phospholipid content of alveolar membranes, leading to production of less inflammatory mediators.[142,143] Clinical research outcomes have shown significant improvement in oxygenation status and reduced ventilator and intensive care unit days with the specialized product.[144] Compared to traditional methods of treating ARDS patients, use of a specialized, medical nutritional product to improve patient condition and decrease length of intensive care stay provides unparalleled cost benefits for specific nutritional components/ingredients.

Medium-chain triglycerides (MCTs) are beneficial as a fat source for those people suffering from malabsorption or maldigestion disorders. MCTs aid gastric emptying because they are absorbed by the intestinal tract without emulsification by bile acids.[145] The small intestine has a greater capacity to absorb MCTs than long-chain triglycerides, and the fatty acids derived from MCTs are transported principally by the portal vein.[146] Due to the ease of MCT absorption, medical nutritionals formulated for use in metabolic stress, acute trauma, and malabsorption conditions such as irritable bowel disease frequently use MCTs in the fat blend. A new use of MCTs involves combining the benefits of MCTs with fish oil into a structured triglyceride (STG). Fish oil is an abundant source of n-3 fatty acids, including EPA and docosahexaenoic acid. EPA functions as a precursor of prostaglandins and leukotrienes that have anti-inflammatory properties.[147] STGs offer absorption characteristics superior to standard oil combinations for delivery of n-3 fatty acids.[148] A fish oil/MCT structured lipid is featured in a medical nutritional formula specific for malabsorption conditions, such as Crohn's disease, due to the potential benefit from increased absorption and anti-inflammatory effects of EPA.[149] Structured lipids are exciting new ingredients that hold much promise for many future medical nutritional product formulations.

Vitamins and Minerals

Vitamins, complex organic compounds present naturally in plant and animal tissue, are cofactors in regulating many metabolic processes.[123] Minerals play an important role in the maintenance of muscle and nerve function, regulation of water balance and metabolism, mineralization of the skeleton, and transformation of energy.[123] Requirements for vitamins and minerals are affected by many factors, including an individual's biochemistry, age, changes in dietary intake, nutritional status, medications, and disease state.[150] Antioxidant vitamins include vitamins C and E, as well as beta carotene, a carotenoid with provitamin A activity. Tocopherols (vitamin E), ascorbic acid (vitamin C), and carotenoids act as scavengers and control potentially damaging reactions by eliminating pro-oxidants and free radicals.[151] During metabolism, the creation of reactive oxygen metabolites, or free radicals, is a normal process and is normally reduced by antioxidants so that oxidative damage is minimal. Injury, infection, or disease processes can induce an inflammatory response that occurs due to increases in the neutrophil and macrophage production of free radicals. The body's antioxidant supply during such stress may not be sufficient to quench the free radicals formed, leading to oxidative damage. The antioxidant levels in several specialty medical nutritional formulations are increased with the intention of moderating the free-radical damage due to the inflammatory response.

Future

Over the past 25 years, both consumers and health care professionals have been barraged with recommendations about diet and health. Focus of these recommendations has changed from deficiency symptoms of vitamins and minerals to macronutrient concerns such as limitation of fat and calories. Recent recommendations to increase intake of foods such as plant foods for fiber and phytochemicals are a change from past recommendations to limit intake of foods such as those high in cholesterol and saturated fat. Obtaining proper nutrient intake in various disease states is difficult for patients; often they depend on their physician for appropriate nutritional recommendations. Research providing appropriate clinical data to establish a food/health/disease relationship combined with logical, educational messages is necessary to help health care professionals determine appropriate nutritional choices. Medical foods have the potential to provide great benefit with better compliance than other therapies may be able to provide. Discovery of new phytochemicals and other natural food components with specific physiological benefits is the constant goal for incorporation into new medical foods.

Summary

The importance of infant feeding is best illustrated by studies that demonstrate the long-term effects that early nutrition has on later growth and development. Since infant formulas are fed to millions of infants in the United States and elsewhere, caution needs to be taken before formulation changes are made. The potential and often unknown risk of incorporating new ingredients into formula requires that clinical benefits be clearly demonstrated. Manufacturers cannot precisely duplicate the composition of human milk because of its unique properties and nutrient concentration variability. The most appropriate model for studying modification of the formulation is the breastfed-infant–performance model, which considers how well infant formula matches human milk in delivering beneficial clinical outcomes to the infant. Manufacturers using different philosophies have developed formulas that differ in ingredients and clinical outcomes. It is important for those responsible for making decisions concerning infant feeding to be aware of these differences when selecting the appropriate formula for infants.

Medical foods provide complete nutrition and/or enhance the dietary management of a disease. The development of medical foods depends heavily on scientific substantiation for safety and efficacy of their use. Various methods of tube feeding, "disease-specific" enteral feedings, modular products, and single-nutrient additions to formulas have been developed and refined. Continued scientific research will enable the breakthrough nutrients or nutrient combinations tested for use in the medical foods of today to be the mainstream functional foods of tomorrow.

REFERENCES

1. Dobbing J. Lasting deficits and distortions of the adult brain following infantile undernutrition. In: World Health Organization, ed. *Nutrition, the Nervous System, and Behavior: Proceedings of the Seminar in Malnutrition in Early Life and Subsequent Mental Development.* Washington, DC: World Health Organization; Scientific Publication 251. 1972:15–23.

2. Dobbing J. Vulnerable periods in developing brain. In: Dobbing J, ed. *Brain, Behavior, and Iron in the Infant Diet.* New York, NY: Springer-Verlag; 1990:1–26.

3. Lucas A. Programming by early nutrition in man. In: Bock GR, Whalen J, eds. *The Childhood Environment and Adult Disease.* New York, NY: John Wiley & Sons; 1991:38–55.

4. Lucas A, Morley R, Cole TJ, et al. Early diet in preterm babies and developmental status in infancy. *Arch Dis Child.* 1989;64:1570–1578.

5. Lucas A, Morley R, Cole TJ, et al. Breast milk and subsequent intelligence quotient in children born preterm. *Lancet.* 1992;339:261–264.

6. de Andraca I, Walter T, Castillo, M, et al. Iron deficiency anemia and its effects upon psychological development at preschool age: a longitudinal study. *Nestle Found Nutr Ann Rep.* 1990:53–62.

7. Lozoff B, Jimenez E, Wolf AW. Long-term developmental outcome of infants with iron deficiency. *N Engl J Med.* 1991;325:667–694.

8. Barker DJP. *Mothers, Babies, and Disease in Later Life.* London, England: BMJ Publishing Group; 1994.

9. Benson JD, MacLean WC Jr, Ponder DL, Sauls HS. Modifications of lipids in infant formulas: concerns of industry. In: American Oil Chemists' Society; *Essential Fatty Acids and Eicosanoids.* Sinclair A, Gibson E, eds. Champaign, IL: 1992:218–221.

10. Benson JD, MacLean WC Jr. Composition of infant formula: infusion of new knowledge. In: Picciano MF, Lonnerdal B, eds. *Mechanisms Regulating Lactation and Infant Nutrient Utilization* New York, NY: Wiley-Liss; 1992:359–371.

11. Benson JD, Masor ML. Infant formula development: past, present and future. *Endocr Regul.* 1994;28:9–16.

12. MacLean WC Jr, Benson JD. Theory into practice: the incorporation of new knowledge into infant formula. *Semin Perinatol.* 1989;13:104–111.

13. Martinez GA, Dodd DA. 1981 milk feeding patterns in the United States during the first 12 months of life. *Pediatrics.* 1983;71:166–170.

14. Martinez GA, Krieger FW. 1984 milk-feeding patterns in the United States. *Pediatrics.* 1985;76:1004–1008.

15. Martinez GA, Nalezienski JP. The recent trend in breast-feeding. *Pediatrics.* 1979;64:686–692.

16. Martinez GA, Nalezienski JP. 1980 update: the recent trend in breast-feeding. *Pediatrics.* 1981;67:260–263.

17. Ryan AS. The resurgence of breastfeeding in the United States. *Pediatrics.* 1997;99:e12 (electronic pages).

18. Ryan AS, Pratt WF, Wysong JL, et al. A comparison of breast-feeding data from the National Surveys of Family Growth and the Ross Laboratories Mothers' Surveys. *Am J Public Health.* 1991;81:1049–1052.

19. Ryan AS, Rush D, Krieger FW. Recent declines in breast-feeding in the United States, 1984 through 1989. *Pediatrics.* 1991;88:719–727.

20. Ryan AS, Martinez GA. Breast-feeding and the working mother: a profile. *Pediatrics.* 1989;83:524–531.

21. American Academy of Pediatrics, Committee on Nutrition. The use of whole cow's milk in infancy. *Pediatrics.* 1992;89:1105–1109.

22. Hallberg, L, Rossander-Hulten L, Brune M, Gleerup A. Calcium and iron absorption: mechanism of action and nutritional importance. *Eur J Clin Nutr.* 1992;46:317–327.

23. Dallman PR, Siimes MA, Stekel A. Iron deficiency in infancy and childhood. *Am J Clin Nutr.* 1980;33:86–118.

24. Pizarro F, Yip R, Dallman PR, et al. Iron status with different infant feeding regimens: relevance to screening and prevention of iron deficiency. *J Pediatr.* 1991;118:687–692.

25. Miller V, Swaney S, Deinard A. Impact of the WIC program on the iron status of infants. *Pediatrics.* 1985;75:100–105.

26. Vazquez-Seoane P, Windom R, Pearson HA. Disappearance of iron-deficiency anemia in a high-risk infant population given supplemental iron. *N Engl J Med.* 1985;313:1239–1240.

27. Yip R, Binkin NJ, Fleshood L, Trowbridge FL. Declining prevalence of anemia among low-income children in the United States. *JAMA.* 1987;258:1619–1623.

28. Yip R, Walsh KM, Goldfarb MG, Binkin NJ. Declining prevalence of anemia in childhood in a middle class setting: a pediatric success story. *Pediatrics.* 1987;80:330–334.

29. Yip R, Parvanta I, Cogswell ME, et al. Recommendation to prevent and control iron deficiency in the United States. *MMWR*. 1998;47:1–29.

30. Yip R. The changing characteristics of childhood iron nutritional status in the United States. In: Filer LJ Jr, ed. *Dietary Iron: Birth to Two Years*. New York, NY: Raven Press; 1989:37–56.

31. Ryan AS, Martinez GA, Yip R. Changing patterns of infant feeding in the United States: evidence to support improved iron nutritional status in childhood. In: Hercberg S, Galan P, Dupin H, eds. *Recent Knowledge on Iron and Folate Deficiencies in the World*. Paris: Colleque INSERM, 1990:631–639.

32. Ryan AS. Iron-deficiency anemia in infant development: implications for growth, cognitive development, resistance to infection, and iron supplementation. *Yrbk Phys Anthropol*. 1997;40:25–62.

33. Nutrient Requirements for Infant Formulas. 50 Fed. Reg. 45106–45108 (1985). 21 CFR § 107.

34. Prentice A. Constituents of human milk. *Food Nutr Bull*. 1996;17:305–312.

35. Lucas A, Gibbs, J, Baum JD. What's in breast milk? *Lancet*. May 1977;7:1011–1012.

36. Picciano MF, Guthrie HA, Sheehe DM. The cholesterol content of human milk. *Clin Pediatr*. 1978;17:359–362.

37. Packard VS. Vitamins. In: Stewart GF, Schweigert BS, Hawthorn J, eds. *Human Milk and Infant Formula*. New York, NY: Academic Press; 1982:29–49.

38. Janas LM, Picciano MF. Quantities of amino acids ingested by human milk-fed infants. *J Pediatr*. 1986;109:802–807.

39. Fomon SJ. Requirements and recommended dietary intakes of protein during infancy. *Pediatr Res*. 1991;30:391–395.

40. Munro HN. Amino acid requirements and metabolism and their relevance to prenatal nutrition. In: Wilkinson AW, ed. *Parenteral Nutrition*. London, England: Churchill Livingstone; 1972:34–67.

41. O'Connor DL, Masor ML, Paule C, Benson J. Amino acid composition of cow's milk and human requirements. In: Welch RAS, Burns DJW, Davis SR, Popay AI, Prosser CG, eds. *Milk Composition, Production and Biotechnology*. Wallingford, England: CAB International; 1997:203–213.

42. Chappell JE, Clandinin MT, Kearney-Volpe C. Trans-fatty acids in human milk lipids: influence of maternal diet and weight loss. *Am J Clin Nutr*. 1985;42:49–56.

43. Chen ZY, Pelletier G, Hollywood R, Ratnayake WMN. *Trans* fatty acid isomers in Canadian human milk. *Lipids*. 1995;30:15–21.

44. Greer FR. Formulas for the healthy term infant. *Pediatr Rev*. 1995;16:107–112.

45. O'Brien JS, Fillerup DL, Mean JF. Quantification of fatty acid and fatty aldehyde composition of ethanolamine choline and serine phosphoglycerides in human cerebral gray and white matter. *J Lipid Res*. 1964;5:29–330.

46. Anderson RE. Lipids of ocular tissues. IV. A comparison of the phospholipids from the retina of six mammalian species. *Exp Eye Res*. 1970;10:339–344.

47. Tinoco J, Miljanich P, Medwadowski B. Depletion of docosahexaenoic acid in retinal lipids of rats fed a linolenic acid-deficient, linoleic acid-containing diet. *Biochem Biophys Acta*. 1977;486:575–578.

48. Clandinin MT, Chappell JE, Leong S, et al. Extrauterine fatty acid accretion in infant human brain: implications for fatty acid requirements. *Early Hum Develop*. 1980;4:131–138.

49. Clandinin MT, Chappell JE, Leong S, et al. Intrauterine fatty acid accretion rates in human brain: implications for fatty acid requirements. *Early Hum Develop*. 1980;4:121–139.

50. Crawford MA, Hassam AG, Hall BM. Metabolism of essential fatty acids in the human fetus and neonate. *Nutr Metab*. 1977;21:187–188.

51. Sanders TAB, Naismith DJ. A comparison of the influence of breast feeding and bottle feeding on the fatty acid composition of erythrocytes. *Br J Nutr*. 1979;4:619–623.

52. Ponder DL, Innis SM, Benson JD, Siegman JS. Docosahexaenoic acid status of term infants fed breast milk or infant formula containing soy oil or corn oil. *Pediatr Res*. 1992;32:683–688.

53. Makrides M, Simmer K, Goggin M, Gibson RA. Erythrocyte docosahexaenoic acid correlates with the visual response of healthy, term infants. *Pediatr Res*. 1993;33:425–427.

54. Innis SM, Nelson CM, Rioux MF, King DJ. Development of visual acuity in relation to plasma and erythrocyte w-6 and w-3 fatty acids in healthy term gestation infants. *Am J Clin Nutr*. 1994;60:347–352.

55. Birch EE, Brich DG, Hoffman DR, et al. Breast-feeding and optimal visual development. *J Pediatr Ophthalmol Strabismus*. 1993;30:33–38.

56. Makrides M, Neumann M, Simmer K, et al. Are long-chain polyunsaturated fatty acids essential nutrients in infancy? *Lancet*. 1995;345:1463–1468.

57. Jorgensen MH, Hernell O, Lund P, et al. Visual acuity and erythrocyte docosahexaenoic acid status in breast-fed and formula-fed term infants during the first four months of life. *Lipids.* 1996;31:99–105.

58. Carlson SE, Ford AJ, Werkman SH, et al. Visual acuity and fatty acid status of term infants fed human milk and formulas with or without docosahexaenoate and arachidonate from egg yolk lecithin. *Pediatr Res.* 1996;39:882–888.

59. Auestad N, Montalto MB, Hall RT, et al. Visual acuity, erythrocyte fatty acid composition, and growth in term infants fed formulas with long chain polyunsaturated fatty acids for one year. *Pediatr Res.* 1997;41:1–10.

60. Janowsky JS, Scott DT, Wheeler RE, Auestad N. Fatty acids affect early language development. *Pediatr Res.* 1995;37:310A.

61. Blennemann B, Baggs G, Masor M. An evaluation of formula tolerance after human milk feeding. *Pediatr Res.* 1996;39:1814A.

62. Kunz C, Lonnerdal B. Re-evaluation of the whey protein/casein ratio of human milk. *Acta Pediatr.* 1992;81:107–112.

63. Paule C, Wahrenberger D, Jones W, Kuchan M, Masor M. A novel method to evaluate the amino acid response to infant formulas. *FASEB J.* 1996;10:A554.

64. Nelson SE, Rogers RR, Frantz JA, Ziegler EE. Palm olein in infant formula: absorption of fat and minerals by normal infants. *Am J Clin Nutr.* 1996;64:291–296.

65. Nelson SE, Frantz JA, Ziegler EE. Absorption of fat and calcium by infants fed a milk-based formula containing palm olein. *J Am Coll Nutr.* 1998;17:1–6.

66. Hansen JW, Huston R, Ehrenkranz R, Bell EF. Impact of palm olein (PO) in infant feeding on fat and calcium absorption in growing premature infants. *J Amer Coll Nutr.* 1996;15:526A.

67. Fomon SJ. *Nutrition of Normal Infants.* St. Louis: Mosby-Year Book, Inc, 1993.

68. Lammi-Keefe CJ, Ferris AM, Jensen RG. Changes in human milk at 0600, 1000, 1400, 1800, and 1200 h. *J Pediatr Gastroenterol Nutr.* 1990;11:83–88.

69. Ziegler EE, Fomon SJ. Lactose enhances mineral absorption in infancy. *J Pediatr Gastroenterol Nutr.* 1983;2:288–294.

70. Mimouni F, Campaigne B, Neylan M, et al. Bone mineralization in the first year of life in infants fed human milk, cow-milk formula, or soy-based formula. *J Pediatr.* 1993;122:348–354.

71. Roggero P, Mosca F, Motta G, et al. Sugar absorption in healthy preterm and full-term infants. *J Pediatr Gastroenterol Nutr.* 1986;5:214–219.

72. Carver JD, Walker WA. The role of nucleotides in human nutrition. *Nutr Biochem.* 1995;6:58–72.

73. Gil A, Uauy R. Dietary nucleotides and infant nutrition. *J Clin Nutr Gastroenterol.* 1989;4:145–153.

74. Schlimme E, Raezke K-P, Ott F-G. Ribonucleosides as minor milk constituents. *Z. Ernahrungswisse.* 1991;30:138–152.

75. Gil A, Sanchez-Medina F. Acid-soluble nucleotides of cow's, goat's and sheep's milk, at different stages of lactation. *J Dairy Res.* 1981;48:35–44.

76. Quan R, Barness LA, Uauy R. Do infants need nucleotide supplemented formula for optimal nutrition? *J Pediatr Gastroenterol Nutr.* 1990;11:429–437.

77. Quan R, Uauy R. Nucleotides and gastrointestinal development. *Sem Pediatr Gastr Nutr.* 1991;2:3–11.

78. Gil A, Sanchez-Medina F. Acid-soluble nucleotides of human milk at different stages of lactation. *J Dairy Res.* 1982;49:301–307.

79. Janas LM, Picciano MF. The nucleotide profile of human milk. *Pediatr Res.* 1982;16:659–662.

80. Leach JL, Baxter JH, Molitor BE, et al. Total potentially available nucleosides of human milk by stage of lactation. *Am J Clin Nutr.* 1995;61:1224–1230.

81. Pickering LK, Granoff DM, Erickson JR, et al. Modulation of the immune system by human milk and infant formula containing nucleotides. *Pediatrics.* 1998;101:242–249.

82. Schmidl MK, Massaro, S, Labuza TP. Parenteral and enteral food symptoms. *Food Technol.* 1988;42(7):77–87.

83. Schmidl MK, Labuza TP. Medical foods: IFT status summary. *Food Technol.* 1992;46(4):87–96.

84. Randall HT. Enteral nutrition: tube feeding in acute and chronic illness. *JPEN.* 1984;8:113–136.

85. Bonsmann M, Hardt W, Lorber CG. The historical development of artificial enteral alimentation. Part 1. *Anasthesiol Intensivmed.* 1993;34:207.

86. Greenstein JP, Birnbaum SM, Winitz M, et al. Quantitative nutritional studies with water-soluble chemically defined diets. I. Growth, reproduction, and lactation in rats. *Arch Biochem Biophys.* 1957;72:396.

87. Winitz M, Graff J, Gallagher N, et al. Evaluation of chemical diets as nutrition for man-in-space. *Nature.* 1965;205:741.

88. Winitz M, Seedman DA, Graff J. Studies in metabolic nutrition employing chemically defined diets. I. Extended feeding of normal human adult men. *Am J Clin Nutr.* 1970;23:525.

89. 21 CFR § 105.

90. *Federal Register.* September 8, 1972;37:18229.

91. *Federal Register.* January 19, 1973;38:2125.

92. 21 CFR § 105.

93. Scarbrough FE. Medical foods: an introduction. *Food Drug Cosmet Law J.* 1989;44:463–466.

94. Fisher KD, Talbot JM, Carr CJ. *A Review of Foods for Medical Purposes: Specially Formulated Products for Nutritional Management of Medical Conditions.* Prepared for the Food and Drug Administration under Contract FDA 223–75–2090 by the Life Sciences Research Office. Bethesda, MD: 1977.

95. Forbes AL. An historical overview of medical foods. In: *Development of Medical Foods for Rare Diseases: Proceedings of a Workshop.* Bethesda, MD: Life Sciences Research Office, Federation for American Societies for Experimental Biology; 1991:7–16.

96. 21 CFR §§ 106 & 107.

97. Orphan Drug Act Amendment of 1988. P.L. 100–290. Amendments of Section 526 (a) (1) of the Federal Food, Drug, and Cosmetics Act. 21 USC § 360ee.

98. 21 USC § 360ee (b).

99. 21 USC § 360ee (b).

100. Food and Drug Administration. *Compliance Program Guidance Manual.* Program 7321.002. Washington, DC: Food and Drug Administration; 1989:chap 21, program 7321.002.

101. Nutrition Labeling and Education Act of 1990. P.L. 101–535. 21 USC § 343.

102. 21 CFR § 101.13 (q)(4)(ii).

103. 61 CFR § 60661.

104. 21 CFR § 110.113.

105. Serafino JM. Food manufacturing practices and quality control for medical foods. *Food Drug Cosmet Law J.* 1989;44:523–532.

106. Anderson JD, Moore RA, Moore EE. Enteral feeding in the critically injured patient. *Nutr Clin Pract.* 1992;7:117–122.

107. Mirtallo JM, Powell CR, Campbell SM, et al. Cost-effective nutrition support. *Nutr Clin Pract.* 1987;2:142–151.

108. Rielly JJ, Hull SF, Albert N, et al. Economic impact of malnutrition: a model system for hospitalized patients. *JPEN.* 1988;12:371–376.

109. Smith PE, Smith AE. High-quality nutritional interventions reduce costs. *Health Financ Manage.* 1997;51(8):66–69.

110. Kudsk KA. Gut mucosal nutritional support: enteral nutrition as primary therapy after multiple system trauma. *Gut.* 1994;1:552–554.

111. Lacey JM, Wilmore DW. Is glutamine a conditionally esssential amino acid? *Nutr Rev.* 1990;48:297–309.

112. Souba WW. *Glutamine: Physiology, Biochemistry and Nutrition in Critical Illness.* Austin, TX: RG Landis Co: 1992.

113. Smith RJ, Wilmore DW. Glutamine nutrition and requirements. *JPEN.* 1990;14:94S–99S.

114. Hadfield RJ, Sinclair DG, Houlsworth PE, Evans TW. Effects of enteral and parenteral nutrition on gut mucosal permeability in the critically ill. *Am J Respir Crit Care Med.* 1995;152:1545–1548.

115. Jenson GL, Miller RH, Talabiska DG, et al. A double-blind, prospective, randomized study of glutamine-enriched compared with standard peptide-based feeding in critically ill patients. *Am J Clin Nutr.* 1996;64:615–621.

116. Barbul A. Arginine: biochemistry, physiology, and the therapeutic implications. *JPEN.* 1986;10:227–238.

117. Larrick JW. Metabolism of arginine to nitric oxide: an area for nutritional manipulation of human disease? *J Optimal Nutr.* 1994;3:22–31.

118. Barbul A, Wasserkrug HL, Seifter E, et al. Arginine enhances wound healing and lymphocyte immune response in humans. *Surgery.* 1990;108:331–337.

119. Barbul A, Sisto DA, Wasserkrug HL, Efron G. Arginine stimulates lymphocyte immune response in healthy human beings. *Surgery.* 1981;90:244–251.

120. Daly JM, Renolds J, Thom A, et al. Immune and metabolic effects of arginine in the surgical patient. *Ann Surg.* 1988;208:512–523.

121. Barbul A, Sisto DA, Wasserkrug HL, et al. Nitrogen sparing and immune mechanisms of arginine: Differential dose-dependent responses during post injury intravenous hyperalimentation. *Curr Surg.* 1983;40:114–116.

122. Brown RO, Hunt H, Mowatt-Larssen CA, et al. Comparison of specialized and standard enteral formulas in trauma patients. *Pharmacotherapy.* 1994;14:314–320.

123. Wilmore DW. *The Metabolic Management of the Critically Ill.* New York, NY: Plenum Medical Co; 1977.

124. Southgate DAT. Definitions and terminology of dietary fiber. In: Valhouny GV, Kritchevsky D, eds. *Dietary Fiber in Health and Disease.* New York, NY: Plenum Press; 1982:1–7.

125. Anderson JW, Deakins DA, Floore TL, et al. Dietary fiber and coronary heart disease. *Crit Rev Food Sci Nutr.* 1990;29: 95–147.

126. Wolever TMS, Jenkins DJA. Effect of fiber and foods on carbohydrate metabolism. In: Spiller GA, ed. *Dietary Fiber in Human Nutrition.* 2nd ed. Boca Raton, FL: CRC Press; 1993:111–152.

127. Cummings JH. Consequences of the metabolism of fiber in the human large intestine. In: Vahouny GV, Kritchevsky D, eds. *Dietary Fiber in Health and Disease.* New York, NY: Plenum Press; 1982:9–22.

128. Achord JL. Dietary fiber and the gastrointestinal tract. *Curr Concepts Gastroenterol.* 1981;6(2);10, 16–18.

129. Hidaka T, Hidaka H, Eida T. Proliferation of bifidobacteria by oligosaccharides and their useful effect on human health. *Bifidobacteria Microflora.* 1991;10:65–79.

130. Mitsuko T, Hidaka H, Eida T. Effect of fructooligosaccharides on intestinal microflora. *Nahrung.* 1987;31:427–436.

131. Gibson GR, Roberfroid MB. Dietary modulation of the human colonic microbiota: introducing the concept of prebiotics. *J Nutr.* 1995;125:1401–1412.

132. Miller-Catchpole R. Bifidobacteria in clinical microbiology and medicine. In: Bezkorovainy A, Miller-Catchpole R, eds. *Biochemistry and Physiology of Bifidobacteria.* Boca Raton, FL: CRC Press Inc; 1989:177–200.

133. Wilmore DW. Fat metabolism. In: Ballinger WF, Collins JA, Drucker WR, et al, eds. *Manual of Surgical Nutrition.* Philadelphia, PA: WB Saunders Co; 1975:33–49.

134. American Heart Association. Dietary guidelines for healthy American adults: a statement for physicians and health professionals by the Nutrition Committee, American Heart Association. *Circulation.* 1988;77:721A–724A.

135. Franz MJ, Horton ES, Bantle JP, et al. Nutrition principles for the management of diabetes and related complications. *Diabetes Care.* 1994;17:490–518.

136. Campbell SM, Schiller MR. Considerations for enteral nutrition support of patients with diabetes. *Top Clin Nutr.* 1991; 7:23–32.

137. Brown RO, Heizer WD. Nutrition and respiratory disease. *Clin Pharm.* 1984;3:152–161.

138. McArdle WD, Katch FI, Katch VL. *Exercise Physiology: Energy Nutrition and Human Performance.* Philadelphia, PA: Lea & Febiger; 1981.

139. Wennberg AK, Nelson JL, DeMichele SJ, Campbell SM. *Affecting Clinical Outcomes in Acute Respiratory Distress Syndrome with Enteral Nutrition: The Role of Oxepa.* Columbus, OH: Ross Products Division; 1997.

140. Mancuso P, Whelan J, DeMichele SJ, et al. Effects of eicosopentanoic and gamma-linolenic acid on lung permeability and alveolar macrophage eicosanoid synthesis in endotoxic rats. *Crit Care Med.* 1997;25:523–532.

141. Mancuso P, Whelan J, DeMichele SJ, et al. Dietary fish oil and fish and borage oil suppress intrapulmonary proinflammatory eicosanoid biosynthesis and attenuate pulmonary neutrophil accumulation in endotoxic rats. *Crit Care Med.* 1997;25:1198–1206.

142. Palombo JD, DeMichele SJ, Lydon EE, et al. Rapid modulation of lung and liver macrophage phospholipid fatty acids in endotoxemic rats by continuous enteral feeding with n-3 and gamma-linolenic fatty acids. *Am J Clin Nutr.* 1996;63:208–219.

143. Palombo JD, DeMichele SJ, Lyndon EE, Bistrian BR. Cyclic vs continuous enteral feeding with n-3 and gamma-linolenic fatty acids: effects on modulation of phospholipid fatty acids in rat lung and liver immune cells. *JPEN.* 1997; 21:123–132.

144. Gadek J, DeMichele S, Karlstad M, et al. Specialized enteral nutrition improves clinical outcomes in patients with or at risk of acute respiratory distress syndrome (ARDS): a prospective, blinded, randomized, controlled multicenter trial. *Am J Resp Crit Care Med.* 1998;157:677.

145. Bach AC, Babayan VK. Medium chain triglycerides: an update. *Am J Clin Nutr.* 1982;36:950–962.

146. Holt PR. Medium chain triglycerides: a useful adjunct in nutritional therapy. *Gastroenterology.* 1967;53:961–966.

147. Henderson WR. The role of leukotrienes in inflammation. *Ann Intern Med.* 1994;121:684–697.

148. Kenler AS, Swails WS, Driscoll DF, et al. Early enteral feeding in postsurgical cancer patients: fish oil structured lipid-based polymeric formula versus a standard polymeric formula. *Ann Surg.* 1996;233:316–333.

149. Gussler JD, ed. *Specialized Elemental Nutrition: The Role of Optimental Ready-to-Feed Elemental.* Columbus, OH: Ross Products Division; 1998.

150. Scrimshaw NS, Young VR. The requirements of human nutrition. *Sci Am.* 1976;235:51–64.

151. Sies H, Stahl W, Sundquist AR. Antioxidant functions of vitamins: vitamins E, C, beta-carotene, and other carotenoids. *Ann NY Acad Sci.* 1992;669:7–20.

CHAPTER 7

Dietary Supplements

Mary Ellen Camire

HISTORY

Dietary supplements have traditionally been used to improve the nutritional status of individuals who cannot meet their nutritional needs by diet alone. Illness, pregnancy, and stress all produce increased nutritional needs, while low income and poor eating habits contribute to reduced intakes of essential nutrients. Since the roles of vitamins, minerals, and amino acids in human health were not understood until the first half of the 20th century, the sale of these compounds as supplements is relatively new. However, formulations promoted as health stimulants and tonics were sold widely in the 19th and early 20th centuries, often by traveling salesmen.

Regulation

The Dietary Supplement Health and Education Act of 1994 (DSHEA)[1] dramatically changed the regulation of these products and thus changed the industry that produces these products. The timing of this legislation coincided with a growing interest in alternative medicine and "natural" foods. The DSHEA defined dietary supplements as "a product, other than tobacco, intended to supplement the diet that contains at least one or more of the following ingredients: a vitamin, a mineral, an herb, an amino acid, or a dietary substance for use to supplement the diet by increasing the total dietary intake; or a concentrate, metabolite, constituent, or extract of any of these ingredients." Supplements may be sold in the form of pills, capsules, liquids, or powders. They cannot be represented as conventional foods or used as the sole item in a meal. Supplements must be labeled as "dietary supplements," although the front panel of some product packages states "herbal supplement."

If a supplement contains a new dietary ingredient that was not in the food supply prior to 1994, the DSHEA requires the manufacturer to notify the Food and Drug Administration (FDA) at least 75 days before marketing. Manufacturers must also include in the notification the manufacturer's basis for its conclusion that a dietary supplement containing the ingredient will reasonably be expected to be safe. Companies are not required to wait for a safety determination from FDA before marketing the product. FDA has authority to establish Good Manufacturing Practice regulations for dietary supplements. The DSHEA and related regulations are discussed further in another chapter.

The DSHEA required the formation of a Commission on Dietary Supplement Labels, which held several hearings around the nation and published its final report in 1997. The Act also called for an Office of Dietary Supplements (ODS) to be created within the National Institutes of Health. ODS was directed to conduct and coordinate research and funding for research on dietary supplements; it was also designated to collect and compile research on supplements and to produce a database of such research. In 1999, ODS, in conjunction with the US Department of Agriculture's Food and Nutrition Information Center, released the International Bibliographic Information on Dietary Supplements, a free, searchable database on the Internet (http://odp.od.nih.gov/ods/databases/ibids.html) that contains more than 300,000 citations and abstracts. ODS is also planning to release a series of brief papers on individual supplements.

The DSHEA specified that dietary supplements could bear statements of nutritional support that indicate how the product or its ingredients (1) relate to a classical nutrient deficiency, (2) affect or maintain structure or function in humans, or (3) contribute to general well-being. FDA no longer uses the term *statement of nutritional support* since many statements were not directly related to nutritional needs. The current preferred term is *structure-function claim*, meaning that the product supports the normal function or structure of the body. These claims must be accompanied by the disclaimer: "This statement has not been evaluated by the Food and Drug Administration. This product is not intended to diagnose, treat, cure, or prevent any disease." The disclaimer is ironic, since many consumers take supplements in order to prevent or treat diseases. Manufacturers must notify FDA within 30 days of the first use of a structure-function claim. In January 2000, FDA released Final Rules for claims on dietary supplements.[2,3]

FDA is responsible for ensuring the safety and efficacy of dietary supplements, as well as enforcing dietary supplement health and structure-function claims. In March 1999, dietary supplements were required to bear Supplement Facts panels that are similar to the Nutrition Facts panels mandated for foods. An example of the new labeling requirements for supplements is shown in Figure 7–1.[4] The new label allows ingredients for which no daily value (DV) has been established to list the quantity present.

The Federal Trade Commission is responsible for dietary supplement marketing information in advertisements, promotional materials, and Web pages on the Internet. It has published a guide for companies regarding dietary supplement claims that offers numerous examples of both allowable claims and claims that are not permissible.[5]

Supplement Use

Currently, the US government does not compile statistics on the scope of the dietary supplement industry. According to the *Nutrition Business Journal*'s annual industry overview[6] (Figure 7–2), retail sales in specialty stores accounted for the largest proportion of 1997 supplement sales, but mail order sales are expected to gain market share as Internet sales increase. Consumers can readily find a wide variety of dietary supplement products in grocery, drug, and discount stores, and major pharmaceutical firms are releasing new supplement to capture some of this "mainstream" consumer market.[7]

Health and nutrition surveys are beginning to ask more detailed questions regarding supplement use by consumers; previous surveys asked for information on the use of vitamin and mineral supplements only. Thousands of products are available to consumers, and some consumers use several products daily or weekly. This dizzying array of options makes accurate collection of dietary supplement data difficult. Figure 7–3 shows 1997 estimated sales for different dietary supplement categories. Vitamins lead sales, followed by herbs/botanicals.[6] Amino acids, melatonin, and other hormones are examples of products in the "other supplements" category. In October 1999 FDA released *Dietary Supplement Sales Information. Final Report*,[8] a report on data collected on a sample of 3,000 dietary supplements sold in the United States.

How do consumers feel about dietary supplements? In February 1999, National Public Radio, the Henry J Kaiser Family Foundation, and Harvard University's Kennedy School of Government con-

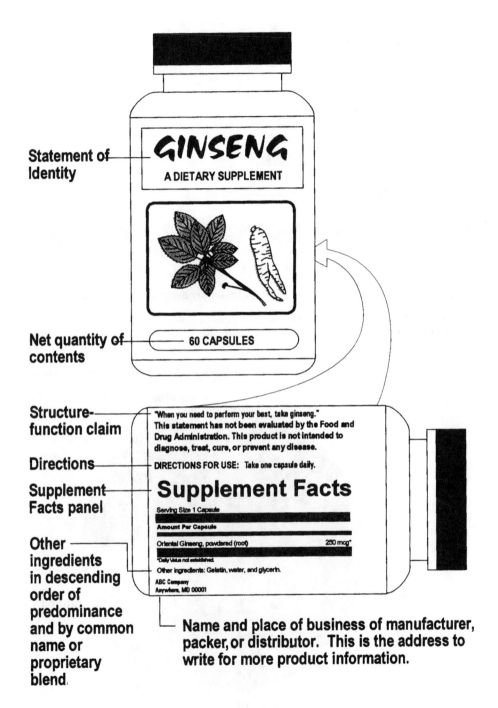

Statement of Identity

Net quantity of contents

Structure-function claim

Directions

Supplement Facts panel

Other ingredients in descending order of predominance and by common name or proprietary blend

GINSENG

A DIETARY SUPPLEMENT

60 CAPSULES

"When you need to perform your best, take ginseng." This statement has not been evaluated by the Food and Drug Administration. This product is not intended to diagnose, treat, cure, or prevent any disease.

DIRECTIONS FOR USE: Take one capsule daily.

Supplement Facts

Serving Size 1 Capsule

Amount Per Capsule

Oriental Ginseng, powdered (root)   250 mcg*

*Daily Value not established.

Other ingredients: Gelatin, water, and glycerin.

ABC Company
Anywhere, MD 00001

Name and place of business of manufacturer, packer, or distributor. This is the address to write for more product information.

Figure 7–1 US dietary supplement label format, effective March 1999. *Source:* Reprinted from P. Kurtzweil, U.S. Dietary Supplement Label Format, Effective March 1999, An FDA Guide to Dietary Supplements, *FDA Consumer*, Vol. 32, No. 5, pp. 28–35, 1998.

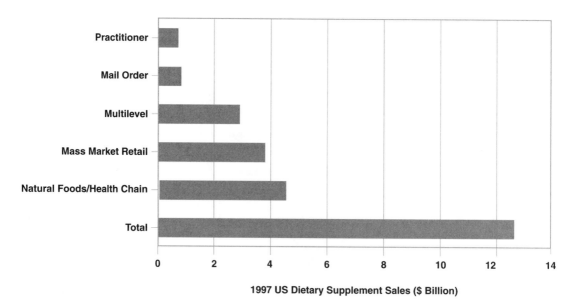

Figure 7–2 1997 US dietary supplement consumer sales channels. *Source:* Reprinted with permission from Annual Industry Overview, *Nutrition Business Journal*, Vol. 3, No. 9, p. 3, Copyright © Nutrition Business International, LLC.

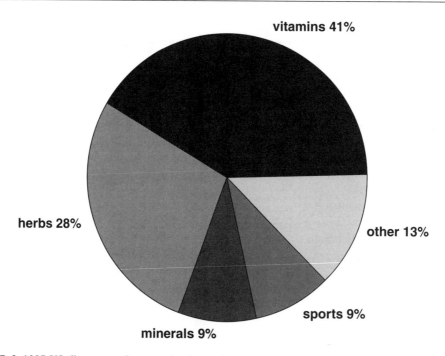

Figure 7–3 1997 US dietary supplement sales by product category. *Source:* Reprinted with permission from Annual Industry Overview, *Nutrition Business Journal*, Vol. 3, No. 9, p. 1, © Nutrition Business International, LLC.

ducted a telephone survey of 1200 Americans over the age of 18 years.[9] Although 55% of the respondents were very or somewhat familiar with supplements other than vitamins and minerals, only 18% used them regularly. The majority of people surveyed who had children said they would not give these products to their children. More than half the respondents felt that supplements were beneficial to health and specifically could help arthritis, colds, and depression. When presented with the structure-function phrase "supports or boosts the immune system," 64% believed that the claim meant disease prevention, and 42% thought it meant it would help a sick person recover. Only 35% stated that the government regulates supplements for safety and efficacy; 48% believed these products are not adequately tested for safety, purity, and dosage accuracy before being sold. Despite these misgivings, 75% said that supplements hurt people sometimes or rarely. Approximately 60% responded that there is inadequate regulation of supplements and supplement claims. Of the 412 persons in the survey who regularly or sometimes used supplements, 72% stated that they would continue to use their supplements even if the government determined that the products were ineffective.

Consumer loyalty to dietary supplements appears to be due, in part, to dissatisfaction with conventional medicine. As a follow-up to a survey on alternative medicine use in 1990, 2055 persons were interviewed by telephone about alternative therapies.[10] Herbal medicine, megavitamin, and homeopathic therapy use had increased significantly ($P \leq 0.001$) (Figure 7–4). Fewer than 40% of the persons who used alternative therapies disclosed this use to their primary care physician. Health insurance did not cover most herbal medicine and some megavitamin therapy, forcing users to pay for these products out of pocket. The researchers estimated that 15 million adults in the United States were taking prescription medications along with herbal and megavitamin supplements.

In a smaller survey of 200 adults visiting a family practice clinic in Wisconsin, Eliason et al[11] found that 52% of the respondents had taken at least one supplement in the past year. The majority of supplements used were vitamin and mineral formulations, and herbs were taken by only 8% of the patients. The most often cited source of information on dietary supplements was the media, including

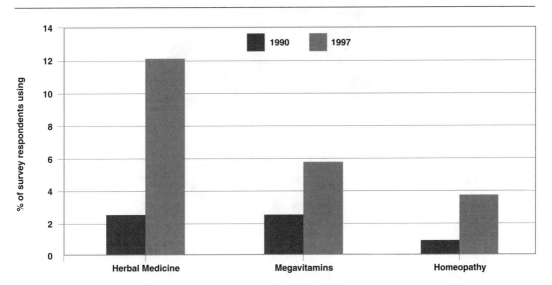

Figure 7–4 Changes in consumer use of alternative medical therapy involving dietary supplements, 1990–1997. *Source:* Reprinted with permission from D.M. Eisenberg et al., Trends in Alternative Medicine Use in the U.S., 1990–1997, *JAMA*, Vol. 280, No. 18, pp.1569–1575, Copyrighted 1998, American Medical Association.

television, radio, and print (Figure 7–5). Surprisingly, aside from physicians, health care professionals such as nurses and dietitians were considered to be a good source of information for only 5% of the patients. Patients with a high school education took significantly fewer supplements than did those with more education.

The demographics of supplement users appear to differ from those of the population as a whole. In two surveys conducted in the early 1990s, supplement use was greater among women, persons with more than high school education, and persons of slim to average weight. The study conducted in Wisconsin also found that supplement users had higher intakes of most nutrients from foods (ie, nutrient intake was adequate) and that persons who had never smoked were more likely to use multivitamins and vitamin E supplements.[12] The Swedish study found that female supplement users tended to have worse perceived health and drank more alcohol than nonuser cohorts.[13]

TYPES

General Categories

Vitamins and minerals in functional foods are addressed in a separate chapter in this book. Amino acids may be used in dietary supplements to produce specific results, such as drowsiness caused by tryptophan. Amino acids are also added to muscle-building products to minimize losses due to digestion. *Dietary fiber* is a term used to describe a variety of products and compounds. Crude, concentrated sources of dietary fiber, such as cereal bran, may produce very different results than a highly purified or even synthetic material. A comprehensive definition of dietary fiber is forthcoming from

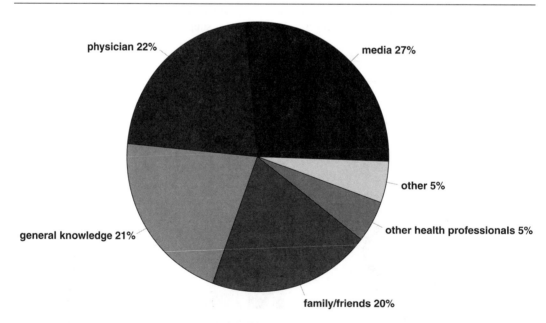

Figure 7–5 Sources of information about dietary supplements for 200 adults visiting a family practice clinic. *Source:* Reprinted with permission from B.C. Eliason et al., Use of Dietary Supplements by Patients in a Family Practice Clinic, *Journal of American Board of Family Practice*, Vol. 9, No. 4, p. 251, © 1996, American Board of Family Practice.

the American Association of Cereal Chemists. Hormones are considered to be dietary supplements if they replace similar chemicals in the body that are lost due to aging or other processes. Abattoir waste such as ovaries and testes may be used as crude sources of hormones.

Herbs, or the preferred term, botanicals, are often unfamiliar to food scientists. While some, such as garlic and rosemary, are also culinary herbs, many products have a history of use only for medicinal purposes. American consumers and health care practitioners have relied upon popular texts such as Tyler's *The Honest Herbal*.[14] In Europe, a committee of the German Department of Health has published many monographs on the use of botanicals. These papers, referred to as the Commission E monographs, have recently been translated into English.[15] An herbal version of the *Physician's Desk Reference* for medication, partly based on the Commission E monographs, is now available.[16] The active chemicals in many products are not definitely known, and there may be no reliable analytical method available for detecting such compounds. The Institute for Nutraceutical Advancement, in Denver, Colorado (http://www.nutraceuticalinstitute.com) has developed methods for detecting the following compounds by high-performance liquid chromatography: flavonol glycosides in *Gingko biloba*, ginsenosides in *Panax* species (ginseng), kavalactones in *Piper methysticum* (kava kava), and a chemical "fingerprint" for *Hypericum perforatum* (St John's wort). Additional methods are under development and will be submitted to the Association of Official Analytical Chemists and the US Pharmacopoeia.

Sports

Improved performance is the objective of consumers who use sports supplements. Katch and coworkers[17] have reviewed the history of sports supplements and offer suggestions for the future. Some dietary supplement ingredients increase energy, while others aid in retention of lean body mass (Table 7–1).

Manufacturers must weigh the ethics of using performance-enhancing components. Caffeine is a restricted chemical that is found naturally in coffee, tea, cola, chocolate, mate, and guarana. The International Olympic Committee (IOC) has set a limit of 12 mg/mL of urine, an amount that can be achieved with three to four mugs of coffee. Although the effect of caffeine on recreational athletes is not established, caffeine doses of 5 to 6 mg/kg of body weight appear to improve performance in short-term, intense activities for trained athletes.[18] Doses over 9 mg/kg could produce illegal urinary levels. Androstenedione has gained popularity as a natural anabolic dietary supplement to build mus-

Table 7–1 Dietary Supplement Ingredients for Sports

Reputed Action	*Supplement Ingredient*
Energy metabolism/sustained energy	Creatine, ribose, deer antler velvet
Resistance to stress	Ginseng (*Panax ginseng* and *P. quinquefolius*), Siberian ginseng or eleuthero (*Eleutherococcus senticosus*)
Electrolyte replacement	Sodium, potassium, chloride, magnesium
Increased lean tissue	Chromium, protein and amino acids, beta-hydroxymethyl butyrate (HMB), creatine, carnitine, choline, yohimbine (from *Pausinystalia yohimbe*), dehydroepiandrosterone (DHEA)

Note: Inclusion of a product in this table does not imply safety or efficacy.

cle and increase strength; record-setting baseball player Mark McGwire's use of the product may have spurred sales in 1998. However, a randomized controlled study of 30 young men revealed an increase in blood estrones rather than testosterone, and the researchers concluded that the supplement did not improve muscle adaptation to resistance training.[19] Sports-specific formulations can be formulated for different age and gender groups; one possibility is a soccer beverage for boys aged 10 to 12 years. At least 30% of the high school athletes questioned in one study used vitamin/mineral supplements, but teens participating in some sports were more likely to use supplements.[20] Teen athletes who use supplements are more likely to believe that these products aid their performance.

Weight Loss

An estimated 24.7% of American adult women are overweight, with another 24.9% being obese.[21] Excess body fat is associated with numerous illnesses, including coronary heart disease, diabetes, and joint problems. Thompson et al[22] estimated the health-related economic cost of obesity to US business to be approximately 5% of total medical care costs. Many more Americans are trying to lose a moderate amount of weight for personal reasons. Individuals with special events such as reunions and weddings looming may be more likely to purchase products to aid in short-term weight loss.

An increasing number of peer-reviewed research papers address the benefits of proprietary mixtures, such as one reported by Hoeger et al[23] that is a patented combination of chromium picolinate, inulin, capsicum, L-phenylalanine, and other "lipotropic" ingredients. Subjects in that study were placed on a 1500-kcal/d diet and were required to walk 45 minutes 5 days per week; those receiving the supplement had significantly different fat mass, but not total weight, compared to those subjects receiving a placebo.

Table 7–2 presents some dietary supplement ingredients that are used in weight loss supplements and the mechanisms by which they are believed to aid in weight reduction. The combination of ephedra or ma huang (*Ephedra sinesis*), a botanical that contains the stimulant ephedrine and related alkaloids, with caffeine or other stimulants may result in fatal heart attacks. In 1997, FDA proposed limits of less than 8 mg per serving and not more than 24 mg daily.[24] Ephedrine use is prohibited by the IOC. Products containing ephedra or its extracts should be labeled with warnings to discourage its use in excess or by individuals at risk for complications.

Table 7–2 Weight Loss Dietary Supplement Ingredients

Reputed Action	Ingredient
Appetite suppressant	Ephedra/ma huang
Increase lean tissue	Chromium, protein and amino acids, beta-hydroxymethyl butyrate (HMB), creatine, carnitine, choline, yohimbine (from *Pausinystalia yohimbe*), dehydroepiandrosterone (DHEA)
Diuretic	Bearberry/uva ursi (*Arctostaphylos uva-ursi*), juniper berries (*Juniperus communis*)
Prevent fat absorption	Chitosan
Prevent fat metabolism	*Garcinia carambogia* hydroxycitric acid

Note: Inclusion of a product in this table does not imply safety or efficacy.

Nervous System Enhancement

Individuals with emotional or sleep problems may not want to call attention to their conditions by seeking prescription medication. Table 7–3 lists popular supplements used to improve or maintain nervous system health. Gingko ranked second in US 1997 herb sales at $250 million, while St John's wort was fifth at $200 million.[6] Tryptophan has gained notoriety due to a number of deaths and illnesses associated with consumption of contaminated L-tryptophan supplements. Promotion of dietary supplement ingredients as alternatives to illegal street drugs has been condemned by supplement industry groups.

Other Conditions

The Council for Responsible Nutrition has issued a paper detailing the benefits of supplementation for reducing risks and their associated costs, for heart disease, birth defects, blindness, and other ailments.[25] Few randomized, double-blind, placebo-controlled clinical studies have been published with sufficient subjects to ease many health professionals' concerns about recommending supplements. Some data exist to support the belief that saw palmetto and isoflavones decrease symptoms of benign prostatic hyperplasia, but these materials do not cure or prevent the condition. Other common supplements and possible structure-function claims are shown in Table 7–4. There is possibly a dietary supplement to aid any condition, but the lack of reliable research makes selection challenging. The *PDR for Herbal Medicine*[16] lists botanicals by indication and therapeutic categories, side effects, and drug interactions.

FORMULAS AND PROCESSES

Dietary supplements are typically sold in the forms of capsules, pills, and extracts. Botanical extracts are usually ethanol based, but water extractions are becoming more commonplace. Very little is known about the efficiency of extraction methods. Many companies develop and test proprietary blends in order to protect their research and development investments.

Adding supplement ingredients to foods may change the process as well as the formulation. Elliott[26] reviewed technical issues associated with addition of antioxidant vitamins to foods and beverages; overages of up to 200% may be necessary. Stirring time may need to be increased. Beverages require soluble ingredients to prevent solids from accumulating on the bottom of the container.

Table 7–3 Dietary Supplement Ingredients That Affect the Nervous System

Reputed Action	Dietary Supplement Ingredient
Improved memory	Gingko (*Gingko biloba*), L-Glycerylphosphorylcholine
Sleep aid	Melatonin, L-tryptophan
Antidepressant	St John's wort, fish oil
Stimulant	Any caffeine-containing plant material, ephedra/ma huang (*Ephedra sinensis*)
Relaxant	Kava kava (*Piper methysticum*), valerian (*Valeriana officinalis*)

Note: Inclusion of a product in this table does not imply safety or efficacy.

Table 7–4 Miscellaneous Dietary Supplement Ingredients and Possible Structure-Function Claims To Accompany Them

Common Name	Latin Name	Part Used	Possible Structure-Function Claim
Botanicals			
Black cohosh	*Cimicifuga racemosa*	Rhizome	Maintains normal female hormone balance
Cranberry	*Vaccinium macrocarpon*	Fruit	Supports urinary tract health
Milk thistle	Silybum marianum	Seeds	Promotes healthy liver function
Echinacea	*Echinacea purpurea*	Flowers*	Stimulates the immune system
Ginger	*Zingiber officinale*	Root	Settles the stomach
Goldenseal	*Hydrastis canadensis*	Rhizomes and roots	Supports normal immunity
Other types			
Colostrum	Not applicable	Not applicable	Immunity-booster
Glucosamine	Not applicable	Not applicable	For joint health

*Roots and other parts of this species and the other major *Echinacea* species may have different effects.

Substances used in powders to be added to water, juice, or milk must also be easily dispersed and stable at the pH of the fluid base. The sticky nature of the sweeteners, gums, and other ingredients poses certain problems for manufacturers. Mixtures deposited on conveyor belts may be difficult to cut with wire knives, and the belts may require chilling to facilitate removal of the bars. Sticky mixtures will also be challenging to remove from molds and dies.[27]

SENSORY CHARACTERISTICS

The addition of dietary supplement ingredients to conventional foods can drastically change the appearance, aroma, and taste of the products. Although relatively little has been published on the impact of botanicals on food sensory properties, ingredient suppliers have been actively examining this issue. The effects of traditional dietary supplement ingredients have been studied in more detail. Table 7–5 presents some common dietary supplements and their effects on food and beverage color.

Table 7–5 Color Changes Due to Dietary Supplement Ingredients

Ingredient Type	Ingredient	Color
Botanical	Bilberry, blueberry	Red to purple
	Spirulina, Chlorella, and other single-cell algae	Green
	Hibiscus	Red
	Many dried herbs and extracts	Brown
Mineral	Iron	Gray
Vitamin	Beta carotene (vitamin A precursor)	Yellow to red
	Riboflavin	Yellow

Note: Inclusion of a product in this table does not imply safety or efficacy.

Unique combinations of dietary supplement ingredients pose new challenges for product development personnel. A lemonade to which bilberry, spirulina, and gingko are added could be a visual disaster, much like a preschooler's first attempt to mix water colors—everything ends up a dull brown! Fish oil and garlic may improve cardiovascular health but would be incompatible with delicate flavors. Draganchuk[28] has proposed three approaches to dealing with off flavors such as green notes from soy and grain ingredients and bitter tastes from many herbs. She suggests: (1) masking with single or multiple flavors, (2) selecting flavors that complement the flavors of functional ingredients, or (3) developing prototypes from the start to avoid flavor issues.

Clydesdale[29] reviewed the color problems encountered with mineral fortification of foods. Solubility, pH, and chemical form (ie, ferrous sulfate vs elemental iron) are critical factors to consider when adding minerals to foods. Undesirable dark colors occur when iron salts interact with phenolic compounds such as anthocyanins in fruit juices and tannins in chocolate. Powdered beverages containing iron salts may not reveal color problems until reconstituted. Iron and other transitional metal ions can also promote lipid oxidation, reducing the nutritional value of polyunsaturated lipids and decreasing vitamins A and C and causing off flavors and odors.

Calcium salts can cause chalky mouth-feel, and as calcium supplementation becomes more common in foods, new forms of calcium have been identified to fill product niches. Calcium-fortified orange juice has become a popular means to increase calcium intake, yet many common forms of calcium are unsuitable for addition to orange juice. Calcium carbonate and calcium hydroxide produce cooked or brown flavors, while calcium chloride decreases juice flavor intensity and adds a brackish flavor at a level of 0.11% addition.[27] Tropicana (Bradenton, FL) originally used tricalcium citrate to fortify its orange juice, but the chalky appearance was believed to be responsible for the product's low market share.[30] The company then procured an exclusive license for 100% juice products from Proctor and Gamble for FruitCal, calcium citrate malate. The organic acids in this salt are found in many fruits and are thus compatible with juice flavors. This formulation change and an aggressive education and marketing program for consumers and health care professionals significantly increased Tropicana Premium sales.

As part of a research project in our laboratory to describe sensory properties of botanicals, we added two types of ginseng extract to orange juice to ascertain whether sensory characteristics are affected by these products. The extracts were manufactured by Draco Natural Products (San Jose, CA), which specializes in spray-dried, aqueous extracts processed at low temperature (less than 90°C). The *Panax ginseng* extract was a tan powder standardized to contain a minimum of 20% ginsenosides, while the Siberian ginseng was a light brown powder made from the roots and rhizomes of *Eleutherococcus senticosus* and standardized to contain 0.8% eleutherosides. Each extract was added to Tropicana orange juice at a concentration of 600 m/L of juice. This concentration is lower than that found in some commercial products, but comparisons are difficult to make since the source and strength of botanical products varies considerably. Benchtop studies indicated that a concentration of 1000 mg of extract per liter caused a pronounced brown color and medicinal taste.

Table 7–6 shows the results of three sequential duo-trio tests conducted with 19 University of Maine students and staff. In the duo-trio test, sensory panelists are given a reference sample and then are asked to match one of two coded samples to that reference. Although this test is not as statistically efficient as the more common triangle test, it is our experience that naïve panelists better understand the task of the duo-trio test, which asks for a "same as the reference" judgment instead of the "which is odd" task for the triangle test. Panelists were not able to significantly discriminate among samples at a probability level of less than 0.05, even though a small majority correctly identified the sample that was the same as the reference in each test. Review of the comments made by panelists who correctly identified the "same" sample indicated that the ginseng-

Table 7–6 Correct Responses in Duo-Trio Tests of Orange Juice with Added Ginseng Extract*

Samples	Panelists	Correct Identification of the Sample the Same as the Reference	Probability Value (P)[†]
Orange juice vs orange juice with *Panax* ginseng	19	10	0.50
Orange juice with *Panax* ginseng vs orange juice with Siberian ginseng	19	12	0.18
Orange juice vs orange juice with Siberian ginseng	19	10	0.50

*600 mg of ginseng extract per liter of juice.
[†]The selected level of statistical significance is $P \leq 0.05$.

containing samples were less sweet and more bitter and tart than plain orange juice. The juice with *Panax* ginseng was said to be more brown and less tart and more sweet than the juice with Siberian ginseng. These subtle differences may become more obvious when the concentrations of the herbal extracts are higher. Also, since the concentration of active compounds differed in each of the two ginseng extracts, other levels of ginsenosides and eleutherosides may have different sensory impacts in beverages.

SPECIFIC EXAMPLES

Herb-Containing Beverage with No Health or Structure-Function Claims

Fresh Samantha Juices (Saco, ME) produces a line of flash-pasteurized refrigerated juices that do not contain juice concentrates or added preservatives. One product is not labeled as a dietary supplement but could easily be mistaken for one. The front label of "Oh Happy Day" bears a turbaned child assuming the lotus yoga position. Underneath the product name, in much smaller type, is the description "lip-smacking fruit beverage with St. John's Wort." Each 16 oz bottle contains two servings and consists of apple juice, bananas, blackberries, orange juice, water, blueberries, lime juice, and St John's wort. Table 7–7 shows the nutritional and herbal content of one bottle.

Table 7–7 Nutrient and Herb Content of Fresh Samantha's "Oh Happy Day" Beverage

Nutrient or Herb	Amount	% Daily Value
Calories	260 kcal	
Dietary fiber	6 g	24
Vitamin A		4
Vitamin C		30
Calcium		4
Iron		12
Riboflavin		16
St John's wort	0.10 g	no DV

Supplement Soup

In 1999, the Hain Food Group, Inc (Uniondale, NY) released a line of herbal supplement soups under the "Kitchen Prescription" brand label. These ready-to-heat canned soups bear a Supplement Facts panel instead of the Nutrition Facts panel required for conventional soups. Two soups—chicken broth with noodles and country vegetable—contain the immune stimulant herb echinacea, and chunky tomato and split pea soups contain the herb St John's wort. These products retail for nearly $2 per can.

The chicken broth with noodles and echinacea exploits two traditional American cold remedies. This soup contains 318 mg of ethanol extract of *Echinacea angustifolia* and *E. purpurea* root per serving; each can yields two one-cup servings. Jamaican ginger extract is also added, but no content per serving is listed. The structure-function claim for this soup is "Support your immune system with a cup of *Kitchen Prescription* today!" Each can also provides 86% of the DV for vitamin A, due to vegetables and carrot powder in the formulation, but a disadvantage to the product is the sodium content—more than 1 g per can.

The chunky tomato with St John's wort soup provides lower quantities of both vitamin A and sodium, but a can also contains more than half the DV for vitamin C. Each can contains 196 mg of St John's wort stem. This quantity is well below the typical dose of 2 to 4 g per day and the levels (3 g/kg of body weight) required to cause photosensitization and other side effects.[16] However, since the flowers and leaves of this herb are most commonly used, the stems may contain far less of the active phytochemicals. Most consumers may not be aware of this difference in quality.

Herbal Teas

One of the most traditional ways to take herbs is in the form of teas. Hot aqueous extracts of herbs are common to many cultures. Celestial Seasonings, Inc, of Boulder, Colorado, initially marketed herbal teas for medicinal use but refocused its product line to emphasize flavored noncaffeinated teas. Today, the company sells selected herb teas as herbal supplements, in addition to mixtures and single-herb supplements. The herbal tea by the name of "Echinacea Cold Season" bears the image of a koala bear snuggled with a warm quilt, tissues, and a pot of tea. According to the Supplement Facts panel, a tea bag is one serving and provides 50% of the DV for vitamin C and 70% of the zinc DV. Each tea bag also provides 470 mg of peppermint (*Mentha X piperita*) leaves, 440 mg of echinacea herb and root, 235 mg of licorice (*Glycorrhiza glabra*) root, and 140 mg of eucalyptus (*Eucalyptus globulus*) leaves. Flavoring is provided by unspecified amounts of chicory root, menthol extract, lavender flowers, and star anise; stevia leaves add sweetness.

Although this product is sold in the conventional tea section in most grocery stores, the front panel clearly states that it is an herbal supplement. The package bears two structure-function claims. In bold type, the first states the echinacea content "to help you feel your cheerful, healthy best through cold season." The second realistically proclaims that there is no cure for the common cold, "but this tea, and maybe a big fluffy quilt, will make your cold just a bit more bearable!"

Celestial Seasonings takes a responsible approach to warning consumers. This tea carries a recommendation in capital letters that the product is for adults only. Consumers are advised not to use the products for more than 6 to 8 consecutive weeks and when needed. The tea is contraindicated for pregnant and nursing women and for persons with allergies to the daisy family, acquired immune deficiency syndrome, human immunodeficiency virus, autoimmune diseases, collagenosis, leukoses, multiple sclerosis, or tuberculosis.

FUTURE

What supplement trends can be expected? The aging US population can expect to live longer than their parents, but age-related illnesses may impair quality of life.[31] Dietary supplements that can reduce risk for heart disease, cancer, and diabetes should appeal to these consumers. Despite the popularity of Viagra, impotence is still a problem for some men, and purchase of supplements provides both anonymity and an excuse to avoid seeking medical help. Since cataracts and macular degeneration decrease vision in older adults, antioxidants and other types of supplements may enjoy a vogue for improving eyesight.

The FDA and the supplement industry must strive to keep dangerous supplements out of the hands of consumers. Stories of tainted and misrepresented supplements are more common in the popular press. Persistent bad publicity may erode consumer confidence in supplements overall. Food manufacturers should weigh the value of adding dietary supplement ingredients to their products. Safe consumption levels are not yet established for many, if not most, of these ingredients. Consumers and health care professionals are finding more resources available to learn of supplement risks and benefits. Undergraduate dietetic programs must now provide basic education about alternative medical practices to their students.

Manufacturers will be faced with decisions that never had to be made before the DSHEA. If dietary supplement ingredients such as botanicals or hormones are added to a food product, should the food be labeled as a food or as a dietary supplement? As the examples in this chapter demonstrate, the path is not clear. Storlie et al[32] examined this controversial issue with two hypothetical products: an oat fiber bar with omega-3 fatty acids and a grape spread with added resveratrol. Labeling these products as dietary supplements offers the opportunity to make structure-function claims but may restrict the location of these products in retail markets. The sheer number of "nutritionally enhanced" products in the market may prevent FDA from scrutinizing these labeling decisions. However, food producers may find themselves shifting into the role of health care providers and thus should consider the motto "Do no harm."

REFERENCES

1. Dietary Supplement Health and Education Act of 1994. U.S. Public Law No. 103–417.
2. Food and Drug Administration. FDA finalizes rules for claims on dietary supplements. FDA Talk Paper. January 5, 2000. (http://vm.cfsan.fda.gov/~/rd)
3. Food and Drug Administration. Regulations on statements made for dietary supplements concerning the effect of the product on the structure or function of the body. Final Rule. *Federal Register*. 2000: 65(4):999–1050. 21 CFR Part 101.
4. Kurtzweil P. An FDA guide to dietary supplements. *FDA Consumer*. 1998;32(5):28–35.
5. Federal Trade Commission. Dietary supplements: an advertising guide for industry. Available on the Internet at http://www.ftc.gov/bcp/conline/pubs/buspubs/dietsupp.htm. Accessed 1998.
6. Annual industry overview. *Nutr Bus J.* 1998;3(9):1–5, 13.
7. Bayer introduces One-A-Day herbal formulas: other pharmaceutical formulas close behind. *Nutr Bus J.* 1998;3(9):6.
8. Muth MK, Domanico JL, Anderson DW, Siegel PH, Bloch LJ. Dietary supplement sales information. Final Report to the FDA. Available on the Internet at http://vm.cfsan.fda.gov/~dms/ds-sales.html. Accessed 2000.
9. National Public Radio, Kaiser Family Foundation, Harvard University Kennedy School of Government. Survey on Americans and dietary supplements. Available on the Internet at http://www.npr.org/programs/specials/survey/front.html. Accessed 1999.
10. Eisenberg DM, Davis RB, Ettner SL, et al. Trends in alternative medicine use in the United States, 1990–1997. *JAMA.* 1998;280:1569–1575.
11. Eliason BC, Myszkowski J, Marbell A, Rasmann DN. Use of dietary supplements by patients in a family practice clinic. *J Am Board Fam. Pract.* 1996;9:249–253.
12. Lyle BJ, Mares-Perlman JA, Klein BEK, Klein R, Greger JL. Supplement users differ from nonusers in demographic, lifestyle, dietary and health characteristics. *J Nutr.* 1998;128:2355–2362.

13. Wallstrom P, Elmstahl S, Hanson BS, et al. Demographic and psychosocial characteristics of middle-aged women and men who use dietary supplements. *Eur J Pub Health.* 1996;6:188–195.

14. Tyler VE. *The Honest Herbal.* 3rd ed. New York, NY: Pharmaceutical Products Press; 1993.

15. Blumenthal M, Busse WR, Goldberg A, et al. *The Complete German Commission E Monographs: Therapeutic Guide to Herbal Medicines.* Austin, TX: American Botanical Council, 1998.

16. Gruenwald J, Brendler T, Jaenicke C. *PDR for Herbal Medicines.* Montvale, NJ: Medical Economics Co; 1998.

17. Katch FI, McArdle WD, Katch VL, Freeman JA. Exercise nutrition: from antiquity to the twentieth century and beyond. In: Wolinsky I, ed. *Nutrition in Exercise and Sport.* Boca Raton, FL: CRC Press; 1998:1–48.

18. Graham TM, Spriet LL. Caffeine and exercise performance. *Sports Sci Exchange.* 1996;9(1):1–5.

19. King DS, Sharp RL, Vukovich MD, et al. Effect of oral androstenedione on serum testosterone and adaptations to resistance training in young men. *JAMA.* 1999;281:2020–2028.

20. Sobal J, Marquart LF. Vitamin/mineral supplement use among high school athletes. *Adolescence.* 1994;29:835–843.

21. Flegal KM, Carroll MD, Kuczmarski RJ, Johnson CL. Overweight and obesity in the United States: prevalence and trends, 1960–1994. *Intl J Obesity.* 1998;22:39–47.

22. Thompson D, Edelsberg J, Kinsey KL, Oster G. Estimated economic costs of obesity to U.S. business. *Am J Health Promot.* 1998;13:120–127.

23. Hoeger WW, Harris C, Long EM, Hopkins DR. Four-week supplementation with a natural dietary compound produces favorable changes in body composition. *Adv Ther.* 1998;15:305–314 .

24. Food and Drug Administration. Dietary supplements containing ephedrine alkaloids; proposed rule. *Federal Register.* 1997;62:30678–30724. To be codified at 21 CFR pt 111.

25. Dickinson A. Optimal nutrition for good health: the benefits of nutritional supplements. Council for Responsible Nutrition. Available on the Internet at http://www.crnusa.org/ben_full.htm, Accessed 1998.

26. Elliott JG. Application of antioxidant vitamins in foods and beverages. *Food Technol.* 1999;53(2):46–48.

27. Matz SA. *Formulating and Processing Dietetic Foods.* McAllen, TX: Pan-Tech International, Inc; 1996.

28. Draganchuk M. Flavors for success. *Nutraceuticals World.* 1999;2(3):60–63.

29. Clydesdale FM. Mineral additives. In: Bauerfeind JC, Lachance PA, eds. *Nutrient Additions to Food: Nutritional, Technological and Regulatory Aspects.* Trumbull, CT: Food and Nutrition Press; 1991:87–107.

30. Green NR. Tropicana Pure Premium with calcium and extra vitamin C: a case study. *J Nutraceuticals, Functional Med Foods.* 1999;1(4):33–41.

31. Sloan AE. The new market: foods for the not-so-healthy. *Food Technol.* 1999;53(2):54–58, 60.

32. Storlie J, O'Flaherty MJ, Hare K. Food or supplement? Choosing the appropriate regulatory path. *Food Technol.* 1998;52(12):62–67, 69.

CHAPTER 8

Dairy Ingredients as a Source of Functional Foods

Gertjan Schaafsma and J.M. Steijns

Increased awareness in many countries of the diet-health relationship has stimulated a trend in nutrition science in which more attention is given to health effects of individual foods. The role of the diet and specific foods in the prevention and treatment of diseases and in improving body functions has become more prominent and active. The existing and expanding knowledge in nutrition science and the purchasing power of the aging population offer the food industry good opportunities to improve existing foods and develop new foods. Traditionally, cow's milk has been considered a basic food in many diets. The milk is rich in a large variety of essential nutrients and other bioactive compounds. Nowadays, modern separation techniques in the dairy industry and enzyme technology offer opportunities to isolate, concentrate, or modify these compounds so that their application in functional foods, dietary supplements, or medical foods has become or becomes possible. This chapter deals with milk components that from a nutritional or medical point of view can be applied to existing or new foods to improve, in a classical sense, their nutritional value or to introduce new characteristics beyond those of nutritional value.

MILK PROTEINS

The proteins in cow's milk can be roughly divided into caseins and whey proteins. The caseins are those (phospho)proteins that precipitate from raw skim milk by acidification to pH 4.6 at 20°C, whereas the whey proteins stay in the milk serum. A similar separation takes place during cheese making when the casein fraction in the milk curdles due to the action of calf rennet or chymosin and the soluble proteins reside in the cheese whey.

The caseins comprise four families of molecules (αs1, αs2, β, and κ) that show genetic polymorphism and post-translational modification with phosphorylation and/or glycosylation. The major whey proteins are α-lactoglobulin and β-lactalbumin, for which also genetic variants are known. The whey fraction also contains substantial amounts of immunoglobulins and serum albumin, which are passed on to the mammary gland and secreted into the milk. Lactoferrin and lactoperoxidase are the best characterized proteins of the growing list of minor protein components in whey; both compounds are now available in commercial amounts. Due to the presence of the endopeptidase plasmin in the milk, β-casein is degraded to N- and C-terminal peptides; the C-terminal peptides, which contain a number of phosphorylated amino acids, precipitate with the caseins during acid treatment, whereas the N-terminal peptides remain in the soluble whey fraction.[1]

The frequencies of the various genetic variants of individual cow's milk proteins have been determined in common breeds (see, eg, Swaisgood[2]). The number of amino acid substitutions may vary from one to eight between variants in a particular family.[1] Within a breed, the production of the individual milk proteins may vary, but it is unlikely that significant differences in amino acid composition will occur between samples of bulk milk.[3]

Table 8–1 lists the amount of the predominant individual casein and whey proteins in milk, their peptide length, and their apparent molecular weight.[1,4–7] In the sections to follow, selected individual minor or major components will be described in further detail and in relation to their envisaged biological role or activity. The reader is referred to several excellent reviews and handbooks for extensive information on the various milk proteins.[1,2,4,5,8–12]

PROCESSING, ISOLATION, AND ENZYMATIC MODIFICATION OF MILK PROTEINS

The manufacturing of casein and its soluble derivatives (caseinates) has been reviewed extensively by Muller[13] and Mulvihill.[14] To destabilize the caseins from skimmed milk, essentially two industrial methods are used. One relies on isoelectric precipitation, whereby the pH is lowered to about 4.6; the other uses proteinases that hydrolyze the κ-casein component in the casein micelle, thus rendering the micelles susceptible to precipitation by the calcium ions naturally present in the milk. The acid precipitation may use mineral acids, like sulfuric acid, or lactic acid. The proteolytic coagulation uses calf rennet or rennet substitutes like chymosin. The curd is then separated from the whey—for example, with centrifuges—and is washed to remove residual whey. Removal of water, drying, tempering and blending, milling, and sieving are the steps that follow. Caseinates are produced by solubilizing the curd with alka-

Table 8–1 Predominant Casein and Whey Proteins in Bovine Milk

Protein	Amount in Milk (g/L)	Number of Amino Acids	Apparent Molecular Weight (Daltons)
Caseins			
αs1-Casein A, B, C, D, E	10.3	199 (B)	22,000–24,000
αs2-Casein A, B, C, D	2.7	207	25,000
β-Casein A$_1$, A$_2$, A$_3$, B, C, D, E	9.7	209 (A$_2$)	24,000
κ-Casein A, B	3.5	169	19,000
C-terminal β-Casein fragments	0.8	Various sizes	Various sizes
Whey proteins			
N-terminal β-casein fragments	0.8	Various sizes	Various sizes
β-Lactoglobulin	3.4	162	18,000
α-Lactalbumin	1.3	123	14,000
Bovine serum albumin	0.4	582	66,000
Immunoglobulins	0.8	Variable	150,000–1,000,000
Lactoferrin	0.02–0.2	689	80,000
Lactoperoxidase	0.03	612	78,000

Source: Data from Barth & Behnke, 1997, Walstra & Jennes, 1984, Eigel et al 1984, Steijns & van Hooydonk, in press, and de Wit & van Hooydonk, 1996.

line solutions to increase the pH toward neutral at elevated temperatures; thus, soluble caseins can be prepared that are rich in sodium, potassium, calcium, magnesium, or some mixture of these, if required.

The cheese and acid whey can be processed in various ways. Concentrating the liquid whey and subsequently drying it simply yields whey powders. Alternatively, the lactose can be removed or the mineral content decreased. Blending with other protein sources yields whey protein blends. Ultrafiltration is used to increase the protein content up to about 80% to produce the so-called whey protein concentrates. The reader is referred to the informative reviews by Marshall[15] and Morr[16] for process details. Recently, much attention has been given to whey protein isolates and fractionation of whey into individual components.[17–19] Whey protein isolates are made from cheese whey and typically have a protein content of 90% or higher. These products are gaining interest because of their bland taste, specific functionality in food products, nutritional value, and functional food use.[20] The industrial production methods for these isolates or enriched fractions are still developing.

Characterization by high-performance liquid chromatography (HPLC) allows for quantification of individual major components from cheese whey, namely α-lactoglobulin, β-lactalbumin, bovine serum albumin, and the κ-casein–derived glycomacropeptide, as shown in Figure 8–1. Screening commercial whey protein isolates with this technique shows that the protein composition and hence production routes, may differ considerably (Table 8–2).

Bioactive proteins like lactoferrin and lactoperoxidase can be purified from milk and whey with ion exchange technology.[21] Immunoglobulins may be extracted from whey by ultrafiltration and/or affinity chromatography,[22] or from milk, after removal of the caseins by precipitation or coagulation, using precipitation or filtration technology.[23]

Within the milk proteins specific peptides have been detected with biological activity, as will be discussed later. These peptides are obtained by enzymatic hydrolysis and subsequent fractionation or enrichment. The reader is referred to the paper of Lahl and Braun[24] that addresses relevant issues for the enzymatic production of protein hydrolysates for food use, like choice of raw materials, process-

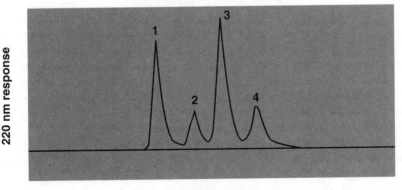

Figure 8–1 HPLC separation of major whey proteins of bovine milk. HPLC separation of proteins present in whey: 1 = BSA, 2 = β-lactoglobulin, 3 = α-lactalbumin, and 4 = GMP; with this method, whey protein isolates can be analyzed with respect to protein composition. Refer to Table 8–2 for the results of analysis of various commercially available whey protein isolates.

Table 8–2 Protein Composition of Some Commercial Whey Protein Isolates (WPIs) as Measured with HPLC

WPI Sample	α-Lactalbumin	β-Lactoglobulin	BSA	GMP
I	22	60	11	0
II	20	43	5	23
III	15	59	21	0

Note: Various commercially available WPIs were screened with HPLC, in the way depicted in Figure 8–1, to measure the amounts of α-lactalbumin, β-lactoglobulin, bovine serum albumin (BSA), and glycomacropeptide (GMP); the amounts were corrected for the differences in response factor for the proteins and are presented as percentages in the product "as is."

ing methods, and quality control. Schmidl and Labuza[25] addressed the use of dairy protein for enteral feeding to enhance nutritional support.

NUTRITIONAL VALUE OF MILK PROTEINS

Essential amino acids and nitrogen are required for growth (synthesis of new body proteins) and for maintenance (compensation of obligatory losses of body nitrogen, mainly caused by oxidation of amino acids). Proteins providing all indispensable amino acids in such amounts that requirements are easily met appeared to have a nutritional value that is higher than that of proteins that are relatively low in one or more of these nutrients. In this regard, proteins like those in milk, egg, and beef were found to be superior, in rat bioassays, to proteins like those in beans, maize, wheat, and soy. It should be realized that the methods currently applied particularly refer to the (essential) amino acid concentration and digestibility of the test protein and do not account completely for other aspects of proteins that are physiologically relevant. Such aspects are, for instance, antibacterial and antiviral properties, effects on nutrient bioavailability, immune stimulation, and hormonal effects of bioactive peptides formed during the digestion of the protein, all of which will be reviewed later in this chapter.

Methods To Measure Protein Quality

It is well accepted that the nutritional value of proteins may differ substantially, depending on their (essential) amino acid composition and digestibility. For many years, bioassays, mainly with rats, have been the methods of choice to assess the nutritional value of proteins. This value has been expressed in parameters like protein efficiency ratio (PER), net protein utilization (NPU), and biological value.

The PER method has been used the most because of its simplicity. The PER of a particular protein is measured in growing rats fed a limited amount (10% w/w) of test or reference protein. Weight gain of the rats is measured and expressed in grams per gram of protein intake.

But despite its popularity and wide application, the PER method has been criticized because of the recognition that the essential amino acid requirement of growing rats differs substantially from that of humans, especially regarding the much higher rat requirement of sulfur-containing amino acids. This was one of the reasons that in 1989 a joint Food and Agricultural Organization/World Health Organization (FAO/WHO) Expert Consultation on Protein Quality Evaluation[26] concluded that protein quality could be assessed adequately by comparing the content of the first limiting essential amino acid of the test protein with the content of that amino acid in a reference pattern of essential amino acids. For this reference pattern, a pattern was taken that was based on the essential amino acid requirements of the preschool child. This comparison was expressed in a so-called amino acid score (AAS). The next step was the correction of the AAS for digestibility of the test protein. For this correction, a factor was applied, called true fecal digestibility, which was measured in a rat assay. The corrected AAS was known as the Protein Digestibility Corrected Amino Acid Score (PDCAAS) and was adopted as the preferred method for measurement of the protein value in human nutrition. The definition of PDCAAS is given below:

$$\text{PDCAAS} = \frac{\text{mg of limiting amino acid in 1 g of test protein}}{\text{mg of same amino acid in reference requirement pattern}} \times \text{true digestibility} \tag{1}$$

True digestibility is measured in a rat experiment and equals the amount of absorbed nitrogen, expressed as a fraction of nitrogen intake and corrected for endogenous nitrogen losses in the feces. It is calculated as follows:

$$\text{True digestibility} = \frac{I(F - f)}{I} \tag{2}$$

where I = nitrogen intake, F = total fecal N excretion, and f = fecal nitrogen excretion on a protein-free diet.

According to Equation 1, PDCAAS values of a protein may exceed the value 1. However, it has been decided[26] that any value above 1 will be truncated to 1.

Now, after 7 years of experience with PDCAAS, it can be concluded that the principle of the method has been widely accepted. On the other hand, questions have been raised in the scientific community, most notably at the Dutch Dairy Foundation's Workshop on Nutrition and Health in 1995, about the validity of the preschool child amino acid scoring pattern, the true fecal digestibility correction, and, most importantly, the truncating of PDCAAS values to 1. Although it is beyond the scope of this section to discuss these three issues in detail, the truncating of PDCAAS values to 1 cannot be accepted in the nutritional evaluation of high-quality proteins, such as milk proteins, since it overlooks the important nutritional characteristic of high-quality proteins that they can balance in mixed diets the amino acid composition of low-quality proteins. A classical and widely known example in this regard is the combination of milk and wheat, where the relatively high lysine concentration of the milk proteins compensates for the low concentration of this essential amino acid in wheat. Thus, it can easily be shown that 1.2 g of casein can balance 1 g of wheat protein, whereas 6.2 g of soy protein is needed to do the same thing. The truncating of PDCAAS values largely eliminates the differences in the power of high-quality proteins to balance the amino acid composition of inferior proteins. The truncating should therefore be dropped, as was concluded at the Dutch Dairy Foundation's Workshop.

Another method to measure the quality of proteins is the performance of balance studies in humans. Such experiments, which are quite time consuming and expensive, establish the minimum amount of a test protein that is required to restore the nitrogen balance in adult subjects after these subjects have been fed on a protein-free diet. The lower this amount, the higher the potency of the protein to satisfy protein needs.

Nutritional Value of Milk Proteins

Table 8–3 shows the FAO/WHO preschool–child amino acid scoring pattern and the essential amino acid contents of a variety of proteins.[9,26,27] The data in Table 8–3 have been used to compute the AAS values, as indicated in Table 8–4.[9,26,28] This table also shows the PER, protein digestibility (PD), and PDCAAS values of the respective proteins. The nutritional value of milk proteins has also been measured in balance studies with humans, as summarized in Table 8–5.

Table 8–3 FAO/WHO Preschool–Child Amino Acid Scoring Pattern and Amino Acid Content (mg/g of Crude Protein) of Several Proteins

	Scoring Pattern	Cow's Milk	Casein	Whey Protein	Beef	Whole Egg	Soy	Wheat
Histidine	19	27	29	22	34	22	—	—
Isoleucine	28	47	57	68	48	54	47	33
Leucine	66	95	104	111	81	86	85	68
Lysine	58	78	83	99	89	70	63	27
Methionine + cystine	25	33	31	48	40	57	24	39
Phenylalanine + tyrosine	63	102	111	73	80	93	97	78
Threonine	34	44	46	80	46	47	38	29
Tryptophane	11	14	14	21	12	17	11	11
Valine	35	64	68	68	50	66	49	43
Total (minus histidine)	320	504	514	569	479	512	414	328

Source: Data from FAO/WHO 1990; Renner 1983; and Sarwar 1984.

Table 8–4 Nutritional Value of Various Proteins, as Expressed by Several Parameters

	PER	AAS	PD	PDCAAS
Milk protein	3.1	1.27	0.95	1.21
Casein	2.9	1.24	0.99	1.23
Whey proteins	3.6	1.16	0.99	1.15
Beef	2.9	0.94	0.98	0.92
Whole egg	3.8	1.21	0.98	1.18
Soy	2.1	0.96	0.95	0.91
Wheat	1.5	0.47	0.91	0.42

Source: Data from FAO/WHO 1990; Renner 1983; and the European Dairy Association 1997.

Table 8–5 Minimum Requirement of Various Proteins (g/kg of Body Weight), as Measured in Nitrogen Balance Studies in Humans

Protein	Requirement
Mixture of egg N and potato N (36:64)	0.369
Mixture of lactalbumin N and potato N (70:30)	0.374
Lactalbumin	0.480
Whole egg	0.500
Potato	0.512
Beef	0.550
Bovine milk	0.568
Soy	0.595
Casein	0.699
Whole wheat flour	0.892

Source: Data from C.A. Barth and U. Behnke, Nutritional Significance of Whey and Whey Components, *Nahrung*, 41, Nr. 1, pp. 2–12, © 1997, VCH Verlagsgesellschaft mbH.

It is clear from this information that there is not always an excellent correlation between the various measures of protein value. One explanation for this is the difference between the amino acid requirements of rats, preschool children, and human adults. Nevertheless, it can be concluded that in all measures milk proteins give an excellent score. With respect to application of milk proteins in functional foods, it is widely recognized that, because of their relatively high concentration of essential amino acids, these proteins have a good potential in foods to balance the amino acid composition of a large variety of plant proteins. This is particularly evident from Table 8–5, which shows that mixtures of plant proteins and milk proteins have a high potential to cover the protein needs in humans. The high concentration of essential or semi-essential amino acids in milk proteins can have special significance under conditions where an increased requirement for these proteins exists.

BIOLOGICAL ACTIVITY OF MILK PROTEINS

Mineral Carriers

The protein micelles in milk contain physiologically significant amounts of calcium and phosphorus, which is made possible by specific structural features of the various casein molecules. It has been suggested that the biological function of the caseins is the protection of the milk gland against calcification by controlling calcium phosphate precipitation.[29] Specific peptide fragments in the β and κ caseins are responsible for this mineral carrier function.

Host Protection

Studies with rats have shown that dairy proteins may offer protection to the host against carcinogens.[30] The animals were treated subcutaneously with the procarcinogen dimethylhydrazine, and the tumor incidence, the total number of tumors per group, and the tumor mass index in the small and large intestine were recorded as a function of the protein composition of the purified diets. Feeding the rats casein or whey protein resulted in significantly fewer tumors per treatment group compared to a red meat or soybean meal. The whey proteins were particularly effective. The mechanism of the protective action was not elucidated, but in the large intestine a positive correlation was found between fecal fat concentration and the tumors per group ($n = 20$; $r^2 = 0.898$; $P = 0.05$). This suggests that free fatty acids and bile acids might be involved in hyperproliferation of colonic epithelial cells, since proteins high in fat were associated with more carcinogens.[31] Also, the liver concentration of glutathione, a natural antioxidant and anticarcinogen, was higher in the dairy groups, which could be attributed to the higher amounts of the sulfur-containing amino acids methionine and, in particular, cysteine, because the latter is the rate-limiting amino acid for glutathione biosynthesis.[32]

Bounous and Amer[33] showed the immunoenhancing effects of dietary whey protein concentrate. With nutritionally adequate diets, the humoral immune response in mice was enhanced with whey protein but not with casein, soy, wheat, corn, egg, beef, or fish protein. The effect was manifested after 2 weeks and persisted for at least 8 weeks.

Cysteine and glutathione (GSH) are important substrates for the immune response.[34] In vitro glutathione supplementation to peripheral blood mononuclear cells (PBMCs) from young and old subjects enhances the interleukin-2 production and the T-cell–mediated mitogenic response.[35] A whey protein concentrate with β-lactoglobulin (56.3%), α-lactalbumin (22.8%), serum albumin (11.1%), lactoferrin (0.7%), and immunoglobulins (9.2%) increased the proliferation in vitro of PBMCs and also the intracellular GSH concentration. On the other hand, tumor cell lines were inhibited in cell proliferation, with concomitant decrease in cellular GSH.[36] This paradoxical situation was explained by assuming negative feedback in GSH synthesis in the tumor cell lines, which had a higher baseline level of GSH. According to the authors, the selective modulation of GSH levels in normal and tumor cells offers opportunities for chemotherapy patients, making tumor cells more vulnerable to chemotherapy and protecting normal tissues against this challenge. Herzenberg et al.[37] presented clinical evidence that GSH depletion in human immunodeficiency virus (HIV) patients was predictive of poor survival. Dröge and Holm[38] reviewed the role of cysteine and GSH in HIV infection and other diseases associated with muscle wasting and immunological dysfunction and concluded that available evidence suggests that the cysteine level is a physiological regulator of nitrogen balance and body cell mass, that the cysteine-mediated regulatory circuit is compromised during catabolism and old age, and that cysteine supplementation is useful in combination with disease-specific treatments. Friedman[39] reviewed the improvement in the safety of foods that have sulfur-containing amino acids and peptides. He pointed out that apart from being precursors for glutathione, SH compounds can conjugate and detoxify xenobiotics and quench potentially toxic free radicals. So it seems that whey proteins that are rich in cysteine are promising candidates for functional foods. The cysteine content of the various whey proteins is listed in Table 8–6.

Lactoferrin and Pathogen Resistance

Lactoferrin is an iron-binding glycoprotein of the transferrin family and has attracted increasing scientific interest since the early 1960s due to its high concentration in human breast milk.[40] The molecule has also been identified in the milks of other mammalian species, like the cow, pig, horse, buffalo, goat, and mouse. On a commercial basis, lactoferrin is isolated from cow's milk. Apart from milks, lactoferrin is generally found in the products of the exocrine glands located in the gateways of the digestive, respiratory, and reproductive systems, like saliva, tears, nasal secretions, and seminal plasma; in blood, lactoferrin is derived from a special group of white blood cells, the neutrophils.[41]

The occurrence of lactoferrin in these biological fluids suggests a role in the nonspecific defense against invading pathogens. In the clinical setting, confirmation for this envisaged protective role comes from the finding of recurrent infections in patients whose neutrophils are not able to produce lactoferrin.[42,43] Recent research suggests that lactoferrin may be a multifunctional immunoregulatory protein.[44]

Most of the proposed biological activities of lactoferrin are related to its excellent iron-binding properties, but non–iron-related activities have been described as well. With respect to its role in pathogen resistance, one can discern the following functions:

- activity against a broad microbial spectrum, including gram-positive and gram-negative bacteria, yeasts, and fungi
- antiviral activity, including activity against cytomegalovirus, influenza, HIV, and rotavirus
- protection of the white blood cells against free iron–catalyzed oxidation reactions
- control of the immune response during infection and inflammation

Table 8–6 Cystein Content of Some Major Whey Proteins

Protein	mmol of Cysteine per Gram of Protein	mg of Cysteine per Gram of Protein
β-Lactoglobulin	0.27	33
α-Lactalbumin	0.56	68
Bovine serum albumin	0.53	64

Source: Data from P. Walstra and R. Jennes, *Dairy Chemistry and Physics*, © 1984, Wiley Interscience.

The antimicrobial activity of lactoferrin includes a variety of working mechanisms:[45]

- Due to the high affinity for iron, bacterial cells become iron deprived and stop growing.
- The microbial membrane is disturbed, and as a consequence the microbial cell loses its integrity and is killed.
- Stimulation of phagocytosis by macrophages and monocytes.

Many in vitro studies have demonstrated these antimicrobial effects. Recent studies also show a potentiating effect of lactoferrin with antibiotics and antifungal agents.[46–48] In vivo results from animal studies are available as well. Bovine lactoferrin was shown to have a bacteriostatic effect in the gut of mice infected with either enterobacters or clostridia.[49,50] In humans, the duration and severity of enteric infections decreased when patients were treated orally with human lactoferrin.[51] Bacterial infections in rainbow trout could be reduced by oral administration of bovine lactoferrin.[52]

Recently, considerable attention has been given to the antiviral activity of lactoferrin. The mechanism of action seems to be the inhibition of the adsorption process of the virus particle to the mammalian cell, through binding to either the host cell or the virus particle. So far, protective effects of lactoferrin have been found for HIV and cytomegalovirus,[53] herpes simplex types 1 and 2,[54] hepatitis C,[55] influenza,[56] and rotavirus.[57] Interestingly, the first in vivo animal studies support the findings from the in vitro work. When mice were pretreated with bovine lactoferrin, the challenge with mouse cytomegalovirus did not result in mortality.[58] The mechanism behind this protective effect seems to be the stimulation of natural killer cell activity. Neutrophils, monocytes, and macrophages are cells of the immune system that kill invading pathogens by oxidant reactions. Since free iron is usually present in areas of inflammation or infection, these oxidant reactions may be accelerated due to the catalytic effort of iron on free-radical production. Lactoferrin will bind the free iron with high affinity and thus function as a local antioxidant protecting the immune cells against the free radicals produced by themselves.[59] Although only the neutrophils degranulate and deliver lactoferrin, monocytes and macrophages have lactoferrin receptors on their cell surface.

One of the most potent stimulants of cytokine activity, compounds produced by immune cells during infection and inflammation to coordinate the defense against pathogens, is the endotoxin lipopolysaccharide (LPS). This microbial membrane–derived LPS is bound and neutralized by lactoferrin, thus preventing the immune response from getting out of control. Not only in vitro studies,[60–62] but also in vivo animal studies,[63–65] have shown this protective effect of lactoferrin. In cats, bovine lactoferrin was used for the treatment of intractable stomatitis; apparently, the phagocytic activity of neutrophils was stimulated.[66] In rainbow trout, bovine lactoferrin also stimulated the phagocytic cells.[67] In mice, bovine lactoferrin induced both the mucosal and the systemic immune response.[68]

One may conclude that lactoferrin makes an important contribution to the host defense system. Lactoferrin eliminates pathogens like bacteria, viruses, and fungi, stimulates and protects cells involved in host defense mechanisms, and controls the cytokine response.

Immunoglobulins

Immunoglobulins are particularly rich in colostrum but decline in concentration during normal lactation.[69] In the cow, the major class of immunoglobulins is lgG1, with concentrations of about 48 g/L in colostrum and 0.6 g/L in normal milk respectively; in human milk the predominant class of immunoglobulins is IgA, with levels of about 17 g/L in colostrum and about 1 g/L in normal breast milk.[70] Cow's milk immunoglobulins have been put forward as possible effective means of preventing or combating diarrhea.[23] Many studies, both animal and human, have been performed up till now with immunoglobulins either from nonimmunized cows or from cows hyperimmunized against specific pathogens. In general, one can say that immunoglobulin preparations from hyperimmunized cows are more effective because the titers of the pathogen-inactivating immunoglobulins are higher than those from nonimmunized cows.[71] The pathogens studied have been *Campylobacter, Clostridium, Shigella, Vibrio,* and *E. coli* species, and also rotavirus. Davidson[72] and Bogstedt et al[71] showed in peer reviews that there were positive-outcome studies with respect to the use of bovine-derived immunoglobulins in humans, both curative and prophylactic, with respect to diarrhea. The potential applications include not only prevention of traveler's diarrhea and infections in day care centers and hospitals but also secondary treatment in immunocompromised subjects, such as acquired immune deficiency syndrome patients and malnourished subjects. Several studies have shown that bovine immunoglobulins may (partially) survive the human digestion tract, which explains why they are still efficacious (see, eg, Roos et al[73] and Kelly et al[74]).

α-Lactalbumin: Source of Tryptophan

Industrial food ingredient manufacturers are focusing nowadays on enriching or purifying α-lactalbumin from cow's milk. The main reason for this development is the desire of infant formula manufacturers to mimic as much as possible the protein composition of breast milk and obtain the same plasma amino acid patterns as are seen in breastfed babies.[75] The biological role of α-lactalbumin in human milk is the interaction with galactosyltransferase to promote the transfer of galactose from UDP-galactose to glucose to form lactose. The human and bovine α-lactalbumins are both 123 amino acids, and their amino acid compositions are much alike.[11] α-Lactalbumin is especially rich in cysteine (Table 8–6) and tryptophan. The human variant has three tryptophan residues, whereas the bovine variant has four; expressed as gram of amino acid residue per 100 g of protein, the levels are 3.97% and 5.25% respectively.

The significance of tryptophan in human nutrition has recently been reviewed by Heine et al.[76] Tryptophan and its metabolites regulate neurobehavioral effects such as appetite, sleeping-waking rhythm, and pain perception.

Bifidogenic Activity

Several studies have shown that cow's milk proteins may stimulate the growth of *Bifidobacterium* species. Petschow and Talbott[77] demonstrated that α-lactalbumin and lactoferrin were potent growth promoters for several *Bifido bacterium* species. Poch and Bezkorovainy[78] showed bifidogenic activity of trypsin digested κ-casein; tryptic digests of the glycosylated part of κ-casein (glycomacropeptide) had no activity. Growth-stimulating activity was lost when the disulfide bridges were modified. These studies suggest that cysteine-rich diets may support the fecal flora. This hypothesis is supported by clinical studies on the fecal flora in infants. Wharton et al[79] summarized their studies on comparison of the fecal flora of breast milk–fed babies to that of infants fed commercial cow's milk–based infant formulas and concluded that whey-predominant infant formulas had higher numbers of bifidobacteria and lactobacilli than casein–dominant formulas; the flora of the whey group was, however, still far from the typical breast milk flora. These authors could not establish benefits for supplementation with lactoferrin. On the other hand, Roberts et al[80] found, in half of the babies, a "bifidus flora" at the age of 3 months when the infant formula was supplemented with 1 mg/mL of bovine lactoferrin.

Other Factors in Milk

Human and bovine milk and colostrum contain minor bioactive components like hormones and growth factors.[81–86] The bovine growth factors have attracted increasing interest since Howarth et al[87] showed that oral administration of a growth factor extract from cheese whey was able to reduce small-bowel damage in methotrexate-treated rats. The application of cow's milk–derived growth factors is still in its infancy because efficacy in humans still remains to be proven. However, given the homology between human and bovine growth factors, some efficacy must be anticipated. Bovine growth factors include, among others, insulin-like growth factor (IGF-1), epidermal growth factor, transforming growth factor (TGF-β), and fibroblast growth factor (acidic and basic).[85] The envisaged (medical) food applications include, among others, gut repair after/during radiation or chemotherapy.

Cow's milk also contains binding proteins for vitamins B_{12}, folate, and riboflavin.[70] The folate-binding protein is present in amounts of about 10 ppm. Parodi[88] has reviewed its role in nutrition and concludes that folate bound to its binding protein is more slowly absorbed throughout the whole small intestine but with greater absorption than free folate and is not available to intestinal folate-requiring bacteria, thus also contributing to the folate economy. Parodi further hypothesizes that, due to the surplus binding capacity of the binding protein, whey protein–containing dairy products will sequester the folates from fruits and vegetables and increase their bioavailability.

The fat globule membrane in bovine milk contains a major glycoprotein called butyrophilin. This component, with a molecular weight of about 65,000 daltons, has been suggested to function as a component of the immune system because it shares structural motifs characteristic of the immunoglobulin superfamily.[89]

BIOLOGICAL ACTIVITY OF MILK PROTEIN–DERIVED PEPTIDES

Within the sequence of amino acids of a given dietary protein, specific peptides may be located with specific biological activity. Animal proteins, in particular milk proteins, are rich in these so-called bioactive peptides.[90–91] The most prominent bioactive peptides from the casein molecules are fairly well characterized and accessible for use in foods, since combinations of specific enzymatic hydrolysis and fractionation technologies are available for their products. These bioactive peptides have been obtained from in vitro digestion with proteolytic enzymes from both animal and microbial origin or from in vivo digestion after feeding the precursor protein.

Casein Phosphopeptides

Of the four major casein molecules, only κ-casein does not contain phosphorylated amino acid residues; the three other casein molecules contain, respectively, 8 to 9 (αs1), 10 to 13 (αs2), and 5 (β) of these phosphorylated serine residues. These phosphoserine-rich amino acid sequences in α- or β-casein are thought to play a crucial role in the protection of the milk gland against calcification by controlling calcium phosphate precipitation. These fragments help to create the thermodynamically stable casein micelles, supersaturated with calcium and phosphate, and thus contribute to the stability of milk during heat processing.[29,92]

By cleavage of casein with enzymes like trypsin and alcalase (microbial enzyme), a number of casein phosphopeptide (CPP) fragments can be obtained in vitro.[93,94] If CPPs exert a function in vivo, one may expect these casein-derived fragments to be relatively resistant to proteolytic breakdown in the intestinal tract by either proteases or phosphatases. A couple of studies have shown that this is indeed the case. Naito and Suzuki[95] fed rats either whole acid–precipitated casein or purified β-casein and subsequently analyzed the contents of the small intestine after an approximate 2.5 hours' digestion of the food. They found evidence for the in vivo formation of CPPs and a relatively slower decomposition rate compared to the non-CPP fragments, as large peptides with a low nitrogen/phosphorus ratio were found only in the distal parts of the small intestine. In later studies, these results were confirmed; furthermore, it was shown that the formation of CPPs in the intestine was able to increase the concentration of soluble calcium.[96–100] Meisel and Frister[101] found that CPP fragments were also released in the intestine of minipigs. The studies in animals trying to find positive effects of dietary CPP on intestinal calcium absorption, calcium balance, or bone formation show controversial results. No significant effects were found by Brommage et al,[102] Pointillart and Gueguen,[103] or Scholz-Ahrens et al.[104] On the other hand, Japanese researchers did find positive data in their extensive studies.[105] One of their main goals is to prevent bone loss by effective utilization of dietary calcium. By removing the ovaries in aged rats, a postmenopausal bone-loss model is obtained with which the effect of CPPs on the metabolism of calcium and phosphorus can be studied during long-term feeding. During 16 weeks of feeding a diet without CPPs but with recommended levels of calcium for nonovariectomized rats, the relative bone mineral density decreased gradually by approximately 18%. On the other hand, rats fed a calcium-CPP complex or $CaCO_3$ and CPPs separately (calcium-free CPPs) did not show this decrease in bone mineral density; the same was true for a non-CPP control group that underwent surgery without subsequent ovary removal.[105] Recently, a rat pup model was used to test the effect of purified CPPs on the passive absorption of both calcium and zinc in an oat-based infant cereal and a soy protein–based infant formula. These diets contain phytate, which reduces the solubility of both zinc and calcium. Compared to the nonsupplemented control diet, the addition of CPPs increased the absorption of calcium by 45% and 10% for the oat- and soy-based diets respectively; the zinc absorption was increased by 39% and 33%, respectively.[106]

Rat studies have also addressed the role of CPPs in increasing the bioavailability of dietary iron.[107] Using young iron-deficient rats, it was shown that the liver iron concentration, expressed in milligrams per gram net weight, was significantly higher in groups fed for 2 weeks on diets enriched with a β-casein–derived CPP-iron hydrolysate or with purified CPP-iron, as opposed to the group fed $FeSO_4$. The iron content in the liver also was higher in the CPP groups when compared to that of a non–iron-deficient control group fed $FeSO_4$.

Heany et al,[108] in a randomized crossover design, studied the effect of a purified CPP preparation (87.5 mg) on the calcium absorption in normal postmenopausal women ($N = 35$). The calcium preparation was labeled with ^{45}Ca and given with a load of 250 mg of calcium during a standard breakfast. No effect was seen when the population was analyzed as a whole; however, when the analysis was limited to women with low calcium absorption values ($N = 17$), the supplementation with CPPs resulted in a significant increase of 5.3% for the mean absorption.

Hansen et al[109] investigated the absorption, in young men and women, of calcium and zinc from rice-based ($N = 11$) or whole-grain–based cereals ($N = 11$) with the labeled isotopes ^{47}Ca and ^{65}Zn. The actual quantity absorbed was calculated from whole-body retention after excretion of the nonabsorbed label via the feces had taken place. CPPs (1 g) increased the actual quantity of absorbed calcium or zinc by approximately 30% when the low-phytate rice-based meal was ingested; CPPs up to 2 g per serving could not increase the absorption of calcium or zinc when the whole-grain cereal, with approximately 10 times' higher phytate content, was tested.

The divalent mineral–binding effect of CPPs can be put to use in applications where one wants to increase the availability for absorption of these minerals from the gut. Drinks with calcium and iron are examples of commercial uses of CPPs; examples can be found especially in the Japanese market. Products for children that incorporate calcium or milk minerals and CPPs in sweets or cookies are found in the Southeast Asian market. As mineral accretion is high during early childhood, incorporation of CPPs provides good solubility and availability for absorption of calcium or zinc and thus is worth considering for infant nutrition. Other possible uses are in calcium-enriched dairy products and natural calcium supplements.

Finally, we would like to point to the potential use in functional food confectionery. It is now known that complexes of calcium, CPPs, and phosphate may reduce caries in a dose-dependent fashion by increasing the level of calcium phosphate in the plaque, thereby influencing the demineralization/

remineralization process[110] or can reduce the adherence of streptococci significantly.[111] Apart from the obvious dental applications, one can foresee potential protective effects of CPPs in carbohydrate-rich products, which tend to promote the growth of the dental microorganisms (eg, *Streptococcus mutans*), causing caries.

Glycomacropeptide

Glycomacropeptide (GMP) is the 64 amino acids long C-terminal residue of κ-casein (169 amino acids), released by rennet or chymosin during cheese making. Thus, GMP ends in the whey fraction. Industrial manufacturing seems easier, however, starting from casein as the raw material because the non-GMP material can be removed by acid (residual) casein precipitation.

Both the genetic variants A and B of GMP are partly glycosylated; the degree of glycosylation is on the order of 40% to 50%, and the pattern is complex, involving galactose, N-acetylgalactosaminitol, and N-acetylneuraminic acid (sialic acid) residues.[112–114]

GMP has been studied extensively for biological activity because of the following hypotheses:

- Due to its rapid formation during digestion, GMP may influence gastric secretions and play a role in the regulation of digestion (see, eg, Yvon et al[112])
- Since parallels have been noted between milk clotting and blood clotting, GMP-derived peptide sequences might provide new opportunities for development or design of bioactive peptides modulating platelet function or thrombosis (see, eg, Fiat et al[115]).
- GMP contains oligosaccharides that may act as growth stimulants for bifidogenic bacteria (see, eg, Idota et al[116]).

The envisaged function of GMP in the regulation of digestion and enhancement of postprandial satiety has, not surprisingly, gained a lot of attention given the increasing number of overweight subjects. Yvon et al[112] concluded from their own work as well as from other studies that GMP seems responsible for the in vivo observed effects on acid gastric secretions and gastric emptying rates. Using an isolated vascularly perfused rat duodenojejunum, Beucher et al[117] were able to show that a specific glycosylated fraction of GMP-A could increase the levels of cholecystokinin (CCK) release about 300% above basal levels. CCK is an important intestinal regulatory peptide known to slow gastric emptying. Available evidence suggests that GMP triggers stimuli from intestinal receptors without being absorbed. Working with this specific fraction of GMP, Corring et al[118] were able to show a similar CCK effect in human volunteers. Maximum increases of stimulation of CCK levels in blood were 350% for 50 g of casein 40 minutes after ingestion, 415% for 50g GMP-free whey at 20 minutes, and 268% for 25 mg of GMP fraction at 20 minutes. Further studies have to show whether GMP added to food products really influences satiety such that subjects will stop eating earlier, but the CCK experiments are encouraging.

Inhibitors of Angiotensin-I Converting Enzyme (ACEI)

Angiotensin-I converting enzyme (ACEI) is a key enzyme involved in the regulation of blood pressure. Due to its activity, two amino acids are removed from angiotensin-I, yielding the octapeptide angiotensin-II, which is a very potent vasoconstrictor. Inhibition of the synthesis of angiotensin-II thus lowers blood pressure. Three peptides from αs1-casein (amino acids 23–34 = C12; 23–27 = C5, and 194–199 = C6) and two peptides from β-casein (177–183 = C7 and 193–202 = C10) have been described. In spontaneously hypertensive rats, these peptides, derived from tryptic hydrolysates of casein, were effective when administered intraperitoneally and orally.[119] The latter suggests that these peptides can pass the intestinal tract and after absorption inhibit the production of angiotensin-II in the blood.

Meisel et al[120] have measured ACEI activities in various milk products, including pasteurized milk, yogurts, quarg, and fresh and ripened cheeses. Low activity was found in products with a low degree of proteolysis, like yogurt, quarg, and fresh cheeses. Ripened cheese contained more activity, but this was dependent on the degree of maturation; above a certain level in cheese maturation, ACEI activity decreased. Thus, dairy products may act as natural functional foods influencing blood pressure; in fact, this means that dairy products may be important vehicles for enrichment with natural ACEIs like those described above (C5–C12).

Other Bioactive Peptides

In bovine β-casein, relatively small peptide derivatives (casomorphins) have been shown to have opioid activity, acting on receptors in the gut and increasing the transit time and enhancing electrolyte absorption.[91,121] There also appear to be immunostimulatory and antithrombotic peptides in the casein molecules (see the review by Schlimme and Meisel[91] for further details).

LACTOSE CARBOHYDRATES

Lactose

The high prevalence of lactase nonpersistency (also called primary adult lactose intolerance) among non-Caucasian populations long overshadowed the beneficial effects of milk as a food or dietary supplement in these populations. Now it is widely recognized that even in the absence of intestinal β-galactosidase, moderate amounts of lactose (up to about 12 g/d, an equivalent of 2 cups of milk) can be tolerated without significant gastrointestinal side effects when the lactose is distributed over the day and taken with meals.[122–124]

The lactose content of bovine milk is about 4.8 g/100 mL. This glucose-galactose disaccharide is digested rather slowly by the mucosal β-galactosidase, and there is good evidence that a part of the lactose escapes digestion and absorption in the small intestine and serves as a substrate for the colonic microflora.[125] The high concentration of lactose in human milk (7 g/100 mL) would explain, at least in part, the stimulation of the development of a so-called *Bifidus* flora, which is considered to be associated with a decrease of the luminal pH and an increased colonization resistance against pathogenic organisms in the intestine of the child.

Studies in experimental animals have consistently shown that lactose stimulates the vitamin D–independent component of the intestinal calcium transport system[126,127] probably because fermentation of undigested lactose decreases the luminal pH and increases calcium solubility. Lactose also stimulates the absorption of magnesium.[127] Studies on the effect of lactose on calcium absorption in humans, however, have yielded conflicting results. In one study,[128] it was found that lactose may enhance calcium absorption in situations where calcium solubility or active vitamin D–dependent calcium absorption is limiting. Lee et al[129] reviewed the importance of non–vitamin D–mediated calcium absorption, which occurs mainly in the ileum. They stated that the bulk of calcium absorption is accomplished in the ileum, a segment with limited capacity for active calcium absorption, being relatively insensitive to the action of calcitriol, the active vitamin D metabolite. Bile salts and lactose were mentioned as examples of agents that can augment vitamin D–independent ileal absorption through the paracellular pathway.

Lactose has been demonstrated to have a lower glycemic index than sucrose or glucose and is therefore considered to be suitable in the diet of diabetics.[130] Lactose is also less cariogenic than other major food sugars, such as glucose, fructose, and maltose.[131]

Lactose-Derived Products

Lactose-derived products include lactulose, lactitol, lactobionic acid, and galacto-oligosaccharides. Lactulose is also present in heated milk in amounts varying between 4 and 200 mg/100 mL.[132] Both lactulose and lactitol are not digestible. They serve as a source of soluble fiber and are widely used in the treatment of constipation and chronic hepatic encephalopathy, acting in a similar way on the gut flora.[133,134] In a recent study lactulose (10 g/d) increased calcium absorption in postmenopausal women.[135] Lactobionic acid forms soluble complexes with minerals, such as calcium. It might therefore increase calcium absorption. Lactobionic acid is not digested in the small intestine but can be fermented by the intestinal flora. It may possess prebiotic properties, but these have not yet been evaluated. The nondigestible galacto-oligosaccharides have been claimed to have prebiotic properties because of their selective stimulation of bifidobacteria in the intestine.[136]

Like lactulose and other nondigestible oligosaccharides, galacto-oligosaccharides have been demonstrated to increase the intestinal absorption of calcium.[137,138]

Other Carbohydrates

In addition to lactose, bovine milk contains minor amounts of glucose and galactose (10–15 mg/100 mL for each monosaccharide), either free or bound to lipids, proteins, or phosphate. Bovine milk also contains at least 30 different oligosaccharides,[10] reaching a total concentration of 10 mg/100 mL. These oligosaccharides include N-acetylneuaraminic acid, neuramino lactose, N-acetylneuramin lactose, and N-acetyl hexoseamine–containing disaccharides. The oligosaccharides are not digested and may have bifidogenic properties, but these have not yet been evaluated in detail.

MILK FAT AND MILK FAT COMPONENTS

From a nutritional point of view, milk fat has a bad image because of its content of saturated fatty acids and cholesterol and its blood cholesterol–raising properties. Dietary guidelines in many Western countries aim at a reduction of the intake of saturated fatty acids to not more than 10% of total energy intake. It is now known that only C12, C14, and C16 raise the blood cholesterol level,[139] which means that 60% of the fatty acids in milk fat do not have this effect. Moreover, the effect of dietary cholesterol on blood cholesterol levels is much less important than that of the C12, C14, and C16 saturated fatty acids.[140] Nevertheless, the cholesterol issue has overshadowed the potential nutritional benefits of milk fat and milk fat components. Only for the last 5 to 10 years have scientists became interested in this area.

Digestion of Milk Fat

The average content of the main fatty acids in milk fat is shown in Table 8–7. This table indicates that milk fat contains approximately 10% (20 mol%) short-chain and medium-chain fatty acids (C4–C10). For several reasons, this is important for the digestion of milk fat. Firstly, preduodenal lipases, and particularly the gastric lipase, demonstrate a preference for these fatty acids. The dominating outer position of C4 to C10 fatty acids on the glycerol molecule in milk fat and the preference of the gastric lipase for these positions[141,142] contribute substantially to the predigestion of milk fat in

Table 8–7 Average Content of the Main Fatty Acids in Milk Fat (%)

Fatty Acid		%
Short-Chain Fatty Acids		
Butyric acid	C4	3.6
Caproic acid	C6	2.3
Medium-Chain Fatty Acids		
Caprylic acid	C8	1.3
Capric acid	C10	2.7
Lauric acid	C12	3.3
Myristic acid	C14	10.7
Myristoleic acid	C14:1	1.4
Long-Chain Fatty Acids		
Pentadecanoic acid	C15	1.2
Palmitic acid	C16	27.6
Palmitoleic acid	C16:1	2.6
Maragaric acid	C17	0.9
Stearic acid	C18	10.1
Oleic acid	C18:1	26.0
Linoleic acid	C18:2	2.5
Linolenic acid	C18:3	1.4

Source: Adapted with permission from E. Renner, *Milk and Dairy Products in Human Nutrition*, © 1983, W-GmbH, Volkswirtschaftlicher Verlag, München, Verlag Th. Mann KG.

the stomach. Second, the partly broken-down particles are subsequently more easily attacked in the duodenum and jejunum by pancreatic lipase.[142] Third, the C4 through C10—and in part also the C12—fatty acids do not have to be incorporated into micelles to be absorbed in the intestine, nor do they need incorporation into chylomicrons to be transported in the blood. Whereas butyric acid can be taken up directly by the gastric and small intestinal mucosa,[143] the medium-chain fatty acids (C6–C12) are transported to the liver via the portal blood flow and can serve as a rapid source of energy upon oxidation. These issues explain the high digestibility of milk fat, which may be particularly useful under conditions of fat digestion disturbances. Since milk fat is the vehicle for fat soluble vitamins A, D, E, and K and also for carotenoids in milk, its digestibility will affect the bioavailability of these nutrients in a beneficial way.

Butyric Acid

It is known that butyric acid, formed during fermentation of carbohydrates in the colon, can be used as a substrate by colonic cells and has a trophic effect on these cells.[144] It is also known that butyrate displays anti–colon cancer properties, probably by enhancing apoptosis of mutant colonic cells.[144] Animal experiments have shown that butyrate also has a trophic effect on the small-intestinal mucosa.[143] So it may be expected that butyrate released from milk fat during digestion will have a trophic effect on both the gastric and small-intestinal mucosal cells.

Anti-Infective Properties of Milk Fat Digestion Products

The liberation of short- and medium-chain free fatty acids in the stomach after milk fat ingestion will contribute to a lowering of the pH, and this may facilitate digestion of proteins and enhance the acid barrier against pathogenic microorganisms. Moreover, it is known that lipid digestion products may display antibacterial and antiviral activity,[142,145] and this especially refers to fatty acids and monoglycerides with chain lengths varying from 8 to 12 carbon atoms. Thus, the specific composition of milk fat and its preduodenal digestion can be associated with anti-infective properties. This interesting area has not yet been explored in much detail.

Minor Components in Milk Fat

In addition to the fat-soluble vitamins and carotenoids, milk fat contains a range of components that may display physiological activities. These components are the conjugated linoleic acids (CLAs), phospholipids, and cerebrosides.

CLA is a mixture of positional and geometrical isomers of linoleic acid. CLA is formed by rumen microorganisms. Milk fat contains 5 to 7 mg of CLA per gram; in much lower amounts, CLA is present in beef, seafood, and processed foods, whereas vegetable oils contain less than 1 mg/g of oil.[146] CLA has been shown to exert a variety of anticancer effects, including the inhibition of the growth of various forms of chemically induced tumors in experimental animals. In vitro experiments have demonstrated cytostatic and cytotoxic effects to human malignant melanoma, colorectal, and breast cancer cells. Tumor growth inhibition is thought to be related to the ability of CLA to inhibit protein and nucleotide biosynthesis. It has been shown that the 9c-, 11t isomer is the only CLA isomer that is incorporated into the phospholipids of mouse forestomach and rat mammary tumors and liver, so this isomer is probably the only one that is biologically active.[147] The potential significance of CLA in human diets as a protective factor against cancer looks promising and justifies further research.

Milk contains a variety of phospholipids (20–50 mg/L), largely present in the milk fat globule and the milk fat globule membrane.[10] The distribution of the main phospholipid fractions in milk is as follows: phosphatidyl ethanolamine, 35%; phosphatidyl choline, 30%, sphingomyelin, 24%; phosphatidyl inositol, 5%; phosphatidyl serine, 2%. Phospholipids are important as membrane constituents and are involved in cell interaction with ions, hormones, and antibodies. The fatty acid composition of milk phospholipids differs from that of the milk fat: their content of unsaturated fatty acids is relatively higher, and short-chain fatty acids are absent. Because of their emulsification properties, it has been suggested that dietary phospholipids may improve lipid absorption in the intestine.[10] The emulsification properties could be used to enhance the bioavailability of fat-soluble bioactive compounds, like fat soluble vitamins and carotenoids, but such effects have not yet been evaluated. Phospholipids are not essential nutrients. Although a large body of literature exists about beneficial effects of phospholipids on the blood lipid profile, a critical review of the literature[148] arrived at the conclusion that there is no conclusive evidence that these effects are specific for the phospholipids, since they can be explained simply by their content of polyunsaturated fatty acids. Milk phospholipids were reported to exert protective effects on the gastric mucosa of humans given aspirin.[149] It has also been suggested that milk phospholipids provide additional protection against pathogenic microorganisms.[150]

Cerobrosides, glycolipids, and glycosphingolipids can be considered as one group of minor lipids in milk.[10] Their physiological significance in human diets (if any) has not yet been explored in sufficient detail to indicate these compounds as functional food ingredients.

MILK MINERALS

Dietary calcium plays an important role in health. In Western diets, dairy products provide about 70% of the recommended daily intake for calcium. It is beyond the scope of this chapter to review the extensive literature on the health effects of dietary calcium. The reader is referred to a recent critical review on dietary calcium and health presented by the International Dairy Federation.[151] This review elaborates on the positive health contribution of dietary calcium, dairy as well as nondairy based, to prevention of osteoporosis, colon cancer, and hypertension. Also, recently published rat studies show the involvement of dietary calcium in resistance against pathogen infections.[152,153]

Recommendations for an increased daily dietary intake of calcium have been made in a number of countries. In the United States, the Food and Drug Administration has advised men and women over 50 years to increase their calcium intakes toward 1200 mg/d. The dairy industry has therefore developed milk mineral preparations enriched in calcium, up to about 25% on a product basis. These are used in a variety of consumer products, like calcium-enriched milk powder supplements, cakes, biscuits, bars, and beverages. Recently, Bonjour et al[154] showed, in a randomized, double-blind, placebo-controlled trial with 8-year-old girls, that food products enriched with a milk calcium extract could increase the rate of bone mass accumulation in these prepubertal girls 1.5% per year.

FUTURE DEVELOPMENTS

We expect, in the next decade, further progress in the isolation, efficacy research, and application of bioactive components from milk and whey. In addition to new research and applications of already commercially available milk components within and outside the food sector, newly isolated milk components or derivatives such as bioactive peptides, immunoglobulins, lactose-derived products, phospholipid fractions, and growth factors will be tested for their efficacy and most likely will be introduced into functional foods, medical foods, and dietary supplements. It is expected that most efficacy trials will be focused on the improvement of intestinal health and enhancement of gastrointestinal functions, including the improvement of the absorption and barrier function of the intestine. We believe that in addition to experiments in vitro and in animal models, more attention will be given to direct studies in humans, either normal healthy subjects or patients, using new techniques.

REFERENCES

1. Eigel WN, Butler JE, Ernstrom CA et al. Nomenclature of proteins of cow's milk: fifth revision. *J Dairy Sci.* 1984;67: 1599–1631.

2. Swaisgood HA. Chemistry of milk protein. In: Fox PF, ed. *Developments in Dairy Chemistry. Vol. 1: Proteins.* London, England: Elsevier Applied Science Publishers: 1982:1–59.

3. Mackenzie DDS. Milk composition as an indicator of mammary gland metabolism. *Proc Nutr Soc NZ.* 1997;22: 126–136.

4. Barth CA, Behnke U. Nutritional significance of whey and whey components. *Nahrung.* 1997;41:2–12.

5. Walstra P, Jennes R. *Dairy Chemistry and Physics.* New York, NY: Wiley Interscience: 1984.

6. Steijns JM, Van Hooydonk ACM. Occurrence, structure, biochemical properties and technological characteristics of lactoferrin. *Br J Nutr.* In press.

7. de Wit JN, Van Hooydonk ACM. Structure, functions and applications of lactoperoxidase in natural antimicrobial systems. *Neth Milk Dairy J.* 1996;50:227–244.

8. Swaisgood HA. Chemistry of the caseins. In: Fox PF, ed. *Advanced Dairy Chemistry. Vol. 1: Proteins.* London, England: Elsevier Applied Science Publishers; 1993:63–110.

9. Renner E. *Milk and Dairy Products in Human Nutrition.* Munich, Germany: W-GmbH, Voolkswirtschaftlicher Verlag, 1983.

10. Renner E, Schaafsma G, Scott KJ. Micronutrients in milk. In: Renner E, ed. *Micronutrients in Milk and Milk Based Food Products.* New York, NY: Elsevier Applied Science: 1989;1–60.

11. Jensen RG, ed. *Handbook of Milk Composition.* New York, NY: Academic Press; 1995.

12. de Wit J. Nutritional and functional characteristics of whey proteins in food products. *J Dairy Sci.* 1998;81:597–608.

13. Muller LL. Manufacture of casein, caseinate and co-precipitates. In: Fox PF, ed. *Developments in Dairy Chemistry Vol. 1: Proteins*. London, England: Elsevier Applied Science Publishers; 1982;315–337.

14. Mulvihill DM. Caseins and caseinates. In: Fox PF, ed. *Developments in Dairy Chemistry*. Vol. 4. London, England: Elsevier Applied Science Publishers; 1989:97–130.

15. Marshall KR. Industrial isolation of milk proteins: whey proteins. In: Fox PF, ed. *Developments in Dairy Chemistry. Vol. 1: Proteins*. London, England: Elsevier Applied Science Publishers; 1982.

16. Morr CV. Whey proteins: manufacture. In: Fox PF, ed. *Developments in Dairy Chemistry*. London, England: Elsevier Applied Science Publishers; 1989:245–284.

17. Fox P, Mulvihill DM. Developments in milk protein processing. *Food Sci Technol Today* 1993;7:152–161.

18. Etzel MR. Whey protein isolation and fractionation using ion exchangers In: Singh RK, Rizvi SH, eds. *Bioseparation Processes in Foods*. New York, NY: Marcel Dekker; 1995;389–415.

19. Hahn R, Schulz PM, Schaupp C, Jungbauer A. Bovine whey fractionation based on cation-exchange chromatography. *J Chromatogr A*. 1998;795:277–287.

20. Horton BS. Commercial utilization of minor milk components in the health and food industries. *J Dairy Sci.* 1995;78:2584–2589.

21. Perraudin JP. Protéines à activités biologiques: lactoferrine et lactoperoxydase: connaissances récemment acquises et technologies d'obtention. *Lait.* 1991;71:191–211.

22. Fukumoto LR, Li-Chan E, Kwan L, Nakai S. Isolation of immunoglobulins from cheese whey using ultrafiltration and immobilized metal affinity chromatography. *Food Res Int*. 1994;27:335–348.

23. Reddy NR, Roth SM, Eigel WN, Pierson MD. Foods and food ingredients for prevention of diarrheal disease in children in developing countries. *J Food Protect*. 1988;51(1):66–75.

24. Lahl WJ, Braun SD. Enzymatic production of protein hydrolysates for food use. *Food Technol*. October 1994;68–71.

25. Schmidl MK, Labuza TP. Nitrogen sources from dairy proteins for enteral feeding. *Dietitians in Nutrition Support.* 1991;13(4):11–16.

26. Food and Agricultural Organization. *FAO/WHO/Expert Consultation on Protein Quality Evaluation*. FAO Food and Nutrition Papers 1990; 51.

27. Sarwar G. Available amino acid score for evaluating protein quality of foods. *J Assoc Off Anal Chem.* 1984;67:623–626.

28. European Dairy Association. *Nutritional Quality of Proteins*. The Netherlands: Maarssen, Van der Hulst; 1997.

29. Holt C. The biological role of casein? *Hannah Res Inst Res Rev.* 1994;60–68.

30. McIntosh GH, Regester GO, Le Leu RK, Royle PJ, Smithers GW. Dairy proteins protect against dimethylhydrazine-induced intestinal cancers in rats. *J Nutr.* 1995;125:809–816.

31. van der Meer R, Govers MJAP, Lapré JA, Kleibeuker JH. Dietary calcium as a possible anti-promotor of colorectal carcinogenesis. In: Serrano R, et al, eds. *Dairy Products in Human Health and Nutrition*. Rotterdam, The Netherlands: Balkema; 1994:439–448.

32. Sen CK. Nutritional biochemistry of cellular glutathione. *J Nutr Biochem.* 1997;8:660–672.

33. Bounous G, Amer MA. The immunoenhancing effect of dietary whey protein concentrate. *Bull Int Dairy Fed.* 1990;253: 44–53.

34. White AC, Thannickal VJ, Fanburg BL. Glutathione deficiency in human disease. *J Nutr Biochem.* 1994;5:218–226.

35. Wu D, Meydani SN, Sastre J, Hayek M, Meydani M. In vitro glutathione supplementation enhances interleukin-2 production and mitogenic response of peripheral blood mononuclear cells from young and old subjects. *J Nutr.* 1994;124:655–663.

36. Baruchel S, Viau G. In vitro selective modulation of cellular glutathione by a humanized native milk protein isolate in normal cells and rat mammary carcinoma model. *Anticancer Res.* 1996;16:1095–1000.

37. Herzenberg LA, De Rosa SC, Dubs JG, et al. Glutathione deficiency is associated with impaired survival in HIV disease. *Proc Natl Acad Sci USA*. 1997;94:1967–1972.

38. Dröge W, Holm E. Role of cysteine and glutathione in HIV infection and other diseases associated with muscle wasting and immunological dysfunction. *FASEB J.* 1997;11:1077–1089.

39. Friedman M. Improvement in the safety of foods by SH-containing amino acids and peptides: a review. *J Agric Food Chem.* 1994;42:3–20.

40. Lönnerdal B, Iyer S. Lactoferrin: molecular structure and biological function. *Ann Rev Nutr.* 1995;15:93–111.

41. Levay PF, Viljoen M. Lactoferrin: a general review. *Haematologica.* 1995;80:252–267.

42. Boxer LA, Coates TD, Haak RA, Wolach JB, Hoffstein S, Baehner RL. Lactoferrin deficiency associated with altered granulocyte function. *New Engl J Med.* 1982;307:404–410.

43. Breton-Gorius J, Mason DY, Buriot D, Vilde J-L, Griselli C. Lactoferrin deficiency as a consequence of a lack of specific granules in neutrophils from a patient with recurrent infections. *Am J Pathol.* 1980;99:413–419.

44. Brock J. Lactoferrin: a multifunctional immunoregulatory protein? *Immunol Today.* 1995;16:417–419.

45. Naidu AS, Arnold RR. Influence of lactoferrin on host-microbe interactions. In: Hutchens TW, Lönnerdal B, eds. *Lactoferrin: Interactions and Biological Functions.* Totowa, NJ: Humana Press; 1997;259–275.

46. Naidu A, Arnold RR. Lactoferrin interaction with *Salmonella* potentiates antibiotic susceptibility in vitro. *Diagn Microbiol Infect Dis.* 1994;20:69–75.

47. Wakabayashi H, Abe S, Okutumi T, Tansho S, Kawase K, Yamaguchi H. Cooperative anti-candida effects of lactoferrin or its peptides in combinations with azole antifungal agents. *Microbiol Immunol.* 1996;40:821–825.

48. Samaranayake YH, Samaranayake LP, Wu PC, So M. The antifungal effect of lactoferrin and lysozyme on *Candida krusie* and *Candida albicans. APMIS.* 1997;105:875–883.

49. Teraguchi S, Ozawa K, Yasuda S, Shin K, Fukuwatari Y, Shimamura S. Bacteriostatic effect of orally administered bovine lactoferrin on intestinal enterobacteriaceae of SPF mice fed bovine milk. *Biosci Biotechnol Biochem.* 1994;58:482–487.

50. Teraguchi S, Shin K, Ozawa K, et al. Bacteriostatic effect of orally administered bovine lactoferrin on proliferation of *Clostridium* species in the gut of mice fed bovine milk. *Appl Environ Microbiol.* 1995;61:501–506.

51. Trüempler U, Straub PW, Rosenmund A. Antibacterial prophylaxis with lactoferrin in neutropenic patients. *Eur J Clin Microbiol Infect Dis.* 1989;8:310–311.

52. Sakai M, Otubo T, Atsuta S, Kobayashi M. Enhancement of resistance to bacterial infection in rainbow trout, Oncorhynchus mykiss (Walbaum), by oral administration of bovine lactoferrin. *J Fish Dis.* 1993;16:239–247.

53. Harmsen MC, Swart PJ, de Béthune M-P, et al. Antiviral effects of plasma and milk proteins: lactoferrin shows potent activity against both human immunodeficiency virus and human cytomegalovirus replication in vitro. *J Infec Dis.* 1995;172:380–388.

54. Marchetti M, Pisani S, Antonini G, Valenti P, Seganti L, Orsi N. Metal complexes of bovine lactoferrin inhibit in vitro replication of Herpes simplex virus type 1 and 2. *BioMetals.* 1998;11:89–94.

55. Yi M, Kaneko S, Yu DY, Murakami S. Hepatitis C virus envelope proteins bind lactoferrin. *J Virol.* 1997;71:5997–6002.

56. Kawasaki Y, Isoda H, Shinmoto H, et al. Inhibition of Kappa casein glycomacropeptide and lactoferrin of influenza virus hemagglutination. *Biosci Biotechnol Biochem.* 1993;57:1214–1215.

57. Superti F, Ammendolia MG, Valenti P, Seganti L. Antirotaviral activity of milk proteins: lactoferrin prevents rotavirus infection in the enterocyte-like cell line HT-29. *Med Microbiol Immunol.* 1997;186:83–91.

58. Shimizu K, Matsuzawa H, Okada K, et al. Lactoferrin-mediated protection of the host from murine cytomegalovirus infection by a T-cell dependent augmentation of natural killer cell activity. *Arch Virol.* 1996;141:1875–1889.

59. Britigan BE, Serody JS, Cohen MS. The role of lactoferrin as an anti-inflammatory molecule. *Adv Exp Med Biol.* 1994;357:143–156.

60. Cohen MS, Mao J, Rasmussen GT, Serody JS, Britigan BE. Interaction of lactoferrin and lipopolysaccharide (LPS): effects on the antioxidant property of lactoferrin and the ability of LPS to prime human neutrophils for enhanced superoxide formation. *J Infect Dis.* 1992;166:1375–1378.

61. Mattsby-Baltzer I, Roseanu A, Motas C, Elverfors J, Engberg I, Hanson LA. Lactoferrin or a fragment thereof inhibits the endotoxin-induced interleukin-6 response in human monocytic cells. *Pediatr Res.* 1996;40:257–262.

62. Wang D, Pabst KM, Aida Y, Pabst MJ. Lipopolysaccharide-inactivating activity of neutrophils is due to lactoferrin. *J Leukoc Biol.* 1995;57:865–874.

63. Zimecki M, Mazurier J, Machniki M, Wieczorek Z, Montreuil J, Spik G. Immunostimulatory activity of lactoferrin and maturation of CD4-, CD8-murine thymocytes. *Immunol. Lett.* 1991;30:119–123.

64. Zagulski T, Lipinski P, Zagulska A, Broniek S, Jarzabek Z. Lactoferrin can protect mice against a lethal dose of E. coli in experimental infection in vivo. *Br J Exp Path.* 1989;70:697–704.

65. Lee WL, Farmer JF, Hilty M, Kim YB. The protective effects of lactoferrin feeding against endotoxin lethal shock in germfree piglets. *Infect Immun.* 1998;66:1421–1426.

66. Sato R, Inanami O, Tanak Y, Takase M, Naito Y. Oral administration of bovine lactoferrin for treatment of intractable stomatitis in feline immunodeficiency virus (FIV)-positive and FIV-negative cats. *Am J Vet Res.* 1996;57:1443–1446.

67. Sakai M, Kobayashi M, Yoshida T. Activation of rainbow trout, Oncorhynchus mykiss, phagocytic cells by administration of bovine lactoferrin. *Comp Biochem Physiol.* 1995;110B:755–759.

68. Debbabi H, Dubarry M, Rautureau M, Tomé D. Bovine lactoferrin induces both mucosal and systemic immune response in mice. *J Dairy Res.* 1998;65:283–293.

69. Kulkarni PR, Pimpale NV. Colostrum: a review. *Indian J Dairy Sci.* 1989;42:216–224.

70. Significance of the indigenous antimicrobial agents of milk to the dairy industry. *Bull Int Dairy Fed.* 1991;264:2–5.

71. Bogstedt AK, Johansen K, Hatta KM, Casswall M, Svensson L, Hammerström L. Passive immunity against diarrhea. *Acta Pediatr.* 1996;85:125–128.

72. Davidson G. Passive protection against diarrheal disease. *J Pediat Gastroenterol Nutr.* 1996;23:207–212.

73. Roos N, Mahé S, Benamouzig R, Sick H, Rautureau J, Tomé D. ^{15}N-labeled immunoglobulins from bovine colostrum are partially resistant to digestion in human intestine. *J Nutr.* 1995;125:1238–1244.

74. Kelly CP, Chetman S, Keates S, et al. Survival of anti-*Clostridium difficile* bovine immunoglobulin concentrate in the human gastrointestinal tract. *Antimicrob Agents Chemother.* 1997;41:236–241.

75. Heine W, Radke M, Wutzke K-D, Peters E, Kundt G. α-Lactalbumin-enriched low-protein infant formulas: a comparison to breast milk feeding. *Acta Paediatr.* 1996;85:1024–1028.

76. Heine W, Radke M, Wutzke K-D. The significance of tryptophan in human nutrition. *Amino Acids.* 1995;9:191–205.

77. Petschow BW, Talbott RD. Response of Bifidobacterium species to growth promoters in human and cow milk. *Pediatr Res.* 1991;29:208–213.

78. Poch M, Bezkorovainy A. Bovine milk α-casein trypsin digest is a growth enhancer for the genus *Bifidobacterium*. *J Agric Food Chem.* 1991;39:73–77.

79. Wharton BA, Balmer SE, Scott PH. Faecal flora in the newborn. *Adv Exp Med Biol.* 1994;357:91–98.

80. Roberts AK, Chierici R, Sawatzki G, Hill MJ, Volpato S, Vigi V. Supplementation of an adapted formula with bovine lactoferrin: 1. Effect on the infant faecal flora. *Acta Paediatr.* 1992;81:119–124.

81. Grosvenor CE, Picciano MF, Baumrucker CR. Hormones and growth factors in milk. *Endocr Rev.* 1992;14:710–728.

82. Koldovsky O, Kong W, Rao RK, Schaudies P. Milk-borne peptide growth factors in human and bovine milk. In: Walker WA, Harmatz PR, Wershil BK. *Immunophysiology of the Gut.* New York, NY: Academic Press; 1993;269–293.

83. Donovan SM, Odle J. Growth factors in milk as mediators of infant development. *Ann Rev Nutr.* 1994;14:147–167.

84. Forsyth IA. The insulin-like growth factor and epidermal growth factor families in mammary cell growth in ruminants: action and interaction with hormones. *J Dairy Sci.* 1996;79:1085–1096.

85. Guimont C, Marchall E, Girardet JM, Linden G. Biologically active factors in bovine milk and dairy byproducts: influence on cell culture. *Crit Rev Food Sci Nutr.* 1997;37:393–410.

86. Pakkanen R. Determination of transforming growth factor-δ2 (TGF-δ2) in bovine colostrum samples. *J Immunoassay.* 1998;19:23–37.

87. Howarth GS, Francis GF, Cool JC, Xu X, Byard RW, Read LC. Milk growth factors enriched from cheese whey ameliorate intestinal damage by methotrexate when administered orally to rats. *J Nutr.* 1996;126:2519–2530.

88. Parodi PW. Cow's milk folate binding protein: its role in folate nutrition. *Aust J Dairy Technol.* 1997;52:109–118.

89. Mather IH, Jack LJW. A review of the molecular and cellular biology of butyrophilin, the major protein of bovine milk fat globule membrane. *J Dairy Sci.* 1993;76:3832–3850.

90. Fiat A-M, Jollès P. Caseins of various origins and biologically active casein peptides and oligosaccharides: structural and physiological aspects. *Mol Cell Biochem.* 1989;87:5–30.

91. Schlimme E, Meisel H. Bioactive peptides derived from milk proteins: structural, physiological and analytical aspects. *Nahrung.* 1995;39:1–20.

92. Holt C, Horne DS. The hairy casein micelle: evolution of the concept and its implications for dairy technology. *Neth Milk Dairy J.* 1996;50:85–112.

93. Adamson NJ, Reynolds EC. Characterization of tryptic casein phosphopeptides prepared under industrially relevant conditions. *Biotechnol Bioeng.* 1995;45:196–204.

94. Adamson NJ, Reynolds EC. Characterization of casein phosphopeptides prepared using alcalase: determination of enzyme specificity. *Enzyme Microb Technol.* 1996;19:202–207.

95. Naito H, Suzuki H. Further evidence for the formation in vivo of phosphopeptide in the intestinal lumen from dietary β-casein. *Agr Biol Chem.* 1974;38:1543–1545.

96. Lee YS, Noguchi T, Naito H. Phosphopeptides and soluble calcium in the small intestine of rats given a casein diet. *Br J Nutr.* 1980;43:457–467.

97. Lee YS, Noguchi T, Naito H. Intestinal absorption of calcium in rats given diets containing casein or an amino acid mixture: the role of casein phosphopeptides. *Br J Nutr.* 1983;49:67–76.

98. Sato R, Noguchi T, Naito H. Casein phosphopeptide (CPP) enhances calcium absorption from the ligated segment of rat small intestine. *J Nutr Sci Vitaminol.* 1986;32:67–76.

99. Sato R, Shindo M, Noguchi T, Naito H. Characterization of phosphopeptide derived from bovine β-casein: an inhibitor to intra-intestinal precipitation of calcium phosphate. *Biochem Biophys Acta.* 1991;1077:413–415.

100. Hirayama M, Toyota K, Hidaka H, Naito H. Phosphopeptides in rat intestinal digests after ingesting casein phosphopeptides. *Biosci Biotechnol Biochem.* 1992;56:1128–1129.

101. Meisel H, Frister H. Chemical characterization of bio-active peptides from in vivo digests of casein. *J Dairy Res.* 1989;56:343–349.

102. Brommage R, Juillerat MA, Jost R. Influence of casein phosphopeptides and lactulose on intestinal calcium absorption in adult female rats. *Lait.* 1991;71:173–180.

103. Pointillart A, Gueguen L. Effect of casein phosphopeptides on calcium and phosphorous utilization in growing pigs. *Reprod Nutr Dev.* 1989;29:477–486.

104. Scholz-Ahrens KE, Kopra N, Barth CA. Effect of casein phosphopeptides on utilization of calcium in minipigs and vitamin-D deficient rats. *Z Ernährungswiss.* 1990;29:295–298.

105. Tschita H, Goto T, Shimizu T, Yonehara Y, Kuwata T. Dietary casein phosphopeptides prevent bone loss in aged ovariectomized rats. *J Nutr.* 1996;126:86–93.

106. Hansen M, Sandström B, Lönnerdal B. The effect of casein phosphopeptides on zinc and calcium absorption from high phytate infant diets assessed in rat pups and Caco-2 cells. *Pediatr Res.* 1996;40:547–552.

107. Ait-oukhatar N, Bouhalab S, Bureau F et al. Bioavailability of casein phosphopeptide bound iron in the young rat. *J Nutr Biochem.* 1997;8:190–194.

108. Heany RP, Saito Y, Orimo H. Effect of casein phosphopeptide on absorbability of co-ingested calcium in normal postmenopausal women. *J Bone Miner Metab.* 1994;12:77–81.

109. Hansen M, Sandström B, Jensen M, Sörensen SS. Casein phosphopeptides improve zinc and calcium absorption from rice-based but not from whole-grain infant cereal. *J Pediatr Gastroenterol Nutr.* 1997;24:56–62.

110. Reynolds EC, Cain CJ, Webber FL, et al. Anticariogenicity of calcium phosphate complexes of tryptic casein phosphopeptides in the rat. *J Dent Res.* 1995;74:1271–1279.

111. Schuepbach P, Neeser JR, Golliard M, Rouvet M, Guggenheim B. Incorporation of casein glycomacropeptide and casein phosphopeptide into the salivary pellicle inhibits adherence of mutans Streptococci. *J Dent Res.* 1996;75:1779–1788.

112. Yvon M, Beucher S, Guilloteau P, Le Huerou-Luron I, Corring T. Effects of casein macropeptide (CMP) on digestion regulation. *Reprod Nutr Dev.* 1994;34:527–537.

113. Coolbear KP, Elgar DF, Ayers JS. Profiling of genetic variants of bovine P-casein macropeptide by electrophoretic and chromatographic techniques. *Int Dairy J.* 1996;6:1055–1068.

114. Minkiewicz P, Slangen CJ, Lagerwerf FM, Haverkamp J, Rollema HS, Visser S. Reversed-phase high-performance liquid chromatographic separation of bovine Kappa casein macropeptide and characterization of isolated fractions. *J Chromatogr A.* 1996;743:123–135.

115. Fiat A-M, Migliore-Samour D, Jollès P, Drouet L, Bal Dit Sollier C, Caen J. Biologically active peptides from milk proteins with emphasis on two examples concerning antithrombotic and immunomedulating activities. *J Dairy Sci.* 1993;76:301–310.

116. Idota T, Kawakami H, Nakajima I. Growth-promoting effects of N-acetylneuraminic acid containing substances on bifidobacteria. *Biosci Biotechnol Biochem.* 1994;58:1720–1722.

117. Beucher S, Levenez F, Yvon M, Corring T. Effects of gastric digestive products from casein on CCK release by intestinal cells in rat. *J Nutr Biochem.* 1994;5:578–584.

118. Corring T, Beaufrere B, Maubois J-L. Release of cholecystokinine in humans after ingestion of glycomacropeptide. Presented at the *International Whey Conference;* October 27–29, 1997; Rosemont, IL.

119. Karaki H, Doi K, Sugano S, et al. Antihypertensive effect of tryptic hydrolysate of milk casein in spontaneously hypertensive rats. *Comp Biochem Physiol.* 1990;96C:367–371.

120. Meisel H, Goepfert A, Günther S. ACE-inhibitory activities in milk products. *Milchwissenschaft.* 1997;52:307–311.

121. Nyberg F, Brant V, eds. β-Casomorphins and related peptides. In: *Proceedings from the First International Symposium on β-Casomorphins and Related Peptides.* Uppsala, Sweden: Fyris-Tryck AB; 1990.

122. Suarez FL, Savaiano DA, Levitt MD. A comparison of symptoms after consumption of milk of lactose hydrolyzed milk by people with self reported severe lactose intolerance. *N Engl J Med.* 1995;333:1–4.

123. Suarez FL, Savaiano D, Arbisi P, Levitt MD. Tolerance to the daily ingestion of two cups of milk by individuals claiming lactose intolerance. *Am J Clin Nutr.* 1997;65:1502–1506.

124. Solomons NW, Guerrero A-U, Torun B. Dietary manipulation of postprandial colonic lactose fermentation: I. Effect of solid foods in a meal. *Am J Clin Nutr.* 1985;41:199–208.

125. Schulze J, Zunft HJ. Lactose, a potential dietary fiber: on the regulation of its microecological efficiency in the intestinal tract. Part 1. Problems, state of knowledge, methodology [in German]. *Nahrung.* 1991;45:849–866.

126. Schaafsma G, Visser WJ, Dekker PR, van Schaik M. Effect of dietary calcium supplementation with lactose on bone in vitamin D-deficient rats. *Bone.* 1988;8:357–362.

127. Schaafsma G. Bioavailability of calcium and magnesium. *Eur J Clin Nutr.* 1997;51(suppl 1):S13–S16.

128. Schuette SA, Yasillo NJ, Thompson CM. The effect of carbohydrates in milk on the absorption of calcium by post-menopausal women. *J Am Coll Nutr.* 1991;10:132–139.

129. Lee DBN, Hu M-S, Kayne LH, Nakhoul F, Jamgotchian N. The importance of non-vitamin D-mediated calcium absorption. *Contrib Nephrol.* 1991;91:14–20.

130. Wolever TMS, Wong GS, Kenshole A, et al. Lactose in the diabetic diet: a comparison with other carbohydrates. *Nutr Res.* 1985;5:1335–1334.

131. Edgar WM. Extrinsic and intrinsic sugars: a review of recent UK recommendations on diet and caries. *Caries Res.* 1993;27(Suppl 1):64–67.

132. Andrews G. Lactulose in heated milk. *Bull Int Dairy Fed.* 1989;238:46–52.

133. Camma C, Fiorello F, Tinè F, Marchesini G, Fabbri A, Pagliaro L. Lactitol in the treatment of chronic hepatic encephalopathy: a meta-analysis. *Dig Dis Sci.* 1993;38:916–922.

134. Blanc P, Daures J-P, Rouillon J-M, et al. Lactitol of lactulose in the treatment of chronic hepatic encephalopathy: results of a meta analysis. *Hepatology.* 1992;15:222–228.

135. Van den Heuvel, EGHM, Muijs TH, van Dokkum W, Schaafsma G. Lactulose stimulates calcium absorption in post-menopausal women. *J Bone Miner Res.* 1999;14:1211–1216.

136. Ito M, Deguchi Y, Miyamori A, et al. Effects of administration of galacto-oligosaccharides on the human faecal microflora, stool weight and abdominal sensation. *Microb Ecol Health Dis.* 1990;3:285–292.

137. Van den Heuvel, EGHM, Muijs TH, van Dokkum W, Schaafsma G. Fructooligosaccharides stimulate calcium absorption in adolescents. *Am J Clin Nutr.* 1999;69:544–548.

138. Chonan O, Watanuki M. Effect of galactooligosaccharides on calcium absorption in rats. *J Nutr Sci Vitaminol.* 1995;41:95–104.

139. Schaafsma G. *The Western Diet with a Special Focus on Dairy Products.* Brussels, Belgium: Institute Danone; 1997:69–79. Danone Chair Monograph 4.

140. Hopkins PN. Effects of dietary cholesterol on serum cholesterol: a meta-analysis and review. *Am J Clin Nutr.* 1992;55:1060–1070.

141. Hamosh M. Oral lipases and lipid digestion during the neonatal period. In: Lebenthal E, ed. *Textbook of Gastroenterology and Nutrition in Infancy.* New York, NY: Raven Press; 1998:445–463.

142. Hamosh M, Bitman J, Wood DL, Hamosh P, Metha NR. Lipids in milk and the first steps in their digestion. *Pediatrics.* 1984;75(Suppl):145–150.

143. Koruda MJ, Rolandelli RH, Zimmaro D, Hastings J, Rombeau JL, Settle RG. Parenteral nutrition supplemented with short chain fatty acids: effect on the small-bowel mucosa in normal rats. *Am J Clin Nutr.* 1990;51:685–689.

144. Mortensen PB, Clausen MR. Short chain fatty acids in human colon: relation to gastrointestinal health and diseases. *Scand J Gastroenterol.* 1996;31(suppl 216):132–148.

145. Isaacs CE, Litov RE, Thormar H. Antimicrobial activity of lipids added to human milk, infant formula, and bovine milk. *J Nutr Biochem.* 1995;6:362–366.

146. Herbei BK, McGuire MK, McGuire MA, Shultz TD. Safflower oil consumption does not increase plasma conjugated linoleic acid concentrations in humans. *Am J Clin Nutr.* 1998;67:332–337.

147. Ip C, Chin SF, Scimeca JA, Pariza MW. Mammary cancer prevention by conjugated dienoic derivative of linoleic acid. *Cancer Res.* 1991;51:6118–6124.

148. Knuiman JT, Beynen AC, Katan MB. Lecithin intake and serum cholesterol. *Am J Clin Nutr.* 1989;49:266–268.

149. Kivinen A, Tarpila S, Salminen S, Vapaatalo H. Gastroprotection with milk phospholipids: a first human study. *Milchwissenschaft.* 1992;47:694–696.

150. Sprong C, Hulstein M, van der Meer R. Phospholipid-rich butter milk decreases the gastro-intestinal survival and translocation of *Listeria* in rats. *Gastroenterology.* 1998;114:A10090.

151. Dietary calcium and health. *Bull Int Dairy Fed.* 1997;322:1–36.

152. Bovee-Oudenhoven I, Termont D, Dekker R, van der Meer R. Calcium in milk and fermentation by yogurt bacteria increase the resistance of rats to *Salmonella* infection. *Gut.* 1996;38:59–65.

153. Bovee-Oudenhoven I, Termont D, Weerkamp AH, Faassen-Peeters MAW, van der Meer R. Dietary calcium inhibits the intestinal colonization and translocation of *Salmonella* in rats. *Gastroenterology.* 1997;113:550–557.

154. Bonjour J-P, Carrie A-L, Ferrari S, et al. Calcium-enriched foods and bone mass growth in prepubertal girls: a randomized, double-blind, placebo-controlled trial. *J Clin Invest.* 1997;99:1287–1294.

CHAPTER 9

Pre- and Probiotics

Dominique Brassart and Eduardo J. Schiffrin

Although functional foods are becoming more and more used all over the world, they are still a category of nutrients for which a consensus definition does not exist. We will consider as functional foods those foods that are designed not only to cover the basic needs in energy, macronutrients (proteins, carbohydrates, lipids), and micronutrients (vitamins, minerals) but also to bring additional nutritional and physiological benefits to consumers. This chapter will discuss how one category of functional foods, probiotics and prebiotics, can positively modify biological and physiological processes in human nutrition and thereby have a beneficial effect on some human pathological conditions or even be used as an adjuvant therapy in the treatment of certain human pathologies. Benefits claimed on product labels and in promotional literature are very diverse, but this chapter will review the scientific evidence for health benefits and place it in the context of human physiology and physiopathology as well as, when possible, explaining underlying cellular and molecular mechanisms. It will focus on gastrointestinal effects of functional foods without discussing possible effects on other organ systems. In fact, we will suggest that functional foods exert their beneficial effects primarily on the gastrointestinal microflora, thereby promoting the health of the host's own tissues.

Mucosal surfaces like the gastrointestinal mucosa are extensive interfaces with an external environment rich in pathogenic microorganisms and are the starting place for most of mammals' infections. Defenses of mucosal surfaces in physiologically colonized organs such as the gastrointestinal tract involve two main mechanisms: (1) antagonisms between commensal bacteria and pathogens and (2) the host mucosal immune response. The latter, in turn, involves two components: (1) antigen-specific defenses, or adaptive immunity, which develops a certain time after the first encounter with a specific antigen/pathogen, and (2) a constitutive first line of defense, called innate immunity, of central importance for eliminating pathogens when they are encountered for the first time.

The hallmark of the adaptive mucosal immune response is the production of antigen-specific secretory immunoglobulins (Ig), such as secretory IgA (sIgA), which is produced in the intestinal lamina propria and is transported through the epithelial layer into the secretions by the polymeric immunoglobulin receptor (pIgR) or secretory component (sc). Once the sIgA reaches the lumen of the intestine,[1] it recognizes and binds specific antigens/pathogens, giving origin to immune complexes that prevent the interaction of the antigens (bacteria, virus, toxins) with the mucosal surface. This process has been called immune exclusion and represents a noninflammatory mechanism of defense. Functional foods, such as pre- and probiotics, may improve mucosal defenses and thus prevent intestinal infections by (1) modifying the composition and/or the metabolic activity of the intestinal microflora so that it can then better antagonize pathogens or (2) modulating the host's immune response against them.

INTESTINAL MICROFLORA

The intestinal microflora can be considered a postnatally acquired organ that is constituted by a large diversity of bacterial cells that perform important functions for the host[2] and is modulated, in structure and function, by environmental factors such as nutrition. During gestation, the fetus develops in the sterile environment afforded by the maternal uterus, but immediately after birth it is exposed to a microbial inoculum derived from the mother's genital tract and feces, her skin microflora, and the environment.[3] Thus, the gut microflora is acquired in the few hours after delivery of a child. The result is the establishment of an intestinal microbial ecology that is very variable at the beginning but will become a more stable system similar to the adult microflora by the end of the breastfeeding period.

The participating factors responsible for the establishment of the steady-state microflora are not fully understood. However, the whole intestinal microenvironment and competition between different bacterial species and genera are of obvious importance. Probably the type of nutrition has a major impact in the modulation of gut ecology; this is of particular importance during the first days of life.

Enterobacteria, streptococci, and clostridia can be recovered from feces as early as the first day of life. By the third day of life, *Bacteroides* bacteria, bifidobacteria, and clostridia have appeared in about half of all infants, regardless of the type of nutrition. Then bifidobacteria become the dominant bacterial genus in breastfed infants; in contrast, formula-fed infants show levels of stool enterobacteria that are higher or similar to those of bifidobacteria.[3,4] At 1 month of age, bifidobacteria become dominant in infants with both types of nutrition, although breastfed infants' microflora are less complex than the microflora of formula-fed infants. These differences in composition are correlated with differences in metabolic activity. Much less putrefactive bacteria, and their associated metabolic activity, are detected in the breastfed population. In fact, the intestinal pH is lower and redox potential is higher. Formula-fed babies exhibit a more putrefactive intestinal microflora (*Bacteroides, Clostridium, Proteus,* etc) characterized by higher pH, lower redox potential, and the presence of ammonium and amines in feces.

At the other end of the life span, it has been observed that lower numbers of bifidobacteria and greater numbers of clostridia, lactobacilli, and enterobacteria occur in the elderly population.[4,5] Modification of the microflora with old age could be the consequence of modifications in the intestinal microenvironment such as redox potential, intestinal blood perfusion, impaired immune function, and change in eating habits and lifestyle. These correlations, very clear in the newborn period but less clear in the elderly, strongly suggest the importance of nutrition in the composition and metabolic activities of the intestinal microflora. Therefore, it is highly possible that this nutritional modulation of the microflora takes place all along the life span, as is also suggested by cultural and epidemiological data.

Since specific components of the intestinal microflora have been associated with beneficial effects on the host, such as promotion of gut maturation and integrity, antagonisms against pathogens, and immune modulation, it would seem logical that the quantity of these components might be enhanced with nutritional interventions.[2] Two kinds of functional food ingredients, prebiotics and probiotics, have become a focus of great interest in the general population, the food industry, and the scientific community because of their potential for improving gut physiology by modifying the physiology of the intestinal microflora.[6] Probiotics and prebiotics have been designed to modify the balance of the existing flora either by administering growth factors specific to a limited number of bacterial species from the colonic flora (in the case of prebiotics) or by directly providing high quantities of microbes belonging to species that are normal inhabitants of the intestinal microflora (in the case of probiotics).

PREBIOTICS AS FUNCTIONAL INGREDIENTS

Recently, the perception of the biological function of the colon has dramatically changed from the traditional view of the colon as just a hindgut for storage and excretion of waste materials with limited absorptive properties to a new view in which the organ, besides its traditional functions, is perceived as a metabolic chamber, thanks to the presence of a rich colonic microbiota that attains levels of 10^{12} bacteria per gram of colonic content. This microbial metabolic activity salvages energy for the host from nutrients that escaped digestion and absorption in the small bowel.[6]

Colonic energy recovery can represent a significant nutritional contribution in diets where "colonic foods" represent an important proportion of the caloric intake. Colonic foods have been defined as "food ingredient[s] entering the colon and serving as substrate for the endogenous bacteria, thus indirectly providing the host with energy, metabolic substrates and essential micronutrients."[6]

Thus, they are foods, usually polysaccharides, whose digestion and absorption are dependent upon bacterial metabolic activities. And even if energy recovery in the colon is of very marginal importance in energy-rich diets such as those in the Western world, metabolites generated during colonic fermentation, such as short-chain fatty acids (SCFAs), can have important regulatory function over host physiology.

In addition to nutrition-related functions, intestinal microflora metabolism helps to preserve the tissular trophism and homeostasis of the intestinal mucosa[7] and to antagonize exogenous microbial pathogens and control the overgrowth of endogenous potential pathogenic bacteria. All of these functions are summarized in Exhibit 9–1.

If the microflora's metabolic activity can regulate the host's physiology, with potential beneficial effects, functional foods can, in turn, modify the microflora's composition and its metabolic activities, thereby indirectly exerting beneficial effects on the host. The main exogenous colonic foods are nondigestible (to humans) carbohydrates; in addition, endogenous mucus is an important substrate for microflora metabolism. Dietary colonic foods include resistant starch, nonstarch polysaccharides, pectins, cellulose, hemicellulose, gums, and nondigestible oligosaccharides.[8]

The proximal (right colon) large bowel has a higher saccharolytic enzyme activity than the distal (left colon) large bowel, which is more proteolytic.[9] The main metabolic end products of colonic fermentation are hydrogen, carbon dioxide, SCFAs, lactate, succinate, ammonia, amines, phenols, and indols.[6,10] Another very important end product of bacterial fermentation is the biomass generated by bacterial growth.

The profile of metabolites generated in colonic fermentation depends on the quality of the microflora and on the availability of specific substrates. Both beneficial metabolites and metabolites that may be detrimental for the host can be produced.

Exhibit 9–1 General Functions of the Intestinal Microflora

- Energy salvage (lactose digestion, SCFA production)
- Modulation of cell growth and differentiation
- Antagonism against pathogens
- Immune stimulation of the gut-associated lymphoid tissue
- Innate immunity against infections
- Production of vitamins
- Reduction of blood lipids

Beneficial Metabolites

Among the beneficial end products of fermentation are SCFAs and lactate. Globally, they modify the general ecology of the lumen, reducing pH and thereby controlling nonacidophilic bacteria. Moreover, these products may improve colonic absorption of calcium, magnesium, and iron. Acetate, propionate, and butyrate are the major SCFAs produced in the colon. Acetate and propionate are absorbed through the epithelium and reach the liver via the portal vein blood. The acetate is also taken up by the muscle and other peripheral tissues.[11] There the SCFAs are metabolized, contributing to the general host energy gain.[9] This gain is estimated to be less than 5% of total energy intake and is dependent on dietary fiber intake. Propionate has been suspected to reduce cholesterol level in blood, but this deserves further confirmation. So far, only one study has shown a decrease of cholesterol in humans.[12] Butyrate is almost totally taken up by the colonocytes and is used as a source of carbon and energy.[9] In addition, it has an important effect on cell growth and differentiation,[13,14] probably by favoring the expression of differentiation markers, as has been demonstrated in vitro.[15,16] Butyrate is also of interest because its absorption is coupled with sodium and water reabsorption, so that it has a potential antidiarrheic effect.[17]

Detrimental End Products

As with other types of nondigestible substrates available for bacterial metabolism, other end products may be generated, such as amines, phenols, ammonia thiols, and potentially toxic products. Colonic food will promote nonspecific growth and metabolic activity of different types of bacteria, including both those with beneficial effects on the host and those with potentially negative effects. In fact, colonic food ingredients available in functional foods may target the specific growth of a "good bacterial" component of the colonic microflora or induce a global modulation of the microflora metabolic activity to generate a desired profile of end metabolic products. So far efforts have been mainly directed toward finding substrates that could selectively promote the growth of beneficial components of the microflora, possibly at the expense of harmful species.

The metabolic selectivity of one substrate for the beneficial components of the microflora is intimately associated with the prebiotic concept. Moreover, bifidobacteria are the sole component of the colonic microflora that has been targeted for specific growth promotion. On the basis of all this, Gibson and Roberfroid[6] have defined prebiotics as "nondigestible food ingredient[s] that beneficially affect the host by selectively stimulating the growth and/or activity of one or a limited number of bacteria in the colon that have the potential to improve host health."

Several comments can be made on this definition. Prebiotics are dedicated to the modulation of the bacterial population of the colon, and they constitute a specific category of what have been called colonic foods. Although the molecular characteristics of prebiotics are as yet undefined, so far only carbohydrates fulfilling the definition are available on the market. In Japan, the first country to have given functional foods a legal status, prebiotic carbohydrates are the largest example in the category.[18] In contrast, in Europe only fructo-oligosaccharides (FOSs) and fiber have a status of promoting bifidobacterial growth. In the near future, however, it is highly possible that other carbohydrates (eg, galacto-oligosaccharides) will also obtain this status. Although the definition is quite broad in terms of the bacterial species that can be modulated, prebiotics today are limited to those that promote bifidobacteria.

It is not clear, from current studies, whether the administration of prebiotics can also result in an increase of fecal biomass and thus of stool frequency and weight (transit regulation). The potential is controversial and, obviously, dose dependent.[19]

FOS selective growth promotion of bifidobacteria has been determined in vitro,[20] in animals, [21] and in clinical studies.[22] This specificity could be due to the production of β-fructosidases by bifidobacteria.[20] Along with bifidobacteria growth, inhibition of potentially pathogenic harmful gram-positive and gram-negative bacteria like *E. coli, Bacteroides* species, *Fusobacterium* species, *C. perfringens, Salmonella* species, *Listeria* species, *Shigella* species, *Campylobacter,* and *Vibrio cholerae* has been reported.[22,23] One explanation for this antimicrobial effect is based on the production of organic acids, mainly acetic and lactic acids, upon fermentation of prebiotics. Those compounds, by acidifying the colonic content, help in controlling the growth of nonacidophilic bacteria in general.[23] Another possibility is that bifidobacteria may produce bacteriocinlike compounds, as has been previously demonstrated for lactobacilli.[24]

An additional benefit of acid production is the acid transformation in the colon of potentially toxic amines in their protonated and nondiffusable form, thus avoiding an increase, for example, in blood ammonia levels.[6] The stimulation of bifidobacteria growth may also result in an increase in availability of group B vitamins and digestive enzymes.[6] Of crucial importance for consumers is the capacity of bifidobacteria to modulate the host's immune response. Bifidobacteria as lactobacilli are able to stimulate both nonspecific and specific immune responses and by so doing play a central role in mucosal defense mechanisms.[2]

Although prebiotics are carbohydrates, they are not readily digested and thus provide few calories. For example, FOSs are believed to have around 1 Kcal/g,[25] making them ingredients of choice for low-calorie products (hypocaloric diets) and low-sugar products (diabetic diets). In addition, FOSs have texturizing properties that allow their use as fat replacers.

In terms of safety, no major concerns have been raised in association with prebiotics. They can be considered basically as macronutrients because they are ingested in relatively high amounts. Some of them are naturally occurring in fruits, herbs, and vegetables (chicory). This is, for example, the case with inulin and oligofructose, which are present in chicory roots, artichokes, and onions.[26] The only side effects of carbohydrates that are present in the type of prebiotics consumed by humans are flatulence and loose stools after the intake of large quantities. The level at which intestinal discomfort is attained is highly variable within normal individuals and, moreover, from one prebiotic to another.

The ability of prebiotics, in particular FOSs, to increase SCFA concentration is difficult to show in stools, probably because SCFAs are mainly produced in the proximal colon and also absorbed in the proximal colon.[19,27]

It has not been reported that prebiotics can change the profile of SCFAs generated during colonic fermentation. In fact, prebiotics can change the composition of the colonic microflora by promoting specifically bifidobacteria, but changes of the microflora metabolic activity are undetectable. It seems worth examining whether, in addition to the beneficial promotion of bifidobacterial growth, prebiotics or other colonic foods can induce a healthier profile of end products as a result of colonic metabolism modulation.

PROBIOTICS AS FUNCTIONAL INGREDIENTS

In contrast to prebiotics, which represent a new concept, probiotic activities have been hypothesized or known since the turn of the century, when Metchnikoff postulated that lactic acid bacteria provided health and longevity benefits.[28] Recently, this concept has been revitalized, and a major scientific effort has been devoted to obtain, with probiotics, health benefits for the general population and for populations at risk of intestinal infections. In addition, recent scientific work has established underlying mechanisms for probiotic protective effects.[29] According to Fuller,[30] a probiotic is "a live microbial feed supplement which beneficially affects the host animal by improving its intestinal microbial balance." This definition clearly mentions live microorganisms, meaning that at the time of administration and in the intestine, bacteria should be alive. Probiotics are not limited to lactic acid bacteria; there are also experimental data suggesting the potential use of yeast strains as probiotics in clinical situations.[31,32] However, yeasts are generally categorized as nonfood formulations and thus do not fall within the scope of this review.

The most often used genera are lactobacilli and bifidobacteria. They can be consumed in fermented foods, such as products similar to yogurt, or in other fermented and nonfermented foods (vegetables, meat, milk-based drinks, etc). Lactobacilli and bifidobacteria are normal inhabitants of the human colonic flora,[4] thus giving a rationale for their use as a component of functional foods. Since the gut microflora is a constitutive component of the intestinal mucosal defenses, probiotics, specifically selected for their protective capacities, give an opportunity to reinforce the gastrointestinal tract anti-infectious defenses. The basis for the improvement of mucosal defenses can be found in a direct capacity to antagonize pathogens and/or in the capacity to modulate the host's defense mechanisms, such as the immune reaction.[2]

Compared to the highly colonized large bowel, the small intestine harbors a poor resident microflora whose barrier effect against pathogens is limited. This might be why most viral and bacterial intestinal infections (eg, *E. coli* O157:H7) target the small bowel. In addition, the majority of gut-associated lymphoid tissue is located in the small bowel,[33] and the immune stimulation, the afferent limb of the immune response, takes place preferentially, if not only, in a definite compartment of the small-intestinal mucosa. In fact, soluble and particulate antigens initiate immune stimulation at lymphoid aggregates that are covered by an epithelium specialized in antigen sampling of the gut content. These anatomical sites are called Peyer's patches and are found only in the small bowel.

If one of the main probiotic functions is to improve intestinal mechanisms of defense, either by direct barrier effect or through immunomodulation, a stringent criterion for probiotic selection will be the capacity to colonize, even transiently, these intestinal compartments. A major goal in the development of probiotic-based functional foods seems to be the establishment, in the upper gastrointestinal tract (small bowel and stomach), of an ecosystem with a stronger barrier effect against pathogens and/or the maintenance of an adequate level of immunomodulation. If this is achieved, the prevention of gastrointestinal infections or even adjuvant therapies for chronic intestinal infections or inflammatory conditions may become possible with probiotics. However, there is no clear agreement on the bacterial features associated with positive probiotic activities. Criteria currently used for isolating and defining probiotic bacteria[34] and specific strains are summarized in Exhibit 9–2.

The human origin requirement seems important, since some effects can be only species specific. In addition, the immune relation between the host and the intestinal microflora can be profoundly affected by this aspect. Foo and Lee[35] demonstrated several years ago that components of the intestinal microflora share common antigenic epitopes with the intestinal mucosa, giving a biochemical basis to the frequently reported immunological tolerance of the host toward its resident bacteria. Adherence to

---

Exhibit 9–2 Criteria for Isolating and Defining Probiotic Bacteria

- Human origin
- Resistance to acidity and bile toxicity
- Adherence to human intestinal cells
- Colonization (even transient) of the human gut
- Antagonism against pathogenic bacteria
- Production of antimicrobial substances
- Immune modulation properties
- Clinically proven health effects (dose-response data)
- History of safe use in humans

intestinal epithelial cells is probably a prerequisite to give bacteria a competitive advantage in the small bowel, where no true stable microflora exists. Adherence to the mucosal surface prevents the displacement and removal of the offensive microorganisms by the intestinal luminal flow and active peristalsis. Several probiotic bacteria have been described as adherent in vitro and in vivo in humans.[36–39] Transient colonization is a process by which a given probiotic can remain at high enough levels to become a dominant species in the small-bowel bacterial ecology. Colonization at a certain level, in a specific intestinal microenvironment, seems required for the achievement of probiotic biological activities.

Antagonism to enteropathogens is a very important property that should be established systematically for any new probiotic claiming functionality. Indeed, it has been demonstrated that lactic acid bacteria can inhibit a wide range of enteropathogens.[24,36,37,40,41]

Immunemodulation (the change of host reactivity toward the challenge of pathogenic microorganisms) is another important feature of a probiotic, with protective functions for the host.[2,42] The combination of ecological and immune functionality is the key to designing functional foods with proven efficacy.

Safety is a major criterion of probiotic selection because probiotics are living bacteria massively administered to human populations. Although lactic acid bacteria have been consumed for many years and the long history of safe use remains the best proof of their safety, additional safety aspects need to be considered all the time. Traditional toxicity tests have been performed with probiotic bacteria. However, new strains require new tests since there is no possible generalization on the safety or potential pathogenicity features among strains. Therefore, colonization and bacterial translocation are important criteria for safety considerations. Excessive deconjugation of bile salts and mucus degradation are also potential risk factors.[43] Platelet aggregating properties that could promote cardiac valve colonization are considered potential risk factors as well.[44] Another important factor that deserves consideration is the antibiotic resistance that probiotics may show and, furthermore, transmit to other components of the microflora. In that sense antibiotic resistance plasmids are of special interest because they may be conjugatively transferred to other strains. Probably the best example for this case is the plasmid transfer from lactobacilli to enterococci that has been detected in vivo.[45]

The clinical efficacy of probiotics as ingredients of functional foods must be proven by scientifically validated work. The clinical evidence is the main support for the credibility of health-related claims.[46] Moreover, legal authorities will request more and more this type of scientific evidence. Foods containing probiotic strains are designed and proposed for the well-being of the general population. Among the great diversity of functional health claims, it is now widely accepted that probiotics may be efficient in preventing intestinal infections.[2] The main interest is to find alternatives to classical antibiotic treatments because of the rapid development of antibiotic resistance. The use of probiotics and possibly prebiotics in a preventive fashion could diminish the need for antibiotic treatments in people at risk of infections, such as hospitalized patients, the elderly, and the immunosuppressed. But it is imperative to use safe probiotic strains. A growing number of vancomycin-resistant *Enterococcus* infections have been recently reported pointing to an important safety issue that needs to be addressed in the field of probiotics.[47]

DIFFERENCES IN PATHOGENICITY AND PROTECTION ACROSS GASTROINTESTINAL REGIONS

For the initiation of an intestinal infectious process, pathogens colonize specific niches in different habitats of the gastrointestinal tract. This selectivity seems to depend upon the expression of specific receptors on mucosal cells, to which they will specifically attach, and the ecology of the gastrointestinal tract at a given site. Mucosal mechanisms of defense differ, in turn, in different habitats of the gastrointestinal tract. For example, stomach acidity renders this organ almost sterile, whereas in the small bowel, Paneth cell– and pancreatic-derived antibacterial peptides, both examples of innate immunity, control the level of colonization in the proximal small bowel.[48] In most of the gastrointestinal tract, the secretory immune system plays a central role in mucosal defenses, although Ig secretion depends on the density and type of local bacterial populations.[49]

It is important to realize that probiotic bacteria, to perform protective functions in the gastrointestinal tract, must also adapt to different microenvironments, especially those that are the site of pathogen colonization or the target organ to induce immunomodulation. Both aspects seem very important in the small bowel; in contrast, attachment to the mucosal surface may be less relevant in the colon, where transit time allows bacterial development and where more general ecological modifications (ie, pH, redox potential, SCFA production) could be highly efficient in controlling intestinal microflora.

Stomach

Normal gastric secretory function creates a very hostile environment for bacterial development, since interdigestive pH varies between 1 and 2. This is a very efficient nonspecific mechanism of defense against infections. In a way, gastric acidity represents the first barrier effect of the gastrointestinal tract against bacterial growth. However, probiotic microorganisms administered to healthy people have to face the same acid stress, and their capacity to survive in such an environment is central to their protective role, since their functions depend on the metabolic interaction with the endogenous microflora.[30]

Hypo- or achlorhydria is a relatively common condition associated to atrophic gastritis in the elderly population[50] and to other pathological or subpathological conditions that require the administration of antacids (Ranitidine, Cimetidine, Omeprazole). Either by the destruction of parietal cells (in atrophic gastritis) or the inhibition of the protonic pump activity (administration of antacids), the gastric acid barrier becomes disrupted. As a consequence of this, colonization of the stomach and the small bowel by potentially pathogenic bacteria is observed. This is the case with *Helicobacter pylori,* which is involved in development of gastritis, gastric and duodenal ulcers, and some gastric cancers.[51] This bacterium, which shows high genetic variability, infects more than 60% of adults over the age of 60 in industrialized countries, and in developing countries most of the individuals are infected at some point during childhood.[52]

In the small intestine, bacterial overgrowth due to the loss of the gastric acid barrier can have important nutritional consequences. Malabsorptive syndromes with diarrhea, steatorrhea, and loss of vitamin absorption are important complications of small-bowel bacterial overgrowth. In addition, gram-negative enterobacteria colonization of the small bowel can promote passage of endotoxin into the blood, triggering an inflammatory/catabolic cascade of events. Another possible outcome of small-intestine bacterial overgrowth is the passage of viable bacteria through the mucosal barrier, a process called bacterial translocation that may well be the triggering event of a septic process of gut origin.[53]

Some lactic acid bacteria have demonstrated their ability to inhibit *H. pylori* through the bactericidal effect of lactic acid.[54] Recently, the inhibitory effect of a whey-based medium fermented by the *L. johnsonii* La1 strain on *H. pylori* gastric colonization has been shown in humans.[55] The mechanism involves lactic acid toxicity but possibly bacteriostatic compounds as well.[24]

Small Bowel

Bacterial attachment is a prerequisite for colonization of the epithelial surfaces of the gastrointestinal tract. This process is mediated by bacterial molecules on the surface of the bacterial cells, called adhesins, that recognize specific proteins or glycoconjugates on the surface of the eukaryotic cell. The regional expression of these receptors gives pathogens a selectivity for rather specific environments of the intestinal mucosa. In the small bowel, the bacteria that are unable to adhere to the epithelial surface will be rapidly removed by intestinal secretions, active peristaltic movements, and shedding of the mucus layer.

To antagonize small-bowel infectious agents, the probiotics must become dominant in an environment where no true stable microflora exists. This will require either continuous administration of the probiotic protective bacteria or the administration of an adhesive strain that could remain attached to the target organ for longer periods.

As mentioned earlier, probiotics can protect the host against pathogens through two major mechanisms: (1) the barrier effect or colonization resistance, and (2) the modulation of the host's own mechanisms of defense.

The barrier effect of the microflora is expressed quite differently in the small bowel and in the colon. The transient presence of the small-bowel flora implies that the barrier effect is much more directed toward an interference of the interactions between the pathogens and their cellular targets.

This can be obtained by basically two mechanisms: direct antagonism between the probiotic and the pathogen and/or the probiotic's modulation of the host's mechanisms of defense, such as IgA secretory activitiy.

Although antagonisms between lactic acid bacteria and several small-bowel pathogens have been shown in vitro and in vivo,[24,36,37] the underlying molecular mechanisms have not been identified, and more than one could be playing a role. For example, certain lactobacilli are able to secrete antibacterial substances active in vivo against the *Salmonella typhimurium* strain.[24]

On the other hand, lactic acid bacteria can increase the specific IgA production against the pathogen.[56] A similar immunoadjuvant effect has been reported in human volunteers after *S. typhi* TY21a oral vaccination.[57]

It is not known whether a direct bacterial-viral antagonism exists or whether the protection that has been observed with lactic acid bacteria in pediatric rotavirus infection is associated only with an immunoadjuvant effect of the probiotic strain.[58,59]

Colon

In the section "Prebiotics," we have discussed the nutritional modulation of the colonic microflora to promote growth of bifidobacteria and, indirectly, to inhibit growth of potentially harmful components of the microflora. Although such effects can be achieved by these means, and potentially are very interesting, only limited clinical evidence of those beneficial effects exists and has been mentioned.

Probably the most clear clinical evidence of the colonic microflora function comes from its disruption following antibiotic treatment. In fact, antibiotic associated diarrhea (AAD) is a frequent complication of antibacterial treatments in hospitalized patients. The limited barrier function of the colonic microflora under antibiotics promotes bacterial overgrowth of opportunistic pathogens such as *Clostridium difficile*. This is a toxin producing strict anaerobe associated with acute episodes of AAD,[60] recurrent colitis, and in very severe cases with pseudomembranous colitis, a life threatening condition. *Bacteroides fragilis* is also frequently associated with AAD.[61] Probiotics *Saccharomyces boulardii* and *Lactobacillus rhamnosus* GG have shown positive results in the prevention and treatment of AAD.

SAFETY OF PROBIOTICS

Probiotic microorganisms have been widely consumed without the observation of major side effects. Very rare pathologies such as endocarditis or septicemia have been described for lactobacilli, and risk factors have been identified.[62–64] Even if anecdotal, infections that may involve probiotic bacteria should be very seriously evaluated, especially for strains that might be used in functional and clinical foods. Another key question of safety is the influence of probiotic bacteria on gut permeability, as well as the absence of inflammatory processes after chronic or acute administration of specific strains.[65]

In conclusion, one should ensure that probiotic microorganisms do not increase the gut permeability that could promote food allergy and inflammatory conditions either local or systemic. This is of particular importance for strains given as clinical and/or therapeutic agents.

SYNBIOTICS, A NEW CONCEPT

Gibson and Roberfroid[6] have proposed a new concept, the *synbiotic*, or combination of both pre- and probiotics to obtain the benefits of both ingredients. They define a synbiotic as a "mixture of probiotics and prebiotics that beneficially affects the host by improving the survival and implantation of live microbial dietary supplements in the gastrointestinal tract, by selectively stimulating the growth and/or activating the metabolism of one or a limited number of health-promoting bacteria, and thus improving welfare." This is an appealing concept that needs to be tested in clinical conditions.

REFERENCES

1. Underdown BJ, Schiff JM. Strategic defense initiative at the mucosal surfaces. *Annu Rev Immunol.* 1986;4:345–417.
2. Brassart D, Schiffrin EJ. The use of probiotics to reinforce mucosal defense mechanisms. *Trends Food Sci Technol.* 1997;9:321–326.
3. Haenel H. Human normal and abnormal gastrointestinal flora. *Am J Clin Nutr.* 1970;23:1433–1439.
4. Mitsuoka T. The human gastrointestinal tract. In: Wood BJB, ed. *The Lactic Acid Bacteria.* New York, NY: Elsevier Applied Science Publishers; 1992:69–114.
5. Bertazzoni Minelli E, Benini A, Beghini AM, Cerutti R, Nardos G. Bacterial faecal flora in healthy women of different ages. *Microb Ecol Health Dis.* 1993;6:43–51.
6. Gibson GR, Roberfroid MB. Dietary modulation of the human colonic microbiota: introducing the concept of prebiotics. *J Nutr.* 1995;125:1401–1412.
7. Koruda MJ, Rolandelli RH, Settle RG, Zimmaro DM, Rombeau JL. Effect of parenteral nutrition supplement with short-chain fatty acids on adaptation to massive small bowel resection. *Gastroenterology.* 1988;95:715–720.
8. Delzenne NM, Roberfroid MR. Physiological effects of nondigestible oligosaccharides. *Lebensm-Wiss Technol.* 1994;27:1–6.
9. Cummings JH. Short chain fatty acids. In: Gibson GR, Macfarlane GT, eds. *Human Colonic Bacteria: Role in Nutrition, Physiology and Pathology.* Boca Raton, FL: CRC Press: 1995:101–130.
10. Macfarlane S, Macfarlane GT. Proteolysis and amino acid fermentation. In: Gibson GR, Mcfarlane GT, eds. *Human Colonic Bacteria: Role in Nutrition, Physiology and Pathology.* Boca Raton, FL: CRC Press; 1995:75–100.
11. Demigné C, Yacoub C, Rérnezy C, Fafournoux P. Effect of absorption of large amounts of volatile fatty acids on rat liver metabolism. *J Nutr.* 1986;116:77–86.
12. Todesco T, Rao VA, Bosello O, Jenkins DJA. Propionate lowers blood glucose and alters lipid metabolism in healthy subjects. *Am J Clin Nutr.* 1991;54:860–865.
13. Boffa LC, Lupton JR, Mariani MR, et al. Modulation of colonic epithelial cell proliferation, histone acetylation, and luminal short chain fatty acids by variation of dietary fiber (wheat bran) in rats. *Cancer Res.* 1992;52:5906–5912.
14. McIntyre A, Gibson PR, Young GP. Butyrate production from dietary fibre and protection against large bowel cancer in a rat model. *Gut.* 1993;34:386–391.
15. Gibson GR, Moeller I, Kagelari O, Folino M, Young GP. Contrasting effects of butyrate on the expression of phenotypic markers of differentiation in neoplastic and non-neoplastic colonic epithelial cells in vitro. *J Gastroenterol Hepatol.* 1992;7:165–172.

16. Kim YS, Gum JR, Ho SB, Deng G. Colonocyte differentiation and proliferation: overview and the butyrate-induced transcriptional regulation of oncodevelopmental placental-like alkaline phosphatase gene in colon cancer cells. In: Binder HJ, Cummings JH, Soergel KH, eds. *Short Chain Fatty Acids*. 3rd ed. Lancaster, England: Kluwer Academic Publishers; 1994:119–134.

17. Scheppach WM, Bartram HP. Experimental evidence for and clinical implications of fiber and artificial enteral nutrition. *Nutrition*. 1993;9:399–405.

18. Clydesdale FM. A proposal for the establishment of scientific criteria for health claims for functional foods. *Nutr Rev*. 1997;55:413–422.

19. Alles MS, Hautvast JGAJ, Nagengast FM, Hartemink R, Van Laere KMJ, Jansen JBMJ. Fate of fructo-oligosaccharides in the human intestine. *Br J Nutr*. 1996;76:211–221.

20. Wang X, Gibson GR. Effects of the in vitro fermentation of oligofructose and inulin by bacteria growing in the human large intestine. *J Appl Bacteriol*. 1993;75:373–380.

21. Howard MD, Gordon DT, Pace LW, Garleb KA, Kerley MS. Effects of dietary supplementation with fructooligosaccharides on colonic microbiota populations and epithelial cell proliferation in neonatal pigs. *J Ped Gastroenterol Nutr*. 1995;21:297–303.

22. Gibson GR, Beatty ER, Cummings JH. Selective stimulation of bifidobacteria in the human colon by oligofructose and inulin. *Gastroenterology*. 1995;108:975–982.

23. Gibson GR, Wang X. Regulatory effects of bifidobacteria on the growth of other colonic bacteria. *J Appl Bacteriol*. 1994;77:412–420.

24. Bernet-Camard M-F, Liévin V, Brassart D, Neeser JR, Servin AL, Hudault S. The human *Lactobacillus acidophilus* strain LA1 secretes a nonbacteriocin antibacterial substance active in vitro and in vivo. *Appl Environ Microbiol*. 1997;63:2747–2753.

25. Roberfroid M, Gibson GR, Delzenne N. The biochemistry of oligofructose, a nondigestible fiber: an approach to calculate its caloric value. *Nutr Rev*. 1993;51:137–146.

26. Van Loo J, Coussement P, De Leenheer L, Hoebregs H, Smits G. On the presence of inulin and oligofructose as natural ingredients in the Western diet. *Crit Rev Food Sci Nutr*. 1995;35:515–552.

27. Cummings JH, Pomare EW, Branch WJ, Naylor CPE, MacFarlane GT. Short-chain fatty acids in human large intestine, portal, hepatic and venous blood. *Gut*. 1987;28:1221–1227.

28. Metchnikoff E. *The Prolongation of Life*. New York, NY: Putnam Sons; 1908.

29. Delneste Y, Donnet-Hughes A, Schiffrin EJ. Functional foods: mechanisms of action on immunocompetent cells. *Nutr Rev*. 1998;56:S93–S98.

30. Fuller R. Probiotics in man and animals. *J Appl Bacteriol*. 1989;66:365–378.

31. Elmer GW, Surawicz CM, Mc Farland LV. Biotherapeutic agents: a neglected modality for the treatment and prevention of selected intestinal and vaginal infections. *JAMA*. 1996;275:870–876.

32. McFarland LV, Bernasconi P. *Saccharomyces boulardii:* a review of an innovative biotherapeutic agent. *Microb Ecol Health Dis*. 1993;6:157–171.

33. Neutra WR, Pringault E, Kraehenbuhl JP. Antigen sampling across epithelial barriers and induction of mucosal immune responses. *Annu Rev Immunol*. 1996;14:275–300.

34. Lee Y-K, Salminen S. The coming of age of probiotics. *Trends Food Sci Technol*. 1995;6:241–245.

35. Foo MC, Lee A. Antigenic cross-reaction between mouse intestine and a member of the autochthonous microflora. *Infect Immun*. 1974;9:1066–1069.

36. Bernet MF, Brassart D, Neeser JR, Servin AL. Adhesion of human bifidobacterial strains to cultured human intestinal epithelial cells and inhibition of enteropathogen-cell interactions. *Appl Environ Microbiol*. 1993;59:4121–4128.

37. Bernet MF, Brassart D, Neeser JR, Servin AL. *Lactobacillus acidophilus* LA1 binds to intestinal cell lines and inhibits cell attachment and cell invasion by enterovirulent bacteria. *Gut*. 1994;35:483–489.

38. Johansson ML, Molin G, Jeppsson B, et al. Administration of different *Lactobacillus* strains in fermented oat meal soup: in vivo colonization of human intestinal mucosa and effect on the indigenous flora. *Appl Environ Microbiol*. 1993;59:15–20.

39. Alander M, Korpela R, Saxelin M, et al. Recovery of *Lactobacillus rhamnosus* GG from human colonic biopsies. *Lett Appl Microbiol*. 1997;24:361–364.

40. Hudault S, Liévin V, Bernet-Camard M-F, Servin AL. Antagonistic activity exerted in vitro and in vivo by *Lactobacillus casei* (strain GG) against *Salmonella typhimurium* C5 infection. *Appl Environ Microbiol*. 1997;63:513–518.

41. Coconnier MH, Liévin V, Bernet-Camard MF, Hudault S, Servin AL. Antibacterial effect of the adhering human *Lactobacillus acidophilus* strain LB. *Antimicrob Agents Chemother.* 1997;41:1046–1052.

42. Marteau P, Rambaud JC. Potential for using lactic acid bacteria for therapy and immunomodulation in man. *FEMS Microbiol Rev.* 1993;12:207–220.

43. Salminen S, von Wright A, Morelli L, et al. Demonstration of safety of probiotics—a review. *Int J Food Microbiol.* 1998;44:93–106.

44. Korpela R, Moilanen E, Saxelin M, Vapaasalo H. Lactobacillus rhamnosus GG (ATCC 53013) and platelet aggregation in vitro. *Int J Food Microbiol.* 1997;37:83–86.

45. Morelli L, Sarra PG, Bottazzi V. In vivo transfer of pAMB1 from Lactobacillus reuteri to Enterococcus fae calis. *J Appl Bacteriol.* 1998;65:371–375.

46. Sanders ME. Effect of consumption of lactic cultures on human health. *Adv Food Nutr Res.* 1993;37:67–130.

47. Gold HS, Moellering RC. Drug therapy - antimicrobial-drug resistance. *N Engl J Med.* 1996;335:1445–1453.

48. Bonan HG. Peptides, antibiotics, and their role in innate immunity. *Annu Rev Immunol.* 1995;13:61–92.

49. Kjell K, Baklein K, Bakken A, Kral JG, Fausa O, Brandttaeg P. Intestinal B-cell isotype response in relation to local bacterial load: evidence for immunoglobulin A subclass adaptation. *Gastroenterology.* 1995;109:819–825.

50. Russell RM, Krasinski SD, Samloff IM, Jacob RA, Hartz SC, Brovender SR. Folic acid malabsorption in atrophic gastritis: possible compensation by bacterial folate synthesis. *Gastroenterology.* 1986;91:1476–1482.

51. Blaser MJ, Parsonnet J. Parasitism by the "slow" bacterium *Helicobacter pylori* leads to altered gastric homeostasis and neoplasia. *J Clin Invest.* 1994;94:4–8.

52. Graham DY, Klein PD, Opekun AR, Boutton TW. Effect of age on the frequency of active *Campylobacter pylori* infection diagnosed by the $\{^{13}C\}$ urea breath test in normal subjects and patients wih peptic ulcer disease. *J Infect Dis.* 1988;157:777–780.

53. Deitch EA. Bacterial translocation: the influence of dietary variable. *Gut.* 1994;35(suppl 1):S23–S27.

54. Midolo PD, Lambert JR, Hull R, Luo F, Grayson ML. In vitro inhibition of *Helicobacter pylori* NCTC 11367 by organic acids and lactic bacteria. *J Appl Bacteriol.* 1995;79:475–479.

55. Michetti P, Dorta G, Brassart D, et al. *L. acidophilus* supernatant as an adjuvant in the therapy of *H. pylori* in humans. *Gastroenterology.* 1995;108:A166.

56. Perdigón G, Alvarez S, Nader de Macias ME, Roux ME, Ruiz Holgado AP. The oral administration of lactic acid bacteria increases the mucosal intestinal immunity in response to enteropathogens. *J Food Protec.* 1990;53:404–410.

57. Link-Amster H, Rocht F, Saudan KY, Mignot O, Aeschlimann JM. Modulation of a specific humoral immune response and changes in intestinal flora mediated through fermented milk intake. *FEMS Immunol Med Microbiol.* 1994;10:55–64.

58. Isolauri E, Juntunen M, Rautanen T, Sillanaukee P, Koivula T. A human *Lactobacillus* strain (*Lactobacillus casei* spp strain GG) promotes recovery from acute diarrhoea in children. *Pediatrics.* 1991;88:90–97.

59. Saavedra JM, Bauman NA, Oung I, Perman JA, Yolken RH. Feeding of *Bifidobacterium* and *Streptococcus thermophilus* to infants in hospital for prevention of diarrhoea and shedding of rotavirus. *Lancet.* 1994;344:1046–1049.

60. Obido RJ, Lyerly DM, Van Tassell RL, Wilkins TD. Proteolytic activity of the Bacteroider fragilis enterotoxin causes fluid secretion and intestinal damage in vivo. *Infect Immun.* 1995;63:3820–3826.

61. Wells CL, Van de Westerlo EMA, Jechorek RP, Feltis BA, Wilkins TD, Erlandsen SL. *Bacteroides fragilis* enterotoxin modulates epithelial permeability and bacterial internalization by HT-29 enterocytes. *Gastroenterology.* 1996;110:1429–1437.

62. Gasser F. Safety of lactic acid bacteria and their occurrence in human clinical infections. *Bull Inst Pasteur.* 1994;92:45–67.

63. Aguirre M, Collins MD. Lactic acid bacteria and human clinical infections. *J Appl Bacteriol.* 1993;75:95–107.

64. Oakey HJ, Harty DWS, Knox KW. Enzyme production and the potential link with infective endocarditis. *J Appl Bacteriol.* 1995;78:142–148.

65. Marteau P, Vaerman J-P, Dehennin J-P, et al. Effects of intrajejunal perfusion and chronic ingestion of *Lactobacillus johnsonii* strain La1 on serum concentrations and jejunal secretions of immunoglobulins and serum proteins in healthy humans. *Gastroenterol Clin Biol.* 1997;21:293–298.

CHAPTER 10

Fats and Oils and Their Effects on Health and Disease

David Firestone

Fats and oils are more than dietary sources of fuel. They play important roles in the structure and function of a host of metabolic and immunological systems and can provide a variety of benefits with their use in medical and designer foods.[1] Metabolic products of fatty acids play important roles in the cardiovascular and immune systems, bone metabolism, and blood clotting and play a role in control of cancer, coronary heart disease (CHD), autoimmune diseases, diabetes, and other disorders. In addition, fats are essential for growth and development, and because they have desirable sensory properties, they are important to the appearance and taste of foods. Nutritionists generally advise consumers to consume a diet with moderate fat content, maintain ideal body weight, and eat a varied, balanced diet. The National Academy of Sciences Food and Nutrition Board's report on diet and health[2] concluded that total amounts and types of fats and other lipids in the diet influence the risk of atherosclerotic cardiovascular disease, certain forms of cancer, and possibly obesity. The report stated that reductions in saturated fatty acids are likely to reduce CHD and that saturated fat together with total fat was associated with higher incidence of and mortality from some types of cancer.

CLASSIFICATION OF LIPIDS

Lipid means "fat" in Greek and refers to three major classes of compounds: *simple* (neutral) lipids, *compound* lipids, and *derived* lipids. Almost all edible fats and oils are simple lipids. They are triacylglycerols (triglycerides) or esters in which a fatty acid is esterified with a glycerol molecule. Mono- or diacylglycerols can also occur if glycerol has two or one, respectively, unesterified hydroxyl group(s). Neutral plasmalogens are found in land and marine animals and consist of glycerol esters of two fatty acids and a vinyl ether. Waxes are esters of alcohols other than glycerol and may be of plant or animal origin. Cholesterol esters are the most widely occurring sterols in animal tissues. Sterol esters are also components of some natural waxes like beeswax and lanolin. Other sterol esters, such as stigmasterol and β-sitosterol, occur in plants. Examples of waxes include ergosterol, which can be isolated from yeast and forms vitamin D_2 under ultraviolet light, and esterified vitamin A, which is found in fish liver oils.

Compound lipids are complex molecular structures that are divided into classes, such as phospholipids, sphingolipids, and ceramides. The most common phospholipids or phosphoglycerides include phosphatidylcholine, phosphatidylethanolamine, phosphatidylserine, and phosphatidylinositol, all of which are widely distributed in plants and animals. Their structures are such that two of the hydroxyl groups of the glycerol are esterified with fatty acids and the third with a phosphate that is attached to a short-chain organic base. Some phospholipids, such as phosphatidylglycerol and diphosphatidyl-glycerol, consist of more than a single glycerol esterified to phosphoric acid. The latter is also called cardiolipin and is found principally in the inner mitochondrial membrane.

Saturated or unsaturated fatty acids are derived from the hydrolysis of triacylglycerols and more complex lipid structures. Saturated fatty acids may have substituents, such as alkyl, keto, hydroxyl, or halide groups, along the hydrocarbon chain. Unsaturated fatty acids may contain one or more double or even triple bonds. Geometric (*cis* and *trans*) and positional isomers can also occur.

Different fats and oils commonly consumed differ widely in fatty acid composition. For instance, the major fatty acid component in coconut oil is lauric (12:0, 45%); in palm oil, it is palmitic (16:0, 50%); in olive oil, it is oleic (18:1, 75%); and in corn oil, it is linoleic (18:2, 58%), the most commonly found diunsaturated fatty acid in seed oils. The notation used refers to number of carbon atoms: number of double bonds. Thus, "18:2" means 18 carbons and 2 double bonds. Marine animals and fish contain fatty acids with longer chains and more unsaturation, such as arachidonic acid (20:4), EPA (eicosapentaenoic acid, 20:5), and DHA (docosahexaenoic acid, 22:6). Most double bonds are usually methylene interrupted—that is, separated by a methylene group:

$$-C = C-CH_2-C = C-$$

but some are conjugated:

$$-C = C-C = C-$$

Rumenic acid, the major isomer of conjugated linoleic acid (CLA), found mainly in dairy products, has two double bonds, a *cis* double bond at carbon C-9 and a *trans* double bond at C-11. Branched-chain fatty acids are primarily of microbial origin. Cyclic fatty acids include those with cyclopropyl, cyclopropenyl, cyclopentyl, cyclopentenyl, cyclohexyl, or cyclohexenyl moeities. Cyclic fatty acid monomers are potentially toxic components of heated vegetable oils, such as those used in deep-frying operations.[3]

ANALYSIS OF LIPIDS

A brief description is presented of useful techniques for lipid analysis. To analyze the fat content of foods or biological matrices by various methods, a preliminary extraction step is usually required. The different lipid components obtained in an analytical extract will depend on the extraction procedure, specifically the solvent used. Nonpolar organic solvents, such as hexane or supercritical carbon dioxide ($SC-CO_2$), are suitable for neutral or simple lipids, which include esters of fatty acids, acylglycerols, and unsaponifiable matter. Complex or polar lipids, such as phospholipids, glycolipids, lipoproteins, oxidized acylglycerols, and free fatty acids, are preferentially extracted in polar solvents like methanol. Solid-phase extraction is particularly useful for complex polar lipids. Quantitative determination of total fat initially involves acid or base digestion of the test sample in order to liberate bound and complex lipids prior to extraction. A method for quantitative measurement of total fat in foods that meets the Nutrition and Labeling Education Act of 1990 requirements[4] involves acid or alkaline hydrolysis and the extraction of fat from various foods with ether, followed by conversion of the fatty acids to their methyl esters (FAMEs) and quantitative measurement of the FAMEs by capillary gas chromatography (GC).

Determination of the position of double bonds is an important aspect of elucidating the structure of fatty acids. This has been accomplished by a combination of complex chemical and separation techniques, and recently by applying gas chromatography–mass spectrometry (GC-MS) to derivatives of fatty acids, such as 4,4-dimethyloxazolines (DMOXs).[5] Classic techniques include ozonolysis and hydrazine reduction. Hydrazine reduction does not alter the double-bond position or configuration.

Thin-layer chromatography (TLC) is a simple chromatographic technique that allows the separation of a mixture of lipids with widely different polarities in a single run. Silica gel G containing calcium sulfate is used for separating cholesterol esters, triacylglycerols, free fatty acids, cholesterol, diacylglycerols, monoacylglycerols, and phospholipids. Silica gel with silver nitrate is employed for separating fatty acids according to the degree of unsaturation and double-bond configuration (*cis* or *trans*). Microcolumn TLC is performed on a silica gel–glass frit coating on a quartz rod. Quantitation is carried out by scanning with a flame ionization detector (FID) and using the commercially available Iatroscan TLC-FID system. Sebedio et al[6] studied the quantitative analysis of FAME geometrical isomers and triglycerides that differed in unsaturation by using silver nitrate-impregnated rods with the Iatroscan system. The authors observed that the system was useful for quantitation of *trans* fatty acids in margarines and partially hydrogenated oils.

Size exclusion chromatography (SEC) and high-performance (HP) SEC are simple separation techniques based on the relative size of lipid molecules. The stationary phase, which consists of cross-linked macromolecules, discriminates among molecules by excluding the larger ones (that will elute first) and allowing the smaller ones to partially or completely diffuse mechanically into pores of appropriate size (eg, 100 or 500 angstroms). The stationary phase usually used for fat analyses consists of copolymers of styrene-divinyl benzene. HPSEC is a powerful technique for certain applications, such as the separation of hydrolytic, oxidative, and thermal products of frying fat.[7] These alteration products include triacylglycerol dimers, oxidized triacylglycerol monomers, diacylglycerols, monoacylglycerols, and fatty acids, including cyclic fatty acids.

HP liquid chromatography (HPLC) is used for separating nonvolatile high–molecular weight lipids in several modes employing either adsorption or partition chromatography. Adsorption chromatography is widely used for separating classes of lipids according to the nature and number of polar functional groups, such as ester bonds, hydroxyl, and phosphate functions. Silver ion chromatography, in which silica gel is impregnated with silver nitrate, is a form of adsorption chromatography in which the separation occurs according to the number and configuration of double bonds in a molecule. Silver ion–HPLC together with chiral chromatography applied to structural analysis of triacylglycerols has been reviewed.[8]

Reverse-phase HPLC, which is based on partition chromatography, is used to separate individual components that belong to one lipid class. In this case, the stationary phase usually is a nonpolar octadecylsilane (C_{18}) bonded phase, while the mobile phase is a more polar solvent system such as methanol-water.[9,10] HPLC-MS is a system with important applications and great potential for lipid analysis.

Gas chromatography (GC) is a powerful separation technique for resolving and quantifying lipids including nonpolar high–molecular weight species such as triacylglycerols. Methyl ester derivatives of fatty acids are almost exclusively used by GC analysts. Complete separation of common fatty acids is usually achieved by using capillary columns with polar liquid phases. These columns can readily separate fatty acids according to chain length and degree of unsaturation. Recent applications of capillary GC to separations of FAMEs were reviewed by Wolff.[11] To obtain structural information about eluting components, infrared and mass spectrometers must be used as detectors for the gas chromatograph. GC was used to determine the total *trans*-18:1 fatty acid content of milk fats.[12]

Gas chromatography–Fourier transform infrared spectroscopy (GC-FTIR) allows on-line measurement of FTIR spectra of analytes eluting from a gas chromatograph to determine double bond configuration of individual geometric FAME isomers in hydrogenated oils[13,14] or cyclic fatty acid monomers in heated oils.[15]

Gas chromatography-mass spectrometry (GC-MS) provides retention time and molecular weight data, as well as structural information (location of double bonds, hydroxyl groups, or alkyl branches) for lipid components.[16] Unsaturated sites can be found with derivatives that can localize the charge of the molecular ion away from the double bonds. Hence, derivatives such as those prepared with DMOX can be used, since they localize the charge predominantly on the nitrogen atom of these functional groups and result in electron ionization (EI) mass spectra that exhibit distinctive fragmentation patterns from which the double bond's location can be easily determined. DMOX derivatives, which have an important advantage of readily chromatographing on the same polar phases used for FAMEs with no loss in resolution, were used to determine by GC-EIMS the molecular weight, position of double bonds along the fatty acid chain, and ring size of monomers in a complex mixture of cyclic fatty acid monomers.[3]

Functional groups in fat molecules give rise to unique absorption bands in the mid-infrared (IR) spectral region (4000–600 wave numbers), making FTIR spectroscopy an important tool for confirmation of structure. A widely used application of dispersive IR or FTIR spectroscopy is the quantitative determination of total *trans* content in FAMEs.[17] Recently, a single-bounce horizontal ATR cell requiring only 50 μL of meat test sample was developed for rapidly quantitating the *trans* fatty acid content of commercial food products.[18,19]

Wide-line nuclear magnetic resonance (NMR) spectroscopy has been used by the edible oils and fats industry for process monitoring and quality control and has been applied to the rapid determination of the oil content of oilseeds and meals and the solid fat content of food fats and the fat content of food products.[20] Low-resolution methods of the American Oil Chemists' Society include procedures for determination of oil content of oilseeds[21] and solid fat content of commercial fats.[22] High-resolution proton NMR and carbon-13 NMR are important tools available for confirming the identity and elucidating the structure of lipid molecules. Unfortunately, they are relatively insensitive and require large amounts of test samples (about 1–100 mg). The usefulness of high-resolution NMR is greatest for determining the structure of individual compounds, but it diminishes for analyzing components of complex mixtures. Applications of this technique include analysis of partially hydrogenated vegetable oils[23] and quantitation of phospholipids in canned peas by ^{31}P-NMR.[24]

EFFECTS ON HEALTH AND DISEASE

Cholesterol Oxidation Products

Cholesterol oxidation products (COPs) are found in foods as a result of food-processing operations. The association of COPs with a wide variety of toxic properties led to debate concerning the health risk of consuming food products containing COPs. Most recent research in the area of cholesterol oxidation has focused on three major areas: (1) toxicological properties of COPs, (2) occurrence of COPs in foods, and (3) effects of food processing and storage conditions on oxidation of cholesterol. Toxicological studies have concentrated primarily on the role of COPs in cardiovascular disease. Food studies have concentrated primarily on developing methodology to determine the levels of COPs in various food products and, using this methodology, to estimate COPs levels in various foods.

Paniangvait et al[25] and Guardiola et al[26] presented overviews of cholesterol oxidation, biological effects of cholesterol oxides, and various problems associated with quantitation of cholesterol and its oxides. Smith and Johnson[27] reviewed the potential role of COPs in atherogenesis. However, in spite of the great deal of work in this area, the relationships between long-term consumption of COPs and human health remain unclear. Most studies report results from in vitro studies using cell cultures incubated in the presence of COPs. While results from these studies are important in determining which COPs exert toxic effects, they can be considered only a preliminary means of screening potentially toxic compounds. It is difficult to correlate in vitro effects with actual dietary exposure, and it is important that follow-up experiments using whole animals be conducted to determine specific effects of dietary exposure to COPs. In cases where whole animal studies have been conducted, the results must be interpreted cautiously since dietary intake is intentionally exceedingly high to produce acute toxicity effects on the animals under study. On the other hand, there have been few low-level feeding studies. Jacobson et al[28] fed cholestane triol to White Carneau pigeons at a level of 0.0002% of the total diet for a period of 3 months. Coronary artery atherosclerosis in the form of lumenal stenosis was observed in the subject pigeons at a significantly greater frequency than in control pigeons fed 0.05% cholesterol with no triol. From these data, the authors concluded that cholestane triol is an atherogenic risk factor in doses approximating US dietary exposure to cholesterol and its oxidation products. However, their estimate of dietary exposure was based only on the relative percentage of COPs in commercially available grades of cholesterol and can be considered only a crude estimate at best. These findings need to be confirmed with additional dietary exposure models as well as with other COPs. Further work remains to be done to establish long-term effects of dietary exposure to cholesterol oxides at levels similar to those in the US diet.

The second major area of cholesterol oxidation research is the investigation of COP levels in foods.[29] Indeed, the occurrence of COPs in numerous foodstuffs has been documented. Of the more than 80 COPs that have been identified,[30] only seven commonly occur in foods.[31] These seven oxides are 7α- and 7β- hydroxycholesterol, cholesterol-5α,6α-epoxide, cholesterol-5β,6β-epoxide, 7-keto-cholesterol, 25-hydroxy cholesterol, and cholestane-3β,5α,6β-triol. Atherogenic properties have been attributed to five of seven COPs commonly found in foods. Only the epoxides fail to exhibit atherogenic properties. However, the epoxides remain a concern, since they are hydrolyzed to the highly atherogenic cholestane triol under certain conditions. Most unprocessed food products contain little or no COPs. Tsai and Hudson[32] reported high levels of cholesterol epoxides (up to 166 ppm) in some commercial spray-dried egg samples. However, most processed foods contain much lower levels of COPs. Spray-drying studies of egg yolk indicated that only lower levels of COPs were formed under conditions approximating typical commercial drying conditions.[33,34] Only in rare instances are the high levels of COPs reported by Tsai and Hudson[32] found in processed foods. Li et al[35] found that fatty acid and tocopherol content had an effect on cholesterol oxidation during storage and heating of food oils.

Experts agree that more quantitative research is required to identify foods that provide lipid oxidation products to the diet of humans.[25,36] The US Department of Agriculture has established guidelines for frying oils in facilities producing fried meats, and many European countries have specific regulations for frying fats and oils.[37] In addition to COPs, frying fats and oils may contain many other classes of potentially harmful oxidation or thermal degradation products. In spite of the many toxicological studies conducted to date, effects of long-term exposure to low levels of COPs remain unknown. An important next step in atherosclerosis research should be animal studies with a variety of combinations of cholesterol and COPs, particularly at levels approximating US dietary intake.

Conjugated Linoleic Acid (CLA)

Pariza et al[38] reported in 1979 that fried ground beef contained what appeared to be a mutagen inhibitor in addition to bacterial mutagens. The partially purified inhibitor obtained from the ground beef fried at 191°C was observed to inhibit initiation of mouse epidermal tumors by 7,12- dimethyl-benz(a)anthracene.[39] With further purification, the anticarcinogen derived from the fried beef was identified as a mixture of conjugated linoleic acid (CLA) isomers (isomers of *cis*-9,*cis*-12-octadecadienoic acid).[40] *Cis*-9,*trans*-11, *trans*-9,*cis*-11, and *trans*-10,*trans*-12 CLA constitute more than 90%

of the isomers in synthetically prepared CLA.[41] CLAs are present in almost all foods but occur mainly in dairy products and foods derived from ruminant animals.[42] The *cis*-9,*trans*-11 isomer was postulated as the biologically active form of CLA because it can be incorporated into biomembranes.[43] Two mechanisms were proposed for formation of CLA in dairy and other ruminant-derived products: (1) free-radical oxidation of linoleic acid due to aging, heat treatment, and protein quality; and (2) hydrogenation/isomerization of linoleic and linolenic acid in the rumen.[41] Free-radical oxidation was postulated as the major mechanism for CLA formation in cheese prior to aging, whereas biohydrogenation was proposed to predominate during the aging process.[44] Additional evidence also suggested that formation of conjugated dienes is both enzymatic by rumen bacteria[45,46] and due to free-radical reactions via lipid peroxidation in biological tissues.[47–49]

CLA is also found in rodent, rabbit, and human tissues.[43,50–52] Plasma CLA was significantly higher in male subjects following cheddar cheese consumption, and the molar ratio of CLA to linoleic acid increased significantly.[51] CLA in normal human serum was uniformly distributed between phospholipids (36%), triglycerides (36%), and cholesterol esters (28%). Only the *cis*-9,*trans*-11 CLA was found in human duodenal bile.[53] It was postulated that this was due to an enzymatic origin of the CLA rather than to a free-radical mechanism. On the other hand, Fritsche et al[54] found two major CLAs, *cis*-9,*trans*-11 and *trans*-9,*trans*-11, plus two less abundant CLAs, *trans*-9,*trans*-11 and *cis*-9,*cis*-11, in human adipose tissue. The presence of CLAs in adipose tissue was confirmed by GC-IR and GC-MS analysis of FAMEs prepared from extracted tissue using potassium methylate transesterification. Both Shantha et al[55] and Kramer et al[56] pointed out that the choice of methylating agent is critical in the analysis of CLAs as FAMEs, since acid-catalyzed methylation results in some loss of CLAs. Also, base-catalyzed procedures do not convert allylic hydroxy oleates (oxidation products of oleic acid) to CLA.[57]

McGuire et al[58] reported that all human milk samples tested ($N = 14$) contained measurable amounts of CLA (*cis*-9,*trans*-11) ranging from 2.2 to 5.4 mg/g of fat (0.02–0.3 mg/g of milk). CLA was detected in only 31% of infant formula samples examined ($N = 16$) at levels significantly lower than in human milk. CLA in human milk is probably due to dietary intake, since additional investigations[59] suggest that humans may not be able themselves to synthesize CLA.

CLA has been reported to be a potent inhibitor of different types of cancer, including mammary cancers,[60–66] and it has also been suggested that CLA may be beneficial in controlling cardiovascular disease. Lee et al[52] observed that rabbits fed an atherogenic diet augmented daily with CLA (0.5 g of CLA/per rabbit per day) experienced a decrease in total and low-density lipoprotein (LDL) cholesterol levels, and examination of the CLA-fed rabbit aortas showed a significant reduction of atherosclerosis. Nicolosi et al[67] also observed that hamsters fed CLA-containing diets had significantly reduced levels of plasma total cholesterol and LDL cholesterol compared to controls, as well as less early atherosclerosis.

Salminen et al[68] studied the dietary and serum CLA levels in a dietary intervention trial in healthy subjects, comparing similar amounts of *trans* monoenoic fatty acids and stearic acid against a dairy fat–based background diet. Evidence was found suggesting that bioconversion of *trans* fatty acids into CLA occurred in humans. Apparently, CLA was formed during the consumption of a diet rich in C18:1 *trans* fatty acids and incorporated into serum lipids. Only 9,11-CLA was identified since the *cis/trans* isomers could not be separated. Katan et al[69] and Salminen et al[68] reported that C18:1 *trans* fatty acids from partially hydrogenated vegetable oils increase serum LDL cholesterol and decrease high-density lipoprotein (HDL) cholesterol levels. These changes were deemed unfavorable for prevention of CHD.

CLA has also been reported to decrease body fat and increase muscle[70,71] and bone mass.[72] Park et al[70] reported that mice fed a CLA-supplemented diet (5% corn oil plus 0.5% CLA) exhibited 60%

lower body weight and 14% increased lean body mass relative to controls. The authors suggested that CLA might also behave like the omega-3 fatty acids of fish oil in modulating eicosanoid pathways and immune response. West et al[73] found that CLA at 1.0% to 1.2% (w/w) in both low- and high-fat diets produced in mice reduced energy intake, increased energy expenditure, and a loss of body fat that ranged from 43% to 88%.

Houseknecht[74] suggested that CLA may also prove to be a useful therapy for prevention and treatment of Type II (non–insulin-dependent) diabetes in human subjects, since CLA normalized impaired glucose tolerance and improved hyperinsulinemia in the prediabetic Zucker Diabetic rat.

A toxicological evaluation of dietary CLA in rats[75] with animals receiving a diet supplemented with 1.5% CLA (50-fold greater than estimated daily intake for teenage boys) indicated a lack of toxicity of CLA. The various chemoprotective properties of CLA were reviewed by Belury.[76] CLA may also be useful in producing leaner meat while reducing feed costs in addition to helping control cancer and cardiovascular disease and regulating fat and protein metabolism.[77]

Phytosterols and Hypercholesterolemia

Phytosterols (plant sterols), structurally related to cholesterol, reduce blood cholesterol and lower LDL cholesterol levels.[78] Dietary sources include corn, soybean, and other vegetable oils. The most abundant plant sterol is β-sitosterol (sitosterol) while sitostanol, the saturated derivative of sitosterol, occurs at negligible levels in plant lipids. Several studies have suggested that phytosterols lower blood cholesterol levels by either (1) inhibiting cholesterol absorption by displacing cholesterol from intestinal micelles or (2) altering the activity of enzymes involved in cholesterol metabolism and excretion.[79–81] In contrast to cholesterol, phytosterols are poorly absorbed.[82,83] Two percent or less of labeled sitostanol or sitosterol was absorbed intestinally in rats from a test meal versus 18% or more cholesterol. Ikeda et al[83] observed that discrimination between cholesterol and sitosterol can be accounted for by greater uptake of cholesterol by the brush border membrane, as well as by a greater rate of cholesterol delivery from the lumenal bile salt micelle to the membrane.

Although plant sterols inhibit cholesterol absorption, the resulting decrease in serum cholesterol level is slight.[84,85] Sitostanol, prepared by hydrogenation of sitosterol, reduced the intestinal absorption of cholesterol and lowered serum cholesterol more effectively than sitosterol.[86,87] In a study of children with severe familial hypercholesterolemia, Becker et al[88] found that sitostanol reduced LDL cholesterol levels by 33% in 3-month trials. In other studies with hypercholesterolemic subjects, dietary sitostanol-ester lowered serum total and LDL cholesterol levels by about 10%, compared to an increase of 0.1% in a control group.[89,90] Serum campesterol, a dietary plant sterol whose levels reflect cholesterol absorption, was decreased 36% with the diet containing sitostanol ester margarine.[90] It was concluded that long-term use of sitostanol ester margarine as a substitute for part of normal dietary fat would be a beneficial dietary measure. It was also noted that sitostanol itself was not absorbed and did not appear to interfere detectably with absorption of fat-soluble vitamins.

Although cholesterol absorption amounts to 30% to 70% in healthy human adults and plant sterol absorption is no more than around 5%, patients with phytosterolemia, a rare inherited sterol absorption/storage disease, who may also have a moderate or pronounced hypercholesterolemia, can be treated with a low-cholesterol, low–plant sterol diet; therapy with cholestyramine; or administration of sitostanol to reduce both cholesterol and phytosterol absorption.[91]

A combination of pravastatin and sitostanol ester (sitostanol ester margarine) was evaluated for control of mild hypercholesterolemia in men with non–insulin-dependent diabetes mellitus.[92] Serum total and LDL cholesterol were lowered 35% and 44%, respectively, compared to levels observed with a control dietary margarine. The sitostanol ester margarine contained 288, 921, 1138, and 11,400 mg of campesterol, campestanol, sitosterol, and sitostanol, respectively, per 100 g of margarine. Sitosterol was hydrogenated to sitostanol and transesterified with rapeseed oil fatty acids and dissolved in the margarine. Sitostanol ester margarine was also used for dietary treatment of children with familial hypercholesterolemia[93] and for reduction of serum cholesterol in postmenopausal women with previous myocardial infarction.[94] The Finnish company Raisio Chemicals patented a margarine for the Finnish market called Benecol, which contains a derivative of sitostanol derived from pine.[95] The Food and Drug Administration has cleared Benecol spread as "safe under the intended conditions of use."[96]

Westrate and Meijer[97] carried out a randomized double-blind placebo-controlled study with margarines fortified with sterol esters from soybean, shea nut, or rice bran oil. Soybean oil sterols were esterified with sunflower oil fatty acids, rice bran oil sterols were esterified to ferulic acid (oryzanol), and shea nut oil sterols were esterified to a mixture of cinnamic acid (69%), acetic acid (25%), and fatty acids (4%). The margarines, with fatty acid composition as close as possible to that of the control margarine, contained 5% to 11% of free phytosterols. The subjects were 100 healthy nonobese normocholesterolemic and mildly hypercholesterolemic volunteers, 13 to 45 years old, with plasma total cholesterol levels below 8 mmol/L at entry to the study. Sterol intake was between 1.5 and 3.3 g/d. None of the margarines induced adverse changes in blood clinical chemistry. Plasma total and LDL cholesterol were reduced by 8% to 13% for margarines enriched in soybean oil sterol esters (soybean sterol margarine) or sitostanol ester (Benecol) compared to the control margarine. The rice bran oil and shea nut oil sterol margarines were not effective in lowering blood cholesterol levels. The lack of effect in lowering blood cholesterol levels may be related to the presence of 4,4'-dimethyl-sterols in rice bran and shea nut oils as well as the lower sterol content and differences in fatty acid composition of these oils. The study demonstrated that sitosterol and other unsaturated plant sterols, when esterified to fatty acids, can be as effective as saturated sterol esters in lowering circulating cholesterol levels in humans. It was concluded that a margarine with sterol esters from soybean oil, mainly esters from sitosterol, campesterol, and stigmasterol, was as effective as a margarine with sitostanol ester in lowering blood total and LDL cholesterol levels without affecting HDL cholesterol concentrations. The effectiveness of esterified sterols might be related to their enhanced solubility in bile salt micelles and the concurrent decrease in solubility of cholesterol in the micelles and therefore lower absorption.[98]

Omega-3 versus Omega-6 Fatty Acids

Dietary polyunsaturated fatty acids (PUFAs) are classified as two families (omega-6, or n-6, and omega-3, or n-3), which have different physiological functions in the regulation of biological processes. The n-6 PUFA linoleic acid (18:2n-6) is converted by desaturation and elongation to arachidonic acid (20:4n-6); n-6 PUFA, but not arachidonic acid, is found primarily in plants. The n-3 family linolenic acid (18:3n-3) is similarly converted to EPA (20:5n-3) and DHA (22:6n-3); linolenic acid is found in plant sources, while EPA and DHA occur in marine oils.[99] Both omega-3 and omega-6 fatty acids are essential for growth and development and maintenance of good health. Interest in the health effects of omega-3 fatty acids has grown in recent years, partly because these fatty acids have been reported to protect against cardiovascular and inflammatory diseases and certain types of cancer[100,101] and because they are essential nutrients in adults as well as children.[102–105]

The PUFAs of the omega-3 and omega-6 series are of clinical interest because they serve as precursors for the synthesis of prostaglandins (PGs), leukotrienes (LTs), and thromboxanes (TXs), which are collectively referred to as eicosanoids. When macrophages, neutrophils, and other cells are stimulated, arachidonic acid (20:4n-6) is processed by cyclo-oxygenase and lipoxygenase to form prostaglandins, thromboxane, leukotrienes, and platelet-activating factor, which are mediators, respectively, of inflammation; thrombosis and bronchoconstriction; inflammation, chemotaxis, and bronchoconstriction; and bronchial hypersensitivity.[106] High concentrations of these molecules are immunosuppressive because of their many effects on lymphocytes and macrophages. The leukotriene LTB4 is an important mediator for endothelial cell adhesion and chemotaxis of neutrophils. LTC4, LTD4, and LTE4, formerly referred to as the slow-reacting substances of anaphylaxis, are potent mediators of capillary and postcapillary venule vasodilation and permeability. The thromboxane TXA2 causes vasoconstriction and platelet aggregation. The n-3 fatty acids act as inhibitors of arachidonic acid metabolism and the related symptoms of over- and unbalanced productions of lipid mediators.

A diet enriched in EPA leads to the incorporation of this fatty acid into the phospholipid fraction of the cell membrane, where it is available as a substrate for eicosanoid production.[106] These PUFAs have been shown to diminish the production of series-2 PG, TX, and series-4 leukotrienes in humans by competitively inhibiting their synthesis from arachidonic acid.[106,107] Because of the position and number of double bonds in EPA, the enzymes of the eicosanoid cascade produce series-3 PG and TX and series-5 leukotrienes, which are generally less metabolically active than the arachidonic acid metabolites. It is through these sequences of events that omega-3 PUFAs derive their anti-inflammatory and immunomodulatory effects.

Reported effects of omega-3 PUFAs are varied, involving a broad range of metabolic, cardiovascular, and immunological conditions. Epidemiological studies have attributed the low incidence of cardiovascular disease in Eskimos to their relatively high dietary intake of omega-3 to omega-6 PUFAs.[108] Fish oil supplementation has been shown to lower systemic blood pressure in patients with mild hypertension in a double-blind controlled crossover study.[109] Fish oil has also been shown to decrease serum triglyceride and cholesterol levels while increasing concentrations of HDL cholesterol.[110] Harris[111] has reviewed the results of available studies of the effects of omega-3 fatty acids from fish oils and plant oils on human serum lipids and lipoproteins. One suggested effect of omega-3 fatty acids from fish and fish oil is the prevention of arrhythmias and the ability to inhibit ventricular fibrillation and resultant cardiac arrest.[101] Connor and Connor[101] recommended that n-3 fatty acids from fish and fish oil be considered an important therapeutic agent in patients with coronary artery disease as well as to prevent coronary disease in highly susceptible people.

Chronic immunologically mediated diseases, including atopic dermatitis, psoriasis, rheumatoid arthritis, ulcerative colitis, and lupus nephritis, have been shown to improve with short-term use of fish oil supplementation.[112] The major risks of this therapy include bleeding, vitamin E deficiency, and vitamin A and D toxicity. The importance of generating biologically less active eicosanoids may also be extended to the critical care setting. Prefeeding guinea pigs with fish oil has been shown to improve survival when the animals are exposed to lethal doses of endotoxin.[113] In addition, the febrile (fever) response to recombinant interleukin-1 has been shown to be blunted with the use of fish oil supplementation.[114] EPAs' ability to compete with arachidonic acid (20:4n-6) for enzymes of prostaglandin synthesis could account for attenuation of the inflammatory response in certain situations, although alteration in the febrile response was not due to changes in arachidonic acid concentration. Febrile response attenuation might be due to competition between the omega-3 fatty acids and arachidonic acid for active sites on cyclo-oxygenase. Animals fed omega-3 PUFAs from fish oil also have a lesser degree of lactic acidemia following endotoxin infusion than those fed omega-6 PUFAs. This is believed to be due to improved microvascular muscle perfusion.[115] Pulmonary infiltrates were less pronounced in the animals that received fish oil. Fish oil also appears to reduce development of fatty infiltration in the liver and to improve glucose utilization in models of septic injury.[116] The consumption of large quantities of fish oil by healthy human volunteers resulted in diminished production of interleukin-1 and tumor necrosis factor by stimulated peripheral blood monocytes in vitro.[117]

This finding suggested one of several mechanisms whereby omega-3 PUFAs might reduce the severity of the injury response.

Many nutritionists note that too much n-6 and not enough n-3 fatty acids are consumed and that the balance between n-6 and n-3 PUFA in the diet is crucial.[105] Accordingly, health authorities recommend that oily (n-3 fatty acid–rich) fish be consumed twice a week to achieve a more balanced intake. Gibson et al[118] recommended that n-6/n-3 (linoleic acid:linolenic acid) ratios of infant formulas should not deviate significantly from the ratios in human milk—that is, no more than 4:1. Okuyama et al[106] proposed that the current dietary n-6/n-3 ratio of 5 or more (25 in modern Western diets) be revised to 2 or less. Sufficient evidence exists for the health message that consumers should choose foods that provide substantial amounts of n-3 fatty acids (fish, products based on n-3–rich seeds, and vegetables) and avoid foods that are very rich in n-6 fatty acids (products based on staple polyunsaturated oils, certain nuts).[119] Food technology is sufficiently advanced that we can now enrich any food with n-3 PUFA. At the forefront of global food enrichment developments are n-3 PUFA enriched foods, including infant and baby follow-on foods. Some of these enriched products include margarines and breads, which have already been introduced in Britain and Europe.[120] These products also include functional foods (nutritional bars, bread) containing algal DHA.[121] Willis et al[122] assessed the nutritional implications of fat composition on the metabolism and health of both infants and adults and examined the new as well as traditional techniques for obtaining fats and oils with specific nutritional properties.

Fats and Coronary Heart Disease (CHD)

Nutritionists often state that eating too much of any dietary component, fat included, is not conducive to good health. While total dietary fat is one of the many factors implicated in development of CHD,[123] excessive consumption of dietary cholesterol and saturated fatty acids such as myristic acid (16:0), which inhibits cholesterol ester formation and increases LDL cholesterol formation, also contributes to the risk of CHD.[122] The Oslo Study[124] suggested that reduction in plasma total cholesterol and to a lesser extent reduction in smoking were correlated with reduced incidence of myocardial infarction and sudden death. Lewis et al[125] reported that dietary studies in the Netherlands involving reductions in fat intake (from 40% energy from fat to 27% energy from fat) resulted in greater than 20% reductions in serum cholesterol and serum triglycerides. Phillipson et al[126] observed that a diet with fish oil rich in n-3 fatty acids was found to be more effective than a diet with vegetable oil rich in n-6 fatty acids in reducing both plasma cholesterol and plasma triglycerides in patients with hypertriglyceridemia.[126] With fish oil, very–low-density lipoprotein levels were dramatically lowered, as were apoprotein E levels, whereas the vegetable oil diet produced a rise in plasma triglyceride levels.

On the other hand, a study begun in 1986 of nearly 45,000 male health professionals who were 40 to 75 years of age and free of known CHD, found no significant association between dietary intake of n-3 fatty acids or intake of fish and risk of coronary disease.[127] The study data suggested that increasing fish intake from one or two servings per week to five or six servings per week does not substantially reduce the risk of coronary disease. However, Connor and Connor[101] observed that n-3 fatty acids of fish and fish oil have potential for the prevention of CHD and treatment of patients with the disease, since the n-3 acids aid in preventing arrhythmias, have antithrombotic effects, retard the growth of the atherosclerotic plaque, promote synthesis of beneficial nitric oxide in the endothelium, lower plasma triglycerides, and have a mild blood pressure–lowering effect in both normal and mildly hypertensive individuals.

There is a widespread view that saturated fatty acids are the major fat components that elevate serum cholesterol levels and contribute to CHD. However, a review of dietary fatty acids pointed out that stearic acid (18:0) cannot be considered cholesterol elevating and that lauric (12:0) and myristic (14:0) acids are of concern only if diets contain palm kernel oil, coconut oil, or dairy products as major dietary constituents.[128] With regard to palmitic acid (16:0), its response depends upon the metabolic status and age of the subjects under study, while older hypercholesterolemic individuals benefit from decreased consumption of palmitic acid and all saturated fatty acids. Also, the relative levels of dietary cholesterol and linoleic acid have a significant bearing on the cholesterolemic response of palmitic acid.

Another review of the effects of individual fatty acids on plasma lipids and lipoproteins[129] maintained that, in general, saturated fatty acids are hypercholesterolemic while unsaturated fatty acids elicit a hypocholesterolemic effect compared to saturated fatty acids. However, it was emphasized that myristic acid is the most potent saturated fatty acid while stearic acid appears to be neutral. Monounsaturated fatty acids exert a neutral or mildly hypocholesterolemic effect, *trans* fatty acids produce effects intermediate to those of the hypercholesterolemic saturated fatty acids and the *cis* mono- and polyunsaturated fatty acids, and polyunsaturated fatty acids elicit the most potent hypocholesterolemic effect.

The relation between dietary intake of specific types of fat, particularly *trans* unsaturated fat, and risk of CHD was studied in women enrolled in the Nurses' Health Study.[130] The findings suggested that replacing saturated and *trans* unsaturated fats with unhydrogenated monounsaturated and polyunsaturated fats is more effective in preventing CHD in women than reducing overall fat intake. Zock and Katan[131] compared the effects of a diet with 8% of energy from *trans*-octadecenoic acid (elaidic acid) on serum lipoproteins with the effects of a diet with 8% of energy from linoleic acid and stearic acid. Replacement of linoleic acid by elaidic acid or stearic acid in the diet caused lower HDL cholesterol and higher LDL cholesterol levels. No differences were found between men and women in the response of serum lipoprotein or apolipoprotein levels to the *trans* fatty acid diet relative to linoleic acid. It was advised that patients at high risk for atherosclerosis might benefit by avoiding excessive intake of *trans* fatty acids and cholesterol. Nevertheless, *trans* fatty acids form only a minor component of normal diets, and marked reductions in *trans* fatty acid intake would be expected to have less of an impact on LDL cholesterol than a sizable reduction in saturated fatty acids and cholesterol.[132]

A working group of the Food and Agricultural Organization and the World Health Organization[133] concluded that high intake of *trans* fatty acids is undesirable and that their consumption should be reduced. However, evidence implicating *trans* fatty acids in the etiology of coronary disease is not consistent, and it has not been shown that *trans* fatty acids elicit an independent effect on risk of heart disease distinct from that of saturated fats or cholesterol.[134] Because *trans* fatty acids provide no nutritional benefit, it has been stated that prudent public policy suggests that their consumption be minimized and that information on the *trans* content of food should be made available to consumers.[135]

A number of studies have reported lipid responses to butter and margarine added to controlled diets. Judd et al[136] reported that within the range of consumption of table spreads in a typical American diet, the fatty acid profile of palatable table spreads (margarines) can be regulated so that if consumed in place of butter (or presumably in place of other fats high in saturated fatty acids), appreciable improvement in blood lipid profiles can be expected for most people. Two table spreads were studied, a blend of liquid soybean and partially hydrogenated soybean oil (oleic acid, 20%; *trans* monoenes, 17%), and a blend of liquid sunflower oil and completely hydrogenated soybean and canola oils (linolenic acid, 1%; linoleic acid, 48%; oleic acid, 20%; *trans* monoenes, 0%). Both table spreads improved blood lipid profiles for total cholesterol and LDL when compared with butter, with greater improvement with 0% *trans* monoene spread.

Other studies pointed to the cardiovascular benefits of fish oils. Dietary supplementation with fish and fish oil was reported to modify platelet and leukocyte function via modulation of eicosanoid synthesis and reduction of plasma triglycerides.[137–139] Vognild et al[140] observed that intake by healthy male and female volunteers of various marine oils caused antithrombotic changes in platelet membranes. A combination of cod liver oil and olive oil appeared to have the best anticoagulant effect.

Parks et al[141] carried out a study to determine whether a program including exercise therapy, stress management, and consumption of a diet containing 10% fat could reduce the oxidative susceptibility of LDL and induce regression of cardiovascular disease in patients with the disease. The oxidative modification hypothesis of atherosclerosis states that LDL particles are altered by oxidation, that the altered particles are taken up by macrophages inside the arterial wall, and that cholesterol-laden macrophages initiate the formation of atherosclerotic plaques.[142] The atherosclerosis-reversal therapy reduced plasma total cholesterol and LDL cholesterol and also reduced LDL oxidation (as measured by assays of conjugated diene and volatiles) in the patients. LDL particle contents of α-tocopherol and β-carotene were increased. The study suggested that a rigorous treatment program, including strict dietary control, may reduce LDL oxidizability, which could be a factor in promoting cardiovascular disease. The role of oxidized phospholipids as a causative agent in atherosclerosis was reviewed recently.[143]

Fats and Cancer

Dietary guidelines developed over the past 30 years have recommended that consumption of total fat and saturated fat be limited as one means of reducing the risk of cancer as well as other chronic diseases, including diabetes.[144] At present, national recommendations are that total fat intake be no more than 30% of total energy and that saturated fat intake be less than 10% of total energy for all persons over the age of 2 years. For most people, the goal is to reduce absolute fat intake but not to compensate by increasing the intake of carbohydrates.

Direct or indirect relationships have been suggested between dietary fat and various cancers including breast cancer.[145] High-fat diets appear to be associated with aggressive prostate cancer, while n-3 fatty acids can retard the progression of prostate cancer. Experimental evidence with animal models suggest that n-6 fatty acids enhance breast cancer invasion and metastasis via eicosanoid production while n-3 fatty acids may exert a suppressive effect.

Breast Cancer

It has been stated that a high percentage of human cancer is attributable to environmental influences.[146] Wynder[147] claimed that one half of all cancers in women and one third of all cancers in men are associated with dietary factors. The earliest suggestion that diet plays a part in breast cancer came from population surveys that indicated that this disease is less common among Asian and African populations than among populations in Europe and North America.[148] Lea[149] stated that consumption of fat was most closely correlated with the incidence of breast cancer. Hems[150] reported that the strongest association was with per capita consumption of fat, sugar, and animal protein. On the other hand, a Canadian study[151] showed a weak relationship between the disease and total dietary fat. Dietary cholesterol intake showed no relation to the disease. Both the Evans County[152] and the Framingham[153] studies, which primarily investigated incidence of heart disease, found no relationship between serum cholesterol and cancer in women.

Recent studies from southern Europe indicated that consumption of olive oil (mostly monounsaturated fat) was inversely related to risk of breast cancer,[154,155] whereas saturated fat (including margarine) and total fat intake appeared to be associated with an elevated risk of the disease. Willett[156] concluded that available epidemiological evidence provided little support for any important relation between intake of either linoleic acid or long-chain n-3 fatty acids from fish and risk of breast cancer. However, Wolk et al[157] stated that a population-based prospective cohort study in Sweden demonstrated a positive association of polyunsaturated fat with the risk of breast cancer compared to an inverse association with monounsaturated fat.

Knekt et al[158] found an inverse relationship between dietary intake of milk and subsequent incidence of breast cancer. Milk fat accounted for the major portion of the total dietary fat ingested by the Finnish study subjects, and it was pointed out that milk fat is a good source of CLA,[42] which has been found to suppress breast cancer cells in animals.[60,159] Ip[160] concluded that CLA is perhaps the most potent anticancer fatty acid in that about 1% or less in the diet is sufficient to produce a significant protective effect. No evidence was found that *trans* fatty acids are associated with an increased cancer risk. The current status of the relationship of dietary fat and human breast cancer was reviewed by Erickson.[161]

Prostate Cancer

Comparisons of populations in different countries have suggested a positive correlation of dietary fat with prostate cancer risk.[162,163] An inverse association was observed between intake of linoleic acid and risk of prostate cancer.[164] On the other hand, a positive association was noted for total fat and saturated fat in men younger than 50. Giovannucci et al[165] also reported that total fat consumption was directly related to risk of prostate cancer, but this was primarily due to animal fat. Fat from dairy products (with the exception of butter) or fish was not related to risk. The authors recommended that intake of animal fat be reduced as a practical means to reduce mortality from prostate cancer. Willett[156] concluded that epidemiological studies support a positive relation between consumption of animal fat and risk of prostate cancer but that vegetable fat is not related to cancer risk. Consumption of linoleic acid or long-chain PUFA did not appear to affect risk of prostate cancer.

Von Holtz et al[166] compared the effect of cholesterol and β-sitosterol on LNCaP human prostate cancer cell growth, differentiation, apoptosis, and sphingomyelin cycle intermediates. Compared with cholesterol, sitosterol decreased growth by 24% and induced apoptosis fourfold. The increase in apoptosis appeared to be associated with alterations in the sphingomyelin cycle. These results suggested a protective role of phytosterols against prostate cancer.

Connolly et al[167] examined the effects of different dietary fatty acids on DU 145 human prostate cell growth in nude mice (the diets included 18% corn oil/5% linseed oil, 18% linseed oil/5% corn oil, or 18% menhaden oil/5% corn oil). Compared to the other two diets, a 30% reduction in tumor growth was observed with the high–menhaden oil diet, possibly due, in part, to a reduction of arachidonic acid available for prostaglandin biosynthesis. In another experiment, with mice injected with DU 145 cells directly into the prostate gland, a high-fat (20% fat) linoleic acid–rich diet resulted in twice the mean tumor weight observed with a low-fat (5% fat) diet. It was concluded that a high-fat, high–linoleic acid diet was associated with increased prostate cancer risk.

Colon Cancer

Consumption of red meat or beef has been related to risk of colon cancer, whereas intake of other major sources of animal fat (dairy products, poultry, fish, vegetable oil) does not increase colon cancer risk.[168] Several studies suggest that long-chain (n-3) fatty acids from marine sources inhibit colon cancer.[169] Lindner[170] found that mice fed a diet high in beef tallow developed the greatest number of 1,2-dimethylhydrazine–induced colon tumors, whereas mice fed fish oil developed the fewest tumors. Monounsaturated fat was strongly correlated with tumors. Saturated fat and n-6 PUFA did not show any correlation with tumor production.

Kim et al[171] found that with male Sprague-Dawley rats, high butter and beef tallow diets (45% of total calories) had a promotive effect on distal colon cell proliferation compared to a high–fish oil diet. A high–corn oil diet had an intermediate effect. These studies suggested that colon cancer proliferation depends on the types of dietary fat and colon site.

Iwamoto et al[172] found that an EPA (20:5n-3) diet (9.5% EPA) suppressed the liver metastasis of ACL-15 rat colon cancer cells injected into male F344 rats, whereas a linoleic acid diet (10% linoleic acid) stimulated metastasis that was greater than that with a palmitic acid diet (9.5% palmitic acid). Awad et al[173] observed that sitostanol but not cholesterol added to the media in the form of sterol cyclodextrin complexes inhibited the growth of a human colon cancer cell line and altered the fatty acid composition of minor phospholipids. The authors speculated that inhibition by sitosterol may have been mediated through the influence of signal transduction pathways involving membrane phospholipids.

Butyric acid, present as butyryldiacylglycerols, a component of milk fat that inhibits cell proliferation and induces cellular differentiation in vitro, was found to act synergistically with all-*trans*-retinoic acid to induce cell differentiation in HL-60 lines and was suggested as a candidate for treatment of human malignancies.[174] Butyric acid also induced apoptosis as well as cell differentiation in human colonic carcinoma cell lines.[175] The antitumor and antiviral activities of interferon in vitro and in vivo were reportedly enhanced by butyric acid.[176] Butyric acid influences cell growth by inhibition of histone deacetylases causing hyperacetylation of the histones H_3 and H_4. This in turn relaxes parts of chromatin structure and exposes damaged portions of DNA to repair enzymes.[177] Fermentation of unabsorbed carbohydrates in the colon by bacteria can lead to the formation of butyric acid and may explain the protective role of butyrate and fiber in colon cancer. Rats fed fiber such as wheat bran, which maintain high butyric acid concentrations in the distal large bowel, had fewer DMH-induced colon tumors than those fed soluble fibers, which did not raise butyric acid levels.[178]

Structured Lipids

Structured lipids are lipids reconstituted by chemical or enzymatic procedures to alter their fatty acid composition and/or the stereochemical positions of the fatty acids in the glycerol molecule.[179] They are produced commercially by chemical hydrolysis of fats and oils and random esterification, or by interesterification catalyzed by alkali metals or alkali metal alkylates at elevated temperatures. They can also be prepared by enzymatic synthesis using specific lipases to catalyze direct esterification or transesterification reactions at low temperatures. Both fatty acid chain length and positional specificity of individual lipases can be exploited to obtain desired products containing specific fatty acids, including essential fatty acids in the right proportions. Babayan[180] described structured lipids with a medium-chain triglyceride (MCT) backbone and containing linoleic acid as suitable for care of critically ill patients.

Reported benefits of structured lipids include enhanced absorption of fatty acids in the *sn*-2 position of triglycerides, reduction in serum triglycerides and LDL cholesterol, improved immune function, prevention of thrombosis, reduced risk of cancer, use as reduced-calorie fat, and use as lipid emulsions for enteral and parenteral feeding.[179,181] Jandacek et al[182] observed that triglycerides with octanoic (caprylic) acid in the *sn*-1 and *sn*-3 positions and long-chain fatty acids (linoleic, etc) in the *sn*-2 position are hydrolyzed and absorbed more efficiently than long-chain triglycerides. Hubbard and McKenna[183] found that structured lipid preparations formulated from MCTs and safflower oil were efficient vehicles for supplementation of linoleic acid to cystic fibrosis patients with decreased plasma and tissue levels of linoleic acid that were attributed to malabsorption of linoleic acid associated with exocrine pancreatic insufficiency.

Although cost is a barrier to widespread use of structured lipids in foods, a number of nutritional products are available for nonclinical uses such as maintaining health or bodybuilding, in addition to medical applications such as impaired gastrointestinal function or food allergies of infants.[184] Structured lipid products developed recently include the Salatrim family, accepted by FDA in 1994 as "generally recognized as safe," providing 5 cal/g and tailored for a range of uses.[185]

MCTs, often the basis for structured lipids, are a source of readily available energy, since they are readily absorbed without the need for pancreatic lipase and are transported directly to the liver, where they are metabolized like carbohydrates. Food applications include use as a carrier for flavors, colors, and vitamins, as an oil coating for dried fruits, as an ingredient of reduced-calorie foods, and as an energy source in special nutritionals.[186] Cheeses made with MCTs as well as margarines and other spreads have been suggested as food items for patients with malabsorption problems.[187] Specialty fats include zero *trans* soybean margarines prepared from interesterified soybean oil–soybean trisaturate blend (80:20) or a blend of 80:20 feedstock with additional 20% liquid soybean oil to produce a softer product.[188] Recent developments in the production of nutritionally functional fats and oils have been reviewed by Willis et al.[122]

CONCLUSIONS

Dietary fats and oils are important sources of energy, provide fatty acids essential for growth and development, and play a variety of roles in health and disease. Specific fatty acids are associated with immune function, blood pressure, CHD, and risk of certain cancers. Interest has grown in recent years in the use of omega-3 dietary supplementation to control autoimmune disease, as anticancer agents, and for blood pressure reduction. Structured and specialty lipids have become important products for patients with special nutritional needs and have the potential of preventing as well as assisting in the treatment of disease.

REFERENCES

1. Schmidl MK. The role of lipids in medical and designer foods. In: McDonald RE, Min DB, eds. *Food Lipids and Health.* New York, NY: Marcel Dekker, Inc; 1996:417–436.

2. National Academy of Sciences. *Diet and Health: Implications for Reducing Chronic Disease Risk.* Washington, DC: National Academy Press; 1989.

3. Mossoba MM, Yurawecz MP, Roach JAG, et al. Elucidation of cyclic fatty acid monomer structures: cyclic and bicyclic ring sizes and double bond positions and configuration. *J Am Oil Chem Soc.* 1995;72:721–727.

4. House SD, Larson PA, Johnson RR, et al. Gas chromatographic determination of total fat extracted from food samples using hydrolysis in the presence of antioxidant. *J AOAC Int.* 1994;77:960–965.

5. Mossoba MM, McDonald RE, Roach JAG, et al. Spectral confirmation of *trans* monounsaturated C_{18} fatty acid positional isomers. *J Am Oil Chem Soc.* 1997;74:125–130.

6. Sebedio JL, Farquharson TE, Ackman RG. Quantitative analysis of methyl esters of fatty acid geometrical isomers using $AgNO_3$ impregnated rods. *Lipids.* 1985;20:555–560.

7. Christopoulou CN, Perkins EG. High performance size exclusion chromatography of monomer, dimer and trimer mixtures. *J Am Oil Chem Soc.* 1989;66:1338–1343.

8. Svensson L, Sisfontes L, Nyborg G, Blomstrand R. High performance liquid chromatography and glass capillary gas chromatography of geometric and positional isomers of long chain monounsaturated fatty acids. *Lipids.* 1982;17:50–59.

9. Sebedio JL, Prevost J, Ribot E, Grandgirad A. Utilization of high-performance liquid chromatography as an enrichment step for the determination of cyclic fatty acid monomers in heated fats and biological samples. *J Chromatogr A.* 1994;659:101–109.

10. Adlof RO, Copes LC, Emken EA. Analysis of monoenoic fatty acid distribution in hydrogenated vegetable oils by silver-ion high-performance liquid chromatography. *J Am Oil Chem Soc.* 1995;72:571–574.

11. Wolff RL. Recent applications of capillary gas-liquid chromatography to difficult separations of positional or geometrical isomers of unsaturated fatty acids. In: Sebedio JL, Perkins EG, eds. *New Trends in Lipid and Lipoprotein Analyses.* Champaign, IL: AOCS Press; 1995:147–180.

12. Molkentin J, Precht D. Optimized analysis of *trans*-octadecenoic acids in edible fats. *Chromatographia.* 1995;41: 267–271.

13. Mossoba MM, McDonald RE, Chen J-YT, et al. Identification and quantitation of *trans*- 9,*trans*-12-octadecadienoic acid methyl ester and related compounds in hydrogenated soybean oil and margarines by capillary gas chromatography/matrix isolation/Fourier transform infrared spectroscopy. *J Agric Food Chem.* 1990;38:86–92.

14. Mossoba MM, McDonald RE. Applications of capillary GC-FTIR. *INFORM.* 1993;4:854–859.

15. Mossoba MM, Yurawecz MP, Lin HS, et al. Application of GC-MI-FTIR spectroscopy to the structural elucidation of cyclic fatty acid monomers. *Am Lab.* 1995;27(14):16K–18K.

16. Christie WW. *Gas Chromatography and Lipids.* Ayr, Scotland: Oily Press; 1989;161–184.

17. Firestone D, Sheppard A. Determination of *trans* fatty acids. In: Christie WW, ed. *Lipid Methodology-One.* Ayr, Scotland: Oily Press; 1992:273–322.

18. Ali LH, Angyal G, Weaver CM, et al. Determination of total *trans* fatty acids in foods: comparison of capillary-column gas chromatography and single-bounce horizontal attenuated total reflection infrared spectroscopy. *J Am Oil Chem Soc.* 1996;73:1699–1705.

19. American Oil Chemists' Society. Isolated *trans* Geometric Isomers Single Bounce-Horizontal Attenuated Total Reflection Infrared Spectroscopic Procedure. In: *Official Methods and Recommended Practices.* 5th ed. Champaign, IL: AOCS; 1998:Recommended Practice Cd 14d-96.

20. Horman I. NMR Spectroscopy. In: Charalambous G, ed. *Analysis of Foods and Beverages.* Orlando, FL: Academic Press: 1984;205–264.

21. Simultaneous determination of oil and moisture contents of oilseeds using pulsed nuclear magnetic resonance spectroscopy. In: *Official Methods and Recommended Practices.* 5th ed. Champaign, IL: AOCS; 1998:Recommended Practice Ak 4-95.

22. Solid Fat Content (SFC) by Low-Resolution Nuclear Magnetic Resonance. In: *Official Methods and Recommended Practices.* 5th ed. Champaign, IL: AOCS; 1998:Recommended Practice Cd 16b-93.

23. Gunstone FD. The composition of hydrogenated fats by high-resolution ^{13}C nuclear magnetic resonance spectroscopy. *J Am Oil Chem Soc.* 1993;70:965–970.

24. Murcia MA, Villalain J. Phospholipid composition of canned peas by ^{31}P-NMR. *J Sci Food Agric.* 1993;61:345–347.

25. Paniangvait P, King AJ, Jones AD, and German BG. Cholesterol oxides in foods of animal origin. *J Food Sci.* 1995;60: 1159–1174.

26. Guardiola F, Codony R, Addis PB, et al. Biological effects of oxysterols: current status. *Food Chem Toxicol.* 1996;34: 193–211.

27. Smith LL, Johnson BH. Biological activities of oxysterols. *Free Radical Biol Med.* 1989;7:285–332.

28. Jacobson MS, Price MG, Shamoo AE, Heald FP. Atherogenesis in White Carneau pigeons: effects of low-level cholestane-triol feeding. *Atherosclerosis.* 1985;57:209–217.

29. Addis PB, Park SW. Cholesterol oxide content of foods. In: Peng SK, Morin JR, eds. *Biological Effects of Cholesterol Oxides.* Boca Raton, FL: CRC Press; 1992:71–88.

30. Smith LL. *Cholesterol Autoxidation.* New York, NY: Plenum Press; 1981.

31. Finocchiaro T, Richardson T. Sterol oxides in foodstuffs: a review. *J Food Protect.* 1983;46:917–925.

32. Tsai LS, Hudson CA. Cholesterol oxides in commercial dry egg products: quantitation. *Food Sci.* 1985;50:229–231.

33. Morgan JN, Armstrong DJ. Formation of cholesterol-5,6-epoxides during spray-drying of egg yolk. *J Food Sci.* 1987;52: 1224–1227.

34. Morgan JN, Armstrong DJ. Quantification of cholesterol oxidation products in egg yolk powder spray-dried with direct heating. *J Food Sci.* 1992;57:43–45.

35. Li SX, Cherian G, Ahn DU, et al. Storage, heating, and tocopherols affect cholesterol oxide formation in food oils. *J Agric Food Chem.* 1996;44:3830–3834.

36. Addis PB, Park SW. Role of lipid oxidation products in atherosclerosis. In: Taylor SL, Scanlan RA, eds. *Food Toxicology: A Perspective on the Relative Risks*. New York, NY: Marcel Dekker, Inc; 1989:297–330.

37. Firestone D, Stier RF, Blumenthal MM. Regulation of frying fats and oils. *Food Technol.* 1991;45:90–94.

38. Pariza MW, Ashoor SH, Chu FS, Lund DB. Effects of temperature and time on mutagen formation in pan-fried hamburger. *Cancer Lett.* 1979;7:63–69.

39. Pariza MW, Hargraves WA. A beef-derived mutagenesis modulator inhibits initiation of mouse epidermal tumors by 7,12-dimethylbenz(a)anthracene. *Carcinogenesis.* 1985;6:591–593.

40. Ha YL, Grimm NK, Pariza MW. Anticarcinogens from fried ground beef: heat altered derivatives of linoleic acid. *Carcinogenesis.* 1987;8:1881–1887.

41. Ha YL, Grimm NK, Pariza MW. Newly recognized anticarcinogenic fatty acids: identification and quantitation in natural and processed cheese. *J Agric Food Chem.* 1989;37:75–81.

42. Chin SF, Liu W, Storkson JM, et al. Dietary sources of conjugated dienoic isomers of linoleic acid, a newly recognized class of anticarcinogens. *J Food Composition Anal.* 1992;5:185–197.

43. Ha YL, Storkson J, Pariza MW. Inhibition of benzo(a)pyrene-induced mouse forestomach neoplasia by conjugated derivatives of linoleic acid. *Cancer Res.* 1990;50:1097–1101.

44. Lin H, Boylston TD, Luedecke LO, Shultz TD. Factors affecting the conjugated acid content of cheddar cheese. *J Agric Food Chem.* 1998;46:801–807.

45. Hughes PE, Hunter WJ, Tove SB. Biohydrogenation on unsaturated fatty acids. *J Biol Chem.* 1982;257:3643–3649.

46. Fujimoto K, Kimoto H, Shishikura M, et al. Biohydrogenation of linoleic acid by anaerobic bacteria isolated from rumen. *Biosci Biotechnol Biochem.* 1993;57:1026–1027.

47. Cawood P, Wickens DG, Iversen SA, et al. The nature of diene conjugation in human serum, bile and duodenal juice. *FEBS Lett.* 1983;162:239–243.

48. Iversen SA, Cawood P, Madigan MJ, et al. Identification of a diene conjugated component of human lipid as octadeca-9,11-dienoic acid. *FEBS Lett.* 1984;171:320–324.

49. Dormandy TL, Wickens DG. The experimental and clinical pathology of diene conjugation. *Chem Phys Lipids.* 1987;45:353–364.

50. Harrison K, Cawood P, Iversen A, Dormandy T. Diene conjugation patterns in normal human serum. *Life Chem Rep.* 1985;3:41–44.

51. Huang Y-C, Luedecke LO, Shultz TD. Effect of cheddar cheese consumption on plasma conjugated linoleic acid concentrations in men. *Nutr Res.* 1994;14:373–386.

52. Lee KN, Kritchevsky D, Pariza MW. Conjugated linoleic acid and atherosclerosis in rabbits. *Atherosclerosis.* 1994;108:19–25.

53. Smith GN, Taj M, Braganza JM. On the identification of a conjugated diene component of duodenal bile as 9Z, 11E-octadecadienoic acid. *Free Radical Biol Med.* 1991;10:13–21.

54. Fritsche J, Mossoba MM, Yurawecz MP, et al. Conjugated linoleic acid (CLA) isomers in human adipose tissue. *Z Lebensm Unters Forsch A.* 1997;205:415–418.

55. Shantha NC, Decker EA, Hennig B. Comparison of methylation methods for the quantitation of conjugated linoleic acid isomers. *J AOAC Int.* 1993;76:644–649.

56. Kramer JKG, Fellner V, Dugan MER, et al. Evaluating acid and base catalysts in the methylation of milk and rumen fatty acids with special emphasis on conjugated dienes and total *trans* fatty acids. *Lipids.* 1997;32:1219–1228.

57. Yurawecz MP, Hood JK, Roach JAG, et al. Conversion of allylic hydroxy oleate to conjugated linoleic acid and methoxy oleate by acid-catalyzed methylation procedures. *J Am Oil Chem Soc.* 1994;71:1149–1155.

58. McGuire MK, Park Y, Behre RA, et al. Conjugated linoleic acid concentrations of human milk and infant formula. *Nutr Res.* 1987;17:1277–1283.

59. Herbel BK, McGuire MK, McGuire MA, Shultz TD. Safflower oil consumption does not increase plasma conjugated linoleic acid concentrations in humans. *Am J Clin Nutr.* 1998;67:332–337.

60. Ip C, Chin SF, Scimeca JA, Pariza MW. Mammary cancer prevention by conjugated dienoic derivatives of linoleic acid. *Cancer Res.* 1991;51:6118–6124.

61. Zu H-X, Schutt HA. Inhibition of 2-amino-3-methylimidazo[4,5-f]quinoline-DNA adduct formation in CDF$_1$ mice by heat-altered derivatives of linoleic acid. *Food Chem Toxicol.* 1992;30:9–16.

62. Shultz TD, Chew BP, Seaman WR, Luedecke LO. Inhibitory effect of conjugated dienoic derivatives of linoleic acid and β-carotene on the in vitro growth of cancer cells. *Cancer Lett.* 1992;63:125–133.

63. Shultz TD, Chew BP, Seaman WR. Differential stimulatory and inhibitory responses of human MCF-7 breast cancer cells to linoleic acid and conjugated linoleic acid in culture. *Anticancer Res.* 1992;12:2143–2146.

64. Ip C, Scimeca JA, Thompson HJ. Conjugated linoleic acid: a powerful anticarcinogen from animal fat sources. *Cancer.* 1994;74:1050–1054.

65. Ip C, Lisk DJ, Scimeca JA. Potential of food modification in cancer prevention. *Cancer Res.* 1994;(suppl);54: 1957S–1959S.

66. Durgam VR, Fernandes G. The growth inhibitory effect of conjugated linoleic acid on MCF-7 cells is related to estrogen response system. *Cancer Lett.* 1997;116:121–130.

67. Nicolosi RJ, Rogers EJ, Kritchevsky D, et al. Dietary conjugated linoleic acid reduces plasma lipoproteins and early aortic atherosclerosis in hypercholesterolemic hamsters. *Artery.* 1997;22:266–277.

68. Salminen I, Mutanen M, Jauhiainen M, Aro A. Dietary *trans* fatty acids increase conjugated linoleic acid levels in human serum. *Nutr Biochem.* 1998;9:93–98.

69. Katan MB, Zock PL, Mensink RP. *Trans* fatty acids and their effects on lipoproteins in humans. *Ann Rev Nutr.* 1995;15: 473–493.

70. Park Y, Albright KJ, Liu W, et al. Effect of conjugated linoleic acid on body composition in mice. *Lipids.* 1997;32: 853–858.

71. Dugan MER, Aalhus JL, Schaefer AL. Kramer JKG. The effects of conjugated linoleic acid on fat to lean repartitioning and feed conversion in pigs. *Can J Anim Sci.* 1997;77:723–725.

72. Li Y, Watkins BA. Conjugated linoleic acids alter bone fatty acid composition and reduce ex vivo bone prostaglandin E$_2$ biosynthesis in rats fed n-6 or n-3 fatty acids. *Lipids.* 1998;33:417–425.

73. West DB, Delany JP, Camet PM, et al. Effects of conjugated linoleic acid on body fat and energy metabolism in the mouse. *Am J Physiol.* 1998;275:R667–R672.

74. Houseknecht KL, VandenHeuvel JP, Moya-Camerena SY, et al. Dietary conjugated linoleic acid normalizes impaired glucose tolerance in the Zucker fatty rat. *Biochem Biophys Res Comm.* 1998;244:678–682.

75. Scimeca JA. Toxicological evaluation of dietary conjugated linoleic acid in male Fischer 344 rats. *Food Chem Toxicol.* 1998;36:391–395.

76. Belury MA. Conjugated dienoic linoleate: a polyunsaturated fatty acid with unique chemoprotective properties. *Nutr Rev.* 1995;53:83–89.

77. Haumann BF. Conjugated linoleic acid offers research promise. *INFORM.* 1996;7:152–159.

78. Ling WH, Jones PJH. Dietary phytosterols: a review of metabolism, benefits and side effects. *Life Sciences.* 1995;57: 195–206.

79. Ikeda I, Sugano M. Some aspects of mechanism of inhibition of cholesterol absorption by β-sitosterol. *Biochem Biophys Acta.* 1983;732:651–658.

80. Ikeda I, Tanabe Y, Sugano M. Effects of sitosterol and sitostanol on micellar solubility of cholesterol. *J Nutr Sci Vitaminol.* 1989;35:361–369.

81. Laraki L, Pelletier X, Mourot J, Debry G. Effects of dietary phytosterols on liver lipids and lipid metabolism enzymes. *Ann Nutr Metab.* 1993;37:129–133.

82. Hassan AS, Rampone AJ. Intestinal absorption and lymphatic transport of cholesterol and β-sitostanol in the rat. *J Lipid Res.* 1979;20:646–653.

83. Ikeda I, Tanaka K, Sugano M, et al. Discrimination between cholesterol and sitosterol for absorption in rats. *J Lipid Res.* 1988;29:1583–1591.

84. Grundy SM, Ahrens EH Jr, Davignon J. The interaction of cholesterol absorption and cholesterol synthesis in man. *J Lipid Res.* 1969;10:304–315.

85. Lees AM, Mok HYI, Lees RS, et al. Plant sterols as cholesterol-lowering agents: clinical trials in patients with hypercholesterolemia and studies of sterol balance. *Atherosclerosis.* 1977;28:325–338.

86. Heinemann T, Leiss O, von Bergmann K. Effect of low-dose sitostanol on serum cholesterol in patients with hypercholesterolemia. *Atherosclerosis.* 1986;61:219–223.

87. Heinemann T, Kullak-Ublick G-A, Pietruck B, von Bergmann K. Mechanisms of action of plant sterols on inhibition of cholesterol absorption: comparison of sitosterol and sitostanol. *Eur J Clin Pharmacol.* 1991;40(suppl 1); S59–S63.

88. Becker M, Staab D, von Bergmann K. Treatment of severe familial hypercholesterolemia in childhood with sitosterol and sitostanol. *J Pediatr.* 1993;122:292–296.

89. Vanhanen HT, Blomqvist S, Ehnholm C, et al. Serum cholesterol, cholesterol precursors, and plant sterols in hypercholesterolemic subjects with different apoE phenotypes during dietary sitostanol ester treatment. *J Lipid Res.* 1993;34:1535–1544.

90. Miettinen TA, Puska P, Gylling H, et al. Reduction of serum cholesterol with sitostanol-ester margarine in a mildly hypercholesterolemic population. *N Engl J Med.* 1995;333:1308–1312.

91. Lutjohann D, von Bergmann K. Phytosterolemia: diagnosis, characterization and therapeutical approaches. *Ann Med.* 1997;29:181–184.

92. Gylling H, Miettinen TA. Effects of inhibiting cholesterol absorption and synthesis in cholesterol and lipoprotein metabolism in hypercholesterolemic non-insulin-dependent diabetic men. *J Lipid Res.* 1996;37:1776–1785.

93. Gylling H, Siimes MA, Miettinen TA. Sitostanol ester margarine in dietary treatment of children with familial hypercholesterolemia. *J Lipid Res.* 1995;36:1807–1812.

94. Gylling H, Radhakrishnan R, Miettinen TA. Reduction of serum cholesterol in postmenopausal women with previous myocardial infarction and cholesterol malabsorption induced by dietary sitostanol ester margarine. *Circulation.* 1997;96:4226–4231.

95. Finnish margarine lowers cholesterol. *Chem Britain.* 1996;32(9):11.

96. FDA clears cholesterol-lowering Benecol spread as safe. *Food Chem News.* 1999;41(14):13–14.

97. Westrate JA, Meijer GW. Plant sterol-enriched margarines and reduction of plasma total- and LDL-cholesterol concentrations in normocholesterolaemic and mildly hypercholesterolaemic subjects. *Eur J Clin Nutr.* 1998;52:334–343.

98. Jones PJH, Ntanios F. Comparable efficacy of hydrogenated versus nonhydrogenated plant sterol esters on circulating cholesterol levels in humans. *Nutr Rev.* 1998;56:245–248.

99. Firestone D. *Physical and Chemical Characteristics of Oils, Fats, and Waxes.* Champaign, IL: AOCS Press; 1999.

100. Simopoulos AP. Omega-3 fatty acids in health and disease and in growth and development. *Am J Clin Nutr.* 1991;54:438–463.

101. Connor SL, Connor WE. Are fish oils beneficial in the prevention and treatment of coronary artery disease? *Am J Clin Nutr.* 1997;66(suppl):1020S–1031S.

102. Holman RT, Johnson SB, Hatch T. A case of human linolenic acid deficiency involving neurological abnormalities. *Am J Clin Nutr.* 1982;35:617–623.

103. Bjerve KS. Alpha-linolenic acid deficiency in adult women. *Nutr Rev.* 1987;45:15–19.

104. Bjerve KS. Omega-3 fatty deficiency in man: implications for the requirement of alpha-linolenic acid and long-chain omega-3 fatty acids. In: Simopoulos AP, Kifer RR, Martin RE, Barlow SM, eds. *Health Effects of Omega-3 Polyunsaturated Fatty Acids in Seafoods. World Rev Nutr Diet.* 1991;66:133–142.

105. Haumann BF. Nutritional aspects of n-3 fatty acids. *INFORM.* 1997;8:428–447.

106. Okuyama H, Kobayashi T, Watanabe S. Dietary fatty acids: the n-6/n-3 balance and chronic elderly diseases: excess linoleic acid and relative n-3 deficiency syndrome seen in Japan. *Prog Lipid Res.* 1997;35:409–457.

107. Lee TH, Hoover RL, Williams JD, et al. Effect of dietary enrichment with eicosapentaenoic and docosahexaenoic acids on in vitro neutrophil and monocyte leukotriene generation and neutrophil function. *N Engl J Med.* 1985;312: 1217–1224.

108. Bang HO, Dyerberg J. Plasma lipids and lipoproteins in Greenlandic west coast Eskimos. *Acta Med Scand.* 1972; 192:85–94.

109. Radack K, Deck C, Huster G. The effects of low doses of n-3 fatty acid supplementation on blood pressure in hypertensive subjects. *Arch Intern Med.* 1991;151:1173–1180.

110. Sirtori CR, Gatti E, Tremoli E, et al. Olive oil, corn oil, and n-3 fatty acids differently affect lipids, lipoproteins, platelets and superoxide formation in Type II hypercholesterolemia. *Am J Clin Nutr.* 1992;56:113–122.

111. Harris WS. n-3 Fatty acids and serum lipoproteins: human studies. *Am J Clin Nutr.* 1997;65(suppl);1645S–1654S.

112. Yetiv, JZ. Clinical applications of fish oil. *JAMA.* 1988;260:665–670.

113. Mascioli E, Leader L, Flores E, et al. Enhanced survival to endotoxin in guinea pigs fed IV fish oil emulsion. *Lipids.* 1988;23:623–625.

114. Pomposelli JJ, Mascioli EA, Bistrian BR, et al. Attenuation of the febrile response in guinea pigs by fish oil enriched diets. *JPEN.* 1989;13:136–140.

115. Pomposelli JJ, Flores EA, Blackburn GL, et al. Diets enriched with n-3 fatty acids ameliorate lactic acidosis by improving endotoxin-induced tissue hypoperfusion in guinea pigs. *Ann Surg.* 1991;213:166–176.

116. Ling MK, Istran N, Colon E. Effect of fish oil on glucose metabolism in the interleukin-1 alpha (IL-1) treated rat. *JPEN.* 1992;15(suppl):215.

117. Endres S, Ghorbani R, Kelly VE, et al. The effect of dietary supplementation with n-3 polyunsaturated fatty acids on the synthesis of interleukin-1 and tumor necrosis factor by mononuclear cells. *N Engl J Med.* 1989;320:265–271.

118. Gibson RA, Makrides M, Neumann MA, et al. Ratios of linoleic acid to α-linolenic acid in formulas for term infants. *J Pediatr.* 1994;125(suppl); S48–S55.

119. Cleland LG, James MJ. Diet and arthritis: rheumatoid arthritis and the balance of dietary n-6 and n-3 essential fatty acids. *Br J Rheumatol.* 1997;35:513–514.

120. Newton IS. Food enrichment with long-chain n-3 PUFA. *INFORM.* 1996;7:169–180.

121. Becker CC, Kyle DJ. Developing functional foods containing algal docosahexaenoic acid. *Food Technol.* 1998;52:68–71.

122. Willis WM, Lencki RW, Marangoni AG. Lipid modification strategies in the production of nutritionally functional fats and oils. *Crit Rev Food Sci Nutr.* 1998;38:639–674.

123. Gurr MI. Dietary lipids and coronary heart disease: old evidence, new perspective. *Prog Lipid Res.* 1992;31:195–243.

124. Hjermann I, Velve-Byre K, Holme I, Leren P. Effect of diet and smoking intervention on the incidence of coronary heart disease. *Lancet.* 1981;2:1303–1310.

125. Lewis B, Hammett F, Katan M, et al. Toward an improved lipid-lowering diet: additive effects of changes in nutrient intake. *Lancet.* 1981;2:1310–1313.

126. Phillipson BE, Rothrock DW, Conner WE, et al. Reduction of plasma lipids, lipoproteins, and apolipoproteins by dietary fish oils in patients with hyperglyceridemia. *N Engl J Med.* 1985;312:1210–1216.

127. Ascherio A, Rimm EB, Stampfer MJ, et al. Dietary intake of marine n-3 fatty acids, fish intake, and the risk of coronary disease among men. *New Engl J Med.* 1995;332:977–982.

128. Khosla P, Sundram K. Effects of dietary fatty acid composition on plasma cholesterol. *Prog Lipid Res.* 1996;35:93–132.

129. Kris-Etherton PM, Yu S. Individual fatty acid effects on plasma lipids and lipoproteins: human studies. *Am J Clin Nutr.* 1997;65(suppl);1628S–1644S.

130. Hu FB, Stampfer MJ, Manson JE, et al. Dietary fat intake and the risk of coronary heart disease in women. *N Engl J Med.* 1997;337:1491–1499.

131. Zock PL, Katan MB. Hydrogenation alternatives: effects of *trans* fatty acids and stearic acid versus linoleic acid on serum lipids and lipoproteins in humans. *J Lipid Res.* 1992;33:399–410.

132. Mensink RP, Katan MB. *Trans* monounsaturated fatty acids in nutrition and their impact on serum lipoprotein levels in man. *Prog Lipid Res.* 1993;32:111–122.

133. Food and Agricultural Organization, World Health Organization. *Fats and Oils in Human Nutrition: Report of a Joint Expert Consultation.* Rome, Italy: FAO; 1994.

134. Shapiro S. Do *trans* fatty acids increase the risk of coronary artery disease? A critique of the epidemiologic evidence. *Am J Clin Nutr.* 1997;66(suppl):1011S–1017S.

135. Ascherio A, Willett WC. Health effects of *trans* fatty acids. *Am J Clin Nutr.* 1997;66(4 suppl):1006S–1010S.

136. Judd JT, Baer DJ, Clevidence BA, et al. Effects of margarine compared with those of butter on blood lipid profiles related to cardiovascular disease risk factors in normolipemic adults fed controlled diets. *Am J Clin Nutr.* 1998;68:768–777.

137. Knapp HR, Reilly IAG, Allesandrini P, FitzGerald GA. In vivo indices of platelet and vascular function during fish-oil administration in patients with atherosclerosis. *N Engl J Med.* 1986;314:937–942.

138. Leaf A, Weber PC. Cardiovascular effects of n-3 fatty acids. *N Engl J Med.* 1988;318:549–556.

139. Kinsella JE, Lokesh BR, Croset M, et al. Dietary n-3 polyunsaturated fatty acids: effects on membrane enzyme activities and microphage eicosanoid synthesis. In: Chandra RK, ed. *Health Effects of Fish and Fish Oils.* St Pohn's, Newfoundland: ARTS Biomedical Publishers and Distributors; 1989:81–126.

140. Vognild E, Elvevoll EO, Brox J, et al. Effects of dietary marine oils and olive oil on fatty acid composition, platelet membrane fluidity, platelet responses, and serum lipids in healthy humans. *Lipids.* 1998;33:427–436.

141. Parks EJ, German JB, Davis PA, et al. Reduced oxidative susceptibility of LDL from patients participating in an intensive atherosclerosis treatment program. *Am J Clin Nutr.* 1998;68:778–785.

142. Zock PL, Katan MB. Diet, LDL oxidation, and coronary artery disease. *Am J Clin Nutr.* 1998;68:759–760.

143. Itabe H. Oxidized phospholipids as a new landmark in atherosclerosis. *Prog Lipid Res.* 1998;37:181–207.

144. Lichtenstein AH, Kennedy E, Barrier P, et al. Dietary fat consumption and health. *Nutr Rev.* 1998;56(5):S3–S28.

145. Rose DP. Dietary fatty acids and cancer. *Am J Clin Nutr.* 1997;66(suppl): 998S–1003S.

146. Doll R. Strategy for detection of cancer hazards to man. *Nature.* 1997;256:589–596.

147. Wynder EL. Nutrition and cancer. *Fed Proc.* 1976;35:1309–1315.

148. Doll R. The geographical distribution of cancer. *Br J Cancer.* 1969;23:1–8.

149. Lea AJ. Dietary factors associated with death rates from certain neoplasms in man. *Lancet.* 1966;2:332–335.

150. Hems G. Association between breast-cancer mortality rates, child-bearing and diet in the United Kingdom. *Br J Cancer.* 1980;41:429–437.

151. Miller AB, Kelly A, Choi NW, et al. A study of diet and breast cancer. *Am J Epidemiol.* 1978;107:499–509.

152. Kark JD, Smith AH, Hames CG. The relationship of serum cholesterol to the incidence of cancer in Evans County, Georgia. *J Chronic Dis.* 1980;33:311–322.

153. Williams RR, Sorlie PD, Feinlieb M, et al. Cancer incidence by levels of cholesterol. *JAMA.* 1981;245: 247–252.

154. Martin-Moreno JM, Willet WC, Gorgojo L, et al. Dietary fat, olive oil intake and breast cancer risk. *Int J Cancer.* 1994;58:774–780.

155. Trichopoulou A, Katsouyanni K, Stuver S, et al. Consumption of olive oil and specific foods groups in relation to breast cancer risk in Greece. *J Natl Cancer Inst.* 1995;87:110–116.

156. Willett WC. Specific fatty acids and risks of breast and prostate cancer: dietary intake. *Am J Clin Nutr.* 1997; 66(suppl):1557S–1563S.

157. Wolk A, Bergstrom R, Hunter D, et al. A prospective study of association of monounsaturated fat and other types of fat with risk of breast cancer. *Arch Intern Med.* 1998;158:41–45.

158. Knekt P, Jarvinen R, Seppanen R, et al. Intake of dairy products and the risk of breast cancer. *Br J Cancer.* 1996;73:687–691.

159. Visonneau S, Cesano A, Tepper SA, et al. Conjugated linoleic acid suppresses the growth of human breast adenocarcinoma cells in SCID mice. *Anticancer Res.* 1997;17:969–993.

160. Ip C. Review of the effects of *trans* fatty acids, oleic acid, n-3 polyunsaturated fatty acids and conjugated linoleic acid on mammary carcinogenesis in animals. *Am J Clin Nutr.* 1997;66(suppl); 1523S–1529S.

161. Erickson KL. Dietary fat, breast cancer, and nonspecific immunity. *Nutr Rev.* 1998;56:S99–S105.

162. Armstrong B, Doll R. Environmental factors and cancer incidence and mortality in different countries with special reference to dietary practices. *Int J Cancer.* 1975;15:617–631.

163. Rose DP, Boyar AP, Wynder EL. International comparisons of mortality rates for cancers of the breast, ovary, prostate and colon, and per capita food consumption. *Cancer.* 1986;58:2363–2371.

164. Kaul L, Heshmat MY, Kovi J, et al. The role of diet in prostate cancer. *Nutr Cancer.* 1987;9:123–128.

165. Giovannucci E, Rimm EB, Colditz GA, et al. A prospective study of dietary fat and risk of prostate cancer. *J Natl Cancer Inst.* 1993;85:1571–1579.

166. von Holtz RL, Fink CS, Awad AB. β-Sitosterol activates the sphingomyelin cycle and induces apoptosis in LNCaP human prostate cancer cells. *Nutr Cancer.* 1998;32:8–12.

167. Connolly JM, Coleman M, Rose DP. Effects of dietary fatty acids on DU 145 human prostate cancer cell growth in athymic nude mice. *Nutr Cancer.* 1997;29:114–119.

168. Giovannucci E, Goldin B. The role of fat, fatty acids, and total energy intake in the etiology of human colon cancer. *Am J Clin Nutr.* 1997;66(suppl):1564S–1571S.

169. Klurfeld DM, Bull AW. Fatty acids and colon cancer in experimental models. *Am J Clin Nutr.* 1997;66(suppl); 1530S–1538S.

170. Lindner MA. A fish oil diet inhibits colon cancer in mice. *Nutr Cancer.* 1991;15:1–11.

171. Kim D-Y, Chung K-H, Lee JH. Stimulatory effects of high-fat diets on colon cell proliferation depend on the type of dietary fat and site of the colon. *Nutr Cancer.* 1998;30:118–123.

172. Iwamoto S, Senzaki H, Kiyozuka Y, et al. Effects of fatty acids on liver metastasis of ACL-15 rat colon cancer cells. *Nutr Cancer.* 1998;31:143–150.

173. Awad AB, Chen Y-C, Fink CS, Hennessey T. β-Sitosterol inhibits HT-29 human colon cancer cell growth and alters membrane lipids. *Anticancer Res.* 1996;16:2797–2804.

174. Chen Z-X, Breitman TR. Tributyrin: a prodrug of butyric acid for potential clinical application in differentiation therapy. *Cancer Res.* 1994;54:3494–3499.

175. Heerdt BG, Houston MA, Augenlicht LH. Potentiation by specific short-chain fatty acids of differentiation and apoptosis in human colonic carcinoma cell lines. *Cancer Res.* 1994;54:3288–3294.

176. Pouillart P, Cerutti I, Ronco G, et al. Enhancement by stable butyrate derivatives of antitumor and antiviral actions of interferon. *Int J Cancer.* 1992;51:596–601.

177. Smith PJ. n-Butyrate alters chromatin accessibility to DNA repair enzymes. *Carcinogenesis.* 1986;7:423–429.

178. McIntyre A, Gibson PR, Young GP. Butyrate production from dietary fiber and protection against large bowel cancer in a rat model. *Gut.* 1993;34:386–393.

179. Akoh CC. Structured lipids: enzymatic approach. *INFORM.* 1995;6:1055–1061.

180. Babayan VK. Medium chain triglycerides and structured lipids. *Lipids.* 1987;22:417–420.

181. Kennedy JP. Structured lipids: fats of the future. *Food Technol.* 1991;45:76–83.

182. Jandacek RJ, Whiteside JA, Holcombe BN. The rapid hydrolysis and efficient absorption of triglycerides with octanoic acid in the 1- and 3-positions and long-chain fatty acids in the 2-position. *Am J Clin Nutr.* 1987;45:940–945.

183. Hubbard VS, McKenna MC. Absorption of safflower oil and structured lipid preparations in patients with cystic fibrosis. *Lipids.* 1987;22:424–428.

184. Haumann BF. Structured lipids allow fat tailoring. *INFORM.* 1997;8:1004–1011.

185. Scarbrough FE. FDA: structured lipids pose challenge. *INFORM.* 1996;7:191–194.

186. Megremis CJ. Medium-chain triglycerides: a nonconventional fat. *Food Technol.* 1991;45:108–114.

187. Babayan VK, Rosenau JR. Medium-chain-triglyceride cheese. *Food Technol.* 1991;45:111–114.

188. List GR, Pelloso T, Orthoefer F, et al. Preparation and properties of zero *trans* soybean oil margarines. *J Am Oil Chem Soc.* 1995;72:383–384.

CHAPTER 11

The Soybean as a Source of Bioactive Molecules

John J.B. Anderson and Sanford C. Garner

HUMAN CONSUMPTION OF SOY FOODS

The use of soybeans (soya beans) and soy products for human consumption has been very limited in Western nations, but it has been historically substantial in many Asian nations, especially China and Japan. In modern Japan, the mean daily intake of soy in diverse foods has been estimated to approach 30 g (dry weight) per person.[1] Only in the last two decades or so has there been a trend toward increasing amounts of soy in the diets of Westerners.[2] Approximately 5 g/d is the estimated per capita consumption in the United States in the late 1990s, largely because of the mixing of soy flour with other flours in the production of breads, other cereal products, and other foods.[2] This trend may, however, be taking a significant upsurge at the beginning of the 21st century because of advances in our knowledge of the health-promoting and disease-preventative effects of many of the non-nutrient molecules found in soy beans and their commercial products. The soy foods most commonly consumed in the United States are tofu, soy milk, soy sauce, miso, and tempeh,[2] but nontraditional uses of soy in breads and other baked goods are increasing greatly. Asian countries will continue to consume large quantities of soy protein (and the many bioactive/components, such as isoflavones within the soy foods), but their consumption patterns are likely to shift more toward Western foods, especially meats, as the availability of these foods and the per capita purchasing power of Asian nations continue to improve.

This review of the bioactive molecules commonly found in soy and soy products focuses on a few nutrient molecules, such as vitamin K and lecithin, and on non-nutrients, such as phenolic acids, isoflavones, tannins, saponins, phytic acid, phytosterols, protease inhibitors, and lectins. The composition of soy and its major food products is covered first. Earlier reviews of these topics may be found in reports by Liener,[3] Garcia et al,[4] Messina and Barnes,[5] and Messina and Erdman.[6]

COMPOSITION OF SOYBEANS AND SOY PRODUCTS

Like most other legumes, soy and soy products contain a host of bioactive molecules that have effects on human tissues; the potency of these effects from processed products, especially products with increased protein content, may differ considerably from the effects of these molecules in soy flour because of increased concentrations. These molecules are typically closely associated spatially with one or another of the macronutrients—proteins, fats, and carbohydrates—in the soybean.

Soybeans, but not necessarily all their products, contain good amounts of protein, dietary fiber, thiamin, riboflavin, niacin, folate, vitamin K, and the minerals, iron, zinc, phosphorus, calcium, and magnesium. In addition, the oil fraction contains phospholipids, including phosphatidylcholine (lecithin). Some fermented soy products may contain vitamin B_{12} because of its synthesis by the culturing microorganisms. Tofu is especially high in calcium when prepared by precipitation with a calcium salt; conversely, tofu is high in magnesium when precipitated with a magnesium salt.

The bioavailability of minerals in soy foods depends on the degree of processing, especially of phytic acid (see below).

Table 11–1 gives the amounts of the major constituents: proteins, fats, carbohydrates, and ash of soy. On a dry weight basis, approximately 40% of soybeans consists of protein. Several different protein molecules are present, but much of it—80% or more—is storage protein, as opposed to proteins involved in structural support and enzymatic/metabolic processes. Glycinins, classified as globulins, constitute 70% to 80% of the storage protein. The protease inhibitors represent an important, though small, percentage of total proteins in soy, and these molecules are located close to the main storage proteins. For optimal nutritional value, the protein in soy must be heated to eliminate protease inhibitors, lectins, and other undesirable factors.[7]

Another 20% of soybeans, on a dry weight basis, is the oil. Soy oil is fairly unsaturated, and it contains four different phospholipids (see below under "Lecithin"). Approximately 50% of the fat in soy oil is linoleic acid, 20% is monounsaturated fatty acids, 20% is saturated fatty acids, and the remaining 10% is α-linolenic acid (7%) and other elongated polyunsaturated fatty acids. The lecithins (phospholipids) and other fat-soluble molecules, including pigments and flavors, are removed in the processing of crude soy oil.

Carbohydrates constitute about 35% of the dry weight of soybeans. The major components are polysaccharides (amyloses and amylopectins) and indigestible fiber, but the flatulence-producing disaccharides raffinose and stachyose also exist in soybeans. Soybeans are rich in dietary fiber of several types.

The remaining 5% of the dry weight of soybeans consists of ash (minerals) and the bioactive organic molecules, several of which are reviewed in this report.

The major non-nutrient molecules found in soy that have potential human health benefits, including polyphenols (phenolic acids, isoflavones, tannins, and saponins), phytic acid, protease inhibitors, and lectins, are reviewed in the following sections along with two important nutrient components of soy, lecithin, and vitamin K. The approximate amounts of some of these bioactive molecules in soy flour, most of which are not considered nutrients per se, are listed in Table 11–2.[7–12] The phenolic molecules and protease inhibitors are closely associated with the proteins in the natural products because of their purported roles as antioxidants or inhibitors of agents that degrade proteins. Phytic acid, which contains six atoms of phosphorus, is found in the water-soluble components of the soybean.

The phospholipids, including lecithin (phophatidylcholine), are located in the fat or oils of the soybean, the typical location also of fat-soluble vitamins—that is, vitamins E and K. Finally, the minerals zinc and iron are typically found bound to phytic acid and proteins within the soybean and its products. (The minerals, vitamin E, and other vitamins are not covered in this review.)

Table 11–1 Approximate Percentages of the Major Constituents of the Soybean (Dry)

Soybean Component	Approximate %
Protein	40
Oil (fat)	20
Carbohydrate	35
Ash (mineral)	5

Table 11–2 Composition of Selected Non-Nutrient Bioactive Molecules in Soybeans or Flour

Bioactive Molecules	Source	Usual Amount	Reference
Absorbable*			
Phenolic acids	Bean	0.03 g/100 g	10
Saponins	Flour	0.50 g/100 g	11
Isoflavones (total)	Flour	0.20 g/100 g	8
Phytic acid	Flour	1.7 g/100 g	12
Nonabsorbable			
Protease (trypsin)			
inhibitors			
Kunitz	Flour	1.1–19.6 mg/g	9
Bowman-Birk	Flour	<5.0 mg/g	9

*Tannins are very low in soybeans (no data available).
Source: Data from references 7–12.

DIGESTION AND ABSORPTION OF LIPID-SOLUBLE MOLECULES

A generic overview on the digestion and absorption of the lipid-soluble bioactive molecules in soy within the small intestine provides a background for understanding the utilization of these molecules.

After being freed from other parts of the bean by digestive processes, the lipid-soluble molecules are absorbed by the small intestine as part of the micelles that are created by the three components secreted in bile fluid, namely bile salts, lecithin, and cholesterol. These lipids, some of which have additional enzymatic modifications that yield other metabolic products within the gut lumen, become incorporated in micelles and enter the intestinal absorbing cells at the brush border membrane, and they complete their absorption at the basolateral membrane (exit step) of these cells by entering the lymphatic drainage as part of chylomicrons. Within the cytosol of the absorbing cells of the small intestine, the lipid-soluble molecules are incorporated in the newly forming chylomicrons, which carry them to extrahepatic cells of the body and eventually to the liver via the residual chylomicron remnants. In addition, these same molecules in the remnants, once distributed via blood to the liver, can be recycled by becoming part of the bile fluid and are then secreted back into the gut lumen where they can be reabsorbed by the enterohepatic circulation (Figure 11–1). The final excretion of most of these molecules occurs in the stool, but the more water-soluble molecules—conjugates of glucuronides and sulfates—are preferentially excreted in the urine.

NUTRIENT MOLECULES

Although soy contains a variety of macro- and micronutrients, two micronutrients are reviewed here because of their potentially important biological properties. Soybean oil is the major source of phosphatidylcholine (lecithin) in the human diet. Soy is also a significant dietary contributor of vitamin K, a nutrient that is present in few food sources.

Lecithin

Lecithin (phosphatidylcholine) is a major component of crude soybean oil: it represents 41.7% of the phospholipids or 8.57 g/100 g.[13] The term *lecithin* also has a broader meaning to the food science field as it includes the four major phospholipids found in food, namely, phosphatidylcholine, phosphatidylserine, phosphatidylethanolamine, and phosphatidylinositol (Figure 11–2); soybeans,

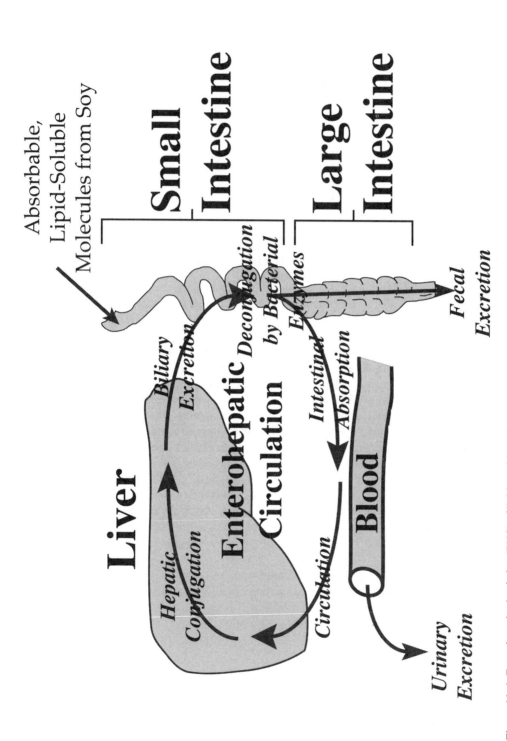

Figure 11–1 Enterohepatic circulation (EHC) of lipid-soluble molecules. Lipid-soluble molecules from foods, including soy products, are absorbed in the small intestine and reach the liver via the circulation. These molecules may be conjugated in the liver and secreted in the bile, from which they can be reabsorbed, returning from the gut to the liver (conjugation may also occur in the gut cells). This process of EHC may recycle a single molecule as many as 5 to 10 times before it is excreted in the feces or urine.

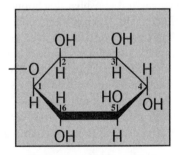

Phosphatidylcholine

Phosphatidylethanolamine

Phosphatidylserine

Phosphatidylinositol

Figure 11–2 Structures of the four main phospholipids. The major phospholipids, phosphatidylcholine (PC), phosphatidylethanolamine (PE), phosphatidylserine (PS), and phosphatidylinositol (PI), share a common structure, with two fatty acid chains esterified to a glycerol molecule. The base molecule is attached to carbon-3 by a phosphate group.

however, have little or no phosphatidyserine. Lecithin has well-known emulsifying properties for lipids that make it widely used by the food industry in a variety of processed foods.[14]

Food Consumption

Soy oils are readily available in the marketplace and are consumed as such, and soy milks are increasingly being consumed. Since phosphatidylcholine is in these foods and others, it is consumed in adequate amounts by the population as a whole. Choline deficiency is practically unheard of in humans, except possibly in newborn babies; so the intake of phosphatidylcholine presents little or no problem.

Structure

The structures of phosphatidycholine and other major phospholipids are shown in Figure 11–2. They contain glycerol, two fatty acids, a phosphate, and a basic component. The other phospholipids have structures like phosphatidylcholine's except for the difference in their base: choline versus serine, ethanolamine, or inositol. Phosphatidic acid, the parent molecule of phospholipids, is formed by the joining of 3-glycerol-phosphate and a diglyceride; then the base (choline or other) is linked to the phosphate group of phosphatic acid to form phosphatidylcholine (or other phospholipids). These biochemical steps are illustrated in Figure 11–3.

Digestion/Absorption

Phosphatidylcholine is hydrolyzed by phospholipases (water soluble) at the surfaces of the lipid-soluble micelles, which contain the water-soluble components (oxygen-containing groups) at the water-lipid interfaces and the lipid-soluble components of the membranes buried in the micelles. The enzymatic cleavage typically occurs at the C3 bond of glycerol, leaving a 3-lyso-diglyceride that remains in the micelle. The choline cleaved from the parent phospholipid is absorbed as an amino acid by a specific carrier mechanism for choline and other amino acids with similar structures. Within the absorbing epithelial cells of the small intestine, phosphatidylcholine and other phospholipids are synthesized from their absorbed components with the use of cellular energy (ATP); these lipids are then taken up in newly forming chylomicrons and transferred passively across the basolateral membrane to lymphatic lacteals.

Circulation/Distribution/Storage

The uptake of resynthesized phosphatidylcholine and other phospholipids into nascent chylomicrons in intestinal epithelial absorbing cells not only allows the second step of absorption to occur but also allows these molecules to be distributed to all cells of the body.

The storage of phosphatidylcholine is really a functional form of the molecule, since it serves, in part, as a component of membrane phospholipids. Very few of these molecules could be considered as stores per se, but they are temporarily in storage during circulation in various lipoproteins, especially very–low-density lipoprotein (VLDL) and low-density lipoprotein (LDL). Choline, on the other hand, may be stored in small amounts in the liver and other tissues as constituents of other labile molecules that can donate choline to phosphatidic acid, when needed, for the synthesis of cellular phosphatidylcholine.

Effects of Lecithin on Cells

Phosphatidylcholine is used by cells for the synthesis of other molecules as well as in its intact form for insertion in membranes.[15] These essential roles depend on a ready availability of choline in the diet. Lecithin has roles in each of the following[15]:

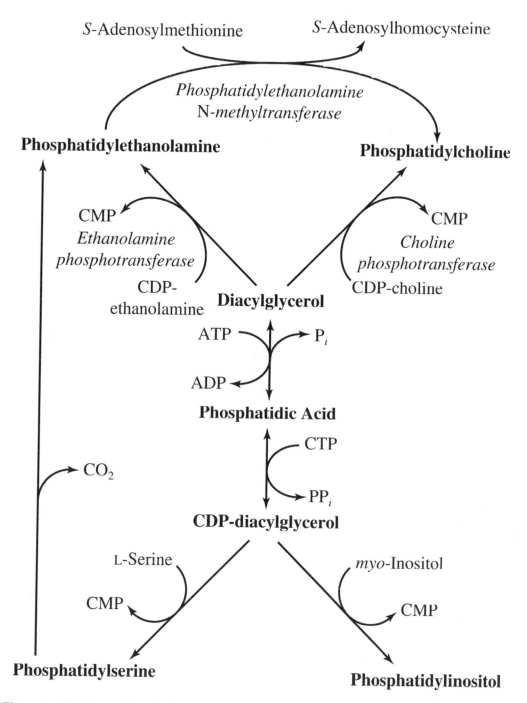

Figure 11–3 Synthesis of phospholipids. The major phospholipids are synthesized from either diacylglycerol or CDP-diacylglycerol with donation of the base group to form the final phospholipid products. In addition, PS can be converted to PE by loss of a carbon dioxide molecule, and PE can be converted to PC by addition of methyl groups via phosphatidylethanolamine N-methyltransferase.

- *Membrane phospholipids*—Phosphatidylcholine is one of the three most common phospholipids found in cell membranes. The metabolism of this molecule in membranes serves as a source of second messengers.
- *Second messengers in cells*—The metabolites of phosphatidylcholine are released to the cytosol, where they act as second messengers that either stimulate or inhibit other steps of metabolism of the cell.
- *Brain molecules*—Choline derived from phosphatidylcholine has a role in brain cells in the resynthesis of new phosphatidylcholine molecules needed for membranes, especially in the membranes of the Schwann (sheath) cells that envelop the axons of neurons.
- *Acetylcholine*—Acetylcholine is the neurotransmitter secreted by neurons at the skeletal muscle junctions that initiates muscular contraction. Choline is a critical component of this neurotransmitter.
- *Breast milk*—Human breast milk contains fairly high amounts of phosphatidylcholine, which is used by suckling infants for new membrane synthesis and other needs.
- *Programmed cell death or apoptosis*—Choline apparently serves in a protective role, allowing cells to die rather than to turn into malignant cells (cancers) when they undergo mutations.

Summary

Phosphatidylcholine may represent a highly important molecule for human health because it provides an essential molecule, choline, that is used for many functions in human tissues. While choline does not have a Recommended Dietary Allowance (RDA), it is classified as an Adequate Intake nutrient; it was listed in 1998 by the Food and Nutrition Board[16] as an essential nutrient for humans. Further information on the amounts of choline required to conduct its functional roles in human tissues will be needed before an RDA is assigned to choline across the life cycle.

Vitamin K

Vitamin K is essential for the synthesis of several proteins, including at least four coagulation factors, three bone matrix proteins, and several other recently identified proteins in various body tissues. In addition to its role in clotting, other functions of vitamin K, an essential micronutrient with an RDA, are emerging.[17–21]

Soy serves as one of the major sources of vitamin K, second only to dark green leafy vegetables and other legumes.[20]

Human Consumption

The consumption of vitamin K, aside from vitamin supplements, is exclusively from plant sources,[18] and soy foods provide good amounts of the natural vitamin. Vegetarians have no trouble obtaining sufficient amounts of vitamin K, but omnivores typically consume fewer servings of dark greens and legumes, including soy products. Concern has been expressed that meat-eating populations may ingest insufficient amounts of this vitamin,[20] so that they may be at risk for inadequate carboxylation of bone matrix proteins and hence the development of osteoporosis.

Structures

Two major food-derived molecules of vitamin K exist: vitamin K_1, or phylloquinone, and vitamin K_2, or menaquinone. The structures of these two forms of vitamin K are illustrated in Figure 11–4. Vitamin K_1 is found only in plant foods, but it has a limited distribution in the plant kingdom. Vitamin

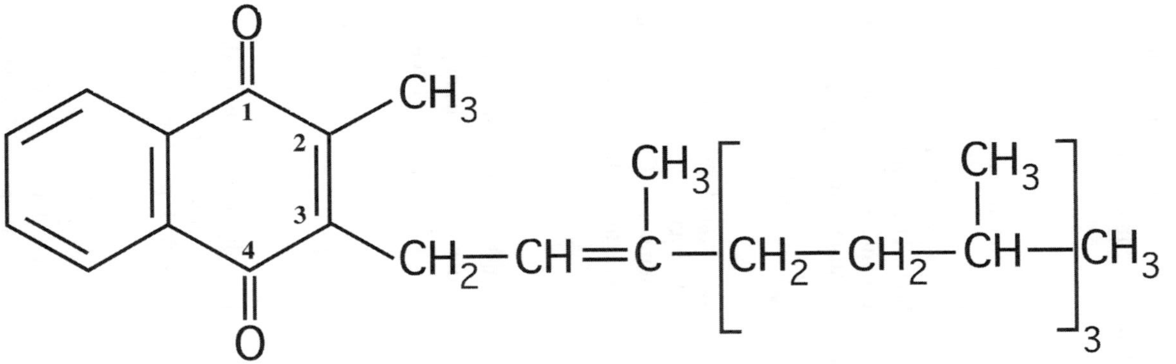

Vitamin K₁ (Phylloquinone)

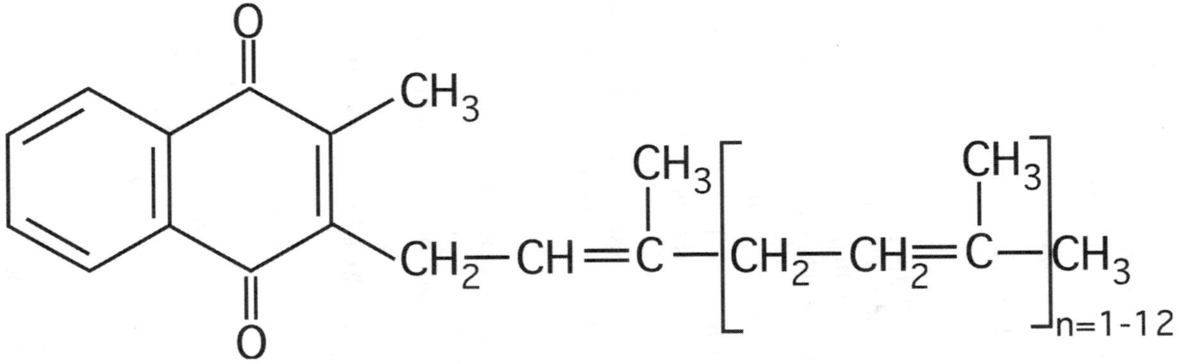

Vitamin K₂ (Menaquinone)

Figure 11–4 Structures of two forms of vitamin K. Vitamin K may be consumed from dietary sources either as vitamin K_1, phylloquinone, the naturally occurring material found in plant foods, or as vitamin K_2, menaquinone, which is formed by bacterial action in fermented food, such as miso, or in the human gut.

K_2, a product found in intentionally fermented foods, results from bacterial synthesis. Vitamin K_2 is also produced in small amounts by bacterial flora in the human intestines.

Digestion/Absorption

The digestion and release of vitamin K from soy foods involves removal of this fat-soluble vitamin from the food components that contain it in the natural state. Absorption of vitamin K is similar to that of other lipid-soluble molecules—that is, via micelles at the luminal membrane of the absorbing epithelial cells of the small intestine (brush border).

Circulation/Distribution/Storage

Chylomicrons made by the intestinal absorbing cells take up absorbed vitamin K along with other lipid-soluble vitamins, triglycerides, phospholipids, and cholesterol-esters.

The storage of vitamin K in the body is extremely limited for a fat-soluble vitamin, and it has been estimated that only about 1 day's supply of the vitamin is stored in the body. The liver ($\sim$150 μg) and bone (40 μg) are considered the major storage sites, while the vitamin K content of fat tissue is practically nil. Approximately 10 μg of vitamin K circulates in blood in association with lipoproteins, primarily chylomicron remnants, and with proteins, like albumin.

Post-Translational Effects of Vitamin K

Enzymatic post-translational modification of proteins containing glutamic acid residues (Glu) involves the action of vitamin K (Figure 11–5). Carboxylase enzymes insert a second carboxyl group at the gamma (γ) carbon of selected glutamate (Gla) residues in these proteins. The Glu residues are therefore converted to Gla residues with the addition of the new carboxyl group. These metabolic events occur in association with the microsomal membranes.

Summary

Vitamin K is provided in good amounts by soybeans and soy products. Because of generally inadequate intakes of vitamin K by meat eaters, especially when coupled with poor vegetable consumption, potential adverse effects on bone tissue may place individuals at increased risk of osteoporotic fractures.

NON-NUTRIENT MOLECULES—ABSORBABLE

Polyphenols

Several types of polyphenols exist in soybeans. Many or perhaps all of them act as antioxidants against free-radical oxygen species in the cells, and many act as phytoalexins, or molecules that repel animal predators. Table 11–3 lists some of the phenolic molecules found in soybeans.[3,4,7,22] Polyphenols as a whole have been recently reviewed.[23]

Phenolic Acids

The term *phenolic acids* covers a large group of organic molecules found naturally in soy and its derivative products but only in small quantities ($\sim$1–10 mg/100 g). The classification of molecules containing phenol groups is complex and includes several types of molecules (by their common names) covered in this chapter, namely isoflavones, saponins, and tannins, in addition to phenolic acids. The phenolic acids, simple molecules such as the organic solvent benzoic acid, are included in this broad class of polyphenols because of diverse phenol groups within them. Also, typically each of these phenol groups contains one and sometimes two hydroxyl groups, which increase their solubility

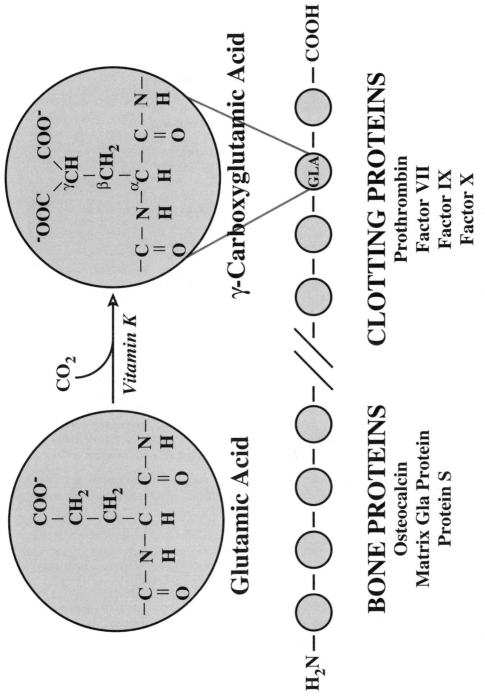

Figure 11–5 Post-translational modifications of proteins by vitamin K. The physiological function of vitamin K is to serve as a cofactor in the post-translational modification of glutamic acid (Glu) residues in several bone proteins and clotting factors to form γ-carboxyglutamic acid (Gla) residues. Without these modifications, the proteins are not functional.

Table 11–3 Variety of Polyphenolic Molecules in Soybeans

Class	Examples
Phenolic acids	p-Hydroxybenzoic acid
	Vanillic acid
	Gentisic acid
	o-Coumaric acid
	Syringic acid
	Sinapic acid
	Phenolic acid
	Caffeic acid
Isoflavones (aglycones)	Genistein
	Daidzein
	Glycetein
Saponins (triterpenoid alcohols) (aglycones)	Soyasapogenol A
	Soyasapogenol B
	Soyasapogenol E
Tannins (hydrolyzable)	Tannic acid
	D-Catechin (flavan-3-ol)
	Gallic acid
	Anthocyanins

Source: Data from references 3, 4, 7, and 22.

in aqueous solutions. They are also found in the soybean spatially associated with proteins. All of the phenolic molecules are thought to be derived from the amino acid phenylalanine, as illustrated in Figure 11–6.[7]

Total phenolic acids are lower in soy flours than in the flours of many other plant foods, but soy flours contain a variety of these molecules, many of which have important health benefits or potential benefits.[24] See Table 11–2 for the amounts of phenolic acids in unprocessed soybeans.

Human Consumption.   Human consumption of phenolic acids in Western nations is very low because of low soy consumption. In Asian nations, the consumption of soy products is not so low, but still only small amounts of phenolic acids are ingested per day.

Structures.   The general structures of phenolic acids are related to syringic acid, which is shown in Figure 11–6. These molecules are largely responsible for the color and flavor of the beans. Little is known about their functions in soybeans, but the water solubility of many of these molecules (low molecular weight) allows them to be associated with proteins.

Digestion/Absorption/Storage.   Little is known about either the digestion or the absorption of the phenolic acids, but it is presumed that they are released from food proteins very similarly to isoflavones and then absorbed by both mechanisms for water-soluble and lipid-soluble molecules, depending on the specific structure of each molecule.

The more lipid soluble of these molecules are carried first by chylomicrons from the gut and then by chylomicron remnants to the liver. The more water-soluble molecules, perhaps the majority of the phenolic acids, are carried in blood from the gut and delivered first to the liver via the hepatic portal vein (upper intestinal collection route) from the small intestine before distribution to other tissues in the body.

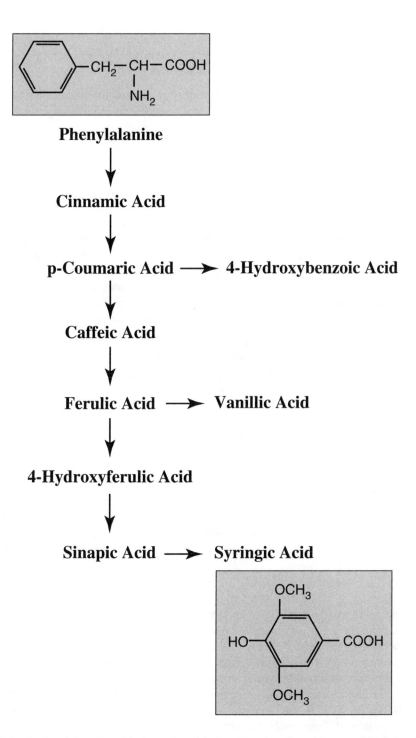

Figure 11–6 Synthesis of phenolic acids from phenylalanine. The phenolic acids are considered to be derived from the amino acid phenylalanine. Illustrated here is one pathway that gives rise to a series of phenolic acids, including syringic acid.

Effects in Cells/Tissues.   The phenolic acids act as antioxidants, but no scientific evidence based on human investigations is available. They may serve as antioxidants that protect especially the protein fraction of soy. These molecules may also have roles in preventing cancer.

Summary.   Phenolic acids need much more investigation to establish their functional roles in humans and to extend beyond their major properties in the soybean as organic solvents, hence as color and flavor amplifiers.

Isoflavones

The isoflavones represent one type of polyphenol, sometimes referred to as flavonoids (see above under "Phenolic Acids"); these molecules are also referred to as phytoestrogens because of their ability to interact with estrogen receptors in cells. The generic chemical structures of flavones and isoflavones are illustrated in Figure 11–7. Their ring structures are the same as in flavones, but the orientation of the B ring is different: in flavones the B ring is linked to carbon 2 of the middle (or C) ring, while in isoflavones it is attached to carbon 3 (see Figure 11–7). The nonsteroidal phytoestrogenic isoflavones in soybeans have elicited great interest in the last decade because of their potentially protective or preventive properties against several major chronic diseases, including cardiovascular diseases, cancers, and osteoporosis.[25–28] The content of isoflavones in soy flour is listed in Table 11–2. Setchell[29] has recently written a general review of the isoflavones.

Human Consumption.   Rough estimates of mean intakes of isoflavones in Western populations are less than 10 mg/d per person, based on assumptions of the estimated daily soy consumption of 5 g or less per person and a range of 0.5 to 3.0 mg/g of soy flour or concentrate.[29] Asian populations may consume as much as 10 times this estimate or more.

Structures.   The structures of the major isoflavones in comparison to estradiol are shown in Figure 11–8. The reason for the comparison of the isoflavones to estradiol relates to the binding properties of these molecules, namely their affinity for the estrogen receptor in mammalian cells. Estrogen receptors combine with true estrogens such as estradiol to initiate responses in reproductive and other tissues. Phytoestrogens behave similarly: they act as estrogen agonists but may also act as estrogen antagonists.[29]

Digestion/Absorption.   The majority of isoflavones in soy and soyfoods exist as glycosides or glycones that have a glucose molecule attached to them. When a glucosidase enzyme attacks the glycosides in the lumen of the small intestine, the single glucose molecule is removed and the aglycone remains. Figure 11–9 diagrams this reaction. Both glycones and aglycones can be measured by high-performance liquid chromatography.[30,31]

The aglycones are apparently preferentially absorbed by the small intestine as part of the micelles created by the action of bile. These aglycone molecules, some of their further metabolic products made in the gut, and a small percentage of undigested glycones complete their absorption by entering the lymphatic drainage as part of chylomicrons. The circulation of the isoflavones in blood is complex, partly because of the lipid solubility of these molecules and partly because they also bind via weak attractive forces with proteins in the circulating blood. The serum concentrations were shown to rise slowly in the blood of human subjects, as part of a feeding study, after a single soy meal.[32] In addition, these same molecules, eventually distributed via blood to the liver, can be recycled by becoming part of the bile fluid and recycling as part of the enterohepatic circulation (see Figure 11–1). The final excretion of these different molecules occurs in the stool and the urine.

Results of a human study demonstrated that the aglycones genistein and daidzein had almost equal bioavailability for absorption.[32] Genistein was found to be highly bioavailable in rats and was shown to participate in the enterohepatic cycle.[33]

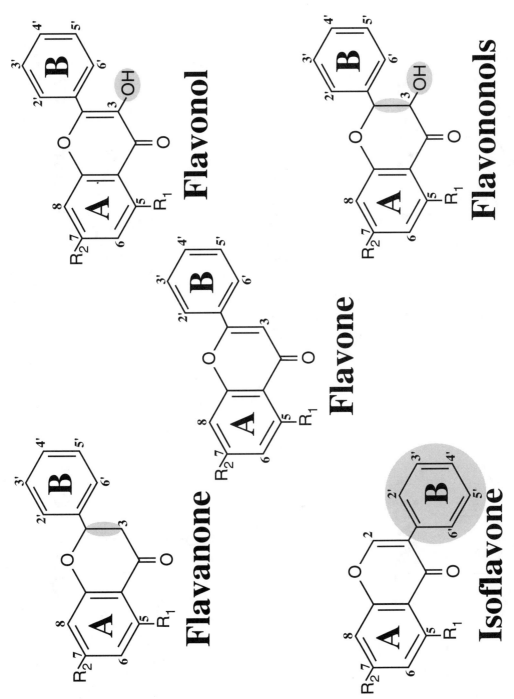

Figure 11–7 Generic chemical structures of five classes of flavonoid molecules. The structures of five major classes of flavonoid molecules are illustrated. The structural difference in comparison to the flavones is highlighted by shading of the distinctive modification in each class.

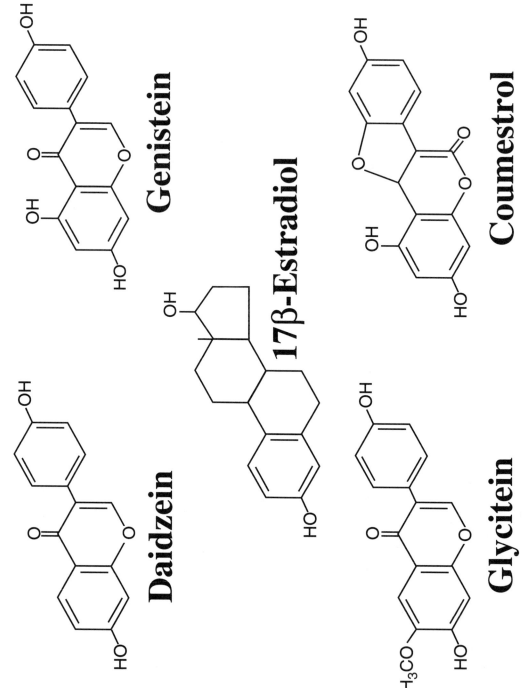

Figure 11–8 Structures of major isoflavones in comparison to estradiol. The major aglycone isoflavones of soy are illustrated. The isoflavones are drawn with the B ring up instead of down as in Figure 11–7 to highlight the similarity to estradiol. Coumestrol belongs to a second major class of phytoestrogens, the coumestans, found in soy, alfalfa, and a few other plant foods.

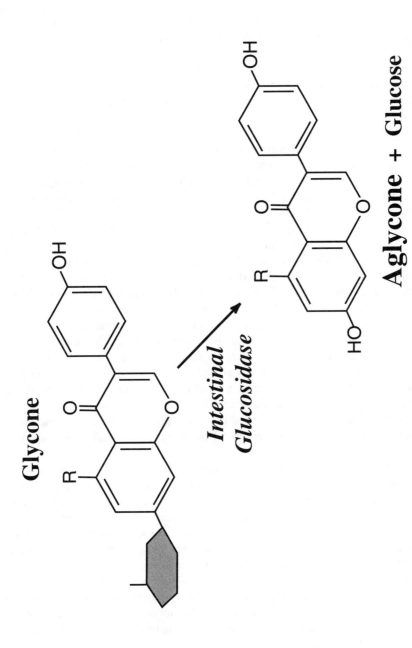

Figure 11–9 Enzymatic conversion of glycones to aglycones. Glucosidase enzymes in the small intestine attack glycoside bonds in the isoflavone glycones derived from soy to yield an aglycone molecule plus a single molecule of glucose.

Effects of Isoflavones on Cells.   The action of the isoflavones on cells, as weak estrogen agonists or as estrogen antagonists, has become an area of great interest to researchers of osteoporosis, cardiovascular diseases, and cancer. The actions, agonistic or antagonistic, differ according to cell type and the quantity of each type of estrogen receptor (ER) in these cells. Types of ERs are discussed below.

Bone cells, especially osteoblasts, are able to respond to isoflavones and other phytoestrogens as they do to estrogens from the blood. In these cells, isoflavones tend to act like the true estrogens, such as estradiol. The isoflavones can combine with ERs in the cytosol and then move into the nucleus where the isoflavone-ER complex activates the genomic estrogen-response element. The outcome is synthesis of the same protein molecules that are produced following stimulation by estradiol.[34] Estrogenic (agonist) effects help to maintain bone mass and density, whereas if reproductive tissues, ie, breast and uterus, are stimulated by estrogen in this way, the risk of cancers of these reproductive tissues may be increased.

Although the interaction of phytoestrogens with ERs may explain many of the beneficial effects of these molecules on bone cells—bone formation and inhibition of bone resorption (degradation)—isoflavones also act as antioxidants by protecting cellular proteins, unsaturated fatty acids, and nucleic acids from reactive oxygen species,[35] and as tyrosine kinase inhibitors that may reduce cell proliferation and the risk of carcinogenesis.[36] These diverse actions of isoflavones may also support the healthy functions of bone and other cells.

Cancer cells have been shown to be inhibited following treatment with soy isoflavones. Dietary supplementation of mice with isoflavones significantly reduced metastases of cancer cells (melanoma).[37] The consistency of experimental findings that demonstrate beneficial effects in inhibiting the proliferation of cancer cells under different conditions has not been good, so the mechanisms of protection against cancer by isoflavones cannot yet be established with any confidence.[38]

Liver parenchymal cells conduct the business of coordinating energy metabolism for the entire body and directly handle many of the detoxifying and waste removal mechanisms of the body.[39,40] In addition, cholesterol is synthesized by the liver for distribution by VLDLs. Whether isoflavones affect cholesterol synthesis has not been established. Since the total circulating cholesterol level is reduced, while high-density lipoprotein cholesterol is not affected by isoflavones,[41] it is presumed that these bioactive, nonsteroidal molecules must also have a direct effect on liver cells. This postulated action, however, requires experimental verification. Blood lipids (total cholesterol, and triglycerides in VLDLs) were shown to be significantly reduced in moderately hypercholesterolemic patients who consumed isoflavones in their diets, independent of other sources of isoflavones, which were excluded, for a period of 9 weeks.[42]

A meta-analysis has shown that soy protein has a significant cholesterol-lowering action in human subjects.[43] The conclusion of this meta-analysis has been confirmed by a large population study of Japanese men and women in which soy consumption and serum total cholesterol concentration were inversely related.[44] The cholesterol-lowering mechanism had been speculated to be based on the protein/amino acids in soy or another factor, such as isoflavones, associated with the protein fraction of soy,[45] but it now appears that the isoflavones are largely, if not totally, responsible for the reduction of serum cholesterol concentration, at least in subjects with moderate hypercholesterolemia.[42]

Arterial tissue cells, mainly endothelial cells[46] and smooth muscle cells,[47–49] have been demonstrated to be responsive to isoflavones, and vascular reactivity is enhanced by these molecules. In addition, isoflavones inhibit the progression of pathological changes in atherosclerotic lesions in monkeys.[41] The scavenging of free radicals by isoflavones may be another mechanism that helps delay the damage of arterial walls from the atherosclerotic process.[35]

Epithelial cells in general with the potential to become cancer cells also are inhibited by isoflavones.[5] The mechanisms postulated for this effect are presumed to be the same as for the beneficial effects of these molecules on osteoblast cells, including tyrosine kinase inhibition and antioxidation (see the discussion of bone cells above).

Epithelial cells of reproductive tissues, especially mammary glands, ovaries, and testes, also contain ERs, so that they too are subject to the actions of isoflavones. In fact, the discovery of phytoestrogens in clover, a legume, resulted from findings of excessive stimulation of reproductive tissues of grazing sheep in Australia, which resulted in low yields of lambs.[50]

Mechanisms of Action of Isoflavones.   The mechanisms of action of isoflavones can be divided into genomic and nongenomic actions, as shown in Figure 11–10.

ERs represent the classical (genomic) mechanism of action of estrogens and other nonsteroidal molecules with estrogenlike effects.[51] Membrane receptors are a second way (nongenomic) in which isoflavones as well as true estrogens can activate most human cell types.[38]

Summary.   The isoflavones from soybeans have estrogenlike properties in human cells—hence their common name of phytoestrogens. They act as weak estrogen agonists or as weak estrogen antagonists, but usually in different tissues. For example, the weak agonistic effects of the isoflavones result in beneficial outcomes in osteoblast and vascular cells, but the antagonistic influences may occur in reproductive tissues. Because of their differential effects on ERs in different cell types, isoflavones may be considered as selective estrogen receptor modulators, similar to drugs like tamoxifen and raloxifene. Further investigations are needed to clarify the cell-specific responses to isoflavones.

Saponins

Saponins, also known as triterpenoids, are classified as polyphenols. They are abundant in soybeans and soy products, more than in any other legume, including various beans and chickpeas.[7] Approximately 5% of soybeans, by dry weight, is the saponin content.[5] Saponins have modest hypocholesterolemic effects: they lower total cholesterol in animal models by approximately 5% to 10% over a period of several months.[5] The amount of saponins in soy flour is listed in Table 11–2.

Human Consumption.   Human intakes of saponins are typically very low because most saponins are largely, but not totally, removed in processing. Because saponins also have a bitter taste, soyfoods containing them are not considered palatable. Their content in soy flour is slightly greater than twice that of isoflavones in soy flour (see Table 11–2). At high levels of consumption, saponins may have toxic effects, but dose-response relationships have not been adequately addressed.[7]

Structures.   A typical structure of a saponin is shown in Figure 11–11. Soyasaponin I is the most abundant component, but two other forms of saponins also exist in soybeans.[52] These glycosides generate aglycones in the gut, namely sapogenins much as the isoflavone glycosides do. The most common aglycone is known as sapogenin B (or soyasapogenol), as shown in Figure 11–11.

Digestion/Absorption.   The saponins are not thought to be absorbed in the digestive tract to any significant extent because they form large complexes of three-dimensional stacks within micelles that greatly increase the total mass and size of the micelles in the small intestine. These mixed micelles containing saponins with bile acids cannot be absorbed efficiently at the brush border because of their great size, and the cholesterol in these large micelles is thereby partially prevented from being absorbed. Saponins also interact with bile acids and thereby render them unavailable for reabsorption from these micelles and their recycling as part of the enterohepatic circulation.[53] In both cases, more cholesterol from that circulating in blood would be required for bile acid production in the liver, which in turn would produce a net lowering of total cholesterol concentration in blood.

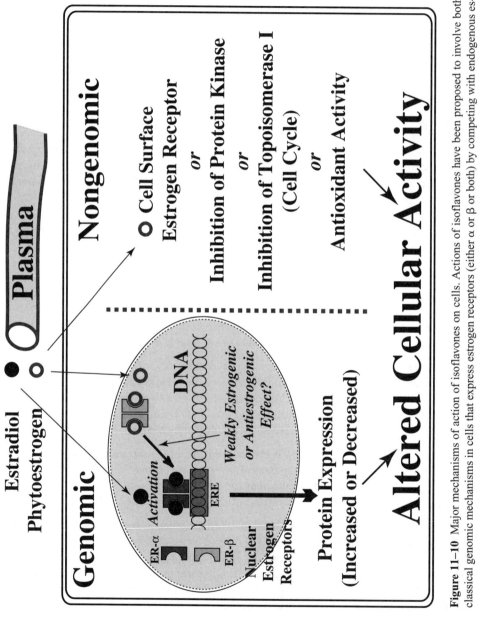

Figure 11–10 Major mechanisms of action of isoflavones on cells. Actions of isoflavones have been proposed to involve both classical genomic mechanisms in cells that express estrogen receptors (either α or β or both) by competing with endogenous estrogens for occupancy and nongenomic mechanisms that may include either binding to a putative cell-surface estrogen receptor or affecting intracellular enzymes that regulate the cell cycle or protein phosphorylation. *Source:* Reprinted with permission from S.C. Garner and J.J.B. Anderson et al., in *Balliere's Clinical Endocrinology and Metabolism,* H. Adlercreutz, ed., figure 3, © 1999, Balliere Tindall, a W.B. Saunders Company.

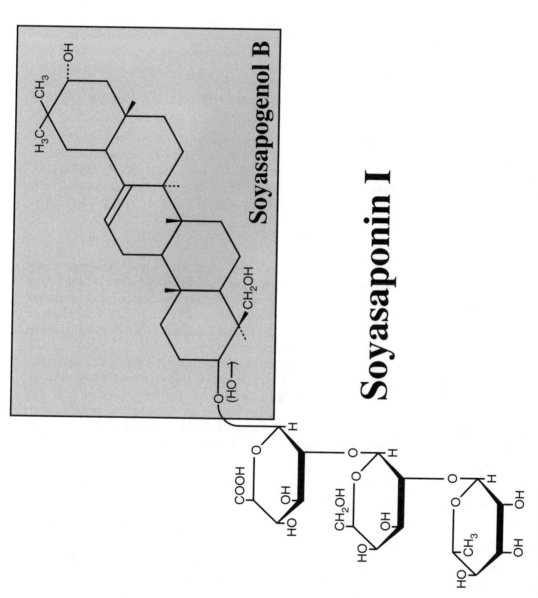

Figure 11–11 Typical structure of a saponin. Soyasaponin I is the most common saponin in soy foods. Removal of the three sugar molecules yields an aglycone, soyasapongenol B.

Effects in Cells/Tissues.   The major effect of saponins within the lumen of the small bowel is their solubilizing of cholesterol (or forming a water-insoluble complex of the two types of molecules) and other sterols and steroids, which moves them down the gastrointestinal tract to elimination. The net effect of a saponin-rich diet—prevention of cholesterol absorption and bile acid reabsorption—is to lower serum cholesterol and decrease the risk of atherosclerotic disease. Insufficient numbers of human investigations of the effects of saponin consumption on blood cholesterol concentrations in subjects with modest total cholesterol levels have been conducted to generate a consensus on the benefit of saponins to human health.

Saponins may also have a role in cancer prevention, but the mechanism is not established.[5] One idea is that saponins favorably modulate the immune system and thereby improve destruction of cancer cells; another is that the saponins serve as antioxidants, a role similar to the one they serve in the soybean. The saponins apparently have no direct effect on the mucosal cells lining the gastrointestinal tract, and they are not considered toxic at modest levels of intake.

Summary.   Saponins, commonly found in soybeans and many processed soy products, hold promise because of their potential role in lowering serum total cholesterol concentrations, especially in individuals who have levels that are not particularly high—that is, levels between 200 and 240 mg/dL. The roles of saponins in preventing cancer require further investigation. Saponins apparently have no toxic or antinutrient effects except insofar as they lower total cholesterol and possibly depress the absorption of fat-soluble vitamins.

Tannins

The tannins represent a special subclass of phenolic acids or polyphenolic molecules, but only small amounts of tannins exist in soy.[4] These molecules are best known for their tanning capability in leather preparation. Two classes of tannins exist: condensed tannins (proanthocyanidins), exemplified by flavonoid polymers, and hydrolyzable tannins, characterized by polymers of gallic acid. The taste of flavonoid-based condensed (oligomeric) tannins, also known as anthocyanidins, is not appreciated by humans because these molecules tend to be bitter and astringent.[22,54] Also, these molecules have toxic properties.[22,55] Plants are thought to make tannins to repel birds, insects, and other animals eating them.[55] In addition to these antipredatory defenses, tannins also have water-binding and antioxidative properties. Tannins bind strongly to proteins, and they can precipitate proteins, including digestive enzymes, within the gastrointestinal tract. Because of these properties, including their inhibitory effect on the growth of intestinal bacteria, tannins may influence human health in different ways. A recent review of tannins focuses on the health effects of these molecules.[56]

Human Consumption.   Extremely low intakes of tannins occur because of the low content of these molecules in soy flour and their partial removal by processing. If present in large amounts, the bitter taste of these molecules would otherwise not encourage human use.

Structures.   Both hydrolyzable and condensed tannins exist under the broad class of phenolic acids. These molecules have a molecular weight greater than 500 daltons.

Digestion/Absorption.   The digestion and absorption processes are similar to those of phenolic acids and isoflavones. Condensed tannins have deleterious effects within the gastrointestinal tract, namely, enzyme inactivation, because they bind to all proteins, including enzymes, and can cause damage to the mucosal membrane surface.[22] Any absorbed tannins or their metabolites may contribute to this tissue damage and further damage other cells of the body following absorption because of their protein-binding property.[22] In addition, in some unknown manner, tannins reduce the metabolic utilization of absorbed amino acids, possibly by being absorbed themselves through tannin-eroded segments of the intestinal mucosa in experimental animal models.[57]

Circulation/Distribution/Storage.   Blood distribution of the tannins is similar to that of other phenolic molecules. Under unusual conditions, such as perforation of the gastrointestinal tract, tannic acid, one type of tannin, has been identified in the water-soluble fraction of blood.

Effects in Cells/Tissues.   Tannins have the capability of precipitating proteins, both within the gut lumen when released from their proteins in soy following digestion and within cells after membrane transfer to the cytosol.[22]

Summary.   Tannins represent an insignificant amount of the polyphenolic molecules consumed in soy and soybean products. Their role in human function is largely unknown, largely because of their bitter taste in certain components of foods, and the fact that they are thereby avoided.

Phytosterols

β-sitosterol, campesterol, and stigmasterol, which constitute almost 100% of the membrane sterols of soybeans, serve a role in plants much like that of cholesterol in animal membranes. The ratio of these phytosterols in refined soybean oil is approximately 2.5:1:1, in which the total sterols represent 221 mg/100 g of oil.[58] Phytosterols clearly reduce serum total cholesterol concentration by ~10% when dietary plant sources are substituted for animal sources in the usual eating pattern of individuals with elevated serum cholesterol. The structure of β-sitosterol, as compared to cholesterol, is illustrated in Figure 11–12.

Human Consumption

The intake of phytosterols is high in all populations; it is in the range of 300 to 400 mg/d for the Japanese and those who consume a diet rich in vegetables and fruits but considerably less—approximately ~80 to 160 mg/d—among omnivorous Westerners who consume less fruits and vegetables.[5,59] The content of β-sitosterol in the edible portion of the soybean is ~90 mg/100 g of dry weight.[5,59]

Digestion/Absorption/Storage

Because of their lipid solubility, phytosterols are removed in the digestion process from the cell membranes of plant foods. Only small amounts of phytosterols are absorbed, but phytosterolemia may result from excessive absorption.[59] A high correlation exists between cholesterol absorption and the absorption of phytosterols.[59] Once released in the gut lumen by digestive processes, the phytosterols then enter micelles for the entry step of absorption into the absorbing columnar epithelial cells, where they are incorporated into chylomicrons for the exit step of absorption into the lymphatic lacteals. The phytosterols are returned via chylomicrons to the liver, and hepatic cells secrete these lipid-soluble molecules in bile to be excreted as acids. How much they participate in the enterohepatic circulation, as has been shown for cholesterol (Figure 11–1), is not clear. Most of the phytosterols are eliminated in the feces.

Effects in Cells/Tissues

No direct effects of phytosterols on cells have been identified. Whether phytosterols are anticarcinogenic has not yet been established.

Summary

Phytosterols represent a class of non-nutrient molecules that are consumed in large amounts because of their ubiquitous presence in plant cell membranes, but they have no known function. Their major benefit to humankind is indirect; as their consumption from plant foods increases, the intake of

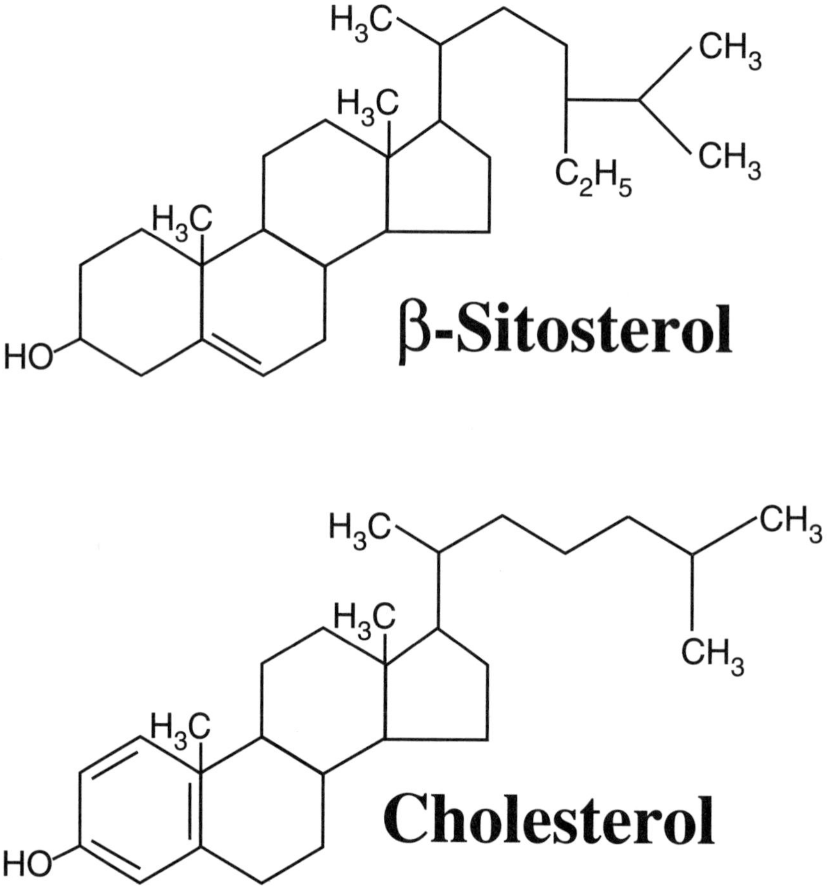

Figure 11–12 Structure of β-sitosterol in comparison to cholesterol. β-sitosterol has a role similar to cholesterol's in plants in the formation of cell membranes.

cholesterol decreases because of the reduced consumption of animal products. The net result of phytosterol substitution for cholesterol in human diets is a significant decline in both total cholesterol and LDL cholesterol. In addition, the phytosterols partially inhibit the absorption of cholesterol when they are present together in the small intestine after a meal.

Phytic Acid and Inositol Phosphates

Phytic acid, also known as *myo*-inositol hexaphosphate, is abundant in soybeans and soy products, especially soy flour (see Table 11–2). Historically, it has been considered an antinutrient, but this view is not consistent with our current knowledge of its actions. In addition to binding various metallic ions to its phosphate groups, phytic acid from soy and soy products may have anticarcinogenic properties.[7] In plants, the phytates serve as a store of phosphorus, as well as several minerals, especially divalent cations.

Human Consumption

Phytates are consumed in Western nations primarily from grains rather than from soy or soy products because of the low acceptance of soy products in the diets of these nations. Phytic acids in the human diet provide a good source of essential minerals from the storage compartments of soy and soy foods, if these are not excessively processed; excessive hydrolysis separates minerals from the phytates by reducing binding of the mineral ions, which then are at least partially discarded. The phytates form complexes with diverse minerals, and this binding may decrease the bioavailability of minerals within the small intestine, especially as the pH within the gut lumen increases from 3.5 in the upper duodenum to alkaline values (>7.0) in the jejunum and ileum. The phytates tend not to release minerals above a pH of 3.5, which means poor bioavailability of minerals bound to phytate exists almost the length of the small intestine—approximately 20 feet—except for the upper duodenum between the pyloric sphincter and the opening of the common bile duct.

A major drawback to human consumption of soy phytate is that it interferes, at least in part, in human subjects with the absorption of zinc,[60] calcium,[61] and other mineral ions. Phytate may also reduce the absorption of iron to a limited extent in humans, but not all researchers are in agreement on this point.[62] Other minerals, such as manganese, may also be reduced in the presence of a high concentration of phytate.[63] The reduction of bioavailability of iron and other divalent cations from soy may also be related to an unknown factor in soy other than phytate.

Structures

The common phytic acid is a hexaphosphate, and it is illustrated in Figure 11–13. Other inositol phosphates may contain from one to five phosphate groups on the inositol ring. Each of these phosphate groups is capable of binding one monovalent or divalent cation, but typically the number of cations bound is less—only three to five cations per phytic acid. Therefore, most phytates (the mineral salts of phytic acid) found in foods contain a mix of minerals, both monovalent and divalent, chelated by strong ionic bonds. Sodium, potassium, calcium, magnesium, zinc, copper, iron, manganese, and other ions have been determined to be associated with food phytates.

Digestion/Absorption/Storage

If phytic acid per se is absorbed, only a small amount of the intact molecule is able to cross the intestinal mucosa. More likely, the simple inositol molecule is absorbed after removal of the phosphate groups and minerals. Phytase enzymes in the soy are probably partly responsible for the hydrolysis of phytates in the gut lumen.[64] Inositol is used biologically, like choline, to synthesize phospholipids in the intestinal absorbing cells prior to their incorporation in chylomicrons for distribution to all cells of the body. The phospholipids are functional molecules and therefore are not stored in the typical meaning of *storage* (see Figure 11–1).

Undigested phytic acids may, however, inhibit the absorption of minerals, such as calcium, zinc, and iron ions, because of low bioavailability of these cations. The inhibition of calcium bioavailability, and hence absorption, has been verified in human subjects.[61]

Effects in Cells/Tissues

Phytic acids per se may be anticarcinogenic, at least in animal models. The mechanisms most involved are second messenger roles of the inositol phosphate metabolic products after their release from cell membranes, as in the case of the choline phosphates (see this chapter under "Lecithin"). The cytosolic actions of inositol phosphates are critical for cell functions because they serve as cell switches that regulate the entry of calcium ions via specific ion channels.[65]

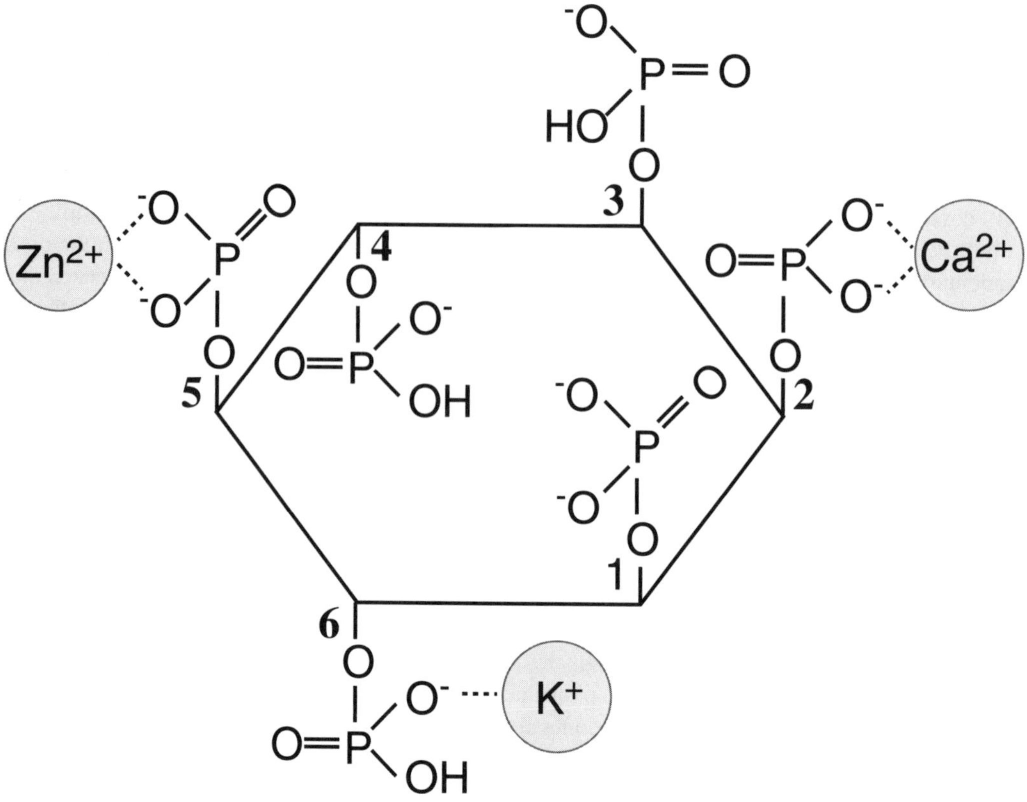

Phytic Acid

Figure 11–13 Structure of phytic acid with ionic binding of cations. Phytic acid consists of an inositol molecule with six phosphate groups. The negative charges on the phosphate groups allow for binding of one or more monovalent (such as K^+ or Na^+) or divalent (such as Ca^{2+}, Cu^{2+}, or Zn^{2+}) ions.

Phytates have been shown to exert anticarcinogenic effects in mice and rats through the action of inositol phosphate 6 (IP6), which similarly inhibits cell proliferation in vitro in rabbit prostate cells.[66,67] IP6 is especially rich in soy flour.[67]

Summary

Soy-derived phytic acid serves two important roles in human health: carrying minerals that are largely, though not totally, absorbed and by exerting an anticarcinogenic action. Since human consumption of soybeans is currently so low in Western nations, it is unlikely that either of these functions has any currently significant impact on health.

NON-NUTRIENT MOLECULES—NONABSORBABLE

The various steps of protein digestion within first the stomach and then the small intestine almost totally inactivate the protease inhibitors and lectins. Any residual molecules that escape even partial digestion must be rare.

Protease Inhibitors

Soybeans contain a broad range of protease inhibitors, including trypsin (Kunitz) inhibitor and chymotrypsin (Bowman-Birk or BBI) inhibitor.[9] Soy is the richest source of these inhibitory proteins in the plant kingdom, and the inhibitors are associated with the storage proteins in the seed. Soaking the beans overnight in water and then the next morning bringing them to a boil and simmering them for an hour or so allows denaturation and inactivation of practically all of these inhibitors. Because raw soybeans contain these inhibitors, plus a few other undesirable products, only cooked products should be consumed. Fermented soy products that are not heated are also safe because the fermentation process permits degradation by microorganisms of the protease inhibitors and other factors that have potentially adverse effects. Trypsin inhibitors have been reported to have a protective role with respect to cancer development.[5,68,69] Table 11–2 lists the estimated amounts of protease inhibitors in soy flour.

Human Consumption

Very low intakes of the protease inhibitors exist in the United States and other Western nations because of low soy consumption. Japanese and other Asian populations, however, consume large amounts of soy—cooked, fermented, or otherwise processed. Heat treatment inactivates these inhibitors, but if soybeans are not heat-treated, the inhibitors, especially BBI, remain, and pancreatic enlargement may follow, as has been shown in experimental animals.[70] No adverse effects of these inhibitor molecules on pancreatic tissue have been observed in the Japanese, whose pancreatic cancer rates are similar to those of the US population, because the soy is appropriately heat-treated.

Structures

The protease inhibitor molecules are proteins that have characteristic three-dimensional shapes.[3]

Digestion

These molecules are freed during digestion from raw soy and raw soy products by digestive enzymes that attack the components of plant cells. Once released into the gut lumen, these inhibitors stimulate cholecystokinin-producing endocrine cells in the gut lining (mucosa) to increase pancreatic secretion, which in turn may lead to pancreatic hyperplasia, hypertrophy, and possibly cancer in the acinar cells.

Because the protease inhibitors are proteins, they are eventually attacked and degraded by gut peptidases. Thus, these inhibitors are not absorbed and do not enter the circulation.

Effects in Cells/Tissues

No effects of protease inhibitors result directly from soy foods, but when BBI is taken in a pill or tablet in large amounts, the protease inhibitors may act to prevent cancer development in rodent models.[68] In vitro studies of BBI suggest that it also may help prevent the development of certain cancers by interfering with malignant transformations.[70] Human trials using an oral preparation have been initiated to determine if BBI can prevent and suppress cancer in high-risk humans.[69]

Summary

Soybean protease inhibitors may have potential roles in the prevention of cancer. These molecules, such as BBI, must first be extracted from raw soy and then purified before they can be functional. Raw soy is not safe to consume, and heat inactivates practically all of these protease inhibitors. Therefore, consumption by pill or another concentrated form would be required to deliver these molecules to tissues, much like a drug, even though extracted BBI would still be derived from a food. Toxicity would have to be presumed to occur at high dosages of protease inhibitors, but no studies with human subjects have been conducted.

Lectins

Lectins, also known as hemaglutinins, exist in minimally processed soy feeds, but they are destroyed by heating, as used in typical food preparation. Lectins have previously been implicated to inhibit growth in animals, especially rats, but more recently investigations have shown that some other component of soy is responsible for the growth inhibition of whole soy meal.[3] These molecules have a high affinity for binding to carbohydrate-containing moieties, mainly with a high specificity for glucose.[3] In vitro, they act as hemaglutinins of red blood cells because of their binding to glucose molecules that are part of complex-carbohydrate recognition molecules extending from the surfaces of cell membranes, and they act similarly by binding to membrane molecules of other cell types, including gastrointestinal mucosal cells.

CONCLUSIONS

The bioactive molecules in soy and soy products have many important roles in promoting health and preventing disease. Several of these components, such as choline, vitamin K, and the minerals, have established essentiality as nutrients, but none of the non-nutritional bioactive molecules is considered essential. Nevertheless, several of these other molecules, especially the isoflavones, have been shown to have significant and positive effects on cardiovascular and skeletal tissues, and possibly on several others. With the exception of the trypsin and BBIs and lectins, the other bioactive molecules discussed here are not inactivated by heat treatment of soy, and they generally remain stable through food-processing steps. More research information is needed on each of these non-nutrient molecules before its utility in the promotion of human health can be fully established.

The most promising of this group of bioactive molecules are the isoflavones, which have been regarded to have beneficial effects in the prevention of cardiovascular diseases, osteoporosis, and some cancers. The mechanistic explanations of these actions are currently being actively investigated.

Two types of molecules in soy have a cholesterol-lowering effect, though by different mechanisms. Since phytosterols substitute for cholesterol when plant foods replace animal foods, adverse effects resulting from the intake of cholesterol may be reduced. Saponins also have promise because of their potential cholesterol-lowering effect.

Two additional molecules in soy have potential anticancer roles. BBI, an important protease inhibitor, may have critical roles in inhibiting the development of cancer in selected tissues. Phytates also have a potential anticarcinogenic effect.

Soy and soy foods also contain a few nutrients in adequate amounts that promote cellular health. Phosphatidylcholine, also known as lecithin, has many cellular roles, and its derivative choline may play a role in defense against cancer by enhancing apoptosis of cells rather than the development of malignant cells. Vitamin K has essential roles in post-translational modification of clotting proteins and bone matrix proteins, and it may have other actions in cancer protection not yet understood. The micronutrients in soybeans that have been reviewed here—choline in lecithin and vitamin K—tend to be deficient in the diets of many individuals who consume few vegetable sources, especially soy and soy products. The consumption of prepared soy foods (sufficiently heated to inactivate lectins and protease inhibitors) may therefore improve the nutritional status of many individuals with respect to these essential micronutrients.

Finally, the full value of the diverse set of molecules, both nutrients and non-nutrients, found in soy foods has not yet been fully appreciated. Much research is needed to expand our understanding of each of these bioactive molecules.

REFERENCES

1. Messina M. Isoflavone intakes by Japanese were overestimated. *Am J Clin Nutr.* 1995;62:645–646. Letter.

2. Golbitz P. Traditional soyfoods: processing and products. *J Nutr.* 1995;125:570S–572S.

3. Liener IE. Implications of antinutritional components in soybean foods. *Crit Rev Food Sci Nutr.* 1994;34:31–67.

4. Garcia MC, Torre M, Marina ML, LaBorda F. Composition and characterization of soybean and related products. *Crit Rev Food Sci Nutr.* 1997;37:361–391.

5. Messina M, Barnes S. The role of soy products in reducing risk of cancer. *J Nat Cancer Inst.* 1991;83:541–546.

6. Messina M, Erdman JW Jr, eds. First International Symposium on the Role of Soy in Preventing and Treating Chronic Disease. *J Nutr.* 1995;125:567S–808S.

7. Anderson RL, Wolf WJ. Compositional changes in trypsin inhibitors, phytic acid, saponins and isoflavones related to soybean processing. *J Nutr.* 1995;125:581S–588S.

8. Wang H-J, Murphy PA. Isoflavone content in commercial soybean foods. *J Agric Food Chem.* 1994;42:1666–1673.

9. DiPietro CM, Liener IE. Soybean protease inhibitors in foods. *J Food Sci.* 1989;54:606–609, 617.

10. Kozlowska H, Zadernowski R, Sosulski FW. Phenolic acids in oilseed flours. *Nahrung.* 1984;27:449–453.

11. Ireland PA, Dziedzic SZ, Kearsley MW. Saponin content of soya and some commercial soya products by means of high performance liquid chromatography of the sapogenins. *J Sci Food Agric.* 1986;37:694–698.

12. Ranhotra GS, Loewe RJ, Puyat LV. Phytic acid in soy and its hydrolysis during breadmaking. *J Food Sci.* 1974;39:1023–1025.

13. Padgette SR, Taylor NB, Nida DL, et al. The composition of glyphosate-tolerant soybean seeds is equivalent to that of conventional soybeans. *J Nutr.* 1996;126:702–716.

14. Szuhaj BF, ed. *Lecithins: Sources, Manufacture and Uses.* Champaign, IL: American Oil Chemists' Society; 1989.

15. Zeisel SH, Blusztajn JK. Choline and human nutrition. *Ann Rev Nutr.* 1994;14:269–296.

16. Food and Nutrition Board, Institute of Medicine. *Dietary Reference Intakes.* Washington, DC: National Academy Press; 1998.

17. Vermeer C, Jie KSG, Knapen MHJ. Role of vitamin K in bone metabolism. *Ann Rev Nutr.* 1995;15:1–22.

18. Booth SL, Pennington JAT, Sadowski JA. Food sources and dietary intakes of vitamin K-1 (phylloquinone) in the American diet. *J Am Diet Assoc.* 1996;96:149–154.

19. Kohlmeier M, Saupe J, Shearer M, et al. Bone health of adult hemodialysis patients is related to vitamin K status. *Kidney Int.* 1997;51:1218–1221.

20. Kohlmeier M, Garris S, Anderson JJB. Vitamin K: a vegetarian promoter of bone health. *Vegetarian Nutr.* 1997;1:53–57.

21. Binkley NC, Suttie JW. Vitamin K nutrition and osteoporosis. *J Nutr.* 1995;125:1812–1821.

22. Deshpande SS, Sathe SK, Salunkhe DK. Chemistry and safety of plant polyphenols. In: Friedman M, ed. *Nutritional and Toxicological Aspects of Food Safety.* New York, NY: Plenum Press; 1984:457–495.

23. Bravo L. Polyphenols: chemistry, dietary sources, metabolism, and nutritional significance. *Nutr Rev.* 1998;56:317–333.

24. Messina M. Modern applications for an ancient bean: soybeans and the prevention and treatment of disease. *J Nutr.* 1995;125:567S–569S.

25. Clarkson TB, Anthony MS, Hughes CL Jr. Estrogenic soybean isoflavones and chronic disease: risks and benefits. *Trends Endocrinol Metab.* 1995;6:11–16.

26. Adlercreutz HC, Goldin BR, Gorbach SL, et al. Soybean phytoestrogen intake and cancer risk. *J Nutr.* 1995;125:757S–770S.

27. Kurzer MS, Xu X. Dietary phytoestrogens. *Ann Rev Nutr.* 1997;17:353–381.

28. Anderson JJB, Garner SC. The effects of phytoestrogens on bone. *Nutr Res.* 1997;17:1617–1632.

29. Setchell KDR. Phytoestrogens: the biochemistry, physiology, and implications for human health of soy isoflavones. *Am J Clin Nutr.* 1998;68:1333S–1346S.

30. Franke AA, Custer LJ, Cerna CM, Narala KK. Quantitation of phytoestrogens in legumes by HPLC. *J Agric Food Chem.* 1994;42:1905–1913.

31. Franke AA, Custer LJ, Tanaka Y. Isoflavones in human breast milk and other biological fluids. *Am J Clin Nutr.* 1998;68:1466S–1473S.

32. King RA, Bursill DB. Plasma and urinary kinetics of the isoflavones daidzein and genistein after a single soy meal in humans. *Am J Clin Nutr.* 1998;67:867–872.

33. Safakianos J, Coward L, Kirk M, Barnes S. Intestinal uptake and biliary excretion of the isoflavone genistein in rats. *J Nutr.* 1997;127:1260–1268.

34. Anderson JJB, Garner SC. Phytoestrogens and bone. In: Adlercreutz H, ed. Phytoestrogens. *Balliere's Clin Endocrinol Metab.* 1998;12:543–557.

35. Ruiz-Larrea MB, Mohan AR, Paganga G, et al. Antioxidant activity of phytoestrogenic isoflavones. *Free Radical Res.* 1987;26:63–70.

36. Akiyama T, Ishida J, Nakagawa S, et al. Genistein, a specific inhibitor of tyrosine specific protein kinases. *J Biol Chem.* 1987;262:5592–5595.

37. Li D, Yee JA, McGuire MH, Murphy PA, Yan L. Soybean isoflavones reduce experimental metastasis in mice. *J Nutr.* 1999;129:1075–1078.

38. Anderson JJB, Anthony M, Messina M, Garner SC. Effects of phyto-oestrogens on tissues. *Nutr Res Rev.* 1999;12;75–116.

39. Gustafsson JA, Mode A, Norstedt G, Eneroth P, Hokfelt T. Central control of prolactin and estrogen receptors in rat liver—expression of a novel endocrine system, the hypothalamo-pituitary-liver axis. *Ann Rev Pharmacol Thera.* 1983;23:259–278.

40. Eisenfeld AJ, Aten RF. Estrogen receptors and androgen receptors in the mammalian liver. *J Steroid Biochem.* 1987;27:1109–1118.

41. Anthony MS, Clarkson TB, Williams JK. Effect of soy isoflavones on atherosclerosis: potential mechanisms. *Am J Clin Nutr.* 1998;68:1390S–1393S.

42. Crouse JR III, Terry JG, Morgan TM, et al. A randomized trial comparing the effect of casein with that of soy protein containing varying amounts of isoflavones on plasma concentrations of lipids and lipoproteins. *Arch Int Med.* 1999;159:2070–2076.

43. Anderson JW, Johnstone BM, Cook-Newell ME. Meta-analysis of effects of soy protein intake on serum lipids in human. *N Engl J Med.* 1995;333:276–282.

44. Nagata C, Takatsuka N, Kurisu Y, Shimizu H. Decreased serum total cholesterol concentration is associated with high intake of soy products in Japanese men and women. *J Nutr.* 1998;128:209–213.

45. Potter SM. Overview of proposed mechanisms for the hypocholesterolemic effect of soy. *J Nutr.* 1995;125:606S–611S.

46. Honore EK, Williams JK, Anthony MS, Clarkson TB. Soy isoflavones enhance coronary vascular reactivity in atherosclerotic female macaques. *Fertil Steril.* 1997;67:148–154.

47. Fujio Y, Fumiko Y, Takahashi K, Shibata N. Responses of smooth muscle cells to platelet-derived growth factor are inhibited by herbimycin-A tyrosine kinase inhibitor. *Biochem Biophys Res Commun.* 1993;195:79–83.

48. Shinokado K, Yokota T, Umezawa K, Sasaguri T, Ogata J. Protein tyrosine kinase inhibitors inhibit chemotaxis of vascular smooth muscle cells. *Arteriosclerosis Thromb.* 1994;14:973–981.

49. Shinokado K, Umezawa K, Ogata J. Tyrosine kinase inhibitors inhibit multiple steps of the cell cycle of vascular smooth muscle cells. *Exp Cell Res.* 1995;220:266–273.

50. Bennetts HW, Underwood EJ, Shier FL. A specific breeding problem of sheep on subterranean clover pastures in western Australia. *Aust Vet J.* 1946;22:2–12.

51. Anderson JJB, Ambrose WW, Garner SC. Biphasic effects of genistein on bone tissue in the ovariectomized, lactating rat model. *Proc Soc Exp Biol Med.* 1998;217:345–350.

52. Oakenfull D. Saponins in food: a review. *Food Chem.* 1981;6:19–40.

53. Sidhu GS, Oakenfull DG. A mechanism for the hypocholesterolaemic activity of saponins. *Br J Nutr.* 1986;55:643–649.

54. Haslam E, Lilley TH. Natural astringency in foodstuffs—a molecular interpretation. *Crit Rev Food Sci Nutr.* 1998;27:1–40.

55. Mangan JL. Nutritional effects of tannins in animal feeds. *Nutr Res Rev.* 1988;1:209–231.

56. Chung KT, Wong TY, Wei C-I, Huang Y-W, Lin Y. Tannins and human health: a review. *Crit Rev Food Sci Nutr.* 1998;38:421–464.

57. Butler LG. Effects of condensed tannin on animal nutrition. In: Hemingway RW, Karchesy JJ, eds. *Chemistry and Significance of Condensed Tannins.* New York, NY: Plenum Press; 1989:301–408.

58. Weihrauch JL, Gardner JM. Sterol content of foods of plant origin. *J Am Diet Assoc.* 1973;73:39–43.

59. Miettinen TA, Tilvis RS, Kesaniemi YA. Serum plant sterols and cholesterol precursors reflect cholesterol absorption and synthesis in volunteers of a randomly selected male population. *Am J Epidemiol.* 1990;131:20–31.

60. Stuart SM, Ketelsen SM, Weaver CM, Erdman JW Jr. Bioavailability of zinc to rats as affected by protein source and previous dietary intake. *J Nutr.* 1986;116:1423–1431.

61. Heaney RP, Weaver CM, Fitzsimmons ML. Soybean phytate content: effect on calcium absorption. *Am J Clin Nutr.* 1991;53:745–747.

62. Beard JL, Weaver CM, Lynch S, et al. The effect of soybean phosphate and phytate content on iron bioavailability. *Nutr Res.* 1988;8:345–352.

63. Davidsson L, Almgren A, Juillerat MA, Hurrell RB. Manganese absorption in humans. *Am J Clin Nutr.* 1995;62:984–987.

64. Harland BF, Narula G. Food phytate and its hydrolysis products. *Nutr Res.* 1999;19:947–961.

65. Mayrleitner M, Schafer R, Fleischer S. IP3 receptor from liver plasma membrane is a (1,4,5)IP3 activated and (1,3,4,5)IP4 inhibited calcium permeable ion channel. *Cell Calcium.* 1995;17:141–153.

66. Shamsuddin AM. Inositol phosphates have novel anticancer function. *J Nutr.* 1995;125:725S–732S.

67. Shamsuddin AM, Yang GY, Vucenik I. Novel anticancer functions of IP6: growth inhibition and differentiation of human mammary cancer cell lines in vitro. *Anticancer Res.* 1996;16:3287–3292.

68. Kennedy AR. The evidence for soybean products as cancer preventive agents. *J Nutr.* 1995;125:733S–743S.

69. Kennedy AR. The Bowman-Birk inhibitor from soybeans as an anticarginogenic agent. *Am J Clin Nutr.* 1998;68:1406S–1412S.

70. Liener IE. Possible adverse effects of soybean anticarcinogens. *J Nutr.* 1995;125:744S–750S.

PART IV

Nutritional Aspects

CHAPTER 12

Dietary Fiber and Its Physiological Effects

Daniel D. Gallaher

Consumption of dietary fiber is now known to produce a number of important physiological effects. Yet until the early 1970s, dietary fiber was considered by most nutritionists as largely inert, useful only as a laxative and therefore of little interest. This view began to change, however, when it was hypothesized that the low levels of dietary fiber consumed in Western societies were responsible for the high incidence of a number of chronic diseases, such as coronary artery disease, appendicitis, diverticular disease, and colon cancer.[1] It was contended that many of the diseases of Western society were, in essence, fiber-deficiency diseases. Efforts to test this hypothesis subsequently created a surge of research effort directed toward understanding the physiological effects of fiber as well as to test its ability to prevent the development of various chronic diseases. This work on dietary fiber continues unabated. As our knowledge of the physiological effects of dietary fiber continues to expand, it is clear that dietary fiber is far from the inert substance it was originally envisioned to be.

RELATIONSHIP TO SPECIFIC DISEASE STATES

Numerous epidemiological studies have examined the association between dietary fiber intake and the incidence of chronic diseases common in Western culture. Colon cancer is the focus of many of these studies. Unfortunately, results have been inconsistent, with both protective effects[2,3] and no protection by high fiber intakes reported.[4] Difficulty in interpreting the epidemiological data, however, arises because it is currently impossible to separate a dietary fiber effect from the effect of other components in fiber-rich foods. Recognizing this problem, Hill[5] concluded that the epidemiological data clearly support a protective effect of cereals and vegetables against colon cancer; however, a role for dietary fiber per se cannot be definitively established. A similar conclusion was reached by Gerber[6] with regard to dietary fiber and breast cancer.

This same problem exists for all epidemiological studies of dietary fiber and disease. Nonetheless, it is of interest that, at least in some studies, an association between decreased disease incidence and increased fiber consumption has been found for a number of other diseases. These include diverticular disease,[7] endometrial cancer,[8] and coronary heart disease.[9] These studies indicate the potential for dietary fiber to reduce the incidence of chronic diseases and provide direction for further studies of the health benefits of increased dietary fiber consumption.

SOURCES OF FIBERS AND THEIR METABOLISM

Dietary fiber is generally agreed to be a constituent of plant materials. Although a number of definitions for dietary fiber have been put forth over the years, one well-known and commonly accepted definition is plant material resistant to hydrolysis by the digestive enzymes of humans.[10] This is a physiological definition, as it relies on the resistance of plant material to enzymatic degradation. Chemical definitions of dietary fiber, such as the sum of nonstarch polysaccharides and lignin, have also been commonly used. Lignin is a highly branched and cross-linked polymer based on oxygenated phenylpropane units. Because it is not a carbohydrate, there has been debate over the years as to whether it should be considered part of the definition of fiber. However, the lignin content of most foods is quite low, so in practical terms, this question is seldom important.

A number of different polymeric plant carbohydrates make up what we commonly refer to as dietary fiber. These include cellulose, the hemicelluloses, β-glucans, pectins, and numerous types of gums[11] (Figure 12–1). Cellulose is a linear polymer of glucose molecules joined by a $\beta(1\rightarrow4)$ linkage. Mammalian amylases are unable to hydrolyze a $\beta(1\rightarrow4)$ linkage, in contrast to the $\alpha(1\rightarrow4)$ linkages that make up starches. Hemicelluloses are highly branched polymers consisting mainly of glucurono- and 4-O-methylglucuroxylans.[12] β-Glucans are mixed-linkage $(1\rightarrow3)$, $(1\rightarrow4)$ β-D-glucose polymers found primarily in cereals, particularly oats and barley.[13] Pectins are a group of polysaccharides composed primarily of D-galacturonic acid,[14] which is methoxylated to variable degrees. Finally, the gums and mucilages represent a broad array of different branched structures. Guar gum, derived from the ground endosperm of the guar seed, is a galactomannan. Other gums, such as gum arabic and gum acacia, have still different structures. Psyllium seed husk is a highly branched arabinoxylan. Thus, from the preceding description, it is clear that dietary fiber is a collection of polymeric carbohydrates that differ significantly in both their composition of sugars and their linkages. It should consequently not be surprising that dietary fibers differ in their physical characteristics as well.

A number of substances, both natural and synthetic, fit into one or more definitions of dietary fiber. These include Maillard reaction products formed during heating of proteins in the presence of reducing sugars, modified celluloses such as methyl cellulose, indigestible animal products such as chitins, and oligosaccharides such as inulin and oligofructose. All of these substances share some characteristics with what is traditionally considered dietary fiber, yet they differ significantly in other ways. None of these substances have yet been universally accepted as dietary fiber, but they nonetheless are sometimes described as such.

Dietary fibers, being indigestible by definition, are not metabolized in the same fashion as the digestible macronutrients. However, most dietary fibers are susceptible to microbial fermentation in the large intestine. The degree to which different fibers are fermented varies widely. For example, cellulose, a glucose polymer, is resistant to fermentation. However, the degree of cellulose fermentation varies considerably, depending on the form. Purified crystalline cellulose is only somewhat degraded (20% or less), whereas the majority of cellulose within a food may be fermented.[15] β-Glucans, another glucose polymer, are highly susceptible to fermentation and are rapidly and completely degraded in the colon. This difference can be ascribed to the tight packing of the glucose chain in cellulose, making it inaccessible to bacterial cellulases. In general, the insoluble fibers, such as cellulose and many hemicelluloses, are resistant to microbial degradation and therefore only partially fermented. In contrast, most soluble types of fiber, such as guar gum, pectins, and β-glucans, are rapidly and completely fermented. However, some fibers considered soluble, such as psyllium, are only slightly fermented, and the highly soluble modified celluloses, such as methylcellulose, are not fermentable at all. Thus, the solubility of a dietary fiber does not equate with its susceptibility to fermentation.

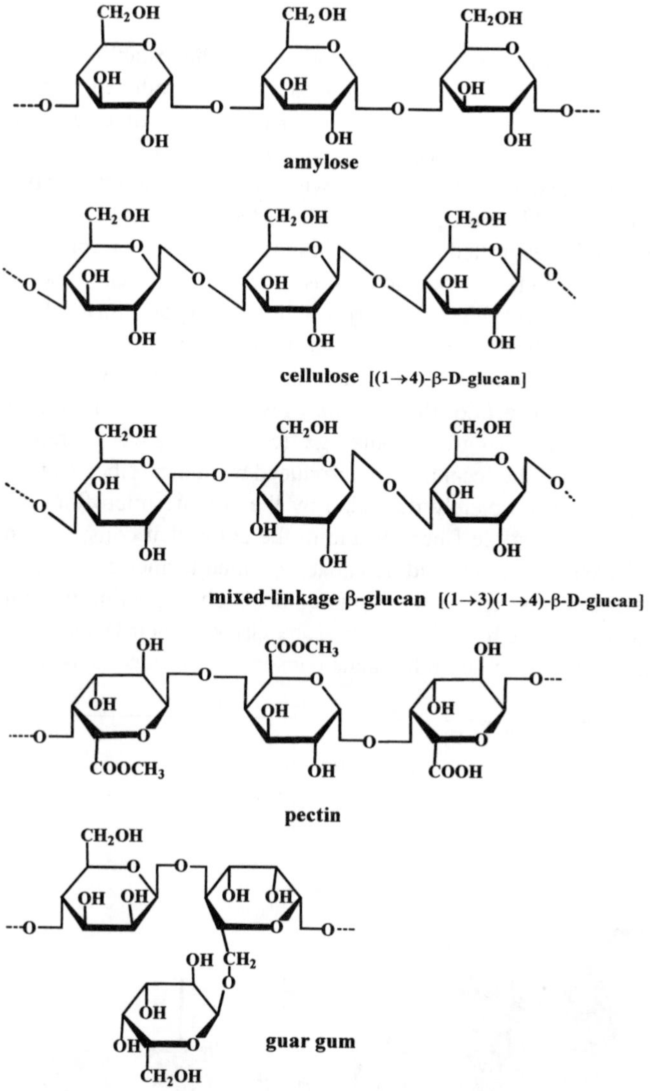

Figure 12–1 Chemical structure of starch (amylose) and several dietary fiber types.

PHYSIOLOGICAL EFFECTS OF FIBER

Dietary fiber, by definition, is not absorbed from the gastrointestinal tract. Yet consumption of dietary fiber has been demonstrated to alter metabolism of carbohydrates, lipids, and proteins, although the magnitude of these changes varies dramatically with the quantity and type of fiber ingested. To understand these actions of dietary fiber, it is important to understand how fiber affects the physical and chemical events within the intestine. Changes within the intestine alter the processes of digestion, absorption, and, in the case of the large intestine, fermentation.

Laxation

A laxative effect is certainly the best known and most firmly established physiological effect of dietary fiber. The laxative effect is related to the fecal bulking produced by a fiber, but not in a linear fashion. At fecal wet weights of 60 g/d or less, transit time, the time required for a consumed substance to be excreted, commonly exceeds 90 hours. As fecal wet weight increases, transit time decreases but reaches a plateau at 40 to 50 hours, when fecal wet weight is between 150 and 200 g/d.[16] Since many individuals within the United States have fecal weights of 120 g or less,[17] increasing dietary fiber intake would be expected to significantly increase laxation for many people.

Regardless of composition, all fiber-rich foods provide some fecal bulking. The degree to which they increase fecal wet weight, however, varies with the type and form of fiber in the food. Wheat bran produces the largest increase in fecal output, followed by fruits and vegetables, gums, and oats and corn.[18] The smallest increment in fecal output is produced by consumption of purified pectin (Figure 12–2). The physical form of the fiber may also influence fecal bulking. For example, coarse wheat bran produces a somewhat greater bulking effect than fine wheat bran.[18]

Several factors influence the fecal bulking produced by dietary fiber. Certain fibers, such as wheat bran and cellulose, are incompletely degraded by the colonic microflora, as discussed above. For these fibers, it is the indigestible fiber residue in the colon that constitutes most of the fecal bulk. These fibers also have some water-holding capacity, which further contributes to the bulk. However, for fibers that are completely fermented in the colon, such as pectin, guar gum, and β-glucans, the fiber itself does not contribute bulk. These fibers, by virtue of their fermentation, produce an increase in the colonic microflora. Since in individuals consuming Western diets, the colonic microflora ac-

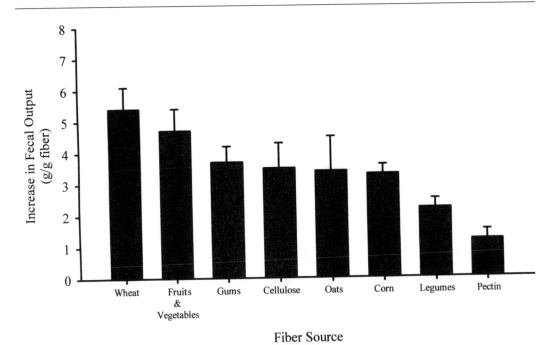

Figure 12–2 Effect of consumption of various dietary fiber sources on fecal output in humans. *Source:* Adapted with permission from J.H. Cummings, The Effect of Dietary Fiber on Fecal Weight and Composition, *CRC Handbook of Dietary Fiber in Human Nutrition*, 2nd ed., pp. 263–349, © 1993, CRC Press.

counts for approximately half the weight of fecal solids,[19] increases in colonic microbial growth can significantly increase the fecal mass. Thus, both highly fermentable and poorly fermentable fibers increase fecal mass, and a larger fecal mass in the colon promotes laxation. However, poorly fermentable fibers such as wheat bran produce a much larger fecal mass and thus are more effective at producing laxation.

Fermentation

Fermentation in the colon leads to the production of a number of end products, including gases such as H_2, CH_4, and CO_2 and short-chain fatty acids (SCFAs). The predominant SCFAs are acetate, propionate, and butyrate. SCFAs are almost entirely absorbed by the colonic mucosa[20] and thus provide a source of energy to the host. For this reason, dietary fiber can be considered to have a caloric value. The exact energy value provided by dietary fiber is difficult to calculate but appears to be approximately 6 kJ/g (1.5 kcal/g) when consumed as part of a mixed meal.[21]

SCFAs are metabolized to different extents at different tissue sites. All three SCFAs undergo some metabolism in the colonic mucosa, although butyrate is a preferred fuel for colonocytes and is extensively metabolized at this site.[22] Most of the remaining propionate and butyrate present in hepatic portal blood is cleared by the liver.[23] Acetate in the peripheral blood is rapidly taken up and oxidized by cardiac and skeletal muscle[24,25] and brain.[26]

Dietary fibers differ both in the quantity and in the profile of SCFAs produced as a result of colonic fermentation. Figure 12–3 shows the total SCFA produced in the cecum of rats fed equal

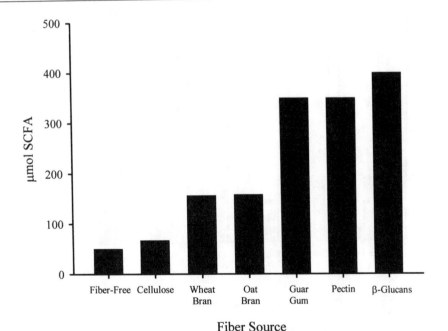

Figure 12–3 Effect of feeding various dietary fiber sources on the quantity of short-chain fatty acids (SCFA) in the cecum of rats. *Source:* Adapted with permission from A. Berggren, I.M.E. Björck, E.M. Nyman, and B.O. Eggum, Short-Chain Fatty Acid Content and pH in Caecum of Rats Given Various Sources of Carbohydrates, *Journal of the Science of Food and Agriculture*, 1993, Vol. 63, p. 401. Copyright Society of Chemical Industry. Reproduced with permission. Permission is granted by John Wiley & Sons Ltd on behalf of the SCI.

quantities of dietary fiber from a number of fiber sources.[27] The quantity of SCFAs produced corresponds roughly to the degree of fermentation of each fiber. That is, purified cellulose, a poorly fermentable fiber, yields the lowest amount of SCFAs, whereas guar gum, pectin, and β-glucans, all highly fermentable fibers, yield the highest. The proportion of SCFAs produced generally follows the pattern of acetate>propionate>butyrate. However, as seen in Figure 12–4, differences in relative proportions do occur. With the exception of pectin, highly fermentable fibers yield relatively less acetate and relatively more propionate and butyrate.[27]

A number of potentially important physiological effects have been ascribed to the increased SCFAs produced by dietary fiber fermentation. The best established of these is a trophic effect on the colonic mucosa. Within physiological relevant ranges, SCFAs stimulate cell proliferation in the large intestine, an effect that is independent of the presence of colonic bacteria or a low lumenal pH.[28] Although the mechanism by which SCFAs produce large intestinal hypertrophy is still a subject of debate, it seems clear that SCFA production is an important factor in the large intestinal enlargement that is typically seen with consumption of fermentable fiber sources.

Specific physiological effects have been attributed to individual SCFAs. The well-established finding that butyrate normalizes the growth of transformed cell lines[29] has led to considerable interest in the possibility that fermentable fibers, by virtue of SCFA production, may have chemoprotective ef-

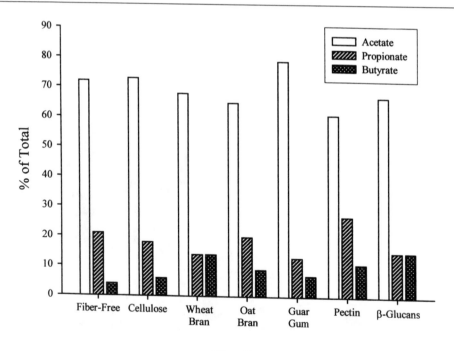

Fiber Source

Figure 12–4 Effect of feeding various dietary fiber sources on the relative proportions of the short-chain fatty acids acetate, propionate, and butyrate in the cecum of rats. *Source:* Adapted with permission from A. Berggren, I.M.E. Björck, E.M. Nyman, and B.O. Eggum, Short-Chain Fatty Acid Content and pH in Caecum of Rats Given Various Sources of Carbohydrates, *Journal of the Science of Food and Agriculture*, 1993, Vol. 63, p. 401. Copyright Society of Chemical Industry. Reproduced with permission. Permission is granted by John Wiley & Sons Ltd on behalf of the SCI.

fects in the colon. However, the trophic effect of butyrate and other SCFAs, described above, has led to the recognition of a paradox regarding butyrate and cell growth.[30] That is, butyrate stimulates the growth of normal cells but inhibits the growth of cancerous cell lines. Further confusing the situation are the inconsistent results of butyrate on colon carcinogenesis in animal models. Medina et al.[31] reported that large intestinal infusion of butyrate decreased colon tumor incidence in carcinogen-treated rats. Similarly, Wargovich et al[32] reported that feeding sodium butyrate reduced the number of colonic precancerous lesions (aberrant crypt foci, ACF). In contrast, Caderni et al[33] found no change in ACF when carcinogen-treated rats were fed butyrate-containing slow-release pellets. Although butyrate shows promise as a chemopreventive agent, the inconsistent results to date indicate that our understanding of the role of butyrate in carcinogenesis is incomplete.

Glucose Metabolism

Modulation of the glycemic response to a glucose load or a meal is one of the most studied physiological effects of dietary fiber. By the late 1970s it was well established that ingestion of certain dietary fibers improved glucose tolerance and decreased postprandial plasma insulin concentrations in normal subjects.[34–36] This effect has subsequently been demonstrated in diabetics as well.[37,38]

Guar gum is the dietary fiber most frequently examined for its ability to modulate the glycemic and insulinemic responses to a meal or glucose load. In most such studies, guar gum has reduced postprandial glucose and insulin response in both normal humans [34,39–44] and animals.[45–48] However, Ryle et al[49] found in normal subjects given guar no effect on plasma glucose at either 1 or 2 hours after a glucose load, and Ellis et al[50] found no consistent effect of guar-containing bread on blood glucose but observed a consistent reduction in serum insulin concentrations compared to the response to bread containing no guar. Nonetheless, the majority of studies find that guar gum reduces the glycemic and or insulinemic response to a meal in subjects or animals in a normal physiological state.

The glycemic response to whole oats, barley, oat bran, and oat gum, a β-glucan–rich extract of oat bran, has been examined in a number of studies. Heaton et al[51] reported that consumption of oats, whether as whole groats, rolled oats, or ground oatmeal, resulted in a reduced glycemic and insulinemic response compared to a wheat-based cereal. However, others report no reduction in the glucose and insulin responses to an oat or barley porridge.[52] Oat bran reduced both the glycemic and insulinemic responses to a porridge meal in normal subjects.[38,42] However, in normolipidemic[53] and hyperlipidemic subjects,[54,55] no change in the glycemic response to a meal was found with oat bran consumption. At present, there are too few studies reported regarding the glycemic effect of whole oats and oat bran to draw conclusions. However, studies using β-glucan–rich concentrates from oats[38,39,42,56,57] or high–β-glucan barley products[52] consistently show improvements in the glycemic response to a meal. These results clearly indicate that β-glucans are the primary component within oats responsible for the improvement in the glycemic response. The inconsistent findings for oat bran probably reflect the low concentration of β-glucans in oat bran, approximately 8% to 10% on a dry weight basis.[58] Oat gum preparations, in contrast, may contain up to 80% β-glucan.

The glycemic response to other dietary fibers has been inconsistent. A number of studies report a decreased glycemic response with psyllium.[59–62] However, other studies have found no effect.[44,63,64] Pectin decreased postprandial glucose concentrations in several studies in humans [35,65] and at one time point in rats,[66] but no effect has also been reported.[67] A number of studies feeding sugar beet pulp have demonstrated improved glucose tolerance[41,63] or a decrease in the area under the glucose curve,[59] but one study reported no effect.[68] Interestingly, a number of studies feeding wheat bran have found either reduced blood glucose concentrations at one or more time points of a glucose tolerance curve, a reduced area under the glucose curve,[69,70] or a reduced area under the insulin curve.[59]

However, other studies have found wheat bran feeding to be without effect on these parameters.[48,67,71] Finally, fiber-rich products from soybeans have been examined. Neither soy polysaccharide (from soy cotyledons) nor soy bran improved glucose tolerance.[72,73]

At present, guar gum and fiber sources high in β-glucans most consistently lead to an improvement in glucose and/or insulin tolerance in normal subjects or animals. For other fibers, either there are too few studies available or the results are too inconsistent to be confident that they improve glucose tolerance.

The improvement in glycemic control found with consumption of certain dietary fibers seems clearly due to a reduced rate of glucose absorption.[34,74] A number of plausible mechanisms have been put forth to explain how fiber slows glucose absorption. These include a slowed rate of glucose uptake within the small intestine, slower gastric emptying, altered small intestinal motility, and a reduced rate of starch digestion. None of these are mutually exclusive, raising the possibility that multiple mechanisms are involved.

Glucose absorption from perfused jejunum in humans was reduced when the perfusate contained guar gum.[75] Similarly, pectin reduced glucose absorption from the small intestine when included in the perfusate, an effect that was concentration dependent.[76] Pectin reduced glucose uptake in perfusion studies in both rats and humans when pectin was present in the perfusate.[77] Thus, the ability of viscous fibers to slow glucose uptake during perfusion of the small intestine appears established.

Numerous studies have examined the effect of dietary fiber on gastric emptying. Table 12–1 presents a summary of such studies.[65,75,78–102] Several different techniques have been employed to determine gastric emptying rate, most commonly the disappearance of a radiolabel from the stomach area or the amount of digesta remaining in the stomach. Solid meals, slurries, and liquid meals have been used, and the emptying rate has been examined in pigs, rats, dogs, and humans. The most common dietary fiber tested has been guar gum, with wheat bran also used in a number of studies. The most striking overall impression from these studies is the consistency with which different dietary fibers reduce the rate of gastric emptying. With the exception of the inconsistent results in pigs by Rainbird[78] and Rainbird and Low,[80] there are few instances in which a dietary fiber did not slow the gastric emptying rate. It is worth noting that several studies report a reduced rate of gastric emptying with wheat bran feeding, a fiber that would not be expected to slow glucose uptake within the small intestine. Since a number of studies have found wheat bran to flatten the postprandial glucose curve, this could be considered evidence supporting a role for delayed gastric emptying as a mechanism for improving glycemic control.

Thus, it appears that dietary fiber improves glucose tolerance primarily by delaying gastric emptying. However, a delay in glucose absorption due to a high viscosity within the small intestine may also contribute to this effect.

Lipid Metabolism

A cholesterol-lowering ability of certain types of dietary fiber has been documented since the early 1960s.[103,104] Guar gum, pectin, psyllium (ie, Metamucil), oats, and other fiber or fiber-rich sources characterized as soluble consistently reduce serum cholesterol.[105] A recent meta-analysis of 67 clinical trials of the effect of soluble dietary fiber on serum cholesterol has confirmed this cholesterol-lowering effect.[106] Many studies of the mechanism by which dietary fiber lowers cholesterol have failed to establish a single mechanism that explains all the observed results. However, the likely explanations are now centered on three phenomena associated with dietary fiber consumption: decreases in cholesterol absorption, fermentation, and increased bile acid excretion.

Table 12–1 Effect of Dietary Fiber on Gastric Emptying Rate in Animals and Humans

Species	Method	Meal Type	Fiber Type	Amount of Fiber	Gastric Emptying Rate*	Reference
Pig	Digesta remaining in stomach	Solid	Guar gum	6%	NC in DM	Rainbird[78]
		Milky drink	Guar gum	10 g/L	NC in DM	
		Glucose drink	Guar gum	20 g/L	↓ in DM @ 0.5 h ↓ in glucose @ 0.5–2 h	
Pig	Digesta remaining in stomach	Solid	Guar gum	2%, 4%, and 6%	↓ in glucose @ 1 h with 6% guar gum diets	Rainbird & Low[79]
Pig	Digesta remaining in stomach	Solid	Pectin	6%	NC in DM, ↓ in glucose @ 2 h	Rainbird & Low[80]
			Carboxymethyl cellulose	4%	↓ in DM @ 2 h, NC in glucose	
			Wheat bran	4%	↓ in DM @ 4 h, NC in glucose	
			Guar gum, low viscosity	4%	NC in DM or glucose	
Pig	Digesta remaining in stomach	Solid	Rolled oats (4% β-glucan)	88.4%	↓ in DM at 0.5–1 h, NC in starch	Johansen et al[81]
			Oat bran (8.6% β-glucan)	96.9%	↓ in DM at 1 h, NC in starch	
			Oat flour (1.6% β-glucan)	83.8%	NC in DM or starch	

continues

Table 12–1 continued

Species	Method	Meal Type	Fiber Type	Amount of Fiber	Gastric Emptying Rate*	Reference
Rat	^{99m}Tc remaining in stomach	Bean slurry	Wheat bran	15 mg	↓	Brown and Greenburgh[82]
Rat	^{99m}Tc remaining in stomach	Bean slurry	Pectin	15 mg	NC	Brown et al[83]
			Guar gum	25–50 mg	↓ at 25 and 50 min	
Rat	Glucose remaining in stomach	Liquid	Guar gums of different viscosities	80 mg	↓ relative to low-viscosity guar gum	Leeds et al[84]
Rat	^{14}C-starch remaining in stomach	Solid	Pectin	5%	↓ at 1 h, NC at 2 or 4 h	Forman and Schneeman[85]
Rat	Glucose remaining in stomach	Liquid	Locust bean gum	110–170 mg	↓ at 2 h	Tsai and Peng[86]
		Slurry	Locust bean gum	49–55 mg	↓ at 2 h	
Dog	^{51}Cr marker in duodenal effluent	Liquid	Guar gum	0.9, 3.0, and 4.5 g	↓ $t_{1/2}$ with greatest amount only	Russell and Bass[87]
			Psyllium	3.0 and 9.0 g	↓ $t_{1/2}$ with greatest amount only	
Human	Dilution in stomach of polyethylene glycol marker	Solid	Guar gum	5 g	↓	Harju[88]
Human	Dilution in stomach of polyethylene glycol marker	Solid	Pectin	10 and 15 g	↓ during initial 2 h postprandially	Flourie et al[89]

Human	Appearance of [24]Na in blood	Liquid	Guar gum	2 g	→	Wilmshurst and Crawley[90]
Human	[99m]Tc remaining in stomach	Solid	Wheat bran, course	14 g	→	Vincent et al[91]
			Wheat bran, fine	14 g	NC	
Human	[99m]Tc remaining in stomach	Liquid	Guar gum	14.5 g	→	Blackburn et al[75]
Human	[99m]Tc remaining in stomach	Liquid	Guar gum	9 g	→	Blackburn et al[92]
Human	[51]Cr remaining in stomach	Liquid	Guar gum	12.5 g	→	Torsdottir et al[93]
Human	[99m]Tc remaining in stomach	Liquid	Guar gum	12 g	→	French and Read[94]
Human	[99m]Tc remaining in stomach	Solid	Ispaghula	3.5 g	NC	McIntyre et al[95]
			Wheat bran	15 g	→	
Human	Difference in plasma glucose concentrations between oral and duodenally instilled meals	Oral meals were solid, instilled meals were a slurry	Guar gum, high viscosity	4.2–4.6 g (oral meal)	→ compared to low-viscosity guar gum	Leclère et al[96]
Human	Ultrasound of gastric antrum diameters	Solid	Mixed	14 g more than control	→	Benini et al[97]

continues

Table 12–1 continued

Species	Method	Meal Type	Fiber Type	Amount of Fiber	Gastric Emptying Rate*	Reference
Human	Gastric contents 30 min postprandially	Liquid	Cellulose	4 g	→	Tadesse[98]
			Pectin, low ester	4 g	→	
Human	[113m]In remaining in stomach	Solid	Guar gum	10 g	NC in half-time or in peak volume (measure of early phase of emptying)	Leatherdale et al[99]
Human	[113m]In remaining in stomach at 30 min	Liquid	Guar gum and pectin	16 g guar + 10 g pectin	→	Holt et al[65]
Duodenal ulcer patients	Dilution in stomach of polyethylene glycol marker	Solid	Guar gum	5 g	→	Harju[88]
Obese humans	[99m]Tc remaining in stomach	Solid	Pectin	15 g	↓ relative to methylcellulose	Di Lorenzo et al[100]
Non–insulin-dependent diabetics	[99m]Tc remaining in stomach	Solid	Guar gum (6 g) + wheat bran (3 g)	9 g	↓	Ray et al[101]
Non–insulin-dependent diabetics	[113m]In remaining in stomach	Solid	Guar gum	10 g	↓ in half-time, NC in peak volume (measure of early phase of emptying)	Leatherdale et al[99]
Non–insulin-dependent diabetics	[51]Cr remaining in stomach	Liquid	Alginate	5 g	→	Torsdottir et al[102]

*Abbreviations used: NC, no change; DM, dry matter; t[1/2], half-life.

Absorption of cholesterol has been examined, either directly or indirectly, for a number of dietary fiber types. Purified pectin and guar gum reduce cholesterol absorption, whereas psyllium does not.[107,108] β-Glucan–rich fiber sources also reduce cholesterol absorption.[109] Hydroxypropyl methyl-cellulose, a modified and highly soluble form of cellulose, has been shown to reduce cholesterol absorption linearly as a function of the logarithm of the intestinal contents' supernatant viscosity that it produces.[110] Overall, it appears that viscous fibers are effective in reducing cholesterol absorption, although the mechanism by which they do so has yet to be established.

As described above, fermentation of dietary fiber leads to production of several SCFAs. Propionate, specifically, has been proposed to mediate the hypocholesterolemic effect of fermentable fibers, such as guar gum and oat bran.[111] Propionate has been found to decrease cholesterol synthesis rates in cultured hepatocytes.[112] Human studies in which propionate was rectally infused, however, found no decrease in serum cholesterol.[60,113] Further, in whole animals, dietary propionate did not reduce cholesterol synthesis.[114] Finally, a reduction in plasma and liver cholesterol has been demonstrated in germ-free rats fed guar gum, thus demonstrating that fermentation is unnecessary to produce the hypocholesterolemic effect associated with this fiber.[115] At present, the current evidence provides little support for a role for cholesterol lowering by propionate.

Bile acids are steroidal compounds, synthesized from cholesterol, that facilitate lipid absorption. They are secreted into the small intestine in response to a meal but are actively reabsorbed in the lower small intestine. However, small quantities of bile acids are continually lost in the feces. The ability of dietary fibers to increase this loss, and thereby to reduce cholesterol levels by stimulating the conversion of cholesterol to bile acids, has been studied extensively. In some human and animal trials, certain dietary fibers increased fecal bile acid excretion. However, there is an inconsistent relationship between increased bile acid excretion and decreased cholesterol levels. For example, Miettinen and Tarpila[116] reported increased bile acid excretion with guar gum and psyllium supplementation in human subjects, with corresponding decreases in serum total cholesterol concentration. Yet Abraham and Mehta[64] reported no change in fecal bile acids during psyllium treatment in humans, despite serum cholesterol reductions. Judd and Truswell[117] reported a 35% increase in bile acid excretion in subjects consuming 125 g/d of rolled oats but no significant reduction in plasma total cholesterol concentrations. Story et al[118] reported that in rats, oat bran and alfalfa reduced liver cholesterol accumulation, whereas corn bran and cellulose did not. However, bile acid excretion was not significantly altered by any of the fibers. Similarly, Gallaher et al[119] found no significant increase in bile acid excretion in rats fed 8% total dietary fiber as oat bran compared to rats fed fiber-free diets. In contrast, Zhang et al[120] noted a significant increase in bile acid excretion in hamsters fed 12% oat bran compared to a fiber-free diet.

Clearly, there is considerable inconsistency in studies examining the effect of dietary fiber on bile acid excretion. Furthermore, several reports indicate that the fiber-induced increases in bile acid excretion are not of sufficient magnitude to completely explain the observed reductions in cholesterol levels.[118] Thus, the role of bile acid excretion in the hypocholesterolemic effect of dietary fibers remains uncertain.

The effect of dietary fiber on triacylglycerol digestion and absorption has also been examined. A number of studies have reported that dietary fibers or fiber-rich foods decrease pancreatic lipase activity in vitro, yet in most cases the lipase activity in vivo is increased.[121] This paradoxical situation points to the importance of studying the effects of dietary fiber in vivo whenever possible. Triacylglycerol absorption may be delayed or impaired in spite of normal or elevated levels of lipase for a number of reasons, including interference with enzyme-substrate interactions, micelle formation, or diffusion of micelles to the intestinal mucosa. The finding that triacylglycerol excretion was increased in ileostomy fluid by 36% by pectin feeding,[122] but was unaltered by wheat bran feeding[123] suggests that viscous fibers may interfere with the process of triacylglycerol absorption. Consistent with this is the finding of a delayed disappearance of triolein from the small intestine in rats fed guar gum or glucomannan, two highly viscous fibers, compared to cellulose.[124] Although decreased triacylglycerol absorption by viscous dietary fibers has been demonstrated, the physiological significance of this phenomenon is uncertain.

Protein Metabolism

As reviewed by Gallaher and Schneeman,[125] dietary fiber generally reduces protein digestibility. This is true for both purified dietary sources and fiber-rich sources. Although the vast majority of these studies have been conducted in rats, comparative studies indicate that the rat is a good model for humans in digestibility studies.[126]

Fiber consumption results in a shift in the pattern of nitrogen excretion. In the majority of studies, there is an increase in fecal nitrogen excretion and a concomitant decrease in urinary nitrogen excretion.[125] The decrease in urinary nitrogen excretion is usually insufficient to compensate for the increased fecal nitrogen excretion, resulting in a slight decrease in nitrogen balance. However, nitrogen balance invariably remains positive, indicating that fiber consumption is unlikely to pose a problem in protein nutriture.

The cause of the increased fecal nitrogen excretion appears to differ for different fiber types. Highly fermentation fibers increase fecal nitrogen due to a large increase in microbial nitrogen. In contrast, the increase in fecal nitrogen that occurs with consumption of poorly fermentable fibers is primarily due to nitrogen associated with the protein bound to the plant cell wall.[127] This indicates that increased fecal nitrogen is due primarily to events in the colon and not the small intestine. This is consistent with studies in vivo that find that dietary fiber generally has no effect or actually increases proteolytic enzyme activity in the small intestine.[121]

DIETARY FIBER AND THE PREVENTION OF DISEASE

The demonstrated physiological effects of dietary fiber, such as improved glucose tolerance, fecal bulking, and lowering of plasma cholesterol, suggest that dietary fiber may reduce the incidence of chronic diseases such as diabetic complications, colon cancer, and heart disease. The epidemiological evidence, as discussed in the section "Relationship to Specific Disease States," is consistent with this expectation. However, there have been few clinical trials examining the effect of increased dietary fiber intake on such conditions due to the expense and difficulty of conducting such trials. No clinical studies have been published that examine the effect of dietary fiber supplementation on the incidence of diabetic complications or of development of atherosclerosis or cancer. The studies that most closely examine the direct effect of dietary fiber on a disease state are those using colonic polyp recurrence as an end point. Colonic polyps are considered precursors to the development of tumors, so their presence or recurrence after removal is considered an excellent marker of risk of tumor development. MacLennan et al[128] examined the effect of a low-fat diet, a high-fiber diet, or the combination of the two on recurrence of adenomatous colonic polyps. Although neither dietary fat reduction nor increased dietary fiber alone significantly reduced polyp recurrence, the combination of the two resulted in a statistically significant reduction in recurrence. There is currently a phase III trial in progress examining the effect of wheat bran supplementation on recurrence of colonic polyps.[129] Since wheat bran is the dietary fiber source that most consistently reduces colon tumors in animal models, this trial is likely to be a definitive study on the reduction in colon cancer risk by dietary fiber.

Clearly, direct evidence (ie, clinical intervention trials with disease end points) that increased dietary fiber consumption reduces the incidence of chronic disease in humans is lacking. Nonetheless, the cumulative weight of evidence from epidemiological studies, studies of physiological effects relevant to disease conditions (eg, cholesterol lowering), and animal studies of these diseases certainly points to an important role in disease prevention.

REFERENCES

1. Burkitt DP, Walker ARP, Painter NS. Dietary fiber and disease. *JAMA.* 1974;229:1068–1074.

2. Negri E, Franceschi S, Parpinel M, et al. Fiber intake and risk of colorectal cancer. *Cancer Epidemiol Biomarkers Preven.* 1998;7:667–671.

3. Haile RW, Witte JS, Longnecker MP, et al. A sigmoidoscopy-based case-control study of polyps: macronutrients, fiber and meat consumption. *Int J Cancer.* 1997;73:497–502.

4. Kato I, Akhmedkhanov A, Koenig K, Toniolo PG, Shore RE, Riboli E. Prospective study of diet and female colorectal cancer: the New York University Women's Health Study. *Nutr Cancer.* 1997;28:276–281.

5. Hill MJ. Cereals, cereal fibre and colorectal cancer risk: a review of the epidemiological literature. *Eur J Cancer Preven.* 1997;6:219–225.

6. Gerber M. Fibre and breast cancer. *Eur J Cancer Preven.* 1998;7(suppl 2):S63–67.

7. Aldoori WH, Giovannucci EL, Rockett HR, et al. A prospective study of dietary fiber types and symptomatic diverticular disease in men. *J Nutr.* 1998;128,714–719.

8. Goodman MT, Wilkens LR, Hankin JH, et al. Association of soy and fiber consumption with the risk of endometrial cancer. *Am J Epidemiol.* 1997;146:294–306.

9. Rimm EB, Ascherio A, Giovannucci E, et al. Vegetable, fruit, and cereal fiber intake and risk of coronary heart disease among men. *JAMA.* 1996;275:447–451.

10. Trowell H. Definition of fiber. *Lancet.* 1974;1:503.

11. Gallaher DD, Schneeman BO. Dietary fiber. In: Ziegler EE, Filer LJ, eds. *Present Knowledge in Nutrition.* 7th ed. Washington, DC: International Life Sciences Institute Press; 1996.

12. Selvendran RR. The plant cell wall as a source of dietary fiber: chemistry and structure. *Am J Clin Nutr.* 1984;39:320–337.

13. Wood PJ. Physicochemical characteristics and physiological properties of oat (1→3), 1→4)-β-D-glucan. In: Wood PJ, ed. *Oat Bran.* St Paul, MN: American Association of Cereal Chemists; 1993:83.

14. Kay RM. Dietary fiber. *J Lipid Res.* 1982;23:221–242.

15. Cummings JH. Consequences of the metabolism of fiber in the human large intestine. In: Vahouny G, Kritchevsky D, eds. *Dietary Fiber in Health and Disease.* New York, NY: Plenum Press;1982:9–22.

16. Spiller GA. Suggestions for a basis on which to determine a desirable intake of dietary fiber. In: Spiller GA, ed. *CRC Handbook of Dietary Fiber in Human Nutrition.* 2nd ed. Boca Raton, FL: CRC Press; 1993:351–354.

17. Marlett JA, Balasubramanian R, Johnson EJ, Draper NR. Determining compliance with a dietary fiber supplement. *J Natl Cancer Inst.* 1986;76:1065–1070.

18. Cummings JH. The effect of dietary fiber on fecal weight and composition. In: Spiller GA, ed. *CRC Handbook of Dietary Fiber in Human Nutrition.* 2nd ed. Boca Raton, FL: CRC Press; 1993:263–349.

19. Stephen AM, Cummings JH. The microbial contribution to human faecal mass. *J Med Microbiol.* 1980;13:45–56.

20. Hoverstad T. Studies of short-chain fatty acid absorption in man. *Scand J Gastroenterol.* 1986;21:257–260.

21. Livesey G. Fiber as energy in man. In: Kritchevsky D, Bonfield C, eds. *Dietary Fiber in Health and Disease.* St Paul, MN: Eagan Press; 1995:46–57.

22. Roediger WEW. Utilization of nutrients by isolated epithelial cells of the rat colon. *Gastroenterology.* 1985;83:424–429.

23. Peters SG, Pomare EW, Fisher CA. Portal and peripheral blood short chain fatty acid concentrations after caecal lactulose instillation at surgery. *Gut.* 1992;33:1249–1252.

24. Lindeneg O, Mellemgaard K, Fabricius J, Lundquist F. Myocardial utilization of acetate, lactate and free fatty acids after ingestion of ethanol. *Clin Sci.* 1964;27:427–435.

25. Lundquist F, Sestoft L, Damgaard SE, Clausen JP, Trap-Jensen J. Utilization of acetate in the human forearm during exercise after ethanol ingestion. *Clin Invest.* 1973;52:3231–3235.

26. Juhlen-Dannfelt A. Ethanol effects of substrate utilization by the human brain. *Scand J Clin Lab Invest.* 1977;37: 443–449.

27. Berggren A, Björck IME, Nyman EM, Eggum BO. Short-chain fatty acid content and pH in caecum of rats given various sources of carbohydrates. *J Sci Food Agric.* 1993;63:397–406.

28. Sakata T. Stimulatory effect of short-chain fatty acids on epithelial cell proliferation in the rat intestine: a possible explanation for trophic effects of fermentable fibre, gut microbes and lumenal trophic factors. *Br J Nutr.* 1987;58:95–103.

29. Kruh J. Effects of sodium butyrate, a new pharmacological agent, on cells in culture. *Mol Cell Biochem.* 1982;42:65–82.

30. Scheppach W, Bartram HP, Richter F. Role of short-chain fatty acids in the prevention of colorectal cancer. *Eur J Cancer.* 1995;31A:1077–1080.

31. Medina V, Afonso JJ, Alvarez-Arguelles H, Hernandez C, Gonzalez F. Sodium butyrate inhibits carcinoma development in a 1,2-dimethylhydrazine-induced rat colon cancer. *JPEN.* 1998;22:14–17.

32. Wargovich, MJ, Chen CD, Jimenez A, et al. Aberrant crypts as a biomarker for colon cancer: evaluation of potential chemopreventive agents in the rat. *Cancer Epidemiol Biomarkers Preven.* 1996;5:355–360.

33. Caderni G, Luceri C, Lancioni L, Tessitore L, Dolara P. Slow-release pellets of sodium butyrate increase apoptosis in the colon of rats treated with azoxymethane, without affecting aberrant crypt foci and colonic proliferation. *Nutr Cancer.* 1998;30:175–181.

34. Jenkins D, Wolever T, Leeds A, et al. Dietary fibers, fiber analogues, and glucose tolerance: important of viscosity. *Br J Med.* 1978;1:1392–1394.

35. Jenkins D, Leeds A, Gassull M, Cochet B, Alberti K. Decrease in postprandial insulin and glucose concentrations by guar and pectin. *Ann Intern Med.* 1977;86:20–23.

36. Gassull M, Goff D, Haisman D, et al. The effect of unavailable carbohydrate gelling agents in reducing the postprandial glycemia in normal volunteers and diabetics. *J Physiol (Lond).* 1976;259:52P–53P.

37. Sels JP, Flendrig JA, Postmes THJ. The influence of guar-gum bread on the regulation of diabetes mellitus type II in elderly patients. *Br J Nutr.* 1987;57:177–183.

38. Braaten JT, Scott FW, Wood PJ, et al. High β-glucan oat bran and oat gum reduce postprandial blood glucose and insulin in subjects with and without type 2 diabetes. *Diabetes Med.* 1994;11:312–318.

39. Braaten J, Wood P, Scott F, et al. Oat gum lowers glucose and insulin after an oral glucose load. *Am J Clin Nutr.* 1991;53: 1425–1430.

40. Kirsten R, Nelson K, Storck J, Hubner-Stiener U, Speck U. Influence of two guar preparations on glucose and insulin levels during a glucose tolerance test in healthy volunteers. *Int J Clin Pharmacol Ther Toxicol.* 1991;29:19–22.

41. Morgan L, Tredger J, Wright J, Marks V. The effect of soluble- and insoluble-fibre supplementation on post-prandial glucose tolerance, insulin, and gastric inhibitory polypeptide secretion in healthy subjects. *Br J Nutr.* 1990;64:103–110.

42. Wood P, Braaten J, Scott F, Riedel D, Poste L. Comparisons of viscous properties of oat and guar gum and the effects of these and oat bran on glycemic index. *J Agric Food Chem.* 1990;38:753–757.

43. McIvor M, Cummings C, Leo T, Mendeloff A. Flattening postprandial blood glucose responses with guar gum: acute effects. *Diabetes Care.* 1985;8:274–278.

44. Jarjis H, Blackburn N, Redfern J, Read N. The effect of ispaghula (Fybogel and Metamucil) and guar gum on glucose tolerance in man. *Br J Nutr.* 1984;51:371–378.

45. Sambrook I, Rainbird A. The effect of guar gum and level and source of dietary fat on glucose tolerance in growing pigs. *Br J Nutr.* 1985;54:27–35.

46. Track N, Cawkwell M, Chin B, et al. Guar gum consumption in adolescent and adult rats: short- and long-term metabolic effects. *Can J Physiol Pharmacol.* 1985;63:1113–1121.

47. Daumerie C, Henquin J-C. Acute effects of guar gum on glucose tolerance and intestinal absorption of nutrients in rats. *Diabetes Metab.* 1982;8:1–5.

48. Cannon N, Flenniken A, Track N. Demonstration of acute and chronic effects of dietary fibre upon carbohydrate metabolism. *Life Sci.* 1980;27:1397–1401.

49. Ryle A, Davie S, Gould B, Yudkin J. A study of the effect of diet on glycosylated haemoglobin and albumin levels and glucose tolerance in normal subjects. *Diabetes Med.* 1990;7:865–870.

50. Ellis P, Apling E, Leeds A, Bolster N. Guar bread: acceptability and efficacy combined. Studies on blood glucose, serum insulin and satiety in normal subjects. *Br J Nutr.* 1981;46:267–276.

51. Heaton K, Marcus S, Emmell P, Bolton C. Particle size of wheat, maize, and oat test meals: effects on plasma and insulin responses and on the rate of starch digestion in vitro. *Am J Clin Nutr.* 1988;47:675–682.

52. Liljeberg H, Granfeldt Y, Björck I. Products based on a high fiber barley genotype, but not on common barley or oats, lower postprandial glucose and insulin responses in healthy humans. *J Nutr.* 1996;126:458–466.

53. Cara L, Dubois C, Borel P, et al. Effects of oat bran, rice bran, wheat fiber, and wheat germ on postprandial lipemia in healthy adults. *Am J Clin Nutr.* 1992;55:81–88.

54. Noakes M, Clifton P, Nestel P, Le Leu R, McIntosh G. Effect of high-amylose starch and oat bran on metabolic variables and bowel function in subjects with hypertriglyceridemia. *Am J Clin Nutr.* 1996;64:944–951.

55. Kestin M, Moss R, Clifton P, Nestel P. Comparative effects of three cereal brans on plasma lipids, blood pressure, and glucose metabolism in mildly hypercholesterolemic men. *Am J Clin Nutr.* 1990;52:661–666.

56. Hallfrisch J, Scholfield D, Behall K. Diets containing soluble oat extracts improve glucose and insulin responses of moderately hypercholesterolemic men and women. *Am J Clin Nutr.* 1995;61:379–384.

57. Wood P, Braaten J, Scott F, et al. Effect of dose and modification of viscous properties of oat gum on plasma glucose and insulin following an oral glucose load. *Br J Nutr.* 1994;72:731–743.

58. Wood P, Anderson J, Braaten J, et al. Physiological effects of β-D-glucan rich fractions from oats. *Cereal Foods World.* 1989;34:878–882.

59. Cherbut C, Bruley des Varannes S, Schnee M, et al. Involvement of small intestinal motility in blood glucose response to dietary fibre in man. *Br J Nutr.* 1994;71:675–685.

60. Wolever TMS, Spadafora P, Eshuis H. Interaction between colonic acetate and propionate in humans. *Am J Clin Nutr.* 1991;53:681–687.

61. Frati-Munari A, Flores-Garduno M, Ariza-Andraca R, Islas-Andrade S, Chavez Negrete A. Effect of different doses of *Plantago* psyllium mucilage on the glucose tolerance test. *Arch Invest Med (Mex).* 1989;20:147–152.

62. Sartor G, Carlstrom S, Schersten B. Dietary supplementation of fibre (Lunelax®) as a means to reduce postprandial glucose in diabetics. *Acta Med Scand.* 1981;656(suppl):51–53.

63. Frape D, Jones A. Chronic and postprandial responses of plasma insulin, glucose and lipids in volunteers given dietary fibre supplements. *Br J Nutr.* 1995;73:733–751.

64. Abraham Z, Mehta T. Three-week psyllium-husk supplementation: effect on plasma cholesterol concentrations, fecal steroid excretion, and carbohydrate absorption in men. *Am J Clin Nutr.* 1988;47, 67–74.

65. Holt S, Heading R, Carter D, et al. Effect of gel fibre on gastric emptying and absorption of glucose and paracetamol. *Lancet.* 1979;1:636–639.

66. Schwartz S, Levine G. Effects of dietary fiber on intestinal glucose absorption and glucose tolerance in rats. *Gastroenterology.* 1980;79:833–836.

67. Wahlqvist M, Morris M, Littlejohn G, Bond A, Jackson R. The effects of dietary fibre on glucose tolerance in healthy males. *Aust NZ J Med.* 1979;9:154–158.

68. Tredger J, Sheard C, Marks V. Blood glucose and insulin levels in normal subjects following a meal with and without added sugar beet pulp. *Diabetes Metab.* 1981;7:169–172.

69. Wolever T, Jenkins D, Leeds A, et al. Dietary fibre and glucose tolerance importance of viscosity. *Proc Nutr Soc.* 1978;37:47A.

70. Molnár D, Dóber I, Soltész G. The effect of unprocessed wheat bran on blood glucose and plasma immunoreactive insulin levels during oral glucose tolerance test in obese children. *Acta Paediatr Hungarica.* 1985;26:75–77.

71. Munoz J, Sandstead H, Jacob R. Effects of dietary fiber on glucose tolerance of normal men. *Diabetes.* 1979;28:496–502.

72. Tsai A, Mott E, Owen G, et al. Effects of soy polysaccharide on gastrointestinal functions, nutrient balance, steroid excretions, glucose tolerance, serum lipids, and other parameters in humans. *Am J Clin Nutr.* 1983;38:504–511.

73. Chiu S, Track N. Guar by-product improves carbohydrate tolerance in rats. *Metabolism.* 1985;34:481–485.

74. Ellis P, Roberts F, Low A, Morgan L. The effect of high-molecular-weight guar gum on net apparent glucose absorption and net apparent insulin and gastric inhibitory polypeptide production in the growing pig: relationship to rheological changes in jejunal digesta. *Br J Nutr.* 1995;74:539–556.

75. Blackburn N, Redfern J, Jarjis, H, et al. The mechanism of action of guar gum in improving glucose tolerance in man. *Clin Sci Mol Med.* 1984;66:329–336.

76. Flourie B, Vidon N, Florent C, Bernie J. Effect of pectin on jejunal glucose absorption and unstirred layer thickness in normal man. *Gut.* 1984;25:936–941.

77. Fuse K, Bamba T, Hosoda S. Effects of pectin on fatty acid and glucose absorption and on thickness of unstirred water layer in rat and human intestine. *Dig Dis Sci.* 1989;34:1109–1116.

78. Rainbird A. Effect of guar gum on gastric emptying of test meals of varying energy content in growing pigs. *Br J Nutr.* 1986;55:99–109.

79. Rainbird A, Low A. Effect of guar gum on gastric emptying in growing pigs. *Br J Nutr.* 1986;55:87–98.

80. Rainbird A, Low A. Effect of various types of dietary fibre on gastric emptying in growing pigs. *Br J Nutr.* 1986;55:111–121.

81. Johansen H, Bach Knudsen K, Sandström B, Skjøth F. Effects of varying content of soluble dietary fibre from wheat flour and oat milling fractions on gastric emptying in pigs. *Br J Nutr.* 1996;75:339–351.

82. Brown N, Greenburgh A. The effects of pectin and wheat bran on the distribution of a meal in the gastrointestinal tract of the rat. *Br J Nutr.* 1994;72:289–297.

83. Brown N, Worlding J, Rumsey R, Read N. The effect of guar gum on the distribution of a radiolabeled meal in the gastrointestinal tract of the rat. *Br J Nutr.* 1988;59:223–231.

84. Leeds A, Bolster N, Andrews R, Truswell A. Meal viscosity, gastric emptying and glucose absorption in the rat. *Proc Nutr Soc.* 1979;38:44A.

85. Forman L, Schneeman B. Dietary pectin's effect on starch utilization in rats. *J Nutr.* 1982;112:528–533.

86. Tsai A, Peng B. Effects of locust bean gum on glucose tolerance, sugar digestion, and gastric motility in rats. *J Nutr.* 1981;111:2152–2156.

87. Russell J, Bass P. Canine gastric emptying of fiber meals: influence of meal viscosity and antroduodenal motility. *Am J Physiol.* 1985;246:G662–667.

88. Harju E. Differences in postprandial acidity and rates of gastric emptying in duodenal ulcer patients and healthy subjects after the addition of guar gum to meals. *S Am J Surg.* 1985;23:151–154.

89. Flourie B, Vidon N, Chayvialle J-A, et al. Effect of increased amounts of pectin on a solid-liquid meal digestion in healthy man. *Am J Clin Nutr.* 1985;42:495–503.

90. Wilmshurst P, Crawley J. The measurement of gastric transit time in obese subjects using [24]Na and the effects of energy content and guar gum on gastric emptying and satiety. *Br J Nutr.* 1980;44:1–6.

91. Vincent R, Roberts A, Frier M, et al. Effect of bran particle size on gastric emptying and small bowel transit in humans: a scintigraphic study. *Gut.* 1995;37:216–219.

92. Blackburn N, Holgate A, Read N. Does guar gum improve post-prandial hyperglycaemia in humans by reducing small intestinal contact area? *Br J Nutr.* 1984;52:197–204.

93. Torsdottir I, Alpsten M, Andersson H, Einarsson S. Dietary guar gum effects on postprandial blood glucose, insulin and hydroxyproline in humans. *J Nutr.* 1989;119:1925–1931.

94. French S, Read N. Effect of guar gum on hunger and satiety after meals of differing fat content: relationship with gastric emptying. *Am J Clin Nutr.* 1994;59:87–91.

95. McIntyre A, Vincent R, Perkins A, et al. Effect of bran, ispaghula, and inert plastic particles on gastric emptying and small bowel transit in humans: the role of physical factors. *Gut.* 1997;40:223–227.

96. Leclère C, Champ M, Boillot J, et al. Role of viscous guar gums in lowering the glycemic response after a solid meal. *Am J Clin Nutr.* 1994;59:914–921.

97. Benini L, Castellani G, Brighenti F, et al. Gastric emptying of a solid meal is accelerated by the removal of dietary fibre naturally present in food. *Gut.* 1995;36:825–830.

98. Tadesse K. The effect of dietary fibre on gastric secretion and emptying in man. *J Physiol.* 1982;332:102P–103P.

99. Leatherdale B, Green D, Harding L, et al. Guar and gastric emptying in noninsulin dependent diabetes. *Acta Diabetica Lat.* 1982;19:339–343.

100. Di Lorenzo C, Williams C, Hajnal F, Valenzuela J. Pectin delays gastric emptying and increases satiety in obese subjects. *Gastroenterology.* 1988;95:1211–1215.

101. Ray T, Mansell K, Knight L, et al. Long-term effects of dietary fiber on glucose tolerance and gastric emptying in noninsulin-dependent diabetic patients. *Am J Clin Nutr.* 1983;37:376–381.

102. Torsdottir I, Alpsten M, Holm G, et al. A small dose of soluble alginate-fiber affects postprandial glycemia and gastric emptying in humans with diabetes. *J Nutr.* 1991;121:795–799.

103. Keys A, Grande F, Anderson JT. Fiber and pectin in the diet and serum cholesterol in man. *Proc Soc Exp Biol Med.* 1961;106:555–558.

104. Wells AF, Ershoff BH. Beneficial effects of pectin in prevention of hypercholesteromia and increase in liver cholesterol in cholesterol-fed rats. *J Nutr.* 1963;74:87–92.

105. Anderson JW, Deakins DA, Floore TL, Smith BM, Whitis SE. Dietary fiber and coronary heart disease. *Crit Rev Food Sci Nutr.* 1990;29:95–147.

106. Brown L, Rosner B, Willett WW, Sacks FM. Cholesterol-lowering effect of dietary fiber: a meta-analysis. *Am J Clin Nutr.* 1999;69:30–42.

107. Kelly JJ, Tsai AC. Effect of pectin, gum arabic and agar on cholesterol absorption, synthesis, and turnover in rats. *J Nutr.* 1978;108:630–639.

108. Fernandez ML. Distinct mechanisms of plasma LDL lowering by dietary fiber in the guinea pig: specific effects of pectin, guar gum, and psyllium. *J Lipid Res.* 1995;36:2394–2404.

109. Osterberg NA, Gallaher DD. The role of cholesterol absorption in the hypocholesterolemic effect of dietary fibers. *FASEB J.* 1994;8:A153.

110. Carr TP, Gallaher DD, Yang C-H, Hassel CA. Increased intestinal contents viscosity reduces cholesterol absorption efficiency in hamsters fed hydroxypropyl methylcellulose. *J Nutr.* 1996;126:1463–1469.

111. Chen W-J, Anderson JW, Jennings D. Propionate may mediate the hypocholesterolemic effects of certain soluble plant fibers in cholesterol-fed rats. *Soc Exp Biol Med.* 1984;75:215–218.

112. Anderson JW, Bridges SR. Plant fiber metabolites alter hepatic glucose and lipid metabolism. *Diabetes.* 1981;30 (suppl 1):532A.

113. Wolever TMS, Brighenti F, Royall D, Jenkins AL, Jenkins DJA. Effect of rectal infusion of short chain fatty acids in human subjects. *Am J Gastroenterol.* 1989;84:1027–1033.

114. Illman RJ, Topping DL, McIntosh GH, et al. Hypocholesterolaemic effects of dietary propionate: studies in whole animals and perfused rat liver. *Ann Nutr Metab.* 1988;32:97–107.

115. Alvarez-Leite JI, Andrieux C, Ferezou J, Riotto M, Vieira EC. Evidence for the absence of participation of the microbial flora in the hypocholesterolemic effect of guar gum in gnotobiotic rats. *Comp Biochem Physiol.* 1994;109A:503–510.

116. Miettinen TA, Tarpila S. Serum lipids and cholesterol metabolism during guar gum, *Plantago* ovata and high fibre treatments. *Clin Chim Acta.* 1989;183:253–262.

117. Judd PA, Truswell AS. The effect of rolled oats on blood lipids and fecal steroid excretion in man. *Am J Clin Nutr.* 1985;34:409–425.

118. Story JA, Watterson JJ, Matheson HB, Furumoto EJ. Dietary fiber and bile acid metabolism. In: Furda I, Bane CJ, eds. *New Developments in Dietary Fiber.* New York, NY: Plenum Press; 1990.

119. Gallaher DD, Locket P, Gallaher CM. Bile acid metabolism in rats fed two levels of corn oil and brans of oat, rye and barley and sugar beet fiber. *J Nutr.* 1992;122:473–481.

120. Zhang J-X, Hallmans G, Adlercreutz H, et al. Effect of oat and rye fractions on biliary and faecal bile acid profiles in Syrian golden hamsters. *Br J Nutr.* 1993;70:525–536.

121. Schneeman BO, Gallaher D. Effects of dietary fiber on digestive enzymes. In: Spiller GA, ed. *CRC Handbook of Dietary Fiber in Human Nutrition.* 2nd ed. Boca Raton, FL: CRC Press; 1993:377–385.

122. Sandberg AS, Andersson H, Hallgren B, Hasselblad K, Isaksson B, Hultén L. The effects of citrus pectin on the absorption of nutrients in the small intestine. *Hum Nutr Clin Nutr.* 1983;37C:171–183.

123. Sandberg A-S, Andersson H, Hallgren G, Hasselblad K, Isaksson B, Hultén L. Experimental model for in vivo determination of dietary fibre and its effects on the absorption of nutrients in the small intestine. *Br J Nutr.* 1981;45:283–294.

124. Ebihara K, Schneeman BO. Interaction of bile acids, phospholipids, cholesterol, and triglycerides with dietary fibers in the small intestine of rats. *J Nutr.* 1989;119:1100–1106.

125. Gallaher D, Schneeman BO. Effect of dietary fiber on protein digestibility and utilization. In: Spiller GA, ed. *CRC Handbook of Dietary Fiber in Human Nutrition.* 2nd ed. Boca Raton, FL: CRC Press; 1993:179–205.

126. Wisker E, Knudsen KE, Daniel M, Feldheim W, Eggum BO. Digestibilities of energy, protein, fat, and non-starch polysaccharides in a low fiber diet and diets containing coarse or fine whole meal rye are comparable in rats and humans. *J Nutr.* 1996;126:481–488.

127. Eggum BO. The influence of dietary fibre on protein digestion and utilization in monogastrics. *Arch Tierernahr.* 1995;48:89–95.

128. MacLennan R, Macrae F, Bain C, et al. Randomized trial of intake of fat, fiber, and beta carotene to prevent colorectal adenomas: the Australian Polyp Prevention Project. *J Natl Cancer Inst.* 1995;87:1760–1766.

129. Martinez ME, Reid ME, Guillen-Rodriguez J, et al. Design and baseline characteristics of study participants in the Wheat Bran Fiber Trial. *Cancer Epidemiol Biomarkers Preven.* 1998;7:813–816.

CHAPTER 13

Food Fortification with Vitamin and Mineral Nutraceuticals

Paul A. Lachance

Nutrition encompasses the sum of the biochemical and physiological processes concerned with the growth, maintenance, and repair of the total organism and/or its constituent organs (eg, liver, kidney, brain).[1] Nutraceuticals are bioactive phytochemical compounds that have disease preventing, health promoting, and/or medicinal properties. Nutrients are nutraceuticals that are essential to life, in the sense that a deficiency disease results from their absence or limited supply. The nutrient nutraceuticals encompass the essential amino acids, sufficient carbohydrate to spare protein, essential omega 3 and 6 fatty acids, vitamins, and minerals. Including the trace minerals, nutrient nutraceuticals number 52 or so chemicals. The non-nutrient nutraceuticals that complement the nutrient nutraceuticals number at least several hundred chemicals.

Preventing disease, especially the chronic diseases that are the leading causes of morbidity and death, requires interventions in the disease process that involve all the nutraceuticals and their interactions. Since as much as 70% of the burden of illness and associated costs (pain, biochemical and physiological incapacities, lost income) is preventable,[2] it can be assumed that nutraceuticals have a significant role.

Nutraceuticals, once ingested, become the intrinsic factors in nutrition and relate to the expression of metabolism dictated by our genetic inheritance. They are capable of blocking or suppressing reactions initiated or stimulated by extrinsic agents, particularly free radicals. Thus, the prevention and thwarting of chronic disease processes require an interplay (ie, synergy) of nutrient and non-nutrient nutraceuticals (Figure 13–1).[3] Such synergy is exemplified in the thwarting of the consequences of untoward oxidations of both metabolic and exogenous origins, which is optimized via the synergy of optimal nutrient nutraceutical antioxidants such as vitamins C and E, vitamin A precursor carotenes, selenium, and possibly folic acid, in concert with the non-nutrient nutraceutical antioxidants such as flavonoids and non–vitamin A precursor carotenes, such as lycopene, lutein, zeaxanthin. The result is the realization of the optimal antioxidant capacity of the tissues.

This chapter concentrates on food enrichment and fortification with vitamins and minerals, the practice of which has resulted in the thwarting of the deficiency diseases of the 20th century. Fortification is a feasible and a suitable method of enhancing public health and is a model practice of nutrification,[4] the enhancement of foods with nutraceutical combinations to prevent illness. Nutrification represents the beginnings of a recognition of the potential, as well as the risks, of complementary medicine in everyday practice, and the consequent expansion of medicine to include it.

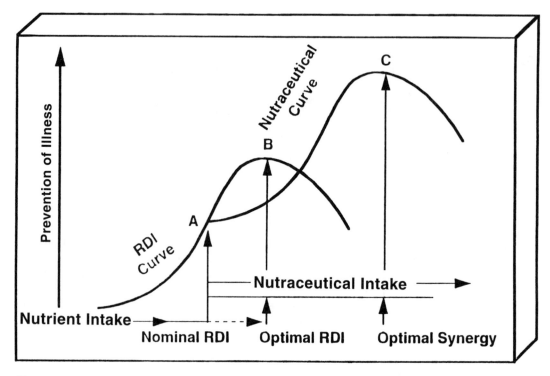

Figure 13–1 Synergy of nutrients and nutraceuticals. Curve A, meeting of nominal nutrient needs; onset of beneficial effects of non-nutrients on biochemistry and physiology. Curve B, meeting of optimal nutrient needs for preventing illness. Curve C, optimal synergy of nutrients and non-nutrients in beneficially affecting biochemistry and physiology to prevent illness. RDI, recommended daily intake.

BRIEF HISTORY OF NUTRIENT ADDITIONS TO FOOD

The Persian physician Melampus, medical advisor to Jason and the Argonauts (4000 BC), prescribed sweet wine laced with iron filings to strengthen the resistance of sailors to spears and arrows.[5] In 1833, Boussingault,[6] a French physician, recommended the addition of iodine to table salt to prevent the development of goiter. However, realistic large-scale addition of nutrients to food could not take place until the 20th century, when chemically defined forms of the nutrient chemicals were economically on hand and could be incorporated into food products in a controlled manner to meet clinically recognized nutrient deficiency conditions.[7] Between the end of World War I and the beginning of World War II, adding iodine to salt became a routine practice, and the incidence of goiter was dramatically reduced. Vitamins A and D were added to margarine and Vitamin D to milk, and the advent of Standard of Identity foods in 1938 brought the opportunity to restore thiamine, riboflavin, niacin, and iron to wheat flour and bread. World War II Food Order No. 1 (an executive order of President Franklin Delano Roosevelt) in 1943 mandated the enrichment of corn, rice, and wheat flours with thiamine, riboflavin, niacin, and iron. All other nutrient additions to foods were optional until January 1998, when folic acid was added to the required nutrients of enrichment. The addition of calcium to cereal grain flours and products has been optional since 1938, but a petition to make it mandatory has been stalled for more than 8 years in the Food and Drug Administration.

In the late 20th century, lifestyle changes such as more geographic and job mobility, increased numbers of divorces, and one-parent homes may have increased the need for food fortification. In Western countries, a majority of women are in the work force. The younger, educated population skips meals and replaces home-cooked meals with fast foods and foods purchased away from home to be eaten on the run or at home several times per week. Seventy percent of decisions concerning the choices for evening meals are made only a few hours earlier. Snacking accounts for as much as 20% of the daily intake of kilocalories. There is a dramatic correlation between the increase in the percentage of food dollars spent away from home and an increase in obesity (Figure 13–2).[8] As these practices continue, individuals become at greater risk for certain nutrient intake inadequacies (calcium, magnesium, zinc, vitamin A, vitamin B$_6$, and ascorbic acid).

PRINCIPLES OF FORTIFICATION

Exhibit 13–1 lists the currently accepted criteria for food fortification.[9] Malnutrition is defined in terms of individual nutrient deficiencies, yet the study of the pathogenesis of disease teaches that several nutrients are involved in the functional solution of a malnutrition problem. Once one solves for the first limiting nutrient of a deficiency, one must consider the next nutrient, which in turn becomes limiting. Although fortification with one micronutrient at a time may help in addressing deficiency, this approach simply postpones realizing the beneficial effects of a more optimal bioavailability of a broad spectrum of nutrients on the biochemical and physiological processes concerned with the body's growth, maintenance, repair, and resistance to stress.

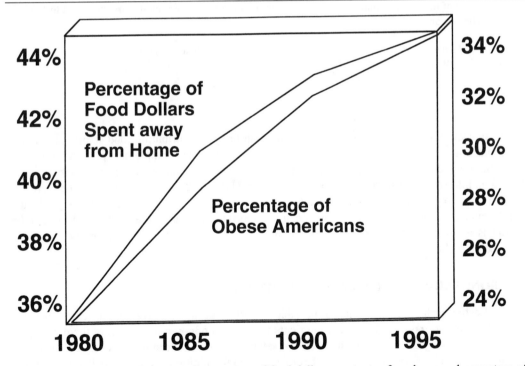

Figure 13–2 Association of changes of percentage of food dollars spent away from home and percentage of obese Americans, 1980–1995. *Source:* Data from USDA and National Restaurant Association, NHANES III (n=15,000) DHHS.

Exhibit 13–1 Criteria for Food Fortification

Food fortification is appropriate when there is a demonstrated need for increasing the intake of essential nutrients in one or more population groups. This may be in the form of actual clinical or subclinical evidence of deficiency, estimated low intakes, or possible risks following changes in eating habits.

The food selected as a vehicle should be consumed by the population at risk.

The intake of the food should be stable and uniform; the lower and upper levels of intake should be known.

The essential nutrient(s) should be present in amounts that are neither excessive nor insignificant, taking into account intakes from other dietary sources.

The amount of the essential nutrient(s) should be sufficient to correct or prevent the deficiency when the food is consumed in normal amounts by the population at risk.

The nutrient(s) added should not adversely affect the metabolism of any other nutrient.

The nutrient(s) added should be sufficiently stable in the food under customary conditions of packaging, storage, distribution, and use.

The nutrient(s) added should be physiologically available from the food.

The nutrient(s) added should not impart undesirable characteristics to the food (changes in color, taste, smell, texture, cooking properties) and should not unduly shorten shelf life.

Technology and processing facilities should be available to permit addition of the nutrient(s) in a satisfactory manner.

The additional cost of the fortification should be reasonable for the consumer.

Methods of measuring, controlling, and/or enforcing the levels of essential nutrients added to foods should be available.

Source: Reprinted from M. Blum, Food Fortification, A Key Strategy To End Micronutrient Malnutrition, *Nutriview*, July 1997, © 1997, Roche Vitamins Ltd, Basel, Switzerland.

The indiscriminate fortification of foods other than protein-contributing food staples and macromolecular sources of energy, such as oils, margarine, and sugar, that complement protein foods needs to be discouraged. Fortifying foods that contribute to the development of obesity, such as casual snacks and desserts, is a highly questionable practice. Unless the micronutrients to be delivered are needed in microgram quantities, it is technologically difficult to fortify a single food ingredient or condiment (as opposed to fortifying a whole, intact food) to serve as an ideal vehicle. Adding iodine to salt is a suitable technology, whereas adding vitamin A (unstable without further coating) to monosodium glutamate (MSG) is unrealistic.

MICRONUTRIENT FORTIFICATION TECHNOLOGY

Fortification technology[10,11] makes it possible to design means of delivering nutrients that can be applied to diverse foods. Adding micronutrients to food (as opposed to delivering them as supplements) significantly decreases the likelihood of safety concerns while simultaneously enhancing nutrient balance and bioavailability. The technology needed for additions to food is well advanced (ie, it is rarely a limiting problem after the initial cost for the equipment). Micronutrients are low cost. A full Recommended Dietary Allowance (RDA) of vitamins in a premix costs about 1¢. Micronutrients are added in milligram and microgram quantities and are "invisible" in that the addition is diluted into the matrix of macromolecules that gives food its structure. Ironically, the attribute of "invisibility" also accounts for malnutrition that occurs because the lack or presence of nutrients cannot be physiologically perceived.

STABILITY OF FORTIFICANTS

Several variables affect the stability of micronutrients in food systems. These are outlined in Figure 13–3. The type of food, solid or beverage, the water activity, the surface area, and the type of processing and packaging involved can either protect the fortificants or exacerbate their sensitivity to the principal parameters affecting stability: temperature, light, oxygen, and pH. Some comparative Arrhenius activation energies in model heated systems from the literature and our laboratory are provided in Table 13–1.[9] One can add overages to correct for the known effects of processing temperature, but temperature extremes during transportation and storage can cause serious deterioration. Interestingly, the sensory properties of the food overall serve as adequate relative indicators. Appropriately selected packaging plays a critical role in preventing exposure to light and oxygen. Alkaline pH, such as that of devil's food cake, is harsh on vitamins such as thiamine. Minerals are very stable to deterioration (in contrast to organic molecules), but they are not inert, and they can contribute to reactive deteriorations (eg, oxidations), which are often detected as rancidity.

The basic technological issues were reviewed by a National Research Council subcommittee on food technology of the food protection committee of the Food and Nutrition Board in 1975.[12] The parameters have not substantially changed. An updated description of state-of-the-art technology by commodity to be nutrified was published in 1991,[10] with more than 5000 citations. It is the first extensive compilation on nutrient addition to foods since the beginnings of the practice in the United

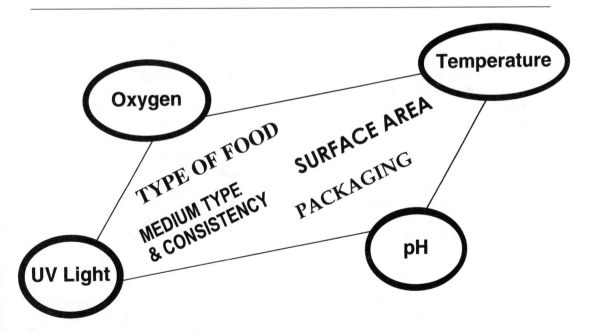

Figure 13–3 Variables affecting nutritive/nutraceutical value.

Table 13–1 Arrhenius Activation Energy in Heated Vitamin Model Systems

Vitamin Model System	Activation Energy (kcal/mol)
Thiamine @ pH 7.5	27
Riboflavin	23
Pantothenic acid	20–27
Pyridoxine @ pH 7.5	23.4
Pyridoxal @ pH 7.5	8.0–10.0
Pyridoxamine @ pH 7.5	31.6
Vitamin B$_{12}$	23
Retinyl palmitate (skim milk)	20
Retinyl palmitate (model)	25

Source: Reprinted from M. Blum, Food Fortification, A Key Strategy To End Micronutrient Malnutrition, *Nutriview*, July 1997, © 1997, Roche Vitamins Ltd, Basel, Switzerland.

States more than 50 years ago. However, an excellent update was published in 1996 by Lofti et al[11] of the Micronutrient Initiative of Canada and the Netherlands. Yet another update is a 1997 special edition of *Nutriview*[9] and the proceedings of the 1997 Symposium on Food Fortification held under the auspices of the Micronutrient Initiative.[13] The Cornell International Institute for Food, Agriculture and Development published "an international research agenda" with an emphasis on food-based approaches to preventing micronutrient malnutrition.[14]

TRENDS AND ISSUES

The RDAs

The RDAs are undergoing changes to include consideration of safe upper levels of micronutrients so that consumers can ingest these substances at levels that prevent deficiency and the pathogenesis of chronic disease but avoid intakes that could have untoward effects.[3] Nutritionists simply will not commit to recommending "optimal" intakes and will deny that safe upper levels may represent such a direction. Importantly, two countries (the United States and Canada) are developing common RDAs. The southeast Asian countries are considering such a unified approach as well.[3] RDAs differ very little between countries, but it is amazing how much time some scientists can spend on slight changes. In terms of food fortification, amounts of micronutrients to add can be selected to ensure intakes approaching the RDAs. It is difficult to predict what effect the Estimated Average Requirement (EAR) will have on applications, but a policy of food fortification might have the objective of at least ensuring intakes approaching the EAR. Classically, food fortification practices have rarely related directly to the RDAs, but rather have been directed to the alleviation of clinical nutrient deficiencies in the practice of public health enrichment.

The Focus of Food Fortification

The international community (eg, Food and Agricultural Organization/World Health Organization, United Nations, The Micronutrient Initiative, OMNI [OMNI/USAID], International Vitamin A Consultive Group continues to focus on the deficiencies of iodine, iron, and vitamin A. In the case of iodine deficiency, considerable inroads have been made with the simple and classic intervention of fortified salt. In the case of iron-deficiency anemia, interventions address the bioavailability of various iron salts when iron itself is available and thus is not limiting. In the United States, iron-deficiency anemia was first addressed by iron enrichment regulation in 1943, but more than 50 years later, the incidence of iron-deficiency anemia remains the same because other nutrients besides iron, such as B_6, folic acid, magnesium, and zinc, are needed for the formation of ferritin and blood, and these are now the limiting nutrients in the American diet. It is proposed that a similar situation exists in developing countries and that the incidence of iron-deficiency anemia attributable to iron deficiency should be reassessed. Interventions for vitamin A deficiency have been plagued by that vitamin's toxicity, which limits the amount and frequency of high-dose supplementation, and by the high instability of this nutrient to light, oxygen, and heat. Only coated forms of the vitamin are stable when incorporated into staples; however, addition to rice requires the formation of imitation rice grains, which must be added at a rate of 1:200 rice granules. Uneducated recipients will screen and discard the synthetic granules if they are not perfectly matched to rice grains in color, shape, and density. Vitamin A added to oils, fats (margarine), dairy products, ready-to-eat cereals, and wheat and corn products has proven to be very stable, whereas vitamin A added to spices, condiments, and flavor enhancers such as MSG has been very unstable in spite of costly research attempts. Vitamin A added to concentrated premixes of sugar requiring the addition of unsaturated oils to cling the vitamin A beadlets to sugar crystals has led to rancidity of the oil in tropical environments. Vitamin A cocrystallized with sucrose is exceedingly stable, but the proprietary technology has limited its use. This limitation will soon expire. A small foundation for advancing this technological approach has been proposed (A. Chen, personal communication, 1997).

Comprehensive Approaches to Food Fortification

The experience in developing and industrialized countries has revealed that food fortification is efficacious and safe. Economic studies by the World Bank confirm that food fortification is a sustainable and cost-effective solution.[15] The intervention, unlike some food and nutrition education, benefits public health even when populations are illiterate or are too poor to have much dietary diversity. In contrast, supplementation approaches fail to meet everyone in need and in developing countries, often require the administration of toxic doses to decrease the dosing frequency and the monitoring needed to realize benefits.

Food fortification of one or more of the staples of the diet has been shown to be a highly effective means of improving public health, as shown by the Santa Maria Cauque trial in Guatemala between 1972 and 1976, in which corn masa flour was fortified with 5% soy flour, lysine, thiamine, riboflavin, niacin, vitamin A, and ferric orthophosphate.[16] The results of the trial were dramatic: a 50% decrease in infant mortality, even in low-birth weight infants, and a 30% decrease in morbidity (diarrhea and upper respiratory infections) (Tables 13–2 to 13–4). This trial, which was funded by USAID and managed by INCAP with the cooperation of Rutgers University, was carried out in a village with years of preintervention control data on birth weights, respiratory infections, and diarrhea incidence for children from birth up to 5 years of age. An internal concomitant control group was based on those who voluntarily elected not to choose the daily addition of fortificant premix to their corn masa, thus taking 40% or fewer daily opportunities. This was compared to those who chose the addition of fortification premix on 60% or more frequent daily occasions during the 3 and 1/2 year trial. These results went unheralded because of changes in fortification policy at the international level that occurred during the time of the trial and thus prevented revealing these significant public health intervention results.

Table 13–2 Infant Mortality in Santa María Cauqué from June 1972 to January 1976, by Frequency of Use of Fortified Masa

| | Mortality Rate/1000 Deliveries* | | | |
| | Deliveries | | Birth Weight | |
Frequency of Use	(N = 304)	All Births	>2500 g	<2500 g
<40%	189	132 (25/189)	112 (13/116)	192 (14/73)
>60%	115	44 (5/115)	42 (3/72)	47 (2/43)

*Rate before fortification (January 1964 to May 1972) for 504 deliveries = 121/1000 (107–133 range).
Source: Urrutia et al., *Corn Fortification: A Field Demonstration Project Model, Final Report,* AID-CSD-3357, U.S. Agency for International Development.

Table 13–3 Mortality Rate of Preschool Children (2–5 Years of Age), in Santa María Cauqué before and after Fortification

| | Before Fortification (January 1964 to May 1972) | | After Fortification (June 1972 to January 1976) Frequency of Fortified MASA Use | | | |
| | | | <40% | | >60% | |
Year of Life	Deaths/Total	Rate/1000	Deaths/Total	Rate/1000	Deaths/Total	Rate/1000
Second	26/454	57	9/187	48	3/126	24
Third	13/389	33	3/167	18	1/118	8
Fourth	5/376	13	5/121	41	0/88	0
Fifth	1/357	3	1/104	10	0/81	0
Total	45/1576	29	18/579	31	4/413	10

Source: Urrutia et al., *Corn Fortification: A Field Demonstration Project Model, Final Report,* AID-CSD-3357, U.S. Agency for International Development.

Table 13–4 Morbidity among Preschool Children in Santa María Cauqué, June 1972 to January 1976, by Frequency of Use of Fortified Masa

| | Age Interval in Months | | | | | | | | | | | |
| | 12–17 (N = 60) | | | 18–23 (N = 67) | | | 24–29 (N = 75) | | | 30–35 (N = 67) | | |
Frequency of Use	(A)	(B)	(C)	(A)	(B)	(C)	(A)	(B)	(C)	(A)	(B)	(C)
<40%	10.5	14.4	6.6	8.4	10.4	6.1	8.9	12.8	5.8	8.3	9.0	3.0
>60%	6.8	7.6	4.6	5.3	9.9	2.7	6.5	6.9	2.9	5.3	8.4	2.4
Benefit days	3.7	6.8	2.0	3.1	0.5	3.4	2.4	5.9	2.9	3.0	0.6	0.6

Note: (A) Percentage of child-days with upper respiratory infection; (B) percentage of child-days with diarrhea; (C) percentage of child-days with combined clinical conditions. Actual number of episodes of illness was unchanged, but episode duration (days) was changed.
Source: Urrutia et al., *Corn Fortification: A Field Demonstration Project Model, Final Report,* AID-CSD-3357, U.S. Agency for International Development.

Along the same lines, the most effective interventions against beri-beri, pellagra, and ariboflavin-iosis have been additions to cereal grain staples (wheat, corn, and rice),[9] and the provision of fortified foods as a component of the Women, Infants, and Children program in the United States has resulted in reduced health care costs (as much as $4.00 saving for each dollar invested).[13] The results of these longitudinal trials are collectively a logical argument for interventions involving fortification with a broad range of nutrients as opposed to interventions that foster one nutrient at a time, via inconsistent and questionable food vehicles.

The adoption of federal policies of food fortification of the significant staple cereal of a country (eg, to corn flour in Colombia, Venezuela, and, pending, in Bolivia) is to be lauded, but even greater public health gains may be achieved if these policies are periodically revised in the light of new knowledge about nutraceutical amelioration of not only deficiency diseases but also chronic diseases. A recent example is the mandatory addition of folic acid to cereal enrichment in the United States in order to decrease the incidence of neural tube defects and, potentially, to lower homocysteine blood levels and slow the progression of cardiovascular disease.

THE FUTURE

Food fortification with vitamins and minerals is an established technology with many scientific questions yet to be examined and resolved. Since many non-nutrient nutraceuticals can interact chemically or physiologically with nutrient fortificants to provide good, bad, or equivocal results, there is a need to initiate research approaches to better define these interactions. Initially, the computational chemical similarities should be methodically studied and advantage taken of existing data. This information could also lead to insights into mechanisms of action. Nutraceutical science must go beyond studying nutrients and explore the potentially crucial role of nutraceuticals in illuminating and combating chronic disease processes.

REFERENCES

1. Lusk G. *The Elements of the Science of Nutrition.* 1928. New York, NY: Johnson Reprint Corp; 1976.

2. Fries JF, Koop CE, Beadle CE, et al. The Health Project. *The Sciences (NYAS).* 1994;34(1):19.

3. Lachance PA. International perspective: basis, need, and application of Recommended Dietary Allowances. *Nutr Rev.* 1998;56,4 pt II: 2–4.

4. Lachance PA. Nutrification: a concept for assuring nutritional quality by primary intervention in feeding systems. *J Agric Food Chem.* 1972;20:522–525.

5. Richardson D. Iron fortification of foods and drinks. *Chem Industry.* 1983;13:498–501.

6. Boussingault M. Memoir sur des salines iodiferes des Andes. *Ann Chem Physique.* 1833;54:163–177.

7. Lachance PA. Nutrient additions to foods. In: Bendich A, Deckelbaum RJ, eds. *Preventive Nutrition: The Comprehensive Guide for Health Professionals.* Totowa: Humana Press Inc; 1997:441–454.

8. National Center for Health Statistics. National Health and Nutrition Examination Survey III (1988–1991). 1994.

9. Blum M. Food fortification: a key strategy to end micronutrient malnutrition. *Nutriview.* July 1997.

10. Bauernfiend JC, Lachance PA, eds. *Nutrient Additions to Food: Nutritional, Technological and Regulatory Aspects.* Trumbull, CT: Food and Nutrition Press Inc; 1991.

11. Lofti M, Venkatesh Mannar NG, Merx RJHM, Naber-van den Heuvel P. *Micronutrient Fortification of Foods: Current Practices, Research, and Opportunities.* Ottawa, Canada: Micronutrient Initiative; 1996.

12. National Academy of Sciences. *Technology of Fortification of Foods.* Washington, DC: National Research Council; 1975.

13. Micronutrient Initiative. Food fortification to end micronutrient malnutrition: state of the art. Proceedings of 2 August 1997 Symposium, Montreal. Ottawa: The Micronutrient Initiative; 1998.

14. Combs GF, Welch RM, Duxbury JM, Uphoff NT, Nesheim MC. *Food-Based Approaches To Preventing Micronutrient Malnutrition: An International Research Agenda.* Ithaca, NY: Cornell International Institute for Food, Agriculture and Development; 1996.

15. *Enriching Lives: Overcoming Vitamin and Mineral Malnutrition in Developing Countries.* Washington, DC: World Bank; 1994.

16. Urrutia J, Mata L, Bressani R, Lachance P. *Corn Fortification: A Field Demonstration Project Model.* Final Report AID-CSD-3357. Guatemala City: INCAP; 1976.

CHAPTER 14

Antioxidants and Their Effect on Health

Lillian Langseth

The idea that certain food ingredients may have health benefits that go beyond basic nutrition lies at the core of the functional food concept. Today, this idea is well accepted by the scientific community, but that was not always the case. As recently as 25 years ago, anyone who proposed that nutrients might have health benefits other than the prevention of deficiency diseases was likely to be dismissed as a quack. A similar fate awaited anyone who was bold enough to suggest that non-nutrient substances in food might be beneficial to health.

One of the first nontraditional ideas about food and health to achieve scientific respectability was the hypothesis that antioxidant nutrients might protect against chronic diseases. At a time when many other ideas about food ingredients and health were still regarded as fanciful notions, this hypothesis was already undergoing serious scientific investigation.

The story of the antioxidant nutrients is of interest not only in itself but also as an illustration of the kind of intense scientific scrutiny that all functional foods must eventually undergo. The progress of research on the antioxidant nutrients provides a "sneak preview" of the developments that are likely to occur when other functional foods and ingredients are subjected to full-scale scientific evaluation. As the antioxidant story illustrates, the road from an intriguing idea to a definitive scientific conclusion is an exciting one—but there may be more than a few bumps along the way.

OXYGEN, FREE RADICALS, AND HUMAN DISEASE

Most people think of oxygen as a benign, life-giving substance, but this is not a completely accurate view. Although oxygen is essential to human life, it is also highly toxic, and it poses a constant threat to the well-being of all living things. The earliest forms of life on earth could not tolerate oxygen, and some microorganisms living today can survive only in oxygen-free environments. Our bodies can tolerate oxygen only because our evolutionary ancestors developed defenses against its toxic effects at the same time that they were evolving the ability to make use of it in the production of energy. Without such defenses, none of the air-breathing creatures on earth could exist.

Free Radicals and Their Effects on Body Cells

It was first proposed in 1954 that most of the damaging effects of oxygen on living organisms could be due to the formation of chemical entities called free radicals.[1] A free radical is an unstable, damaged molecule. It is missing at least one electron, and this deficiency makes it highly unstable.

Free radicals will react with practically any molecule with which they come into contact, causing a kind of chemical havoc. When this occurs in a living organism, it can lead to cell injury and disease.

Table 14–1 lists some of the free radicals and other reactive oxygen species that may be formed in the human body.[2] These highly reactive molecules can cause extensive damage to crucial biological molecules, including proteins, carbohydrates, lipids, and DNA, thus interfering with the normal functioning of body cells. Figure 14–1 illustrates the many mechanisms by which free-radical damage can lead to cell injury.

Free radicals and other highly reactive oxygen compounds are believed to contribute to the causation of a wide variety of human diseases, especially the chronic diseases associated with aging. Some clinical conditions in which oxygen free radicals are thought to be involved include

- cancer
- atherosclerosis and its consequences, including coronary heart disease, stroke, and peripheral arterial disease
- diseases involving inflammation, such as rheumatoid arthritis
- Parkinson's disease
- Alzheimer's disease
- cataract
- macular degeneration
- radiation poisoning
- emphysema
- diseases involving iron overload, such as hemochromatosis

Free radicals and oxidative processes are also believed to be involved in the aging process.[3] During aging, degenerative changes occur in the cells and tissues of the human body. These changes increase vulnerability to disease and limit the life span. Free-radical reactions apparently contribute to this degradation of biological systems over time.

Sources of Free Radicals

Cells produce free radicals as a part of their normal functioning, so everyone is exposed to some extent to these potentially harmful molecules. Some people are also exposed to additional, external sources of free radicals, including cigarette smoke, environmental pollutants, certain drugs, and ultraviolet light or ionizing radiation.

Table 14–1 Some Important Reactive Oxygen Species in Living Organisms

Free radicals	
Hydroxyl radical	$OH^{\bullet}$
Superoxide radical	$O_2^{\bullet -}$
Nitric oxide radical	$NO^{\bullet}$
Lipid peroxyl radical	$LOO^{\bullet}$
Nonradicals	
Hydrogen peroxide	H_2O_2
Singlet oxygen	$^1O_2$
Hypochlorous acid	$HOCl$
Ozone	O_3

Source: Reprinted with permission from L. Langseth, Oxidants, Antioxidants, and Disease Prevention, *ILSI Europe Concise Monograph Series*, p. 1, © 1996, International Life Sciences Institute/ILSI Europe: Washington, DC.

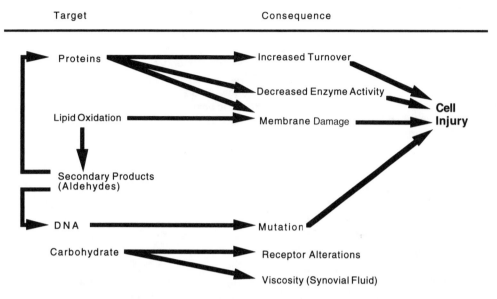

Figure 14–1 Free radical damage. *Source:* Reprinted with permission from *Antioxidant Vitamins Newsletter*, Vol. 1, Numbers 1 and 2, p. 10, Copyright © Roche Vitamins Inc., Parsippany, New Jersey.

Cigarette smoke is a particularly important source of "extra" free radicals. One of the reasons why smoking is so harmful to the human body is that it increases oxidative injury to body cells.[4] Oxidation is believed to play a role in the causation of many of the diseases that are associated with smoking, including lung cancer and other cancers, coronary heart disease, emphysema, and cataract.

THE ANTIOXIDANT DEFENSE SYSTEM

Many defense mechanisms limit the levels of free radicals in the body and inhibit the damage that they cause.[5] Structural defenses, such as membranes and chelating agents, help to prevent contact between free radicals and vulnerable molecules. Enzymes called proteases degrade protein molecules that have been damaged by free radicals, so that they can be replaced with properly functioning proteins. Repair enzymes excise and replace damaged portions of DNA molecules. Some antioxidant enzymes (superoxide dismutase, catalase, and glutathione peroxidase) react with free radicals and inactivate them. Small molecules synthesized in the body, such as glutathione and uric acid, help to stabilize free radicals.

In addition to the defenses listed above, several types of small molecules that cannot be synthesized in the body contribute to the antioxidant defense system. These molecules, which must be obtained from the diet, include several nutrients: vitamin E, vitamin C, and a group of plant pigments called carotenoids (some of which are precursors of vitamin A). In addition, some foods contain nonnutrient antioxidants, most of which are phenolic or polyphenolic compounds—substances that have no known nutritional function but may be important to human health because of their antioxidant actions. Table 14–2 lists some of the non-nutrient antioxidants and their food sources.[2]

The antioxidants found in foods are of special interest to researchers because it is possible to vary people's exposure to these substances by dietary change or the use of supplements. Most of the other components of the antioxidant defense system cannot be modified easily.

Table 14–2 Some Foods That Contain Non-Nutrient Antioxidants

Product	Antioxidants
Soybeans	Isoflavones, phenolic acids
Green tea, black tea	Polyphenols, catechins
Coffee	Phenolic esters
Red wine	Phenolic acid
Rosemary, sage, other spices	Carnosic acid, rosmaric acid
Citrus and other fruits	Bioflavonoids, chalcones
Onions	Quercetin, kaempferol
Olives	Polyphenols

Source: Reprinted with permission from L. Langseth, Oxidants, Antioxidants, and Disease Prevention, *ILSI Europe Concise Monograph Series*, p. 4, © 1996, International Life Sciences Institute/ILSI Europe: Washington, DC.

THE CONCEPT OF OXIDATIVE STRESS

Under normal circumstances, a balance exists in the human body between factors that promote oxidation and the antioxidant defenses that inhibit this process (see Table 14–3).[2] An increase in the production of oxidants or a deficiency in the antioxidant defense system can disturb the balance in the direction of increased oxidation. This situation, in which an imbalance exists between pro-oxidants and antioxidants, is called oxidative stress. Increased intakes of antioxidant nutrients may reduce oxidative stress and help to restore the balance between oxidants and antioxidants in the human body.

THE ANTIOXIDANT NUTRIENTS

Although a variety of substances with antioxidant properties are found in foods, most research has focused on three major antioxidant nutrients: vitamin E, vitamin C, and the carotenoids. Table 14–4 lists the most important food sources of each of these nutrients.[2]

Table 14–3 Pro-Oxidant Factors and Antioxidant Defenses

Pro-Oxidant Factors	Antioxidant Defenses
Inflammation	Vitamin E
Cigarette smoke	Vitamin C
Exercise	β-Carotene
Air pollutants	Other carotenoids
Radiation	Glutathione
Diet high in polyunsaturated fatty acids (PUFAs)	Superoxide dismutases
Carcinogens	Catalase
	Glutathione peroxidases
	Selenium
	Uric acid
	Other antioxidants derived from foods

Source: Reprinted with permission from L. Langseth, Oxidants, Antioxidants, and Disease Prevention, *ILSI Europe Concise Monograph Series*, p. 5, © 1996, International Life Sciences Institute/ILSI Europe: Washington, DC.

Table 14-4 Natural Food Sources of Antioxidant Nutrients

Vitamin E
Best sources: vegetable oils and products made from them, such as margarine and mayonnaise,
    cold-pressed seed oils, wheat germ
Other significant sources: vegetables, fruits, meat/poultry/fish
Vitamin C
Fruits, especially citrus fruits, strawberries, and melons
Vegetables, especially tomatoes, leafy greens, cabbage family vegetables (such as broccoli and cau-
    liflower), potatoes
Carotenoids
Beta carotene: yellow-orange vegetables and fruits, dark green vegetables
Alpha carotene: carrots
Lycopene: tomatoes, watermelon
Lutein and zeaxanthin: dark green leafy vegetables, broccoli
β-Cryptoxanthin: citrus fruits

Source: Reprinted with permission from L. Langseth, Oxidants, Antioxidants, and Disease Prevention, *ILSI Europe Concise Monograph Series*, p. 5, © 1996, International Life Sciences Institute/ILSI Europe: Washington, DC.

Vitamin E is the major fat-soluble antioxidant in all cellular membranes. It also protects polyunsaturated fatty acids against oxidation.

Vitamin C, unlike vitamin E, is a water-soluble substance. Vitamin C is the most important antioxidant in the fluids outside of body cells, and it has many known activities within the cell as well. Vitamin C also has an important synergistic interaction with vitamin E. It regenerates the active form of vitamin E after vitamin E has reacted with a free radical.

The carotenoids are a large group of red to yellow pigments found in plant foods and in the tissues of animals that eat the plants. There are more than 670 naturally occurring carotenoids, many of which have biologically important antioxidant activity. Some of the most important carotenoids found in foods include beta carotene, alpha carotene, lycopene, lutein, and β-cryptoxanthin. Some of these carotenoids (such as beta carotene) can be converted into vitamin A in the body; others (such as lycopene) cannot. All of the carotenoids may be important to health because of their antioxidant activity, regardless of whether they are vitamin A precursors.

ANTIOXIDANTS AND HUMAN DISEASE

The antioxidant nutrients have been a major focus of nutrition research in recent years. For a time, they were regarded by many as magic bullets that could help to prevent virtually all human ills. Both scientists and laypeople were enthusiastic about the prospect of reducing disease risk by increasing the intake of these nutrients—either through widespread supplementation or through the development of food products with an increased antioxidant content.

More recently, however, several major studies of beta carotene and lung cancer have had disappointing results, prompting a reevaluation of the risks and benefits of antioxidant supplementation. When the negative findings of these studies were first announced, many people jumped to the conclusion that antioxidants were worthless and that they would never be of any value in the prevention or treatment of human diseases. But this very negative view of antioxidants is as unjustified as was the earlier overenthusiasm.

Currently, most experts believe that antioxidant nutrients will still prove to be of benefit in preventing or postponing the onset of some degenerative diseases, even though not all of the hypothesized links between antioxidants and disease will prove to be valid. Each such association must be subjected to a thorough, careful scientific evaluation before any definite conclusions can be reached.

TYPES OF RESEARCH STUDIES ON FOOD COMPONENTS AND DISEASE

To understand the current status of scientific knowledge of antioxidant nutrients, it is necessary to know a little about the types of studies that researchers conduct when investigating a possible link between a food component and the prevention of a human disease.

Research often starts with animal experiments, usually involving rats or mice. Animal experiments are convenient and relatively inexpensive (in comparison with research on humans), and they can involve procedures that would be unacceptable if applied to human research subjects (such as deliberate exposure to toxic chemicals). It is important to remember, however, that the fact that something occurs in an animal model does not necessarily mean that it also occurs in humans. There are important biological differences between animal species (eg, rats and mice can synthesize their own vitamin C, but humans cannot). In addition, the ways that diseases are produced in animal models are different from the ways that they occur naturally.

Second, researchers often conduct in vitro (test-tube) biochemical experiments to complement animal research. Experiments of this type allow scientists to investigate biochemical mechanisms in detail, but great caution must be exercised in extrapolating their results to the human situation.

A third line of evidence is epidemiological research—observational studies of human populations. In some types of epidemiological studies, groups of people are compared with one another. For example, scientists might compare the per capita intake of a particular food in different countries with the rates of a disease in those countries to see if any relationship is evident. This is a rather crude form of research, since population groups differ from one another in many ways and it is difficult to pinpoint the effects of any one specific factor.

In more sophisticated types of epidemiological studies, researchers compare individuals rather than groups. For example, they might interview people who have cancer and those who do not in order to find out whether their lifelong dietary habits have differed. Or they might collect information on the dietary habits of a group of healthy people, wait for several years (or even decades), and then compare those individuals who develop cancer during the follow-up period to those who remain healthy.

These types of epidemiological investigations are observational in nature. The researchers observe people and collect data, but they do not deliberately modify the study participants' exposure to factors that might increase or decrease their risk of a disease. For example, although researchers conducting an observational study of vitamin E and heart disease would certainly want to find out whether any of the study participants had ever used vitamin E supplements, they would not ask any of the participants to take (or refrain from taking) those supplements.

Observational epidemiology can provide many useful clues about the relationship between dietary factors and disease risk. Because this type of research involves real people rather than artificial experimental models, the results are directly applicable to the human situation. However, observational epidemiology also has a very important limitation. It can demonstrate only associations between dietary factors and the risk of a disease; it cannot prove a cause-and-effect relationship. Observational epidemiology may tell us that people who have high intakes of a specific nutrient are less likely to develop a particular disease, but it cannot tell us whether the nutrient actually prevents the disease.

For true proof of a cause-and-effect relationship, scientists must conduct a more definitive type of study called a randomized controlled trial. The researchers recruit a large group of volunteers and randomly assign each of them either to receive or not to receive the treatment or preventive measure that is under investigation. Ideally, such research should be conducted in a double-blind fashion, meaning that neither the study participants nor the researchers who work with them know who is receiving the active intervention until after the study is completed. This is usually accomplished by giving some participants placebos—treatments that look like the real thing but do not contain the active agent. For example, in a randomized controlled trial of beta carotene supplementation, some people would be given tablets or capsules that actually contain beta carotene, while others would be given identical-appearing tablets or capsules with no beta carotene in them.

Randomized controlled trials are the ultimate test of a hypothesis about a link between a dietary component and a disease. Unlike the other types of research described above, they can provide a definitive answer to the question of whether a substance helps to prevent a disease. However, controlled trials cannot be undertaken lightly. Researchers conduct this type of study only when it is very clearly justified—that is, when other types of research have provided them with good reason to believe that the intervention they are proposing to test will be effective.

Randomized controlled trials are huge projects. They are extremely expensive and time consuming, involving large numbers of people. For example, a controlled trial of beta carotene supplementation in Finnish men (to be discussed in the next section) involved 29,000 men, took more than 8 years to complete, and cost $43 million.

Controlled trials also involve special ethical considerations that do not apply to other types of research. In a controlled trial, people are asked to do something that they would not otherwise have done and to allow themselves to be assigned—by pure chance—to one test group or another. There is always a possibility that those assigned to the control group might miss out on an important health benefit (if the intervention being tested proves to be effective) or that those assigned to the treatment group might face a health risk (if the intervention has unanticipated side effects). It is unethical to ask people to expose themselves to these risks without a very compelling justification.

ANTIOXIDANTS AND CANCER: THE RISE AND FALL OF BETA CAROTENE

One of the first ideas about antioxidants and disease prevention to receive major research attention was the proposition that beta carotene might prevent cancer—especially lung cancer. This hypothesis was based on two complementary lines of evidence. First, animal studies and in vitro experiments had shown that beta carotene has several biological activities that may be relevant to cancer prevention.[6,7] Second, a very substantial body of epidemiological research had shown a strong, consistent inverse association between the risk of lung cancer and the consumption of fruits and vegetables—the principal dietary sources of beta carotene.[8,9]

By the mid-1980s, the evidence linking beta carotene with a reduced risk of lung cancer was considered sufficiently compelling to warrant large-scale controlled trials. Three major trials of beta carotene supplementation were started at that time, one in Finland and two in the United States.

The Finnish trial was the first to be completed.[10] In this trial, the researchers recruited more than 29,000 middle-aged or elderly male smokers and randomly assigned them to receive supplements of beta carotene, vitamin E, both supplements together, or placebo for 5 to 8 years. The study specifically focused on smokers because they have a very high risk of lung cancer.

Unfortunately for the study participants, the supplements were not effective in preventing lung cancer; in fact, a few more cases of lung cancer occurred among the men receiving beta carotene than among those receiving placebos. Thus, beta carotene supplementation may actually have been harmful (although the adverse effect was relatively small). These findings were unexpected, and they came as an absolute shock to the scientific community.

At the time that these surprising results were announced, many observers suspected that they might be a statistical fluke or that they were attributable to some unique characteristic of the Finnish population. However, in 1996, similar findings were announced in another study of smokers, this time in the United States.[11] Again, people who received beta carotene supplements showed a small increase, rather than a decrease, in lung cancer risk.

Also in 1996, a third trial, involving US physicians, showed no effect of beta carotene supplementation (even when continued for as long as 12 years) on the risk of lung cancer.[12] In this instance, no adverse effect was observed (probably because very few of the doctors were smokers, and the harmful effect of beta carotene seems to be limited to smokers).

Why did beta carotene fail to prevent lung cancer in these trials, even though the previous research had seemed so promising? There are several possible explanations. First, the reductions in cancer risk associated with diets high in beta carotene may be attributable to other substances found in carotene-rich fruits and vegetables rather than to beta carotene itself. Second, beta carotene may be of value only in the earliest stages of carcinogenesis, which occur at relatively young ages—much younger than the ages of the people included in the trials. Third, the doses of beta carotene used in the trials, all of which were substantially greater than those normally obtained from the diet, may have had effects different from those that occur at lower doses.

Scientists are still trying to determine why beta carotene may have promoted lung cancer in the two trials involving smokers. It is possible that the increase in lung cancer risk results from a direct interaction between cigarette smoke and beta carotene.[13] Oxidative agents in the smoke may attack the beta carotene molecule, producing harmful products. Or perhaps there is some other explanation. No one knows for sure.

What everyone does know, however, is that beta carotene is not the hoped-for magic bullet that would offset the effects of cigarettes and protect smokers against lung cancer. The controlled trials clearly demonstrated (after great effort and expense) that beta carotene does not work. This does not mean, however, that beta carotene may not be of value in the prevention of other diseases. It also does not mean that all antioxidants have been proven worthless in the prevention of cancer. It simply means that a specific scientific hypothesis that was once viewed with great enthusiasm has been shown to be wrong.

Currently, most research on antioxidants and cancer is focusing on substances other than beta carotene. One intriguing possibility is that some antioxidant nutrients may protect against prostate cancer. The results of one controlled trial showed that vitamin E supplementation is associated with a reduction in prostate cancer risk,[14] while a second trial showed a similar benefit for supplementation with selenium (a mineral that plays an important role in the body's antioxidant defense system).[15] Both of these trials, however, were designed primarily to investigate other scientific questions; the favorable results for prostate cancer were unexpected. Further research is needed to confirm the apparent benefits of the antioxidants, to determine the mechanisms of action, and to pinpoint the optimum intake of these two nutrients for the purpose of minimizing prostate cancer risk.

VITAMIN E AND HEART DISEASE

One of the most exciting current topics in antioxidant research is the prospect that supplementation with vitamin E may be of value in the prevention of coronary heart disease, the leading cause of death in the United States and most other industrialized nations.

There are at least two very plausible mechanisms by which vitamin E might protect against heart disease. One way is by protecting blood lipoproteins (the proteins that carry cholesterol through the bloodstream) against oxidation.[16,17] Recent research has shown that lipoproteins such as low-density lipoprotein, the so-called "bad" cholesterol, are harmful to the arteries when they are oxidized. Long-term exposure to oxidized lipoproteins contributes to the causation of atherosclerosis, the underlying disorder of the arteries that predisposes an individual to heart attacks and strokes. Since vitamin E can inhibit the oxidation of lipoproteins, it may retard the development of atherosclerosis.

A second mechanism by which vitamin E might protect against heart disease involves the inhibition of blood clotting, a process that is involved in the initiation of a heart attack.[18] This is also the mechanism by which aspirin helps to prevent heart attacks. In fact, there is evidence that the actions of aspirin and vitamin E may be complementary, since each agent inhibits blood clotting in a different way.[19]

A variety of studies in human populations have linked increased intakes or blood levels of vitamin E with a reduced risk of heart disease:

- Two large observational epidemiological studies conducted by researchers from the Harvard School of Public Health showed that middle-aged men and women who took single-entity vitamin E supplements (100 IU/d or more) had substantially lower risks of heart disease than those with lower vitamin E intakes.[20,21] A similar association was also found in a study of elderly Americans.[22]
- A major controlled trial in the United Kingdom showed that when patients with existing heart disease were given supplements of 400 or 800 IU/d of vitamin E for about a year and a half, the number of new nonfatal heart attacks decreased by 77%.[23]
- A study from the University of Southern California indicated that men with existing heart disease who were taking cholesterol-lowering drugs showed less progression of the narrowing in their coronary arteries if they also took vitamin E supplements.[24]
- A controlled trial in Finland (see the previous section) showed a small but statistically meaningful decrease in the risk of angina among men who had been randomly assigned to take supplements of vitamin E at a dose of 50 mg/d (a relatively low dose in comparison with those investigated in other studies).[25]
- A cross-cultural study of 16 European population groups showed that the risk of heart disease was higher in populations with low blood levels of vitamin E.[26]
- An observational study of middle-aged men in Quebec showed that the use of vitamin supplements in general was associated with a significant reduction in the risk of death from heart disease and that the effect of supplemental vitamin E was stronger than that of any other vitamin.[27]
- In a small controlled trial conducted in Japan, 161 middle-aged people were randomly assigned to receive either 100 mg or 3 mg of vitamin E daily for 6 years.[28] Seven new cases of coronary disease occurred during the trial; all seven occurred in the 3-mg control group. Coronary disorders were significantly more frequent in this group than in the 100-mg group.

In most of the studies listed above, protective effects of vitamin E were observed primarily among people who took relatively high doses of the vitamin—usually, 100 mg/d or more. Intakes this high cannot be obtained from natural food sources alone. Thus, if future research confirms that vitamin E is protective against heart disease, it may be necessary to fortify foods with this vitamin or use supplements to take advantage of its effects.

Although quite a lot of scientific evidence points toward a protective effect of vitamin E against heart disease, a cause-and-effect relationship has not been definitively established. No large-scale randomized, controlled trials of high-dose vitamin E supplementation in healthy people have yet been completed. (The Japanese trial mentioned above was very small, the British trial involved people who were already ill, and the Finnish trial involved a low dose of the vitamin.) Thus, the link between vitamin E and heart disease must be regarded as promising rather than conclusive.

ANTIOXIDANTS AND DEGENERATIVE DISEASES OF THE EYE

Many older people develop degenerative diseases of the eye that may be due at least in part to oxidative damage. One such disease is cataract, which results from progressive increases in the opacity (cloudiness) of the lens of the eye. Once a cataract starts to develop, it may become progressively worse over the years, often reaching the point where it must be removed surgically. More than a million cataract operations are performed in the United States every year, at a total cost of more than $3 billion.[29] If there were some way to prevent cataracts from occurring, many of these surgeries—and the medical bills that go with them—could be avoided.

Another important age-related disorder of vision that may be linked to oxidative stress is macular degeneration (also called maculopathy). This disease of the retina can cause permanent visual impairment, sometimes progressing to total blindness. Unlike cataract, macular degeneration cannot be treated effectively, so the need for preventive measures is especially urgent.

During the 1990s, scientific studies have indicated that higher intakes of antioxidant nutrients or vitamin supplements may be associated with reduced risks of both cataract and macular degeneration. For example:

- A large study of US women showed that the risk of cataract was 45% lower among those who used vitamin C supplements for 10 years or more.[30]
- A study of male US physicians showed that the risk of cataract was 25% lower in those who took multivitamins than in those who did not.[31]
- Another US study also showed that regular use of multivitamins was associated with decreased risk of all types of cataract. In this study, supplement users had less than two thirds the risk of cataract seen in nonusers. High intakes of vitamins C and E and carotene, considered individually, were also associated with decreased risks of some types of cataract.[32]
- Also in the United States, a group of older people were followed for several years, during which they had annual eye examinations. Regular users of multivitamin supplements or vitamin E supplements had a decreased risk of developing clouding in the lenses of their eyes during the follow-up period.[33]
- In a study in Finland, people with low blood levels of vitamin E and beta carotene had an increased risk of developing cataracts.[34]
- A study in Canada showed that nonusers of vitamin C supplements were 3.3 times more likely than users to develop cataracts. Similarly, nonusers of vitamin E supplements were 2.3 times more likely than users to develop cataracts.[35]
- A study in the United States showed that low blood levels of vitamin C were associated with an 11-fold increase in the risk of one type of cataract, while low blood carotenoid levels were associated with a 7-fold increase in the risk of a different type of cataract. Low dietary intakes of vitamin C and low intakes of fruits and vegetables were also associated with increased cataract risk in this study.[36]
- A study in Finland that focused on the early stages of development of lens clouding showed that individuals with low blood vitamin E levels had an increased risk of progression of lens clouding during the 3-year study period.[37]
- A large study conducted in five areas of the United States showed relationships between macular degeneration and antioxidants, especially carotenoids.[38,39] People with moderate or high blood levels of carotenoids had a markedly decreased risk of macular degeneration when compared to those with low blood carotenoid levels. High dietary intakes of carotenoids were also associated with reduced risk, with the strongest effect for lutein/zeaxanthin (a carotenoid that is found in leafy dark-green vegetables and also in the retina of the eye).

- In another study of macular degeneration, people with low blood levels of lycopene (a carotenoid found in tomatoes) were found to have twice the risk of macular degeneration observed in those with high lycopene levels.[40]

Scientists are continuing to investigate the relationship between antioxidants and degenerative eye diseases. If future research confirms the evidence already available, vitamin supplements or improvements in diet may prove to be a useful tool for helping to decrease the risks of cataract and macular degeneration. It is important to note, however, that all of the studies described above were observational studies. No randomized controlled trials have demonstrated a beneficial effect. In fact, in the large trial in Finland in which smokers were given supplements of vitamin E and/or beta carotene for about 6 years, the supplements had no noticeable effect on the risk of cataract.[41] Therefore, the evidence for a protective effect of antioxidant nutrients or vitamin supplements against cataract and macular degeneration cannot be considered definitive. Additional research will have to be done before any recommendations about dietary change or supplementation can be made.

VITAMIN E SUPPLEMENTATION IN ALZHEIMER'S DISEASE AND OTHER NEUROLOGICAL DISORDERS

Until very recently, Alzheimer's disease was thought to be untreatable, but this is no longer the case. Well-designed studies have identified several agents that may be of value in slowing the progression of disability in patients with Alzheimer's disease. One of these agents is vitamin E.

In a controlled trial in the United States, patients with moderately severe Alzheimer's disease were randomly assigned to receive 2000 IU/d of vitamin E or placebo for 2 years.[42] Functional loss (as assessed by the time to the occurrence of any of the following: death, institutionalization, loss of the ability to perform basic activities of daily living, or severe dementia) occurred later among patients receiving vitamin E than among those receiving placebo. The difference between the two groups was a matter of months—a small benefit but a promising one.

The results of this trial support the concept that oxidative stress plays a role in the causation of Alzheimer's disease. Oxidation may also be involved in other neurological disorders, and antioxidants may be of value in their treatment. For example, oxidative processes are believed to be involved in the causation of tardive dyskinesia (a movement disorder that may result from the long-term use of antipsychotic drugs), and vitamin E supplementation may be helpful in decreasing the symptoms of this problem.[43,44] However, vitamin E is not necessarily useful in the treatment of all chronic neurological disorders. For example, a controlled trial in patients with Parkinson's disease showed no benefit from high-dose vitamin E supplementation.[45] This negative result was a disappointment, since there is substantial evidence that oxidative processes are involved in Parkinson's disease, and it had been hoped that vitamin E, as an antioxidant, would help to slow the progression of this disease.

VITAMIN C AND THE COMMON COLD

Although most antioxidant research has focused on the prevention of chronic, serious diseases, the antioxidant nutrients may have other uses as well. For example, there is scientific basis for the common belief that vitamin C may be of value in the prevention and treatment of colds.

Many controlled trials have been conducted in human volunteers to evaluate the effect of vitamin C on colds. In general, these trials have shown that people who regularly take 1000 mg/d or more of vitamin C have shorter and milder colds than those who do not.[46,47] It is unclear from the results of the trials, however, whether vitamin C supplementation can actually prevent colds from occurring. In some trials, participants who took vitamin C had fewer colds than those taking placebos, but in other trials, no such difference was found.[46]

It is possible that vitamin C supplementation may prevent colds only in certain high-risk groups of people. A recent reanalysis of data from previously completed studies showed that vitamin C supplementation reduced the number of colds experienced by people who exercise heavily, such as marathon runners.[48] Another analysis associated vitamin C supplementation with a reduced risk of colds in men and boys from population groups with relatively low vitamin C intakes,[49] but not in those with better diets.

SAFETY OF ANTIOXIDANT NUTRIENTS

The safety of antioxidant nutrients is of the utmost importance if these nutrients are to be included in functional foods. Fortunately, there appear to be few concerns about the safety of vitamin E, vitamin C, and the carotenoids, even when they are consumed in high doses. As with all dietary supplements, however, individuals with chronic medical problems and those who are taking medication should consult with a physician before taking antioxidant supplements.

Beta Carotene

The principal side effect of beta carotene is hypercarotenodermia, a reversible, harmless yellow discoloration of the skin that is especially noticeable on the palms of the hands and the soles of the feet. Beta carotene is not otherwise toxic.[6] Because two controlled trials showed an increase in lung cancer risk in smokers who took beta carotene supplements,[10,11] it is usually recommended that smokers avoid these supplements. However, smokers who have taken beta carotene supplements in the past need not be overly concerned. The increase in lung cancer risk associated with long-term use of high-dose beta carotene supplements is about 20%. This is trivial in comparison with the roughly ten-fold (1000%) increase in the risk of lung cancer associated with smoking itself.

Other Carotenoids

Little is known about the safety of high doses of carotenoids other than beta carotene. Until more research is completed, it is prudent for individuals who wish to increase their intake of carotenoids such as lycopene or lutein to do so by eating more fruits and vegetables rather than by taking supplements.

Vitamin C

Although there have been many unconfirmed reports of harmful effects associated with the use of vitamin C supplements, there is actually no convincing scientific evidence of any adverse effects at doses up to 1000 mg/d.[50–52] Because vitamin C promotes the absorption of iron from food, patients with iron overload diseases such as hereditary hemochromatosis or thalassemia major should consult a physician before using vitamin C supplements.

Vitamin E

Vitamin E is considered safe under most circumstances, even at doses as high as 800 mg/d.[52–55] However, since high doses of vitamin E may inhibit blood clotting, their use may be undesirable in some circumstances (eg, shortly before or after surgery). Because of possible adverse effects, patients who are taking anticoagulant drugs (blood thinners) and those with the hereditary eye disease retinitis pigmentosa should consult a physician before taking vitamin E supplements.

SUMMARY

Extensive scientific evidence supports the concept that free radicals and oxidative processes contribute to the causation of a wide variety of diseases, including the major chronic diseases of aging. They may also play a role in aging itself.

In the late 1980s and early 1990s, there was great enthusiasm, among both scientists and laypeople, about the possibility that the antioxidant nutrients could help to prevent a wide variety of diseases. However, as research on antioxidants has progressed, it has become evident that not all of the proposed uses of these nutrients in the prevention or treatment of disease will prove to be effective.

At the present time (late 1998), the most promising areas of antioxidant research are those involving vitamin E. A growing body of scientific evidence indicates that vitamin E may be of value both in preventing the onset of coronary heart disease and in slowing the progression of this disease in people who have already developed heart problems. Vitamin E also appears to have a small but meaningful beneficial effect in patients with Alzheimer's disease, and it may protect against prostate cancer.

Other potential links between antioxidants and disease are also under investigation. For example, antioxidant nutrients and vitamin supplements may be of value in preventing or postponing the onset of cataracts. Some antioxidants, especially carotenoids such as lutein, may protect against macular degeneration. Vitamin C appears to be useful in reducing the severity and duration of the common cold, and it may even reduce the frequency of colds under some circumstances.

The progress of research on the antioxidant nutrients illustrates the importance of randomized controlled trials in the evaluation of relationships between food ingredients and human diseases. Two very plausible hypotheses about antioxidants and disease—the hypothesis that beta carotene supplementation reduces the risk of lung cancer and the hypothesis that vitamin E supplementation slows the progression of Parkinson's disease—have been disproved by controlled trials.

Vitamin E, vitamin C, and beta carotene all have excellent safety records, even when consumed in high doses. Nevertheless, individuals who have chronic health problems and those who are taking medication should consult with a physician before taking antioxidant supplements (or any other kind of dietary supplement).

Much remains to be learned about the relationship between antioxidants and human health. Despite some setbacks, the antioxidant nutrients continue to be one of the most exciting and promising topics in nutrition research.

REFERENCES

1. Aruoma OI, Kaur H, Halliwell B. Oxygen free radicals and human diseases. *J Roy Soc Health*. 1991;111:172–177.

2. Langseth L. *Oxidants, Antioxidants, and Disease Prevention*. Brussels, Belgium: ILSI Europe; 1996.

3. Harman D. Free radical theory of aging: dietary implications. *Am J Clin Nutr*. 1972;25:839–843.

4. Morrow JD, Frei B, Longmire AW, et al. Increase in circulating products of lipid peroxidation (F2-isoprostanes) in smokers. *N Engl J Med*. 1995;332:1198–1203.

5. Ames BN, Shigenaga MK, Hagen TM. Oxidants, antioxidants, and the degenerative diseases of aging. *Proc Natl Acad Sci USA*. 1993;90:7915–7922.

6. Gerster H. Anticarcinogenic effect of common carotenoids. *Int J Vit Nutr Res*. 1993;63:93–121.

7. Gerster H. β-Carotene, vitamin E and vitamin C in different stages of experimental carcinogenesis. *Eur J Clin Nutr*. 1995; 9:155–168.

8. Block G, Patterson B, Subar A. Fruit, vegetables, and cancer prevention: a review of the epidemiological evidence. *Nutr Cancer.* 1992;18:1–29.

9. van Poppel G. Carotenoids and cancer: an update with emphasis on human intervention studies. *Eur J Cancer.* 1993; 29A:1335–1344.

10. Alpha-Tocopherol, Beta Carotene Cancer Prevention Study Group. The effect of vitamin E and beta carotene on the incidence of lung cancer and other cancers in male smokers. *N Engl J Med.* 1994;330:1029–1035.

11. Omenn GS, Goodman GE, Thornquist MD, et al. Effects of a combination of beta carotene and vitamin A on lung cancer and cardiovascular disease. *N Engl J Med.* 1996;334:1150–1155.

12. Hennekens CH, Buring JE, Manson JE, et al. Lack of effect of long-term supplementation with beta carotene on the incidence of malignant neoplasms and cardiovascular disease. *N Engl J Med.* 1996;334:1145–1149.

13. Mayne ST, Handelman GJ, Beecher G. β-Carotene and lung cancer promotion in heavy smokers: a plausible relationship? *J Natl Cancer Inst.* 1996;88:1513–1515.

14. Heinonen OP, Albanes D, Virtamo J, et al. Prostate cancer and supplementation wth alpha-tocopherol and beta carotene: incidence and mortality in a controlled trial. *J Natl Cancer Inst.* 1998;190:440–446.

15. Clark LC, Dalkin B, Krongrad A, et al. Decreased incidence of prostate cancer with selenium supplementation: results of a double-blind cancer prevention trial. *Br J Urol.* 1998;81:730–734.

16. Reaven PD, Khouw A, Beltz WF, Parthasarathy S, Witztum JL. Effect of dietary antioxidant combinations in humans: protection of LDL by vitamin E but not by β-carotene. *Arterioscler Thromb.* 1993;13:590–600.

17. Jialal I, Fuller CJ, Huet BA. The effect of α-tocopherol supplementation on LDL oxidation: a dose-response study. *Arterioscler Thromb Vasc Biol.* 1995;15:190–198.

18. Calzada C, Bruckdorfer KR, Rice-Evans CA. The influence of antioxidant nutrients on platelet function in healthy volunteers. *Atherosclerosis.* 1997;128:97–105.

19. Steiner M, Glantz M, Lekos A. Vitamin E plus aspirin compared with aspirin alone in patients with transient ischemic attacks. *Am J Clin Nutr.* 1995;62:1381S–1384S.

20. Stampfer MJ, Hennekens CH, Manson JE, Colditz GA, Rosner B, Willett WC. Vitamin E consumption and the risk of coronary disease in women. *N Engl J Med.* 1993;328:1444–1449.

21. Rimm EB, Stampfer MJ, Ascherio A, Giovannucci E, Colditz GA, Willett WC. Vitamin E consumption and risk of coronary heart disease in men. *N Engl J Med.* 1993;328:1450–1456.

22. Losonczy KG, Harris TB, Havlik FJ. Vitamin E and vitamin C supplement use and risk of all-cause and coronary heart disease mortality in older persons: the established populations for epidemiologic studies of the elderly. *Am J Clin Nutr.* 1996; 64:190–196.

23. Stephens NG, Parsons A, Schofield PM, et al. Randomised controlled trial of vitamin E in patients with coronary disease: Cambridge Heart Antioxidant Study (CHAOS). *Lancet.* 1996;347:781–786.

24. Hodis HN, Mack WJ, LaBree L, et al. Serial coronary angiographic evidence that antioxidant vitamin intake reduces progression of coronary artery atherosclerosis. *JAMA.* 1995;273:1849–1854.

25. Rapola JM, Virtamo J, Haukka JK, et al. Effect of vitamin E and beta carotene on the incidence of angina pectoris. *JAMA.* 1996;275:693–698.

26. Gey KF, Puska P, Jordan P, Moser UK. Inverse correlation between plasma vitamin E and mortality from ischemic heart disease in cross-cultural epidemiology. *Am J Clin Nutr.* 1991;53:326S–334S.

27. Meyer F, Bairati I, Dagenais GR. Lower ischemic heart disease incidence and mortality among vitamin supplement users. *Can J Cardiol.* 1996;12:930–934.

28. Takamatsu S, Takamatsu M, Satoh K, et al. Effects on health of dietary supplementation with 100 mg d-α-tocopheryl acetate, daily for 6 years. *J Int Med Res.* 1995;23:342–357.

29. Taylor A. Cataract: relationships between nutrition and oxidation. *J Am Coll Nutr.* 1993;12:138–146.

30. Hankinson SE, Stampfer MU, Seddon JM, et al. Nutrient intake and cataract extraction in women: a prospective study. *Br Med J.* 1992;305:335–339.

31. Seddon JM, Christen WG, Manson JE, et al. The use of vitamin supplements and the risk of cataract among US male physicians. *Am J Public Health.* 1994;84:788–792.

32. Leske MC, Chylack LT Jr, Su S. The lens opacities case-control study: risk factors for cataract. *Arch Ophthalmol.* 1991;109:244–251.

33. Leske MC, Chylack LT Jr, He Q, et al. Antioxidant vitamins and nuclear opacities: the longitudinal study of cataract. *Ophthalmology.* 1998;105:831–836.

34. Knekt P, Heliovaara M, Rissanen A, Aromaa A, Aaran R. Serum antioxidant vitamins and risk of cataract. *Br Med J.* 1992;305:1392–1394.

35. Robertson JM, Donner AP, Trevithick JR. Vitamin E intake and risk for cataracts in humans. *Ann NY Acad Sci.* 1989;570:372–382.

36. Jacques PF, Chylack LT Jr. Epidemiologic evidence of a role for the antioxidant vitamins and carotenoids in cataract prevention. *Am J Clin Nutr.* 1991;53:352S–355S.

37. Rouhiainen P, Rouhiainen H, Salonen JT. Association between low plasma vitamin E concentration and progression of early cortical lens opacities. *Am J Epidemiol.* 1996;144:496–500.

38. Eye Disease Case-Control Study Group. Antioxidant status and neovascular age-related macular degeneration. *Arch Ophthalmol.* 1993;111:104–109.

39. Seddon JM, Ajani UA, Sperduto RD, et al. Dietary carotenoids, vitamins A, C, and E, and advanced age-related macular degeneration. *JAMA.* 1994;272:1413–1420.

40. Mares-Perlman JA, Brade WE, Klein R, et al. Serum antioxidants and age-related macular degeneration in a population-based case-control study. *Arch Ophthalmol.* 1995;113:1518–1523.

41. Teikari JM, Virtamo J, Rautalahti M, Palmgren J, Liesto K, Heinonen OP. Long-term supplementation with alpha-tocopherol and beta carotene and age-related cataract. *Acta Ophthalmol Scand.* 1997;75:634–640.

42. Sano M, Ernesto C, Thomas RG, et al. A controlled trial of selegiline, alpha-tocopherol, or both as treatment for Alzheimer's disease. *N Engl J Med.* 1997;336:1216–1222.

43. Egan MF, Hyde TM, Albers GW, et al. Treatment of tardive dyskinesia with vitamin E. *Am J Psychiatry.* 1992;149:773–777.

44. Elkashef AM, Ruskin PE, Bacher N, Barrett D. Vitamin E in the treatment of tardive dyskinesia. *Am J Psychiatry.* 1990;147:505–506.

45. Parkinson Study Group. Effects of tocopherol and deprenyl on the progression of disability in early Parkinson's disease. *N Engl J Med.* 1993;328:176–183.

46. Hemilä H. Vitamin C and the common cold. *Br J Nutr.* 1992;67:3–16.

47. Hemilä H, Herman ZS. Vitamin C and the common cold: a retrospective analysis of Chalmers' review. *J Am Coll Nutr.* 1995;14:116–123.

48. Hemilä H. Vitamin C and common cold incidence: a review of studies with subjects under heavy physical stress. *Int J Sports Med.* 1996;17:379–383.

49. Hemilä H. Vitamin C intake and susceptibility to the common cold. *Br J Nutr.* 1997;77:59–72.

50. Rivers JM. Safety of high-level vitamin C ingestion. *Ann NY Acad Sci.* 1987;498:445–454.

51. Hathcock JN. Safety of vitamin and mineral supplements. In: Bendich A, Butterworth CE, eds. *Micronutrients in Health and in Disease Prevention.* New York, NY: Marcel Dekker, Inc; 1991.

52. Garewal HS, Diplock AT. How "safe" are antioxidant vitamins? *Drug Safety.* 1995;13:8–14.

53. Bendich AB, Machlin LJ. The safety of oral intake of vitamin E: data from clinical studies from 1986 to 1991. In: Packer L, Fuchs J, eds. *Vitamin E in Health and Disease.* New York, NY: Marcel Dekker, Inc; 1993.

54. Meydani SN, Barklund MP, Liu S, et al. Vitamin E supplementation enhances cell-mediated immunity in healthy elderly subjects. *Am J Clin Nutr.* 1990;52:557–563.

55. Meydani SN, Meydani M, Blumberg JB, et al. Assessment of the safety of supplementation with different amounts of vitamin E in healthy older adults. *Am J Clin Nutr.* 1998;68:311–318.

PART V

Safety and Efficacy

CHAPTER 15

Assessment of Safety and Efficacy of Functional Foods and Ingredients

Seppo Salminen and Jorma Ahokas

It is important for the scientist to improve the understanding of the role of specific foods in the health and well-being of consumers. The nutritional sciences are changing from the traditional concept of providing adequate nutrition to a new concept of optimizing nutrition. This has led to the development of science-based functional foods aimed at promoting health and adding value to foods supplementing a nutritionally balanced diet.

It is difficult to define functional foods. The approach taken in Europe by the European Union (EU) concerted action expert group coordinated by International Life Sciences Institute (ILSI), was to define a food as "functional if it is demonstrated to beneficially affect one or more target functions in the body in a way that is relevant either to promoting the state of well-being and health and/or reduction of the risk of disease."[1] By this definition, functional foods are foods in which harmful components have been taken away or beneficial components have been added to produce a health benefit to a significant portion of consumers. Thus, for instance, removing or degrading lactose in dairy products may result in a functional food with health benefits for lactose-intolerant subjects. Adding dietary fiber or fructo-oligosaccharides such as inulin to a non–fiber-containing food may improve the health benefits resulting in a functional food. However, the health benefits of a functional food have to be demonstrated in properly conducted clinical or nutritional studies in humans. Additional in vitro and animal studies can be used to support the human studies and to explain mechanisms.[1] It should be noted that at present the above definition does not apply in the United States since current United States food laws prohibit any disease claims, whether scientifically established or not, for foods and dietary supplements.

There are general scientific guidelines for testing the safety and clinical efficacy of functional foods and ingredients. These include target functions in the body and specifications for conducting proper clinical studies and safety studies. Sometimes functional foods also fall into the category of novel foods in Europe and this requires different clinical and safety assessment.[2]

The health benefits of functional foods must be scientifically documented and validated in well-planned studies, preferably by several laboratories. At the same time, the safety date of new functional foods must also be carefully documented.

REQUIREMENTS FOR GOOD CLINICAL STUDIES

Clinical and nutritional studies should be well designed and should follow the guidelines set for pharmaceutical research. However, food poses a set of problems that are not present for pharmaceutical preparations. In nutritional studies, defects can often be found in study design or in the description of study protocol or product details. Often the studies have not been blinded or randomized, or the placebo control has been difficult to arrange. The criteria for planning nutritional and clinical studies with foods or food components are:

1. Each component, additive, or microbial strain should be carefully described, complete chemical analysis given, and the origin or the manufacturer clearly identified.
   - Microbial strains should be clearly identified with proper methodology.
   - Extrapolation of data from closely related components or microbial strains is not acceptable for documenting health benefits.
   - Well-defined study products and study populations should be chosen and described.
2. Double-blind, placebo-controlled, randomized human studies are preferred, with crossover study design if reasonable.
3. Results should be confirmed by at least two independent research groups.
4. Publication of results in peer-reviewed journals is encouraged.

A summary of the European approach in functional foods and their assessment is given in ILSI's consensus document report.[1]

LEGAL REQUIREMENTS FOR SAFETY ASSESSMENT

Functional foods must be safe according to all standards of assessing food risk. But the concept of risk versus benefit cannot be applied as straightforwardly as it is for drugs; new concepts and new procedures may need to be developed and validated.

No clear guidelines for the safety assessment of functional foods exist. In addition, the regulations vary from country to country or do not exist at all. Often a case-by-case approach must be adapted. According to Huggett and Verschuren,[3] assessing the safety of the amount of food or its component(s) needed for functionality requires evidence that is equally applicable to all major groups in the population, including those who are indulging in behavior that may be expected to compromise the anticipated benefits of the functional food. It may involve postmarketing surveillance and monitoring, including the effects on the whole diet. Further, safety evaluation of micronutrients must take into account potential adverse effects of low intakes (clinical deficiency) as well as effects from intakes that are too high (clinical toxicity). Sometimes this margin is narrow, in which case calculating the safety limit by the arithmetic or geometric means of the LOAEL and Recommended Dietary Allowance is a more appropriate approach.[4]

A more traditional toxicology testing approach may be used to assess the safety of phytochemicals, the daily intake of which remains low, but it is not necessarily ideal for new functional food components, which may account for a rather larger percentage of the total food intake. The classical "dose-effect relationship" cannot be easily applied to them because it may lead to physiological/nutritional disturbances that are irrelevant to safety assessment. If a functional food is considered a novel food, the decision tree leading to an effective package of toxicity testing will be driven by the principle of "substantial equivalence" of the tested food with traditional food.[2]

Protocols for human nutrition studies need to be developed, including, in some cases, postmarketing surveillance. Even though the design of clinical studies as used in drug development can serve as a model, specific protocols and specific safety criteria relevant to functional foods and nutritional studies of these foods may be needed. It may also be necessary to identify specific target groups of individuals who might present higher or lower susceptibilities to potential adverse effects and to consider that the effects of functional foods might be positive in some target groups and negative in others. Finally, the long-term consequences of the interactions between functional food components and functions in the body and the interactions between components must be carefully monitored.

The EU's Novel Food Regulation Directive introduces a statutory premarket approval system for novel foods across the entire EU, and in some cases this may apply also to functional foods. Novel foods are foods that have not previously been consumed to a significant degree; they include foods containing or obtained from genetically modified organisms. The regulation envisages an initial safety assessment at member state level, although centralized procedures are available to resolve any objections between member states. To ensure a consistent approach to the safety assessment of novel foods, the European Commission has published a series of guidelines to accompany the regulation. The United Kingdom already has a well-developed system for assessing the safety of novel foods dating back to 1983.[5] Our case examples later in this chapter include a novel functional ingredient assessed in the United Kingdom. Several European countries now have novel food committees, and some also have assessment expert groups for functional food health effects. A similar process is followed by the Australia and New Zealand Food Authority.

The diversity of novel foods and novel ingredients covered by the EU regulations is such that a complete checklist approach to safety evaluation is impossible at the moment. Rather, a case-by-case approach is suggested, taking into account the composition of the functional or novel food, its intake, and its role in the diet and the intended target group.[2] The so-called SAFEST approach provides a means of focusing the safety evaluation on the nutritional and toxicological aspects of a functional or novel food that are of particular concern. Using this approach, novel foods as well as functional foods can be assigned to one of three classes on the basis of certain background information. For those novel foods that can be shown to be in SAFEST class 1, namely those foods that are substantially equivalent to a traditional counterpart, no further information is required to demonstrate safety. For those novel foods in SAFEST class 2, namely those sufficiently similar to a traditional counterpart or differing from it only in particular, well-defined characteristics, the evaluation will focus on those differences. For novel foods that are not in class 1 or class 2, extensive testing of the whole food will be required. The testing should follow a scientifically based hierarchical approach involving literature reviews, chemical analysis, appropriate in vitro and in vivo tests, and, if necessary, confirmation of safety and nutritional value in humans. Examination of the causes of any adverse effects reported by consumers after the novel food or ingredient has been approved and is introduced into the market may provide additional reassurance of safety.[2,3]

In the United States, similar approaches have been taken under different legislation. Pharmacokinetic data are not routinely required for food safety evaluation, and there is an increasing need to understand more about a chemical's mechanism of action to support other scientific evidence. The advent of novel foods has made conventional toxicological methods inappropriate. For example, noncaloric fat substitutes that may constitute a large portion of the daily diet cannot be adequately tested in conventional animal studies. These substances may require human nutritional studies or clinical studies similar to those used for drugs.[6,7] A short summary of the safety assessment situation is also given in Bellisle et al[8] and The European Consensus Document.[1]

The case examples presented below concern assessment of the safety and efficacy of three different substances that can all be used either as ingredients for functional foods or as functional foods themselves. These are xylitol, a five-carbon noncariogenic sugar alcohol; *Lactobacillus* strain GG (a probiotic *Lactobacillus* strain); and fructo-oligosaccharides such as inulin.

XYLITOL

Xylitol is a naturally occurring five-carbon sugar alcohol that can be found in many fruits and berries. It is most abundant in yellow plums and greengages, but it is also present in many other fruits and vegetables.[9] Xylitol can be produced by acid hydrolysis of plant material and hydrogenation of the resulting xylose to xylitol. Commercially, it is produced from birch wood chips, but it can also be produced by microbial methods. Xylitol has been documented to be noncariogenic and anticariogenic as a sweetener in chewing gum, other functional foods, and candies. Recent studies have indicated that xylitol may have prebiotic properties and other effects on health.

Clinical Efficacy

Noncariogenic and Anticariogenic Properties

The noncariogenic properties of xylitol have been well documented in studies conducted worldwide. In most studies both a noncariogenic and an anticariogenic property have been described. A comprehensive review is available in the summary report on the Turku Sugar Studies.[10] More recently, the results were confirmed in Michigan in 6-year-old children[11] and in 10-year-old children.[12] A summary of the Michigan trials conducted between 1986 and 1995[13] indicates that xylitol is more effective than sorbitol and cariologically safer than sorbitol. Even sorbitol was significantly less cariogenic in chewing gum when compared to sucrose. The findings suggest that the use of xylitol chewing gum can be considered a valuable tool in caries prevention and also in stabilization of caries in all age groups. Most recently, it was shown that subjects consuming sucrose chewing gums during 40 months and thereafter xylitol-containing chewing gum for 16 months decreased their DFMS index (delayed, filled, and missing surfaces) from 10.9 to 9.3 during the xylitol period ($P = 0.0013$). The reduction of this score was attributed to the stabilization of the caries process during xylitol intake.[14] Similar results were also reported by Virtanen and coworkers,[15] indicating that the best protection was obtained for teeth that erupted during the xylitol intervention. Edgar[16] has reviewed all xylitol and sorbitol studies on dental caries and concluded that xylitol has useful anticaries properties when compared to sorbitol and other sugar substitutes.

Generally, xylitol is used in most countries as a noncariogenic sugar substitute in foods such as chewing gum and hard candies.

Effects on Otitis Media

Xylitol is known to reduce caries by inhibiting the growth of *Streptococcus mutans*. On the basis of studies of its properties, it has been shown to prevent the adhesion of some pathogenic bacteria to mucosal cells[17,18] and also to have a significant effect on intestinal microflora.[19] It has therefore been hypothesized that xylitol could also affect the growth of other nasopharyngeal bacterial flora.[20] This could be of significance in considering respiratory infections caused by such bacteria. In vitro studies using the addition of xylitol to the growth medium have shown that 1% and 5% xylitol reduced the growth of alpha-hemolytic streptococci, including *S. pneumoniae*. It reduced slightly the growth of beta-hemolytic streptococci but not that of *Haemophilus influenzae* or *Moraxella catarrhalis*. The inhibitory growth pattern was similar to that previously reported with *S. mutans*.[20] Earlier, effects on both intestinal and oral microflora composition were demonstrated.[19,21]

A large clinical study was conducted to examine whether xylitol may have clinical importance in the prevention of acute otitis media caused by *S. pneumoniae*.[22] In a double-blind randomized trial with xylitol administered in chewing gum at 11 day care nurseries, 306 day care children were enrolled, with 149 children receiving sucrose chewing gum (76 boys; mean [SD] age 4.9 [1.5] years) and 157 receiving the xylitol chewing gum (80 boys; 5.0 [1.4] years) for 2 months at the level of 8.4 g/d. Most of the subjects had had problems with recurrent acute otitis media. The occurrence of acute otitis media and antimicrobial treatment received during the intervention and nasopharyngeal carriage of *S. pneumoniae* were maintained. The results indicated that during the 2-month monitoring period at least one event of acute otitis media was experienced by 31 out of the 149 (20.8%) children who received sucrose compared with 19 out of 157 (12.1%) of the children receiving chewing gum containing xylitol (difference, 8.7%; 95% confidence interval, 0.4% to 17.0%; $P = 0.04$). Significantly less antimicrobials were prescribed to those receiving xylitol, and 29 of 157 (18.5%) children receiving xylitol had at least one period of treatment versus 43 of 149 (28.9%) of those receiving sucrose (difference, 10.4%; confidence interval, 0.9% to 19.9%; $P = 0.032$). The carriage rate of *S. pneumoniae* varied from 17.4% to 28.2%, with no difference between the groups. Two children in the xylitol group experienced diarrhea, but no other adverse effects were noted among the xylitol users.[22]

A similar study was reported with 857 children recruited from day care centers and randomized to five treatment groups receiving "Wither" control syrup, xylitol syrup, control chewing gum, xylitol chewing gum, or xylitol lozenges. During the study, it was observed that xylitol decreased acute otitis media, whether given in chewing gum, lozenge, or syrup.[23] The authors concluded that xylitol, when given in syrup or chewing gum form, decreased acute otitis media in children.[23] The available studies indicate that xylitol preparations, when consumed regularly, prevent the acute otitis media and decrease the need for antimicrobials.[23] Further studies in different countries should be conducted to assess the safety of these supplements.

Safety Assessment of Xylitol

Xylitol has been assigned an E-number in the EU food additive directives. It is one of the few components that can be consumed in large quantities either as a polyol sweetener or as a noncariogenic sugar substitute additive. The safety of xylitol has been assessed in large-scale toxicological studies in animals, and there are epidemiological surveillance data available from long-term human consumption. The World Health Organization Joint Expert Committee on Food Additives expert group has assigned an acceptable daily intake value for xylitol. The only known side effect of xylitol consumption is transient diarrhea when large amounts of the polyol are consumed without prior adaptation.[19,24] Most people can daily consume about 15 g without any side effects. When gradual adaptation takes place, daily consumption figures of more than 100 grams per person have been reported without any ill effects.[10,13]

Xylitol can also be assessed as a natural prebiotic substance and the natural intake does have a gut microflora–modifying effect. Thus, the safety of xylitol in functional food use has been well characterized and documented both as a component from natural sources (fruit such as plums and berries) and as a natural ingredient or additive in foods.[9]

LACTOBACILLUS STRAIN GG (ATCC 53103)

Lactobacillus GG is a natural *Lactobacillus* strain of human origin isolated from a healthy human intestinal microflora. It was originally isolated because of its acid and bile tolerance, its production of antimicrobial substances effective against pathogenic bacteria, and its ability to adhere to human intestinal mucus and to colonize temporarily the human intestinal tract.[25,26]

Clinical Efficacy of *Lactobacillus* GG in Specific Target Populations

Treatment of Acute and Rotavirus Diarrhea in Children

Lactobacillus GG has been reported effective in the treatment of rotavirus diarrhea. It repeatedly reduces the duration of diarrhea to about half in children with rotavirus diarrhea. It has also been reported to be effective in the treatment of watery diarrhea in several studies in Asia with favorable results on colonization.[27,28] A placebo-controlled, randomized, and double-blinded multicenter study is underway in Europe on the efficacy of the strain.

Comparison of Different Probiotic Preparations.   When different lactic acid bacteria were compared for their effects on the immune response to rotavirus in children with acute rotavirus gastroenteritis, differences between various strains were observed.[29,30] Serum antibodies to rotavirus, total number of immunoglobulin-secreting cells, and specific antibody-secreting cells (sASCs) to rotavirus were measured at the acute stage and at convalescence. Treatment with *Lactobacillus* GG was associated with an enhancement of IgA sASC to rotavirus and serum IgA antibody level at convalescence. It was therefore suggested that *Lactobacillus* GG promotes systemic and local immune response to rotavirus, which may be of importance for protective immunity against reinfections.[29] In another study on immunological effects of viable and heat-inactivated lactic acid bacteria, *Lactobacillus* GG was administered as a viable preparation during acute rotavirus gastroenteritis.[30] This resulted in a significant rotavirus-specific IgA response at convalescence. The heat-inactivated *Lactobacillus* GG was clinically as efficient as the viable, but the IgA response was not detected. This result suggests that the viability of the strain is critical in determining the capacity of lactic acid bacteria to induce immune stimulation. Also, in a comparison study of different lactic acid bacteria for the treatment of rotavirus diarrhea, it was shown that *Lactobacillus* GG was most effective, while preparations containing *Streptococcus thermophilus* and *Lactobacillus bulgaricus* or *Lactobacillus rhamnosus* did not have any effect on the duration of diarrhea.[30] A recent review considered *Lactobacillus* GG as an effective treatment and as a potential rotavirus vaccine adjuvant to improve the efficacy of the vaccine.[31] Further studies in Pakistan, Thailand, and Peru confirmed the efficacy of *Lactobacillus* GG in different environmental and hygienic conditions.[27,32] More recently, Guandalini and coworkers[33] reported on a study coordinated by the European Society for Pediatric Gastroenterology, Hepatology, and Nutrition (ESPHGAN) concerning treatment of rotavirus diarrhea with *Lactobacillus* GG. In this study, which had 15 European study centers, the efficacy of *Lactobacillus* GG in the treatment of rotavirus diarrhea in children was verified.[33]

Possible Mechanisms.   The mechanisms by which *Lactobacillus* GG–fermented dairy products reduce the duration of infantile diarrhea are unclear. One explanation is the intrinsic lactase activity of many fermented products. However, *Lactobacillus* GG lacks lactase activity.[25,34] Alternatively, lactic acid bacteria may prevent the growth of bacterial pathogens by competing for the attachment sites or nutrients or by producing antibacterial substances. Several species of lactic acid bacteria, especially *Lactobacillus* GG, can potentiate the immune response to enteric pathogens in humans and animals when given orally.[29,35]

It has been suggested that the enhanced immune response seen in infants with rotavirus diarrhea could somehow be associated with the restoration of gut mucosal barrier function by *Lactobacillus* GG.[36,37] In a suckling rat model of rotavirus infection, introduction of cow's milk to the diet amplified intestinal permeability to proteins. This barrier dysfunction was alleviated by the administration of *Lactobacillus* GG with the cow's milk. The authors also suggested that rotavirus diarrhea is biphasic, with an initial viral diarrhea followed by bacterial diarrhea. The use of probiotics in the treatment is thought to alleviate or eliminate the bacterial diarrhea, as indicated by decreased fecal urease levels, thus shortening the duration of diarrhea.[36] The efficacy of *Lactobacillus* GG in the treatment of acute diarrhea in children was verified in a multicenter study organized by ESPHGAN. Analysis of data from a total of 287 children in 12 centers in 10 countries showed a statistically significant reduction of diarrhea in *Lactobacillus* GG–treated children (reduction in duration of diarrhea from 72 to 58 hours). In children receiving *Lactobacillus* GG, only 2.8% of all patients had diarrhea lasting longer than 6.5 days, while in the placebo group the figure was 11.4%.[33]

It has been concluded that *Lactobacillus* GG is a strain that can temporarily colonize the human gastrointestinal tract and that it is effective in reducing the duration of acute-onset diarrhea. The mechanisms include clearing the watery phase of diarrhea, preventing secondary bacterial diarrhea following rotavirus, reducing rotavirus shedding, balancing intestinal permeability and barrier mechanisms, and enhancing the immune response of the host.

Antibiotic-Associated Diarrhea

Siitonen et al[38] investigated the efficacy of a *Lactobacillus* GG preparation in preventing antibiotic-associated diarrhea. Healthy human volunteers receiving erythromycin had significantly lower incidence of diarrhea if they took *Lactobacillus* GG yogurt than a control group taking erythromycin and pasteurized normal yogurt. Other side effects of erythromycin, such as abdominal distress, stomach cramps, and flatulence, were also less common in the *Lactobacillus* group than in the group taking pasteurized yogurt. Fecal counts of *Lactobacillus* GG indicated that these organisms colonized the bowel in spite of erythromycin treatment. In an American study by Young et al[39] it was demonstrated that *Lactobacillus* GG is also effective in preventing antibiotic-associated diarrhea in children receiving antibiotics for an acute infection of the lower or upper respiratory tract.

Traveler's Diarrhea

A large placebo-controlled double-blind study of a freeze-dried *Lactobacillus* GG preparation was conducted with 820 Finnish tourists traveling to two destinations in southern Turkey.[40] The overall incidence of diarrhea was 43.8%. The total incidence of diarrhea in the travelers to both destinations was 46.5% in the placebo group and 41.0% in the *Lactobacillus* GG group (not a significant difference). *Lactobacillus* GG appeared to be effective in reducing the occurrence of traveler's diarrhea in one of the two destinations, with a protection rate of 39.5%. No side effects were recorded for either the test preparation or the placebo.[40]

More recently, in a double-blind placebo-controlled study involving 245 American subjects traveling to Mexico, the risk of having diarrhea on any given day was 3.9% for those receiving *Lactobacillus* GG and 7.4% for those receiving placebo ($P = 0.054$). The protection rate was 47% and did not vary as a function of age or gender. However, the effect appeared to be more pronounced in travelers with a prior history of traveler's diarrhea.[41] Although efficacy has been demonstrated in two studies, further studies may be required to assess the effects of *Lactobacillus* GG against different infectious diarrhea–inducing microorganisms.

Treatment of Food Allergy

Most recently, *Lactobacillus* GG has been used in studies on treatment of food allergy in infants and adults. Successful immunological changes have been reported in infants with atopic dermatitis and adults with milk hypersensitivity.[42,43]

Food allergy is defined as an immunologically mediated adverse reaction against dietary antigens. The immaturity of the immune system and the gastrointestinal barrier may explain the peak prevalence of food allergies in infancy.[44] In food allergy, intestinal inflammation[45] and disturbances in intestinal permeability[46] and antigen transfer[47] occur when an allergen comes into contact with the intestinal mucosa. During dietary elimination of the antigen, the barrier and transfer functions of the mucosa are normal.[45–47] It has therefore been concluded that impairment of the intestine's function is secondary to an abnormal intestinal immune response to the offending antigens.

Atopic dermatitis is a common and complex, chronically relapsing skin disorder of infancy and childhood. Hereditary predisposition is an important denominator of atopic dermatitis, and hypersensitivity reactions contribute the expression of this predisposition.[48] The relationship between environmental allergens and exacerbation of atopic dermatitis is particularly apparent in infancy, so that dietary antigens predominate and allergic reactions to foods are common.[44]

The mechanisms of the immune-enhancing effect of *Lactobacillus* GG are not entirely understood, but they may relate to antigen transport in the intestinal mucosa. Therefore, the effect of *Lactobacillus* GG on the gut mucosal barrier was investigated in a suckling rat model. Rat pups were divided into three experimental feeding groups to receive a daily gavage of cow milk, or *Lactobacillus* GG with cow's milk, while controls were gavaged with water. At 21 days, the absorption of horseradish peroxidase across patch-free jejunal segments and segments containing Peyer's patches was studied in Using chambers. Gut immune response was indirectly monitored by the ELISPOT method of αASC to β-lactoglobulin. Prolonged cow milk challenge increased macromolecular absorption, whereas *Lactobacillus* GG stabilized the mucosal barrier, with a concomitant enhancement of antigen-specific immune defense and proportional transport across Peyer's patches. These results indicate that there is a link between stabilization of nonspecific antigen absorption and enhancement of the antigen-specific immune response. They further suggest that the route of antigen absorption is an important determinant of the subsequent immune response to the antigen.[38]

Safety Assessment of *Lactobacillus* Strain GG (ATCC 53103)

Lactobacillus strain GG (ATCC 53103) is one of the most extensively studied probiotic lactic acid bacteria strains. Its safety assessment covers traditional toxicity tests and studies on the safety of the strain in both in vivo and in vitro conditions. Several studies in different clinical conditions and human volunteer studies, as well as epidemiological surveillance, indicate that the strain is safe for human consumption even in large amounts. At present *Lactobacillus* GG has the most extensive safety assessment record of any probiotic strain.

Before their incorporation into products, probiotic strains of lactic acid bacteria should be carefully assessed and tested for the safety and efficacy of their proposed use. As yet, no general guidelines exist for the safety testing of probiotics. However, outlines have been proposed for safety assessment of lactic acid bacteria.[49,50] Different aspects of the safety of probiotic bacteria can be assessed using a panel of in vitro methods, animal models, and human subjects with epidemiological studies following large-scale human exposure via food products.

Most dairy strains of lactic acid bacteria have a long history of safe use, and most are considered commensal microorganisms with no pathogenic potential. Lactic acid bacteria have a ubiquitous presence in intestinal epithelium and the human gastrointestinal tract, and their traditional use in fermented foods and dairy products without significant problems attests to their safety. Members of the genus *Lactobacillus* are most commonly given "safe" or "generally regarded as safe" status. Members of the genera *Streptococcus* and *Enterococcus* contain many opportunistic pathogens and have been indicated as vehicles in the transfer of antibiotic resistance. Thus, enterococci and streptococci, excluding *Streptococcus thermophilus,* are not recommended for probiotic use.

The safety of *Lactobacillus* GG (ATCC 53103) has been verified using in vitro methods, animal models, and human subjects. No general guidelines exist for the safety testing of probiotics. Safety assessment procedures and recommendations reported earlier[50–52] have been followed in the studies with *Lactobacillus* GG.

In Vitro Studies

The local effects of lactic acid bacteria on the intestine are commonly measured by their in vitro ability to adhere to human intestinal cell lines and to degrade protective intestinal mucus. These tests provide an indirect measure of the potential of lactic acid bacteria to invade intestinal cells and to damage the protective glycoproteins of the intestinal mucus.

A large number of adhesion studies have been conducted with different strains of lactic acid bacteria, most often using the Caco-2 cell line as a model. *Lactobacillus* GG has shown no invasive properties in this test system, even though the strain is in most studies strongly adherent to this cell line.[53,54]

Probiotic strains that do not degrade intestinal mucus or its glycoproteins are likely to be noninvasive. In a recent study, commercial probiotic strains including *Lactobacillus* GG were shown to be inactive in mucosal degradation injury.[55] In earlier studies, some fecal bifidobacteria were found to degrade mucus.[56]

Production of antimicrobial compounds and inhibition of pathogen growth by lactic acid bacteria has been assessed in vitro. Data from these tests support the safety of *Lactobacillus* GG and indicate that the strain decreases intestinal pH and reduces the numbers of pathogenic bacteria in the intestinal tract, thus protecting the host.[53,54,57]

Lactobacillus isolates from cases of infective endocarditis have some properties in common, including platelet aggregation, binding of fibronectin and fibrinogen, and production of glycosidases and proteases, which are postulated as factors in the pathogenesis of endocarditis. A recent report has verified that platelet aggregation is not demonstrated with *Lactobacillus* GG, although some laboratory strains carry these properties.[58] Also, assessment of clinical strains has shown that platelet aggregation and adhesion properties may not be common virulence factors.[59]

Animal Studies

To further explore the acute oral toxicity of viable lactic acid bacteria, an acute toxicity study was conducted on three different strains of bacteria. *Lactobacillus* strain GG (ATCC 53103) was used as a test compound in such studies.[52,60]

There were no treatment-related deaths or treatment-related adverse effects in any of the groups, and no treatment-related signs of toxicity were observed. Although extrapolation of oral acute toxicity values from animals to humans has limited value, the levels observed in this study would correspond to a dose of more than 420 g of washed, freeze-dried bacteria for a 70-kg human. The normal daily intake from fermented milks varies from 10^9 to 10^{10} cfu/mL (usually 10^6 to 10^9 cfu/mL in fermented milks), corresponding to 150 to 500 g of yogurt with live or viable cultures, while the lactic acid bacteria and propionic acid bacteria content of Swiss cheese and Edam cheese varies in the range of 10^6 to 10^7 cfu/g of cheese. Thus, with an average daily consumption of 20 g of cheese and 400 g of yogurt products, the intake of lactic acid bacteria would be on the order of 10^{10} viable bacteria, or about 0.5 to 2 g of freeze-dried cells.[52]

In lethally irradiated mice, *Lactobacillus* GG did not cause bacteremia. Rather, oral *Lactobacillus* GG intake was reported to prolong survival.[61] In another study, *Lactobacillus* GG was observed to slow down the development of dimethyl hydrazine–induced colon tumors in rats on a high-fat diet.[62] Similarly, *Lactobacillus* GG was shown to decrease alcohol toxicity in mice, reducing plasma endotoxin levels.[63] Animal studies on immunocompromised mice have shown that *Lactobacillus* GG has no adverse effects on this animal model either.[64,65]

A large amount of data from clinical trials or studies in human volunteers also attests to the safety of *Lactobacillus* GG. These studies have included short-term trials in more than 2000 healthy normal volunteers; studies on prevention and treatment of acute diarrhea in premature infants[66] and infants;[67] studies of older children with diarrhea;[35] studies on immune effects;[30] and studies of patients with severe intestinal infections.

All available data indicate that no harmful effects have been observed in controlled clinical studies with *Lactobacillus* GG. On the contrary, during treatment of intestinal infections, beneficial effects have been observed, including stabilization of the gut mucosal barrier, prevention of diarrhea, and amelioration of infant and antibiotic-associated diarrhea and intestinal inflammation.

Postmarketing Surveillance

Case reports from the literature of lactic acid bacteria in association with clinical infection in humans have recently been analyzed in two reviews.[68,69] Both reviews conclude that considering their widespread consumption, lactic acid bacteria appear to have very low pathogenic potential. Two recent Finnish studies[69,70] confirm that the number of infections associated with lactic acid bacteria is small. In the first study, genetic methods (16S rRNA) were used to characterize and identify lactic acid bacteria isolated from blood cultures of bacteremic patients in southern Finland.[69] The results showed that a newly introduced probiotic strain in fermented milks was not associated with infections and that the total number of infections caused by lactobacilli was extremely low. In a further study, lactobacilli isolated from bacteremic patients between 1989 and 1994 were compared to common dairy or pharmaceutical strains.[70] From a total of 5192 blood cultures of patients with septicemia, 12 were positive for lactobacilli, an incidence of 0.23%. None of the clinical cases could be related to lactobacilli strains used by the dairy industry. In both studies, patients with lactic acid bacteria bacteremia had other severe underlying illnesses.

Conclusions on the Safety of *Lactobacillus* GG

Before *Lactobacillus* GG was incorporated into food products, its safety was carefully assessed, and the strain was tested for the safety and efficacy of the proposed use. The reported procedures and follow-up have been

- assessment of the intrinsic properties of *Lactobacillus* GG, using, for example, adhesion factors, antibiotic resistance (no plasmids, thus no plasmid transfer potential), and enzyme profile in comparison to dairy and intestinal strains
- assessment of the safety and effects of the metabolic products of *Lactobacillus* GG
- assessment of the acute toxicity of ingestion of extremely large amounts of *Lactobacillus* GG
- estimation of the in vitro invasiveness of *Lactobacillus* GG, using cell lines and human intestinal mucus degradation, and assessment of infectivity in animal models, including lethally irradiated animals
- determination of the efficacy of ingested *Lactobacillus* GG, as measured by dose response (minimum and maximum dose required, consequent health effects), and assessment of the effect of massive probiotic doses on the composition of human intestinal microflora
- careful assessment of side effects, with none reported during human volunteer studies or clinical studies in various disease-specific states
- epidemiological surveillance of people ingesting large amounts of *Lactobacillus* GG– or *Lactobacillus* GG–fermented dairy products

In conclusion, it is clear that *Lactobacillus* GG is safe and useful in its intended uses as a functional food or ingredient. All safety and toxicity studies support the safety of *Lactobacillus* GG, and it is currently one of the best documented strains as far as safety and efficacy are concerned (Table 15–1). The strain was also approved by Japanese functional food authorities (in 1996) and by the United Kingdom Advisory Committee on Novel Foods and Processes (in 1992).

Table 15–1 Examples of Studies Reporting on Efficacy and Safety of *Lactobacillus* GG

Type of Study	Reported Efficacy and Safety
In Vitro Studies	
Invasiveness	Adhesion but no invasion, prevention of pathogen invasion
Mucus degradation	Mucus adhesion, but no mucus degradation observed
Antimicrobial production	Prevention of pathogen growth, no effects on other lactobacilli or bifidobacteria
Animal Studies	
DMH-induced tumor formation	Tumor development delayed in animals
Alcohol-induced liver damage	Prevention and alleviation of liver damage in experimental alcoholic liver disease
Clinical Studies	
Prevention of traveler's diarrhea	Prevention of traveler's diarrhea, no side effects
Antibiotic-associated diarrhea	Prevention of antibiotic-associated diarrhea and reduced duration of diarrhea, no harmful effects or side effects
Infant diarrhea, rotavirus diarrhea	Prevention and treatment of acute diarrhea and rotavirus diarrhea, enhancement of natural immune system, no side effects or harmful changes
Colonization of preterm infants and infants	Colonization observed, balancing intestinal microflora, no harmful effects or side effects
Treatment of food allergy in infants	Alleviation of symptoms of food allergy in infants, no side effects
Epidemiological studies	
Infection potential/genetic methods	No *Lactobacillus* GG–related infections in normal or immune-compromised subjects

PREBIOTIC FRUCTO-OLIGOSACCHARIDES

Fructo-Oligosaccharides as Prebiotics

Among the components likely to be used in functional foods, prebiotics and probiotics have interesting properties, and some are already recognized and used as food ingredients. A prebiotic has been defined as "a non-digestible food ingredient that beneficially affects the host by selectively stimulating the growth and/or activity of one or a limited number of bacteria in the colon, and thus improves host health."[49]

To be classified as a prebiotic, a food ingredient (1) should be neither hydrolyzed nor absorbed in the upper part of the gastrointestinal tract and (2) should selectively stimulate the growth of potentially beneficial bacteria in the colon; in addition, (3) it may repress pathogen growth and virulence and induce systemic effects that can be beneficial to health.

The first criterion for classifying a food component as prebiotic is resistance to hydrolysis, digestion, and absorption in the upper part of the intestinal tract.[71] Resistance to hydrolysis and digestion in vitro, in breath hydrogen tests, and in experiments with rats, healthy volunteers, and ileostomy patients in vivo have all confirmed that fructo-oligosaccharides do indeed meet the criterion and can be classified as nondigestible oligosaccharides.[71–73]

The second criterion for classification as a prebiotic is colonic fermentation.[71] That fructo-oligosaccharides are quantitatively fermented by the microbes of the colon is supported by many experiments in vivo both in animals and in humans but also by in vitro batch cultures using human fecal slurries as inocula. A health-promoting consequence of such large fermentation is increased fecal biomass and consequently stool weight and/or stool frequency.[72,73] The increase in fecal wet weight is on the order of 1.0 to 1.5 g/g of prebiotic and is equivalent to the increase reported to soluble fibers such as pectin or guar gum.[73]

Colonic fermentation leading to a selective stimulation of the growth of one population of potentially health-promoting bacteria, namely bifidobacteria, has been reported for the fructo-oligosaccharides obtained by enzymatic synthesis and for oligofructose and inulin.[72] In addition, the experiments reported by Gibson et al[72] demonstrate that the overgrowth of bifidobacteria is accompanied by a reduction in the number of other populations like bacteroides, clostridia, and fusobacteria, thereby leading to a major modification in the composition of the colonic microflora.

The prebiotics that are identified today and have served to introduce the concept[71] are carbohydrates that resist digestion but are quantitatively fermented in the colon. They are obtained either by hot water extraction from plants and partial hydrolysis of the extracted molecules or by enzymatic (osyl transferases) synthesis from a disaccharide. The prebiotics that are available in Europe at present or that are being developed belong to a group characterized by the major osyl monomer of which they are composed, namely the fructosyl-type prebiotics or fructo-oligosaccharides.

Nutritional Effects of Prebiotics in the Gastrointestinal Tract

Colonic fermentation produces short-chain fatty acids (SCFAs), mainly acetate, propionate, and butyrate, and lowers the colonic pH. The real ratio of the different SCFAs produced *in vivo* in the human colon is still unclear. The fermentation of inulin, however, has been reported to produce four times more butyrate than wheat and oat brans or pea and carrot fibers, an effect that, if confirmed, could be of great interest, since it is hypothesized that butyrate could play a role in maintaining epithelial cell function and in preventing colon carcinogenesis.[73]

Through their colonic fermentation, carbohydrate-type prebiotics have part of their energy salvaged via the absorption of SCFAs. As a consequence, it is generally accepted that their digestible energy is on the order of 1 to 2 kcal/g (4–9 KJ/g). This property justifies their use as bulking ingredients and sugar substitutes, and, in the case of inulin, as a fat replacer.

Safety of Fructo-Oligosaccharides

Prebiotics are basically macronutrients because they are ingested in relatively high amounts. Their safety evaluation cannot be performed in the same way as food additives are evaluated but can be done on a case-by-case basis.[2]

When fructo-oligosaccharides have been evaluated by European health authorities, these prebiotics have been recognized as safe because inulin and oligofructose are found as natural components in a large variety of fruits, cereals, and vegetables and consequently are consumed regularly as part of the normal diet. For many years, there has been widespread and common knowledge of their high natural occurrence and consumption as human and animal foods. Their average consumption in the normal diet has been evaluated at several grams per day.[74]

Studies conducted to evaluate potential toxic effects of oligofructose in animals and humans have revealed no significant adverse effects. The only side effects noted have been the occurrence of flatulence and soft stools after ingestion of large quantities. Excessive consumption of nondigestible oligosaccharides may indeed cause intestinal discomfort such as flatus, bloating, colic, borborygmi, or even loose stools at high doses.[49] These effects are comparable to the ones observed with other dietary fibers. The practical use levels of prebiotics (typically 2–3 g per serving) are far below the amounts at which intestinal discomfort and/or laxative effects occur.

CONCLUSIONS

The studies presented here on three different functional food ingredients provide further confirmation of their safety and efficacy. For each component, research efforts had a new area of functional food research to explore when the studies were initiated. However, in all cases, good clinical and nutritional studies have been conducted, and the safety and efficacy of the functional ingredients have been demonstrated. In each case, the safety has been assessed in relation to the target groups and target functions of the particular food or ingredient. Thus, these examples provide a scientific basis for the use of functional foods and backing for "health and nutritional" claims on such products.

REFERENCES

1. Diplock AT, Agget P, Ashwell M, Bornet F, Fern E, Roberfroid M. Scientific concepts of functional foods in Europe: consensus documents. *Br J Nutr.* 1999;81(suppl 1):S1–S27.

2. Jonas DA, Antignac E, Antoine JM, et al. The safety assessment of novel foods: guidelines prepared by ILSI Europe Novel Food Task Force. *Food Chem Toxicol.* 1996;34:931–940.

3. Huggett A, Verschuren P. The safety assurance of functional foods. *Nutr Rev.* 1996;54(suppl 2):132–140.

4. ILSI. *Addition of Nutrients to Foods: Nutritional and Safety Considerations.* 1990.

5. Tomlinson N. The EC Novel Foods Regulation: a UK perspective. *Food Additives Contaminants.* 1998;15:1–9.

6. Clydesdale F. A proposal for the establishment of scientific criteria for health claims for functional foods. *Nutr Rev.* 1997;55:413–422.

7. Scheuplein RJ. Information needed to support hazard identification and risk assessment of toxic substances. *Toxicol Lett.* 1995;79:23–28.

8. Bellisle F, Diplock A, Hornstra G, et al. Functional food science in Europe. *Br J Nutr.* 1998;80(suppl 1):S1–S193.

9. Washüttl S, Riederer P, Bancer E. Qualitative and quantitative study of sugar alcohols in several foods. *J Food Sci.* 1973;38:1262–1264.

10. Scheinin A, Mäkinen KK. Turku Sugar Studies I–XXI. *Acta Odontol Scand.* 1975;33(suppl 70):5–348.

11. Mäkinen KK, Hujoel P, Bennett C, Isotupa K, Mäkinen P, Allen P. Polyol chewing gums and caries rates in primary dentition: a 24-month cohort study. *Caries Res.* 1996;30:408–417.

12. Mäkinen KK, Chen C, Mäkinen P, et al. Properties of whole saliva and dental plaque in relation to 40-month consumption of chewing gums containing xylitol, sorbitol, or sucrose. *Caries Res.* 1996;30:180–188.

13. Mäkinen KK, Mäkinen PL, Pape HR, et al. Conclusion and review of the Michigan xylitol program (1986–1995) for the prevention of dental caries. *Int Dent J.* 1996;46:22–34.

14. Mäkinen KK. Review of xylitol studies. *Oral Dis.* 1998;4:226–230.

15. Virtanen J, Bloigu R, Larmas M. Timing of first restorations before, during and after preventive xylitol trial. *Acta Odontol Scand.* 1996;54:211–216.

16. Edgar WM. Sugar substitutes, chewing gum and dental caries—a review. *Br Dent J.* 1998;84:29–32.

17. Naaber P, Lehto E, Salminen S, Mikelsaar M. Inhibition of adhesion of *Clostridium difficile* to Caco-2 cells. *FEMS Immunol Med Microbiol.* 1996;14:205–209.

18. Naaber P, Mikelsaar S, Salminen S, Mikelsaar M. Bacterial translocation, intestinal microflora and morphological changes of intestinal mucosa in experimental models of *Clostridium difficile* infection. *J Med Microbiol.* 1998;47:1–8.

19. Salminen S, Salminen E, Bridges JW, Marks V, Koivistoinen P. Gut microflora interactions with xylitol in rats, mice and man. *Food Chem Toxicol.* 1985;23:985–990.

20. Kontiokari T, Uhari M, Koskela M. Effect of xylitol on growth of nasopharyngeal bacteria in vitro. *Antimicrob Agents Chemother.* 1995;39:1820–1823.

21. Mäkinen KK, Scheinin A. The administration of the trial and control of the dietary regimen. *Acta Odontol Scand.* 1975;33(suppl 70):105–127.

22. Uhari M, Kontiokari T, Koskela M, Niemela M. Xylitol chewing gum in prevention of acute otitis media: double blind randomised trial. *Br Med J.* 1996;313:1180–1184.

23. Uhari M, Kontiokari T, Niemelä M. A novel use of xylitol sugar in preventing acute otitis media. *Pediatrics.* 1998;102:879–884.

24. World Health Organization. Food Additive Series XXI. WHO; 1987.

25. Goldin B, Gorbach S, Saxelin M, Barakat S, Gualteri L, Salminen S. Survival of *Lactobacillus* species (strain GG) in the human gastrointestinal tract. *Dig Dis Sci.* 1992;37:121–128.

26. Salminen S. Functional dairy foods with *Lactobacillus* GG. *Nutr Rev.* 1996;54(suppl 2):99–101.

27. Raza S, Graham SM, Allen SJ, Sultana S, Cuevas L, Hart CA. *Lactobacillus* GG promotes recovery from acute nonbloody diarrhea in Pakistan. *Pediatr Infect Dis J.* 1995;14:107–111.

28. Sheen P, Oberhelman RA, Gilman RH, Cabrera L, Verastegui M, Madico G. Short report: a placebo-controlled study of *Lactobacillus* GG colonization in one-to-three-year-old Peruvian children. *Am J Trop Med Hyg.* 1995;52:389–392.

29. Kaila M, Isolauri E, Soppi E, Virtanen E, Laine S, Arvilommi H. Enhancement of the circulating antibody secreting cell response in human diarrhea by a human *Lactobacillus* strain. *Pediatr Res.* 1992;32:141–144.

30. Majamaa H, Isolauri E, Saxelin M, Vesikari T. Lactic acid bacteria in the treatment of acute rotavirus gastroenteritis. *J Pediatr Gastroenterol Nutr.* 1995;20:333–338.

31. Rabalais GP. Recent advances in the prevention and treatment of diarrheal diseases. *Curr Opin Infect Dis.* 1996;9:210–213.

32. Pant AR, Graham SM, Allen SJ, Harikul S, Sabchareon A, Cuevas L, Hart CA. *Lactobacillus* GG and acute diarrhoea in young children in the tropics. *J Tropic Pediatrics.* 1996;42:162–165.

33. Guandalini S. Probiotics in the treatment of diarrheal diseases in children. *Gastroenterol Int.* 1998;11(suppl)1:87–90.

34. Gorbach SL. Lactic acid bacteria and human health. *Ann Med.* 1990;22:37–41.

35. Isolauri E, Juntunen M, Rautanen T, Sillanaukee P, Koivula T. A human *Lactobacillus* strain (*Lactobacillus casei* sp. strain GG) promotes recovery from acute diarrhea in children. *Pediatrics.* 1991;88:90–97.

36. Isolauri E, Kaila M, Arvola T, et al. Diet during rotavirus enteritis affects jejunal permeability to macromolecules in suckling rats. *Pediatr Res.* 1993;33:548–553.

37. Isolauri E, Majamaa H, Arvola T, Rantala I, Virtanen E, Arvilommi H. *Lactobacillus casei* strain GG reverses increased intestinal permeability induced by cow milk in suckling rats. *Gastroenterology.* 1993;105:1643–1650.

38. Siitonen S, Vapaatalo H, Salminen S, Gordin A, Saxelin M, Wikberg R, Kirkkola A-M. Effect of *Lactobacillus* GG yoghurt in prevention of antibiotic associated diarrhea. *Ann Med.* 1990;22:57–60.

39. Young RJ, Whitney DB, Hanner TL, Antonson DL, Lupo JV, Vanderhoof JA. Prevention of antibiotic-associated diarrhea using *Lactobacillus* GG. *Gastroenterol Int.* 1998;11(suppl 1):86.

40. Oksanen P, Salminen S, Saxelin M, et al. Prevention of travellers' diarrhoea by *Lactobacillus* GG. *Ann Med.* 1990;22:53–56.

41. Hilton E, Kolakowski P, Smith M, Singer C. Efficacy of *Lactobacillus* GG as a diarrheal preventative in travelers. *J Travel Med.* 1997;4:41–43.

42. Isolauri E. Immunological effects of probiotics and their clinical applications in pediatric patients. *Gastroenterol Int.* 1998;11(supp 1):83–85.

43. Pelto L, Salminen S, Lilius E-M, Nuutila J, Isolauri E. Milk hypersensitivity—key to poorly defined gastrointestinal symptoms in adults. *Allergy.* 1998;53:307–310.

44. Isolauri E, Turjanmaa K. Combined skin pric and patch testing enhances identification of food allergy in infants with atopic dermatitis. *J Allergy Clin Immunol.* 1996;97:9–15.

45. Majamaa H, Isolauri E. Evaluation of the gut mucosal barrier: evidence for increased antigen transfer in children with atopic eczema. *J Allergy Clin Immunol.* 1996;97:181–187.

46. Jalonen T. Identical intestinal permeability changes in children with different clinical manifestations of cow's milk allergy. *J Allergy Clin Immunol.* 1991;88:737–742.

47. Heyman M, Desjeux JF. Significance of intestinal protein transport. *J Pediatr Gastroenterol Nutr.* 1992;15:48–57.

48. Sampson HA, McCaskill CC. Food hypersensitivity and atopic dermatitis: evaluation of 113 patients. *J Pediatr* 1985;107:669–675.

49. Salminen S, von Wright A, Morelli L, et al. Demonstration of safety of probiotics—a review. *Int J Food Microbiol.* 1998;44:93–106.

50. Salminen S, von Wright A. Current human probiotics—safety assured? *Microb Ecol Health Dis.* 1998;10:68–77.

51. Donohue DC, Salminen S. Safety of probiotic bacteria. *Asia Pacific J Clin Nutr.* 1996;5:25–28.

52. Donohue D, Deighton M, Ahokas JT, Salminen S. Toxicity of lactic acid bacteria. In: Salminen S, von Wright A, eds. *Lactic Acid Bacteria.* New York, NY: Marcel Dekker, Inc; 1993:307–313.

53. Coconnier MH, Klaenhammer TR, Kerneis S, Bernet MF, Servin A. Protein mediated adhesion of *Lactobacillus acidophilus* BG2F04 on human enterocyte and mucus secreting cell lines in culture. *Appl Environ Microbiol.* 1992; 58:2034–2039.

54. Elo S, Saxelin M, Salminen S. Attachment of *Lactobacillus casei* strain GG to human colon carcinoma cell line Caco-2: comparison with other dairy strains. *Lett Appl Microbiol.* 1991;13:154–156.

55. Ruseler van Embden J, van Lieshout L, Gosselink M, Marteau P. Inability of *Lactobacillus casei* strain GG, *L. acidophilus* and *Bifidobacterium bifidum* to degrade intestinal mucus glycoproteins. *Scand J Gastroenterol.* 1995;30:675–680.

56. Ruseler van Embden J, van Lieshout L, Marteau P. No degradation of intestinal mucus glycoproteins by *Lactobacillus* strain GG. *Microecol Ther.* 1995;25:304–309.

57. Silva M, Jacobus NV, Deneke C, Gorbach SL. Antimicrobial substance from a human *Lactobacillus* strain. *Antimicrob Agents Chemother.* 1987;31:1231–1233.

58. Korpela R, Moilanen E, Saxelin M, Vapaatalo H. *Lactobacillus* GG (ATCC 53103) and platelet aggregation in vitro. *Int J Food Microbiol.* 1997;37:83–86.

59. Kirjavainen PV, Crittenden RG, Donohue DC, et al. Adhesion and platelet aggregation properties of bacteremia-associated lactobacilli. *Infect Immun.* 1999;67:2653–2655.

60. Donohue D, Deighton M, Salminen S, Marteau P. Safety of probiotic bacteria. In: Salminen S, von Wright A, eds. *Lactic Acid Bacteria: Microbiology and Functional Effects.* New York, NY: Marcel Dekker, Inc; 1998.

61. Dong MY, Chang TW, Gorbach SL. Effect of feeding *Lactobacillus* GG on lethal irradiation in mice. *Diagn Microbiol Infect Dis.* 1987;7:1–7.

62. Goldin B, Gualtieri L, Moore R. The effect of *Lactobacillus* GG on the initiation and promotion of dimethylhydrazine-induced intestinal tumors in the rat. *Nutr Cancer.* 1996;25:197–204.

63. Nanji AA, Khettry U, Sadrzadeh SMH. *Lactobacillus* feeding reduces endotoxemia and severity of experimental alcoholic liver (disease). *Proc Soc Exp Biol Med.* 1994;205:243–247.

64. Wagner RD, Pierson C, Warner T, et al. Biotherapeutic effects of probiotic bacteria on candidiasis in immunodeficient mice. *Infect Immun.* 1997;65:4165–4172.

65. Wagner RD, Warner T, Roberts L, Farmer J, Balish E. Colonization of congenitally immunodeficient mice with probiotic bacteria. *Infect Immun.* 1997;65:3345–3351.

66. Millar MR, Bacon C, Smith SL, Walker V, Hall MA. Enteral feeding of premature infants with *Lactobacillus* GG. *Arch Dis Child.* 1993;69:483–487.

67. Sepp E, Mikelsaar M, Salminen S. Effect of administration of *Lactobacillus casei* strain GG on the gastrointestinal microbiota of newborns. *Microb Ecol Health Dis.* 1993;6:309–314.

68. Aguirre M, Collins MD. Lactic acid bacteria and human clinical infection. *J Appl Bacteriol.* 1993;75:95–107.

69. Saxelin M, Rautelin H, Salminen S, Mäkelä P. The safety of commercial products with viable *Lactobacillus* strains. *Infect Dis Clin Pract.* 1996;5:331–335.

70. Saxelin M, Rautelin H, Chassy B, Gorbach SL, Salminen S, Mäkelä P. Lactobacilli and septic infections in southern Finland during 1989–1992. *Clin Infect Dis.* 1996;22:564–566.

71. Gibson GR, Roberfroid MB. Dietary modulation of the human colonic microbiota: introducing the concept of prebiotics. *J Nutr.* 1995;125:1401–1412.

72. Gibson GR, Beatty ER, Wang X, Cummings J. Selective stimulation of bifidobacteria in the human colon by oligofructose and inulin. *Gastroenterology.* 1995;108:975–982.

73. Roberfroid MB. Functional effects of food components and the gastrointestinal system: chicory fructo-oligosaccharides. *Nutr Rev.* 1996;54:517–520.

74. Van Loo J, Coussement P, De Leenheer L, Hoebregs H, Smits G. On the presence of inulin and oligofructose as natural ingredients in the Western diet. *Crit Rev Food Sci Nutr.* 1995;35:525–552.

PART VI

Regulatory Issues

CHAPTER 16

U.S. Government Regulation of Food with Claims for Special Physiological Value

Peter Barton Hutt

Government regulation of food and drugs extends back throughout recorded history.[1] This chapter focuses on the US regulatory status of food for which claims are made for special physiological value under the Federal Food, Drug, and Cosmetic Act of 1938 (FD&C Act or 1938 Act)[2] and the Federal Trade Commission Act of 1914 (FTC Act).[3] U.S. regulatory policy regarding these products can be approached either chronologically or analytically. Neither approach, by itself, is fully explanatory. Accordingly, this chapter first reviews the chronological development of federal law and then analyzes current regulatory requirements and prohibitions.

TERMINOLOGY

In discussing regulation of food under the FD&C Act, it is extremely important to use the correct statutory and regulatory terminology. The term *functional food* and numerous similar terms have no legal or regulatory meaning and no other accepted definition and therefore are not used in this chapter. These are marketing terms that may well serve to convey the nature of the new public interest in the nutrient content and physiological value of food, but they have no place in discussion of the regulation of food. The correct statutory and regulatory terminology is used throughout this chapter and in the summary table set forth in Appendix 16–A.

HISTORICAL OVERVIEW

Although federal legislation to regulate food and drugs was continually considered in the United States from 1879 on, the narrow view of the Supreme Court on federal jurisdiction over interstate commerce[4] forced Congress to delay enactment of national legislation until the first decade of the 20th century.

The Federal Food and Drug Act of 1906

The Federal Food and Drug Act of 1906[5] defined *food* to include all articles used for food and defined *drug* to include all substances intended to be used for the cure, mitigation, or prevention of disease. The jurisdiction of the 1906 Act was limited to the package or label of the article. It prohibited the adulteration or misbranding of any food or drug in interstate commerce. There were no requirements for affirmative labeling of food and no provisions specifically dealing with nutrition or any other physiological properties of food.

The Federal Trade Commission Act of 1914

The FTC Act broadly declares all unfair methods of competition and unfair or deceptive acts or practices affecting commerce to be unlawful. The FTC Act applies to all marketed products, including food and drugs, and to all methods of competition, including representations made in the label, labeling, and advertising for any product. Thus, the broad jurisdiction of the FTC under the FTC Act overlaps the narrower jurisdiction of the federal laws that specifically regulate food and drugs.

The Wheeler-Lea Act of 1938

Not long after enactment of the 1906 Act, the U.S. Department of Agriculture (USDA), which was responsible for implementation of the law at that time,[6] concluded that the statute should be amended in a number of important respects.[7] When the Roosevelt administration introduced a bill in 1933 to replace the 1906 Act,[8] the legislation would have transferred jurisdiction over advertising of food and drugs from the FTC to the Food and Drug Administration (FDA), which had become responsible for implementing the law in 1930. The FTC strongly resisted this approach, and a classic intergovernmental turf battle ensued. After five years of bitter fighting between the two agencies, Congress resolved the dispute by granting primary jurisdiction over the advertising of food and drugs to the FTC under the Wheeler-Lea Act of 1938,[9] which amended the FTC Act, and primary jurisdiction over the label and labeling of these products to FDA under the FD&C Act, which replaced the 1906 Act. Under a September 1971 Memorandum of Understanding,[10] FDA has primary responsibility for labeling and the FTC has primary responsibilities for advertising.[9]

The Federal Food, Drug, and Cosmetic Act of 1938

The FD&C Act repealed and replaced the 1906 Act. It made a number of important changes in the way that food and drugs are regulated. For present purposes, the most important provisions of the 1938 Act may be briefly summarized as follows:

- Congress determined that it was the representations made for the product, not its inherent properties, that would determine its proper classification as a food or a drug.[11]
- The definition of a food was unchanged from the 1906 Act.[12]
- The definition of a drug was expanded to include any article (except food) intended to affect the structure or any function of the human body.[13]
- Congress determined that if an article was represented both as a food and to prevent or treat disease it would be classified as both a food and a drug but that if it was represented to affect the structure or function of the human body it would remain solely a food.[14]
- As in the 1906 Act, the adulteration or misbranding of a food or a drug was declared to be unlawful.[15]
- As a result of the Elixir Sulfanilamide drug disaster that occurred in the fall of 1937, any "new drug" (i.e., a drug not generally recognized as safe) was prohibited from marketing until a new drug application (NDA) was submitted to and reviewed by FDA.[16]
- Both food and drugs were required to bear specified affirmative labeling. Food was required to show the name of the food, the ingredients in descending order of predominance, the net quantity of contents, and the name and address of the manufacturer.[17]
- FDA was given discretionary authority to issue binding standards of identity for any food.[18]
- FDA was given discretionary authority to promulgate regulation governing the information concerning vitamin, mineral, or other dietary properties of any food that is "represented for special dietary uses" in order fully to inform purchasers as to its value for those uses.[19]

These basic provisions of the 1938 Act have not been changed in the intervening sixty years. Congress has amended the FD&C Act on more than 100 occasions, however, with the result that what was originally a short, simple, and direct statute has become a lengthy, complex, and highly obscure statute. As Congress has added new provisions to the FD&C Act, for the most part it has failed to make any attempt to reconcile them with those that already exist in the law. Provisions that were once of substantial importance have become either secondary or completely obsolete, even though they remain unchanged. The following chronology traces the enactment of the amendments to the FD&C Act since 1938 that are of direct importance to the subject matter of this chapter.

The Vitamin-Mineral Amendments of 1976

In response to what FDA concluded to be unwarranted formulation and labeling claims for vitamin-mineral products, beginning in the early 1950s the agency sought for two decades to bring enforcement action, and then to adopt regulations, designed to control these practices.[20] Following the promulgation of final regulations intended to limit the formulation and labeling of vitamin-mineral products in 1973,[21,22] the dietary supplement industry attacked them in two different ways: by appeal to the courts and to Congress. The industry prevailed in both places. Part of the regulations was overruled by a U.S. Court of Appeals[23,24] and part was overruled by Congress in the Vitamin-Mineral Amendments of 1976.[25] As a result, FDA withdrew all of the regulations[26] and abandoned this approach.

The Vitamin-Mineral Amendments of 1976 prohibited FDA from imposing maximum limits on the levels of safe vitamins and minerals that can be used in dietary supplements, from classifying vitamins or minerals as drugs solely because a particular level is exceeded, and from limiting the combination or number of any safe vitamin or mineral in a dietary supplement. At that time, it was the most humiliating legislative defeat in the history of FDA.

The Infant Formula Act of 1980

FDA regulated infant formula as special dietary food for forty years under Section 403(j) of the FD&C Act. In mid-1979, three soy protein–based infant formulas were recalled because of an inadequate level of a nutrient, resulting in harm to very young children. Congress responded by enacting the Infant Formula Act of 1980,[27] which directly established in the FD&C Act the minimum and in some cases the maximum levels of every nutrient that must be included in an infant formula.[28] FDA has used its authority under this statute to establish detailed requirements for all aspects of infant formula content, manufacturing, and claims.[29]

The Orphan Drug Act Amendments of 1988

Beginning in the late 1960s, FDA administratively recognized a new category of medical food products labeled for the dietary management of disease that fell between the categories of food and dietary supplements on one side and drugs on the other side.[30] FDA imposed the food requirements of the FD&C Act on these products but not the drug requirements.[31] FDA officially recognized this category for the first time in exempting medical food from the requirements of nutrition labeling in 1973,[32,33] but a definition of medical food was not established by Congress until it was included as part of the Orphan Drug Act Amendments of 1988.[34] Under the Nutrition Labeling and Education Act of 1990,[35] medical food was exempted from the requirements of nutrition labeling, nutrient descriptors, and disease prevention claims.[36] Concerned that the 1988 congressional definition of medical food might be interpreted to expand this exemption too broadly, FDA adopted a narrower definition in the implementing regulations promulgated in January 1993.[37] In November 1996, FDA published an Advance Notice of Proposed Rulemaking, soliciting comments to initiate a reevaluation of its approach to the regulation of medical food.[38]

The Nutrition Labeling and Education Act of 1990

FDA initiated nutrition labeling in the early 1970s[39–41] on the basis of a provision in the FD&C Act stating that a food is misbranded if it fails to reveal facts material in the light of other representations in the labeling or advertising.[42] When FDA failed to amend nutrition labeling requirements to keep pace with developing scientific knowledge, however, Congress took the matter into its own hands and enacted the Nutrition Labeling and Education Act of 1990. Under the 1990 Act, Congress required FDA to promulgate new regulations governing nutrition labeling, nutrient descriptors, and disease prevention claims for food. After a fierce political battle, these regulations were promulgated in January 1993.[43]

The Dietary Supplement Health and Education Act of 1994

When FDA promulgated its regulations implementing the provision authorizing disease prevention claims under the Nutrition Labeling Education Act of 1990, the agency ignored a congressional invitation to establish separate standards and requirements for dietary supplements,[44] instead using the new statutory authority to deny all such claims for this category of food products,[45–47] and threatening to impose stringent food additive requirements on important dietary supplement ingredients.[48] The dietary supplement industry again displayed its formidable political power and obtained enactment of the Dietary Supplement Health and Education Act of 1994[49] over the strong objection of FDA. The 1994 Act was an even more overwhelming and humiliating defeat for FDA than the Vitamin-Mineral Amendments of 1976. The 1994 Act broadly defined dietary supplements; explicitly authorized structure-function claims and related nutritional support claims for dietary supplements; exempted the dietary ingredients in dietary supplement products from the food additive requirements of the FD&C Act and substituted more flexible food safety provisions that place the burden on FDA to demonstrate the lack of safety; and exempted literature reprints from the labeling provisions of the FD&C Act, thereby establishing substantial limitations on FDA authority over dietary supplements.

The Food and Drug Administration Modernization Act of 1997

Both the conventional food industry and the dietary supplement industry concluded that the limitations on disease prevention claims for food established by FDA under the Nutrition Labeling and Education Act of 1990 were far too severe. The industry was successful in obtaining a provision as part of the Food and Drug Administration Modernization Act of 1997[50] expanding permissible disease prevention claims for food products to include authoritative statements by other federal health agencies and the National Academy of Sciences.[51] A parallel provision was also enacted for nutrient descriptors.[52] Both of these provisions were enacted by Congress over the strong objection of FDA.

TEN STATUTORY FOOD CATEGORIES UNDER THE FD&C ACT

The chronological development of the FD&C Act tells part, but not all, of the story. It is equally important to segregate each statutory food category under the current provisions of the FD&C Act, to analyze the specific requirements that pertain to each category, and to distinguish among the different categories for purposes of determining applicable regulatory requirements. This section of the chapter is not a comprehensive review of all labeling and formulation requirements but deals only with those aspects of FDA regulation that directly relate to claims about the special physiological value of a food.

Category No. 1: Food

The category of food—now often referred to as conventional food—is defined under the FD&C Act to include all articles used for food. After struggling with this definition, the courts have determined that food consists of articles that are consumed primarily (but not exclusively) for taste, aroma, or nutritive value.[53] Normally, claims for special physiological value would reclassify a conventional food into one of the other statutory categories. FDA has specifically provided in regulations, however, that claims that an essential nutrient will prevent a known deficiency disease may be made for conventional food without reclassifying the product either into a different food category or as a drug.[54] FDA has also permitted conventional food claims for dietary guidance that do not relate a specific food to a specific disease and general claims about good health and well-being.[55] Thus, even the conventional food category may be the subject of claims for special physiological value.

Category No. 2: Food for Special Dietary Use

This food category was of major importance between 1938 and 1972. During that time, all regulations governing the labeling of the nutrient content of food were promulgated pursuant to this statutory authority. In 1972, however, FDA abandoned the use of this provision and based its nutrition labeling on the broader misbranding provisions of the FD&C Act.[56] As FDA has established regulations governing nutrient descriptors under the Nutritional Labeling and Education Act of 1990, moreover, the comparable regulations that once were based upon this statutory authority have been revoked or transferred. At present, only hypoallergenic food, infant food, and weight control food are still subject to special dietary use regulations.[57] Thus, this category of food is rapidly becoming obsolete.

It is anomalous that the statutory phrase used to describe this category of food—"food for special dietary use"—is undoubtedly the best terminology available to cover the diverse group of products that fall within the broad class of food with claims for special physiological value. Because the special dietary use category has now become largely obsolete under the FD&C Act, however, it is doubtful, as a practical matter, that this terminology could be resurrected, even though it represents the most comprehensive and easily understandable description of these products.

Category No. 3: Food Intended To Affect the Structure or Any Function of the Human Body

The term *drug* is defined in basically two ways: an article intended to prevent or treat disease and an article (other than food) intended to affect the structure or any function of the body.[58] Quite clearly, any explicit disease claim automatically classifies a food as a drug. Because of the statutory exclusion of food from the structure-function portion of the drug definition, however, structure-function claims may lawfully be made for a food without classifying the food as a drug.

For many years, FDA took the position that any structure-function claim was an implied drug claim that would render a food illegal. Although this position was declared "untenable" by a court in the starch blocker cases,[59] the food industry is very conservative and thus generally has not pursued structure-function claims for conventional food products.

In the final regulations promulgated by FDA to implement the Nutrition Labeling and Education Act in January 1993, FDA made an initial step to bring the claims permitted for conventional food into conformity with the structure-function claims explicitly permitted for dietary supplements under the Dietary Supplement Health and Education Act. The agency stated that a structure-function claim would be permitted for a conventional food if there was no express or implied reference to any dysfunction of or damage to the body or any biological parameter that is a recognized risk factor for disease.[60] Because this discussion in the preamble had no counterpart in the regulation itself and was written so obscurely, it had very little impact at the time.

In the preamble to the FDA regulations promulgated to implement the labeling requirements of the Dietary Supplement Health and Education Act of 1994, FDA explicitly abandoned the position that it had taken earlier in the starch blocker litigation. FDA stated that it was committed to as much parity between dietary supplements and conventional food as was possible within the statute.[61,62] The agency concluded that a proper structure-function claim could be made for a conventional food, such as cranberry juice cocktail, without resulting in the product's being classified as a drug, whether the cranberry juice cocktail was marketed as a conventional food or as a dietary supplement.[63] FDA pointed out that, because Congress changed the law explicitly to permit dietary supplements to be marketed in food form, it is up to the manufacturer of a product, through labeling statements, to determine whether the article is represented as a dietary supplement or as a conventional food. Regardless of which form is taken, the same structure-function claim can be made.

FDA did propose to limit structure-function claims for conventional food to those that directly relate to the nutritive value of the food.[64] There is no statutory or other basis for such a limitation, however, and it is doubtful that this proposal is enforceable.

FDA published proposed regulations in April 1998[65] and promulgated final rules in January 2000[66] to distinguish between a permitted structure-function claim for conventional food and dietary supplements and an illegal disease claim. These regulations are discussed in greater detail in the section of this chapter relating to Category No. 10.

Finally, it should be noted that a conventional food—whether or not it is the subject of a structure-function claim—is fully subject to the food additive requirements of the FD&C Act.[67] If the identical claim is made for the identical product labeled as a dietary supplement, on the other hand, any dietary ingredients in the dietary supplement will be exempt from the food additive requirements and will instead be subject to the more flexible safety provisions of the Dietary Supplement Health and Education Act of 1994.[68] For some dietary ingredients, this distinction may be of substantial importance. On the other hand, although a dietary supplement must bear the statutory disclaimer[69] discussed below and must be the subject of a notification to FDA about any structure-function claim,[70] these two requirements do not apply when structure-function claims are made for conventional food. For some products, this may also be a significant consideration.

Category No. 4: Food Intended for the Prevention or Treatment of Disease

When Congress enacted the FD&C Act in 1938, the legislative history made clear that a disease claim was an unequivocal drug claim and that, if it were made for a food product, the result would be dual classification of the product as both a food and a drug.[71] Until 1990, there was no exception to this rule. Under the Nutrition Labeling and Education Act of 1990 and the Food and Drug Administration Modernization Act of 1997, however, Congress authorized two kinds of disease prevention claims for food that do not result in the food being classified as a drug: those approved by FDA[72] and those approved as authoritative statements by other federal health agencies or the National Academy of Sciences.[73] The food products for which such claims may lawfully be made are discussed under Category No. 9.

For decades, FDA has recognized the legality of combination products (e.g., a product that combines a cosmetic and a drug[74]) where each component is analyzed separately under the FD&C Act. A deodorant and antiperspirant, for example, is a combination of a cosmetic and a drug. Similarly, a dietary supplement and nonprescription drug meeting all applicable requirements of an over-the-counter (OTC) drug monograph[75] could be marketed as a combination product if properly labeled, without reclassifying the dietary supplement component as a drug.

In the early 1970s, FDA sought to reclassify two fat-soluble vitamins, vitamin A and vitamin D, as prescription drugs at any level higher than the upper limit of the U.S. Recommended Dietary Allowance (RDA).[76,77] After protracted litigation, the regulations were declared unlawful because FDA failed to demonstrate an objective intent that high levels of the vitamins are for therapeutic rather than nutritional purposes.[78]

Category No. 5: Medical Food

The category of medical food was initially an administrative creation without statutory authority. It was subsequently defined in the Orphan Drug Act Amendments of 1988[79] and is now reflected in the FD&C Act in the form of statutory exemptions from nutrition labeling, nutrient descriptors, and disease prevention claims for food.[80]

The statutory definition in the Orphan Drug Act requires only the supervision of a physician and the use of the product for specific dietary management of a disease for which distinctive nutritional requirements, based on recognized scientific principles, are established by medical evaluation. FDA has attempted to narrow this definition by regulation, however, to add five limitations not found in the statute itself: (1) a medical food is specifically formulated and processed for feeding a patient and is not a natural food, (2) the patient must have limited or impaired capacity to use ordinary food or must have other special medical needs that cannot be achieved by the modification of the normal diet alone, (3) the nutritional support must be specifically modified for the unique nutrient needs that result from the specific disease involved, (4) the product is intended for use under medical supervision, and (5) the medical supervision must be personal and active.[81] Although these additional nonstatutory requirements are of highly doubtful legality, they have served to reduce the number of medical foods that are marketed, and they indicate a very negative FDA view of these products.

For many years, FDA has considered imposing additional regulatory requirements on medical food by adopting new regulations. The agency asked for public comment on this subject in an Advance Notice of Proposed Rulemaking published in November 1996[82] but has taken no further action.

Category No. 6: Vitamin and Mineral Products

When Congress enacted the Vitamin-Mineral Amendments of 1976[83] to overrule FDA's 1973 regulations, it incorporated in the statute a definition of "special dietary use" that was patterned directly on the FDA administrative definition of this category.[84] As already noted in the discussion of Category No. 2, however, the special dietary use category is now largely obsolete.

The 1976 Amendments were explicitly limited to vitamin-mineral products intended for ingestion in tablet, capsule, liquid, or similar form. At that time, Congress determined that these products could not be marketed in conventional food form. When Congress enacted the Dietary Supplement Health and Education Act of 1994, however, it amended the provisions added by the 1976 Amendments to delete the prohibition against dietary supplements being marketed in conventional food form.[85] For that reason, as will be discussed further below under Category No. 10, dietary supplements may now be marketed either in the traditional tablet or capsule form or in conventional food form, as long as a dietary supplement in food form is not represented as a conventional food.

The 1976 Amendments also added a new provision to the FD&C Act, authorizing FDA to take regulatory action against a vitamin-mineral product if its advertising is false or misleading in a material respect.[86] Before taking any such action, FDA is required to notify the FTC of the proposed action thirty days in advance. If the FTC responds within thirty days that it is taking such action and in fact takes action within sixty days, FDA is precluded from proceeding. Because of the complexity of this provision, and the long-standing relationship between FDA and the FTC, FDA has not used this authority since it was enacted.

The Vitamin-Mineral Amendments of 1976 served their purpose by overruling the FDA regulations. In light of the Dietary Supplement Health and Education Act, however, this category of vitamin and mineral products has now been subsumed under the broader category of dietary supplements and thus is of less importance than it once was.

Category No. 7: Infant Formula

There are only a handful of infant formula manufacturers in the United States, and even if the Infant Formula Act were to be repealed, it is doubtful that other companies would wish to enter this risky field. FDA regulations and enforcement activity with respect to infant formula make this an inhospitable area for product innovation or creative marketing. FDA exercises strict controls over formulation and product claims.

Category No. 8: Food with Claimed Characteristics for Nutrient Levels

Prior to 1990, FDA defined by regulation some, but far from all, nutrient descriptors used in the food industry, usually under the category of special dietary food (e.g., nutrient descriptors for weight control claims).[87,88] Under the Nutrition Labeling and Education Act of 1990, Congress required FDA to complete that job.[89] Nutrient descriptors have now been defined by FDA for virtually all recognized nutrients.[90] Although there is still some controversy about specific definitions for particular nutrient descriptors, this program has largely been successful. In effect, FDA has established a standard dictionary for nutrient terminology. The FTC has made clear that it intends to enforce the same definitions in advertising that FDA enforces in labeling.[91] Thus, this provision has resulted in standard terminology for all marketed food.

The FDA Modernization Act of 1997 authorized nutrient descriptor claims without an FDA regulation that were based on an authoritative statement of a U.S. health agency or the National Academy of Sciences, under specified conditions.[92] Although this authority may eventually become applicable to nutrient descriptors, it has much greater potential for use with disease prevention claims and thus is discussed at greater length under Category No. 9.

Category No. 9: Food with Claimed Characteristics for Disease Prevention and Treatment

From 1938 to 1984, FDA explicitly prohibited all disease claims in food labeling.[93] From 1984 to 1990, FDA relented and permitted this type of claim for food products that met agency guidelines.[94] Congress then provided in the Nutrition Labeling and Education Act that a food disease claim may be made only after, and in accordance with, a regulation promulgated by FDA. FDA has promulgated regulations authorizing the use of disease claims for food.[95]

Under the FDA Modernization Act of 1997, Congress broadened the permitted use of food disease claims to include an authoritative statement of a U.S. health agency or the National Academy of Sciences.[96] The new law requires premarket notification of such claims to FDA at least 120 days before use. This provision was enacted over strenuous objection by FDA. The antipathy of the agency for the provision is clear in its rejection of the first nine submitted authoritative statements,[97] and the narrow interpretation of this provision in a guidance[98] and proposed regulations.[99] Nonetheless, FDA subsequently allowed a disease claim for whole grain foods that was based on a National Academy of Sciences report.[100]

FDA interpreted the FDA Modernization Act provision to apply only to conventional food and not to dietary supplements. In accordance with its general policy of attempting to achieve parity between conventional food and dietary supplements with respect to product claims, however, FDA announced that it would apply this new provision to dietary supplements as well.[101]

FDA's implementation of the provisions regulating disease claims has met a strong constitutional challenge in the courts. In two areas, the courts have found that FDA's policies violate the First Amendment to the U.S. Constitution.

Following FDA promulgation of its final regulations governing food disease claims in January 1993, these regulations were challenged because they failed to contain a time period within which FDA would commit to issuing a final regulation after the date of publication of a proposed regulation for a disease claim. A court agreed that once FDA has proposed to allow a disease claim, an absence of a time frame for the issuance of a final regulation on the matter fails to satisfy First Amendment requirements.[102] FDA therefore promulgated regulations establishing a 270-day time frame between the publication of a proposal and the promulgation of a final regulation, with the possibility of two 90-day extensions.[103]

The second opinion portends a far more fundamental impact, not just on the way that FDA administers the provisions governing disease claims for food products but on the way that FDA regulates all product claims under the FD&C Act. In this case, two dietary supplement marketers challenged the FDA denial of four disease claims: dietary fiber and cancer, antioxidant vitamins and cancer, omega-3 fatty acids and coronary heart disease, and 0.8 mg folic acid and neural tube defects.[104] The court made two determinations. First, the court concluded that FDA had violated the First Amendment by declining to permit the use of the contested claims with disclaimers rather than banning them outright. Second, the court held that FDA had violated the Administrative Procedure Act by failing to provide an adequate definition of the "significant scientific agreement" standard required by the Nutrition Labeling and Education Act to justify approval of a disease claim.[105] The court therefore ordered FDA to reconsider its entire approach to this matter. Because the Solicitor General turned down FDA's request to seek review by the Supreme Court, the decision is final and requires FDA to undertake a fundamental revision of its entire approach to disease claims for all forms of food regulated under the FD&C Act.

Category No. 10: Dietary Supplements

Although vitamin-mineral products were indirectly defined through the definition of special dietary uses in the Vitamin-Mineral Amendments of 1976, there was no definition of a dietary supplement either in the FD&C Act or in FDA regulations until the Dietary Supplement Health and Education Act of 1994. The definition of a dietary supplement[106] contained in the 1994 Act is extremely broad. It includes not only recognized nutrients but other substances that FDA believes to be outside the field of nutrition. Thus, the Dietary Supplement Health and Education Act of 1994 substantially expanded the field of dietary supplements and the scope of "nutrition."[107]

Technically, the FD&C Act has permitted structure-function claims for all food, including dietary supplements, since it was enacted in 1938. The FDA policy to discourage and inhibit such claims resulted, as a practical matter, in very few of those claims ever being made before 1994.

Following the explicit statutory authority for structure-function claims under the Dietary Supplement Health and Education Act of 1994, FDA has sought to limit those claims in two ways. First, it has taken the position that a dietary supplement that contains a natural source of the active ingredient in a previously approved new drug is automatically precluded from marketing in the United States under the 1994 Act. This FDA decision was challenged in the courts and was overturned in a decision handed down in February 1999.[108]

Second, FDA has proposed and promulgated regulations intended to limit the scope of permitted structure-function claims.[109] It proposed to redefine the term *disease* in a very broad way that, if construed and applied literally, would leave very little room for structure-function claims, but withdrew this proposal in the final regulations. It also proposed to determine that implied as well as direct disease claims may not be made for dietary supplements and has indicated that many structure-function claims could be regarded as implied disease claims and strengthened this determination even further in the final regulations. Numerous comments opposing this approach have been filed with FDA, and the ultimate outcome of this rule making is uncertain. These regulations are certain to provoke litigation. In the interim, FDA has sent dozens of letters objecting to structure-function claims submitted to the agency under the requirements of the 1994 Act.[110]

Nonetheless, many of the specific examples used by FDA in its proposed and final regulations included structure-function claims approved by FDA that only a few years earlier would have been regarded by the agency as illegal disease claims. For example, the following cardiovascular claims of Cardiohealth Heart Tabs were listed by FDA as permitted structure-function claims:

- Helps maintain a healthy cholesterol level.
- Helps maintain cardiovascular function and a healthy circulatory system.
- Cardiohealth
- Heart Tabs

Only "lowers cholesterol" was regarded by FDA as an illegal disease claim.

FDA has also sought to limit the form in which dietary supplements may be marketed. In spite of the explicit congressional determination that dietary supplements may be marketed in conventional food form,[111] and the specific FDA recognition in the *Federal Register* that such products as cranberry juice cocktail could be marketed as dietary supplements,[112] the agency has contended that at least some food forms inherently represent the product as a conventional food and thus cannot be used for dietary supplement products.[113] Although FDA has not described or listed product forms that it believes are inherently conventional foods, it has objected to the marketing of a spread form as a dietary supplement but does not appear to have objected to a beverage or bar form as a dietary supplement. There is no statutory provision or legislative history to support the FDA position on this matter.

The 1994 Act also establishes a new medium through which dietary supplement claims may be made without FDA regulation.[114] A publication that is reprinted in its entirety is exempt from FDA regulation as labeling when used in connection with the sale of a dietary supplement to consumers if it meets the following five conditions: (1) it is not false or misleading, (2) it does not promote a particular manufacturer or brand of a dietary supplement, (3) it is displayed or presented so as to provide a balanced view of the available scientific information, (4) it is physically separate from the dietary supplement, and (5) it does not have appended to it any information by any method. The provision also recognizes the right of dietary supplement stores to sell books or other publications as part of their business, without FDA classifying these publications as labeling.

Thus, six years after enactment of the Dietary Supplement Health and Education Act of 1994, the scope of and permitted claims for dietary supplements remain unclear. It may well be some years, and perhaps require additional legislation, before the parameters of this field become better known and understood.

In addition to the changes made in FDA regulation of labeling, the 1994 Act revised the safety requirements for dietary supplements. All dietary supplement ingredients continue to be subject to FDA regulation under the general adulteration provision that prohibits any added poisonous or deleterious substance in food that may render the product injurious to health.[115] For dietary ingredients, however, Congress replaced the premarket approval requirements for food additives[116] with a premarket notification requirement for new dietary ingredients under which FDA, rather than the manufacturer, has the burden of proof.[117]

REGULATION BY THE FEDERAL TRADE COMMISSION

The FTC has a long history of instituting enforcement action against dietary supplement manufacturers for false or misleading advertising claims.[118] From 1984 to 1998, the FTC brought some sixty cases against dietary supplement products. Following the enactment of the Dietary Supplement Health and Education Act of 1994, the FTC has increased its surveillance of dietary supplement advertising to prevent the escalation of unsupported claims.

The FTC has a well-documented policy with respect to all advertising claims. First, advertisers must have adequate substantiation for each objective product claim prior to making that claim. Second, the required substantiation must satisfy the standard of competent and reliable scientific evidence. Third, this level of evidence must be available to support all express and implied claims that the advertisement conveys to consumers.

Because of the increased FTC surveillance of dietary supplement advertising claims and the corresponding increase in dietary supplement cases, the FTC concluded to prepare a special advertising guide for the dietary supplement industry. That guide was released in February 1998[119] and should be consulted by any person involved in advertising dietary supplements. The guide sets forth all of the relevant FTC requirements and includes thirty-six specific examples illustrating the application of these principles.

CONCLUSION

The regulation of dietary supplements both by FDA and the FTC is in a relatively early and uncertain stage. The Dietary Supplement Health and Education Act of 1994 was a direct backlash by Congress against the ill-advised frontal assault by FDA Commissioner David Kessler against the dietary supplement industry. In her March 1999 testimony before the House Committee on Government Reform, new FDA Commissioner Jane Henney signaled a more flexible and cooperative approach to the implementation of the 1994 Act:

> I understand that this statute was passed with bipartisan support and by the hard work of you and others in Congress in developing an appropriate regulatory scheme that would facilitate consumers' access to dietary supplements. It is important that the Agency's implementation of the statute be true to Congressional intent. As I stated during my confirmation process, I am aware that many Americans place great faith in dietary supplements to help them maintain and improve their health and that the scientific evidence documenting the benefits of a number of supplements is increasing. The challenge to FDA is to strike the right balance between preserving consumers' access to both products and information and assuring the safety and proper labeling of all of these products. It is clear, with the benefit of hindsight, that we still have a way to go both in achieving full implementation of DSHEA and in developing a workable regulatory framework. I want to take the opportunity to acknowledge our progress, shortcomings, remaining challenges, and commitment to fully implement the statute.[120]

Particularly because of the recent First Amendment decision, Commissioner Henney has an opportunity to refashion the entire FDA approach to regulating dietary supplements to reflect the balance articulated in this statement to Congress.

REFERENCES

1. Hutt PB, Hutt II PB. A History of Government Regulation of Adulteration and Misbranding of Food, 39 *Food Drug Cosm. L.J.* 2 (1984).
2. 21 U.S.C. 301.
3. 15 U.S.C. 41.
4. Hutt PB. The Transformation of United States Food and Drug Law, 60 *J Ass'n. Food Drug Officials, No. 3,* at 1, 8–9 (September 1996).
5. 34 Stat. 768 (1906).
6. For a history of the government organizations responsible for implementation of federal food and drug law, see Peter Barton Hutt, A Historical Introduction, 45 *Food Drug Cosm. L.J.* 17 (1990).
7. 1917 Report of Bureau of Chemistry 15–16, in Food Law Institute, Federal Food, Drug, and Cosmetic Law Administrative Reports 1907–1949 355, 369–370 (1951).
8. S. 1944, 73d Cong., 1st Sess. (1933).
9. 52 Stat. 11 (1938).
10. 36 Fed. Reg. 18539 (September 16, 1971).
11. S. Rep. No. 361, 74th Cong., 1st Sess. 4 (1935).
12. Section 201(f) of the FD&C Act, 21 U.S.C. 321(f).
13. Section 201(g) of the FD&C Act, 21 U.S.C. 32(g).
14. Id.
15. Section 301 of the FD&C Act, 21 U.S.C. 331.
16. Section 505 of the FD&C Act, 21 U.S.C. 355.
17. Section 403 of the FD&C Act, 21 U.S.C. 343.
18. Section 401 of the FD&C Act, 21 U.S.C. 34.
19. Section 403(j) of the FD&C Act, 21 U.S.C. 34(j).
20. Peter Barton Hutt, Government Regulation of Health Claims in Food Labeling and Advertising, 41 *Food Drug Cosm. L.J.* 3, 52–63 (1986).
21. 38 Fed. Reg. 2143, 2152 (January 19, 1973).
22. 38 Fed. Reg. 20708, 20730 (August 2, 1973).
23. National Nutritional Foods Assn v. FDA, 504 F.2d 761 (2d Cir. 1974).
24. National Nutritional Foods Assn v. Kennedy, 572 F.2d 377 (2d Cir. 1978).
25. 90 Stat. 401, 410 (1976); Section 411 of the FD&C Act, 21 U.S.C. 350.
26. 44 Fed. Reg. 16005 (March 16, 1979).
27. 94 Stat. 1190 (1980).
28. Section 412 of the FD&C Act, 21 U.S.C. 350a.
29. 21 C.F.R. pt. 106 and pt. 107.
30. Hutt, *supra* note 20, at 68–73.
31. 37 Fed. Reg. 18229, 18230 (September 8, 1972).
32. 38 Fed. Reg. 2124, 2126 (January 19, 1973).
33. 38 Fed. Reg. 6951 (March 14, 1973).
34. 102 Stat. 90, 91 (1988), 21 U.S.C. 360ee(b), not codified in the FD&C Act.
35. 104 Stat. 2353 (1990).
36. Sections 403(q)(5)(A)(iv) and 403(r)(5)(A) of the FD&C Act, 21 U.S.C. 343(q)(5)(A)(iv) and 343(r)(5)(A).
37. 21 C.F.R. 101.9(j)(8).
38. 61 Fed. Reg. 60661 (November 29, 1996).
39. 37 Fed. Reg. 6493 (March 30, 1972).
40. 38 Fed. Reg. 2125 (January 19, 1973).

41. *Supra* note 33.

42. Section 201(n) of the FD&C Act, 21 U.S.C. 321(n).

43. For a complete history of the nutrition labeling and related regulations, see Peter Barton Hutt, A Brief History of FDA Regulation Relating to the Nutrient Content of Food, *in* N.R. Shapiro, ed., Nutrition Labeling Handbook Ch. 1, at 1 (Marcel Dekker, 1995).

44. Section 403(r)(5)(D) of the FD&C Act, 21 U.S.C. 343(r)(3)(C) and (D).

45. 58 Fed. Reg. 33700 (June 18, 1993).

46. 58 Fed. Reg. 53296 (October 14, 1993).

47. 59 Fed. Reg. 395 (January 4, 1994).

48. 58 Fed. Reg. 33690 (June 18, 1993).

49. 108 Stat. 4325 (1994).

50. 111 Stat. 2296 (1997).

51. Section 403(r)(3)(C) and (D) of the FD&C Act, 21 U.S.C. 343(r)(3)(C) and (D).

52. Sections 403(r)(2)(G) and (H) of the FD&C Act, 21 U.S.C. 343(r)(2)(G) and (H).

53. NutriLab, Inc. v. Schweiker, 713 F.2d 335, 338 (7th Cir. 1983).

54. 21 C.F.R. 101.14(a)(6).

55. 58 Fed. Reg. 2478, 2479 (January 6, 1993); codified at 21 C.F.R. 101.14(a)(1) and (2).

56. Hutt, *supra* note 20, at 61.

57. 21 C.F.R. pt. 105.

58. Section 201(g)(1) of the FD&C Act, 21 U.S.C. 321(g)(1).

59. American Health Products Co., Inc. v. Hayes, 574 F. Supp. 1498, 1501 (S.D.N.Y. 1983), affirmed on other grounds. American Health Products Co. v. Hayes, 744 F.2d 912 (2d Cir. 1984) (per curiam).

60. 58 Fed. Reg. 2478, 2482 (January 6, 1993).

61. 62 Fed. Reg. 49859, 49861 (September 23, 1997).

62. See also 63 Fed. Reg. 34084–34085 (June 22, 1998).

63. 62 Fed. Reg. at 49860.

64. Id.

65. 63 Fed. Reg. 23624 (April 29, 1998).

66. 65 Fed. Reg. 1000 (January 6, 2000).

67. Sections 201(s) and 409 of the FD&C Act, 21 U.S.C. 321(s) and 348.

68. Sections 201(ff), 402(f), and 413 of the FD&C Act, 21 U.S.C. 321(ff), 342(f), and 350b.

69. Section 403(r)(6)(C) of the FD&C Act, 21 U.S.C. 343(r)(6)(C).

70. Section 403(r)(6) of the FD&C Act, 21 U.S.C. 343(r)(6).

71. *Supra* note 11. FDA published the report of its advisory panel on nonprescription vitamin-mineral drugs in 44 Fed. Reg. 16126 (March 16, 1979) but subsequently withdrew the proposed monograph in 46 Fed. Reg. 57914 (November 27, 1981) and has taken no further action on this matter.

72. Sections 403(r)(3)(A) and (B) of the FD&C Act, 21 U.S.C. 343(r)(3)(A) and (B).

73. Sections 403(r)(3)(C) and (D) of the FD&C Act, 21 U.S.C. 343(r)(3)(C) and (D).

74. E.g. 21 C.F.R. 701.3(d).

75. 21 C.F.R. Part 330.

76. 37 Fed. Reg. 26618 (December 14, 1972).

77. 38 Fed. Reg. 20723 (August 2, 1973).

78. National Nutritional Foods Ass'n v. Mathews, 557 F.2d 325 (2d Cir. 1977). The regulations were subsequently revoked in 43 Fed. Reg. 10551 (March 14, 1978).

79. *Supra* note 34.

80. *Supra* note 36.

81. 21 C.F.R. 101.9(j)(8)(i)-(v).

82. *Supra* note 38.

83. *Supra* note 25.

84. Section 411(c)(3) of the FD&C Act, 21 U.S.C. 350.

85. Section 3(c) of the Dietary Supplement Health and Education Act of 1994. 108 Stat. 4325, 4328 (1994).

86. Section 403(a)(2) and 707 of the FD&C Act, 21 U.S.C. 343(a)(2) and 378.

87. 42 Fed. Reg. 37166 (July 19, 1977).

88. 43 Fed. Reg. 43248 (September 22, 1978).

89. Sections 403(r)(1)(A), 403(r)(2), and 403(r)(3) of the FD&C Act, 21 U.S.C. 343(r)(1)(A), 343(r)(2), and 343(r)(3).

90. 21 C.F.R. pt.101, subpart D.

91. FTC, *Enforcement Policy Statement on Food Advertising* (May 1994).

92. Sections 403(r)(2)(G) and (H) of the FD&C Act, 21 U.S.C. 343(r)(2)(G) and (H).

93. Hutt, *supra* note 20, at 42–44.

94. Peter Barton Hutt & Richard A. Merrill. *Food and Drug Law: Cases and Materials,* 183–187 (2d ed. 1991).

95. 21 C.F.R. pt. 101, subpart E.

96. *Supra* note 50.

97. 63 Fed. Reg. 34084 (June 22, 1998).

98. 63 Fed. Reg. 32102 (June 11, 1998).

99. 64 Fed. Reg. 3250 (January 21, 1999).

100. FDA, *Health Claim Notification for Whole Grain Foods* (July 1999).

101. 63 Fed. Reg. at 34084-34085.

102. Nutritional Health Alliance v. Shalala, 953 F.Supp. 526 (S.D.N.Y. 1997).

103. 62 Fed. Reg. 12579 (March 17, 1997); 62 Fed. Reg. 28230 (May 22, 1997).

104. Pearson v. Shalala, 164 F.3d 650 (D.C. Cir. 1999); rehearing en banc denied, 172 F.3d 72 (D.C. Cir. 1999). Two earlier cases failed for procedural reasons. National Council for Improved Health v. Shalala, 122 F.3d 878 (10th Cir. 1997) (lack of standing); Nutritional Health Alliance v. Shalala, 144 F.3d 220 (2d. Cir. 1998) (not ripe for judicial review).

105. Section 403(r)(3)(B)(i) of the FD&C Act, 21 U.S.C. 343(r)(3)(B)(i).

106. Section 201(ff) of the FD&C Act, 21 U.S.C. 321(ff).

107. FDA has taken the unusual position in two Federal Register preambles that Congress was "inaccurate" in using the term *nutritional* to include all dietary ingredients and thus that the term *statements of nutritional support* as used by Congress in Section 6 of the Dietary Supplement Health and Education Act of 1994 would no longer be used by FDA for dietary ingredients that have no *nutritive value.* 62 Fed. Reg. at 49863. 63 Fed. Reg. 23624, 23625 note 1 (April 29, 1998). Nonetheless, FDA has no authority to overrule the explicit congressional definition and determination that the structure-function claims that may lawfully be made for all dietary ingredients are in fact statements of nutritional support, whether or not FDA classifies an ingredient as "nutritive."

108. Pharmanex, Inc. v. Shalala, 35 F.Supp. 3d 1341 (D. Utah 1999).

109. *Supra* notes 65 and 66.

110. *Supra* note 70.

111. *Supra* note 85.

112. *Supra* note 63.

113. Letter from FDA Director of the Center for Food Safety and Applied Nutrition Joseph A. Levitt to President of McNeil Consumer Products Company Brian D. Perkins (October 28, 1998).

114. Section 403B of the FD&C Act, 21 U.S.C. 343-2.

115. Section 402(a)(1) of the FD&C Act, 21 U.S.C. 342(a)(1).

116. *Supra* note 67.

117. *Supra* note 68.

118. Hutt, *Supra* note 20, at 9–20.

119. FTC Bureau of Consumer Protection, *Dietary Supplements: An Advertising Guide for Industry* (1998).

120. "Dietary Supplement Health and Education Act: Is the FDA Trying To Change the Intent of Congress," Hearings before the Committee on Government Reform, House of Representatives, 106th Cong., 1st Sess. 28, 33–34 (1998).

Ten Statutory Food Categories under the Federal Food, Drug, and Cosmetic Act

This table summarizes ten statutory food categories under the Federal Food, Drug, and Cosmetic Act (FD&C Act), the year(s) they were created and defined, the specific citation to the statutory definition in the FD&C Act, the substance of that definition, and the statutory requirements or prohibitions relating to claims made for these different categories of food on the label or in accompanying labeling.

These ten categories are not mutually exclusive. Some food products fall within more than one category. A food must satisfy all of the requirements for each applicable category.

So-called "functional food" (also referred to as novel/designer/pharma/hypernutritious/therapeutic/medicinal/health/natural health/special nutritional/particular nutritional/specified health/biotechnology/nutraceutical/phytochemical food) is not a statutory food category under the FD&C Act. A "functional food" (by any name) is currently regulated under one or more of the ten statutory categories under the FD&C Act set forth in this table.

All food, regardless of its classification under the FD&C Act, is also subject to regulation by the Federal Trade Commission (FTC) under the FTC Act. Under a Memorandum of Understanding, FDA has primary (but not exclusive) jurisdiction over food labeling, and FTC has primary (but not exclusive) jurisdiction over food advertising.

Statutory Food Category	Year(s) Created/ Defined	Statutory Definition (FD&C Act)	Substance of Definition	Statutory Requirements/ Prohibitions Applicable to Product Claims
1. Food*	1938	§201(f)	Foods are articles used for food or drink and their components.	§403(a)(1) prohibits labeling that is false or misleading in any particular. 21 C.F.R. §101.14(a)(6) provides that claims about diseases from essential nutrient deficiencies are not disease prevention claims and may be made for any food.

continues

*The "food" category includes standardized and nonstandardized food and fortified (enriched) and unfortified (unenriched) food.

Statutory Food Category	Year(s) Created/ Defined	Statutory Definition (FD&C Act)	Substance of Definition	Statutory Requirements/ Prohibitions Applicable to Product Claims
				In 58 Fed. Reg. 2478, 2479 (January 6, 1993), FDA stated that the following are not disease prevention claims and may be made for any food: (1) dietary guidance that does not relate a specific food or substance to a specific disease and (2) general claims about good health and well-being.
	1938	§402(a)(1)	No food may contain an added poisonous or deleterious substance that may render the food injurious to health.	This prohibition applies to all substances in all categories of food.
	1958	§§201(s), 409	A food additive (which is also a food) is any substance that is a component of or affects food and that is not generally recognized as safe (GRAS) and that was not approved for its use by FDA or USDA during 1938–1958 (a prior sanction).	A food additive may not be used prior to FDA promulgation of a food additive regulation for the substance, but a GRAS or prior-sanctioned substance may be used without premarket approval. All food substances in all categories of food must be GRAS, prior-sanctioned, or the subject of a food additive regulation before they can be used, except for the dietary ingredients in dietary supplements. See Category No. 10 below.
2. Food for special dietary use	1938	§403(j)	Purports to be or is represented for special dietary use.	§403(a)(1) prohibits labeling that is false or misleading in any particular.
	1976	§411(c)(3)	A particular use for which a food purports or is represented to be used, including supplying a special dietary need that exists by reason of a physical, physiological, pathological, or other condition; supplying a vitamin, mineral, or other ingredient to	§403(a)(2) prohibits advertising of a vitamin or mineral product that is false or misleading in a material respect. §403(a)(2) prohibits labeling that lists ingredients in a vitamin or mineral product that are not dietary supplement ingredients described in §201(ff)(1) except as part of a

Statutory Food Category	Year(s) Created/ Defined	Statutory Definition (FD&C Act)	Substance of Definition	Statutory Requirements/ Prohibitions Applicable to Product Claims
			supplement the diet; or use as the sole item of the diet.	list of all ingredients of the food in accordance with applicable FDA regulations. §403(j) authorizes regulations governing the labeling of food for special dietary uses. §411(b) permits the labeling and advertising of a vitamin or mineral product to refer to dietary supplement ingredients that are not vitamins or minerals but prohibits prominence or emphasis of ingredients that are not dietary supplement ingredients described in §201(ff)(1). 21 C.F.R. pt 105 establishes regulations governing the labeling of three categories of food for special dietary use: hypoallergenic food, infant food, and weight control food.
3. Food intended to affect the structure or any function of of the human body	1938	§201(g)(1)(C)	The term *drug* is defined as an article (other than food) intended to affect the structure or any function of the human body.	§403(a)(1) prohibits labeling that is false or misleading in any particular. In 58 Fed. Reg. 2478, 2482 (January 6, 1993), FDA stated that a claim that a food or nutrient affects a biological parameter of the body is not an illegal drug or disease prevention (health) claim if there is no express or implied reference to (1) any dysfunction of or damage to the body (other than a reference to an essential nutrient deficiency disease) or (2) any biological parameter that is a recognized risk factor for a disease or health-related condition. In 62 Fed. Reg. 49859, 49860 (September 23, 1997), FDA stated that a "structure or function" claim may be made for any food (e.g., "cranberry

continues

Statutory Food Category	Year(s) Created/ Defined	Statutory Definition (FD&C Act)	Substance of Definition	Statutory Requirements/ Prohibitions Applicable to Product Claims
				products help maintain urinary tract health") if the claim is truthful, not misleading, and derives from the nutritional (rather than from some non-nutritive) value of the food. In 63 Fed. Reg. 23624 (April 29, 1998) and 65 Fed. Reg. 1000 (January 6, 2000), FDA published proposed and final regulations to distinguish between a "structure or function" claim permitted for a food and a "disease prevention or treatment" claim that classifies the product as a drug.
4. Food intended for the prevention or treatment of human disease	1938	§201(g)(1)(B)	The term *drug* is defined as an article intended for use in the diagnosis, cure, mitigation, treatment, or prevention of human disease.	§502(a) prohibits labeling that is false or misleading in any particular. §505 prohibits the marketing of any new drug (including a new claim for an old drug) for which FDA has not approved a new drug application (NDA). Nonprescription vitamin and mineral drugs are regulated under the FDA OTC Drug Review, 21 C.F.R. pt 330. Nonprescription homeopathic drugs are regulated under FDA Compliance Policy Guide 400.400 (formerly 7132.15). Nonprescription traditional Chinese remedies are not subject to any specific written FDA policy. §403(r)(3)(A) authorizes FDA to promulgate regulations establishing disease prevention claims for food that do not result in the food being classified as a drug. See Category No. 9 below.

Statutory Food Category	Year(s) Created/ Defined	Statutory Definition (FD&C Act)	Substance of Definition	Statutory Requirements/ Prohibitions Applicable to Product Claims
				In 63 Fed. Reg. 23624 (April 29, 1998) and 65 Fed. Reg. 1000 (January 6, 2000), FDA published proposed and final regulations to distinguish between a "structure or function" claim permitted for a food and a "disease prevention or treatment" claim that classifies the product as a drug.
5. Medical food	1972 1988	21 U.S.C. §360ee(b)(3) 21 C.F.R. §101.9(j)(8) 21 C.F.R. §101.13(q)(4)(ii) 21 C.F.R. §101.14(f)(2)	Medical food is formulated to be consumed or administered enterally under the supervision of a physician and is intended for the specific management of a disease or condition for which distinctive nutritional requirements, based on recognized scientific principles, are established by medical evaluation.	§403(a)(1) prohibits any labeling that is false or misleading in any particular. §403(q) exempts medical food from nutrition labeling. §403(r) exempts medical food from the requirements for nutrient descriptors and disease prevention claims. 21 C.F.R. §101.9(j)(8) lists five requirements for medical food that narrow the statutory definition.
6. Vitamin and mineral products	1976 1994	§411(c)(1)	A food for a special dietary use that is or contains any natural or synthetic vitamin or mineral and is intended for ingestion in tablet, capsule, or liquid form or, if not intended for ingestion in such form, is not represented as conventional food and is not represented for use as a sole item of a meal or of the diet.	§403(a)(1) prohibits labeling that is false or misleading in any particular. §411(b) permits the labeling and advertising of a vitamin and mineral product to refer to dietary supplement ingredients that are not vitamins or minerals but prohibits prominence or emphasis of ingredients that are not dietary supplement ingredients described in §201(ff)(1). §411(a)(1)(A) prohibits FDA from establishing maximum limits on the potency of a vitamin or mineral for reasons other than safety. §411(a)(1)(B) prohibits FDA from classifying a vitamin or mineral as a drug solely because of potency.

continues

Statutory Food Category	Year(s) Created/ Defined	Statutory Definition (FD&C Act)	Substance of Definition	Statutory Requirements/ Prohibitions Applicable to Product Claims
				§411(a)(1)(C) prohibits FDA from limiting the combinations of vitamins, minerals, and other ingredients. §707 authorizes FDA to take regulatory action against false or misleading advertising of vitamin-mineral products if FTC declines to do so.
7. Infant formula	1980	§201(aa)	A food that purports to be or is represented for special dietary use solely as a food for infants by reason of its simulation of or suitability as a substitute for human milk.	§403(a)(1) prohibits labeling that is false or misleading in any particular. §412 and 21 C.F.R. pts 106 and 107 establish a comprehensive regulatory program for the formulation, manufacture, and labeling of infant formula. §412(h) exempts from the §412 requirements an infant formula represented for an infant with an inborn error of metabolism, low birth weight, or unusual medical or dietary problems.
8. Food with claimed characteristics for nutrient levels (nutrient descriptors)	1990	§403(r)(1)(A)	A food for which a claim characterizes the level of a nutrient.	§403(r)(2)(A) provides that a nutrient descriptor may be used in food labeling only if it uses specific terms defined in FDA regulations and meets the requirements of the substantive definitions of those terms.
	1997	§403(r)(2)(G) and (H)		§403 (r)(2)(G) provides that a food nutrient descriptor claim may be made without an FDA regulation if (1) it is based on an authoritative statement of a U.S. public health agency or the National Academy of Sciences, (2) a premarket notification is submitted to FDA at least 120 days before use, (3) it otherwise complies with the requirements of the FD&C Act, and (4) it accurately represents the authoritative

Statutory Food Category	Year(s) Created/ Defined	Statutory Definition (FD&C Act)	Substance of Definition	Statutory Requirements/ Prohibitions Applicable to Product Claims
				statement in the context of a total daily diet.
				§403(r)(2)(4) provides that FDA may take action to prevent the use of a claim based on an authoritative statement.
				21 C.F.R. pt 101, subpart D, establishes the nutrient descriptors defined by FDA.
9. Food with claimed characteristics for disease prevention and treatment (disease prevention and treatment claims)	1990	§403(r)(1)(B)	Food for which a claim characterizes the relationship of a nutrient to a disease or a health-related condition.	§403(r)(3)(A) provides that a food disease prevention claim may be made only after, and in accordance with, a regulation promulgated by FDA.
	1997	§403(r)(3)(C)		§403(r)(3)(C) provides that a food disease claim may be made without an FDA regulation if (1) it is based on an authoritative statement of a United States public health agency or the National Academy of Sciences, (2) a premarket notification is submitted to FDA at least 120 days before use, (3) it otherwise complies with the requirements of the FD&C Act, and (4) it accurately represents the authoritative statement in the context of a total daily diet.
				§403(r)(3)(D) provides that FDA may take action to prevent the use of a claim based on an authoritative statement.
				21 C.F.R. pt 101, subpart E, establishes the disease claims permitted by FDA for food.
				§201(g)(1) provides that a disease claim for a food that does not comply with §403(r)(3) results in the food being classified as a drug. See Category No. 4.

continues

Statutory Food Category	Year(s) Created/ Defined	Statutory Definition (FD&C Act)	Substance of Definition	Statutory Requirements/ Prohibitions Applicable to Product Claims
				21 C.F.R. §101.14(a)(6) excludes claims about diseases from essential nutrient deficiencies from §403(r)(3)(A).
				In 58 Fed. Reg. 2478 (January 6, 1993), FDA stated that the following are not disease claims: (1) dietary guidance that does not relate a specific food or substance to a specific disease and (2) general claims about good health and well-being.
				In 62 Fed. Reg. 49859, 49860 (September 23, 1997), FDA stated that a "structure or function" claim (see Category No. 3) may be made for any food if it is not a disease claim.
				In 63 Fed. Reg. 23624 (April 29, 1998) and 65 Fed. Reg. 1000 (January 6, 2000), FDA published proposed and final regulations to distinguish between a "structure or function" claim permitted for a food and a "disease prevention or treatment" claim that classifies the product as a drug.
				In 63 Fed. Reg. 34084 (June 22, 1998), FDA published 9 decisions prohibiting requested disease claims based on authoritative statements.
				In 64 Fed. Reg. 3250 (January 21, 1999), FDA proposed regulations for disease claims based on authoritative statements.
				In Pearson v. Shalala, 164 F.3d 650 (D.C. Cir. 1999), the court declared the FDA restrictive policy on food disease claims to be an unconstitutional

Statutory Food Category	Year(s) Created/ Defined	Statutory Definition (FD&C Act)	Substance of Definition	Statutory Requirements/ Prohibitions Applicable to Product Claims
				violation of the First Amendment.
10. Dietary supplement	1994	§201(ff)	A product intended to supplement the diet that bears or contains a dietary ingredient that is a vitamin, a mineral, an herb or other botanical, an amino acid, a dietary substance to supplement the diet by increasing the total dietary intake, or a concentrate, metabolite, constituent, extract, or combination of any of the prior ingredients, and that is intended for ingestion in tablet, capsule, or liquid form or, if not intended for ingestion in such form, is not represented for use as a conventional food or as a sole item of a meal or the diet.	§403(a)(1) prohibits labeling that is false or misleading in any particular. §403(r)(6) permits a statement for a dietary supplement that claims a benefit related to a classical nutrient deficiency disease, describes the role of a nutrient or a dietary ingredient intended to affect the structure or function of the human body, characterizes the documented mechanism by which a nutrient or dietary ingredient acts to maintain such structure or function, or describes general well-being from consumption of a nutrient or dietary ingredient. §403(r)(6) requires the label of a dietary supplement with a nutritional support claim to hear the statement "This statement has not been evaluated by the Food and Drug Administration. This product is not intended to diagnose, treat, cure, or prevent any disease." §403(r)(6) requires the manufacturer of a dietary supplement for which a nutritional support claim is made in labeling to notify FDA at least 30 days after first use. §403B permits the use of publications printed in their entirety if they meet five statutory conditions. In 63 Fed. Reg. 23624 (April 29, 1998) and 65 Fed. Reg. 1000 (January 6, 2000), FDA published proposed and final

continues

Statutory Food Category	Year(s) Created/ Defined	Statutory Definition (FD&C Act)	Substance of Definition	Statutory Requirements/ Prohibitions Applicable to Product Claims
				regulations to distinguish between a "structure or function" claim permitted for a dietary supplement and a "disease prevention or treatment" claim that classifies the product as a drug.
				In Pharmanex, Inc. v. Shalala, 35 F. Supp. 2d 1341 (D. Utah 1999), the court held that a product containing a natural substance that is an active ingredient in a new drug may be marketed as a dietary supplement.
		§413	A new dietary ingredient is a dietary ingredient in a dietary supplement that was not marketed in the United States before October 15, 1994.	At least 75 days before a new dietary ingredient is marketed, unless the ingredient has been present in the food supply as an article used for food in a form in which the food has not been chemically altered, a submission must be made to FDA containing information on which the manufacturer concluded the ingredient is reasonably expected to be safe.
		§201(s)(6)		Dietary ingredients in dietary supplements (but not other ingredients in dietary supplements) are excluded from the food additive provisions in §409.

CHAPTER 17

Regulatory Issues: Europe and Japan

Lorraine Eve

In response to increased consumer awareness over recent years of the link between diet and health, so-called "functional foods" have begun to appear on the market. Functional foods are generally considered to be ordinary foods or food ingredients that have perceived health or medical benefits, usually bearing a claim relating to those benefits. The success of a functional food can depend to a large extent on the regulatory framework in which it is allowed to be marketed. From country to country, the method of regulating the composition and labeling of functional foods varies enormously.

From a legal point of view, the first issue that must be considered in relation to functional foods is the nature of the ingredient: When does a food or ingredient become "functional," and what specific health benefit does it actually confer on the consumer? A whole host of products on the market profess to be "functional," ranging from vitamin- and mineral-enriched products to products containing added fiber ingredients and pre- and probiotics. Is there really a distinction between a vitamin added for fortification and a vitamin added for its perceived functional properties?

The second key issue is the labeling of the product and, in particular, the use of health-related claims. Quite naturally, manufacturers want to draw attention to the key ingredients in their product that confer perceived health benefits on the consumer; however, any such claims must be justified and be capable of being substantiated. In Japan, once a food has gained approval and is recognized as suitable for a specified health use, it may bear an approved statement on its label indicating its specific health benefit. Within Europe, the lack of harmonized European Community (EC) legislation controlling the use of claims in food labeling means that manufacturers may be unsure how far they can go in making health-related claims. The job of the regulator must be to ensure that, while manufacturers have the freedom to convey justified and genuine health messages about the properties of a food, consumers are not misled. Legislation on health claims is therefore intrinsically linked to legislation on functional foods.

Is there a need for specific premarket approval of functional foods? In Japan, the home of functional foods, a foodstuff must undergo strict procedures to be approved and marketed as a "food for specified health use" (FOSHU). In contrast, in Europe, where the market for functional foods is beginning to grow, there are no specific legal controls on the composition of a functional food, and there is even considerable debate as to what makes a food functional.

This chapter considers the legal controls on functional foods in two principal regions, Europe and Japan, and looks at the existing and future regulatory systems. Table 17–1 summarizes the current situation.

Table 17–1 Comparison of Regulatory Approaches to Functional Foods and Health Claims in Various Countries

Country	Standards for Functional Foods	Position on Health Claims
Japan	"Foods for Specified Health Use" (FOSHU) system	Health claims permitted within FOSHU
USA	No regulation	Certain health claims permitted under the Nutritional Labeling Education Act (NLEA); structure-function claims under the Dietary Supplement Health and Education Act (DSHEA)
Codex Alimentarius	No regulation	Draft Recommendations on Health Claims (step 3)
European Community	No regulation	Not permitted
Austria	No regulation	Not permitted
Belgium	No regulation	Draft Code of Practice
Denmark	No regulation	Not permitted
Finland	No regulation	Not permitted
France	No regulation	Opinions of the National Food Council (CNA) and the Inter-Ministerial Commission for the Study of Products Intended for a Particular Diet (CEDAP)
Germany	No regulation—industry guidelines on functional foods	Not permitted
Greece	No regulation—approval required under Food Code	Not permitted
Ireland	No regulation	Not permitted
Italy	No regulation—notification required	Not permitted
Luxembourg	No regulation	Not permitted
Netherlands	No regulation	Advertising/Labeling Code of Practice and Code of Practice on the Scientific Assessment of Health Claims
Portugal	No regulation	Not permitted
Spain	No regulation	Not permitted
Sweden	No regulation	Industry Self-Regulating Guidelines
United Kingdom	No regulation	Joint Health Claims Initiative (JHCI) Code of Practice

EUROPEAN COMMUNITY LEGISLATION

Currently, there is no definition of a "functional food" laid down in harmonized EC legislation, nor are there any specific compositional regulations. Therefore, the individual European Union (EU) member states may draw up their own rules on functional foods, provided these do not present a barrier to trade. The use of varying legal controls will inevitably cause problems in trade; however, as the current trend in European legislation is moving away from so-called "vertical" or compositional requirements and toward more informative labeling, it is unlikely that specific regulations controlling the composition of functional foods will ever be issued at the EC level.

There are, however, certain pieces of horizontal EC legislation that may be applicable to functional foods, such as the general labeling directive, the Regulation on Novel Foods and Novel Food Ingredients, and the Directive on Foods for Particular Nutritional Uses. Therefore, the general provisions laid down in EC legislation and the national rules of the EU member states must be taken into account when considering marketing a functional food in Europe.

Novel Foods

Functional foods or food ingredients may fall under the scope of the EU Novel Foods Regulation and require premarket approval by that route. For example, some functional foods may obtain their specific health-related benefits from ingredients based on totally new compounds that have not previously been consumed as foods within Europe, or they may be produced using newly developed processing methods, including biotechnology; if this is the case, the foods or ingredient are considered as "novel."

Getting regulatory approval for novel foods can be a long, time-consuming process, as well as extremely expensive. EC legislation to harmonize approval of novel foods was agreed on during 1997 in the form of Regulation (EC) No. 258/97 on novel foods and novel food ingredients.[1] It is significant that this legislation takes the form of a regulation as opposed to a directive and therefore was automatically binding in all 15 of the EU member states when it entered into force on May 15, 1997.

A novel food or ingredient is defined as a food or food ingredient that has not yet been used for human consumption to a significant degree within the EC and that falls under one of the following categories:

1. foods and food ingredients containing or consisting of genetically modified organisms
2. foods and food ingredients produced from, but not containing, genetically modified organisms
3. foods and food ingredients with a new or intentionally modified primary molecular structure
4. foods and food ingredients consisting of or isolated from microorganisms, fungi, or algae
5. foods and food ingredients consisting of or isolated from plants, and food ingredients isolated from animals, except for foods and food ingredients obtained by traditional propagating or breeding practices and having a history of safe food use
6. foods and food ingredients to which has been applied a production process not currently used, where that process gives rise to significant changes in the composition or structure of the foods or food ingredients that affect their nutritional value, metabolism, or level of undesirable substances

Foods and food ingredients that fall within the scope of this regulation must not present a danger for the consumer, mislead the consumer, or differ from foods or food ingredients that they are intended to replace to such an extent that their normal consumption would be nutritionally disadvantageous for the consumer. For a novel food to be approved, the person responsible for placing the food on the market must submit a request to the member state in which the product is to be first marketed and also to the commission. An initial assessment is then carried out within the procedure outlined in the regulation.

If a functional food or food ingredient falls under the scope of the Novel Foods Regulation, it may be subject to additional labeling requirements; this is an area that has been of significant interest following the publication of the regulations on genetically modified ingredients and may have similar implications for "functional" novel ingredients.

The regulation requires that additional specific labeling requirements may be applied to foodstuffs to ensure that the final consumer is informed of

1. Any characteristics or food properties such as composition, nutritional value, or nutritional effects, or intended use of the food, that render a novel food or food ingredient no longer equivalent to an existing food or food ingredient. A novel food or food ingredient is deemed to be no longer equivalent if scientific assessment, based upon an appropriate analysis of existing data, can demonstrate that the characteristics assessed are different from those of a conventional food or food ingredient with respect to the accepted limits of natural variation for such characteristics. The labeling of the foodstuff must indicate the characteristics or properties modified, together with the method by which that characteristic or modification was obtained.
2. The presence in the novel food or food ingredient of material that is not present in the existing equivalent foodstuff and may have implications for the health of certain sectors of the population, or give rise to ethical concerns; or
3. The presence of a genetically modified organism.

Examples may include a genetically modified organism present in a probiotic dairy product or a fiber ingredient produced from a novel food source. Where such novel ingredients are used in a functional food, their presence must be indicated clearly in the labeling of the product. Of course, this does not automatically mean that such foods or food ingredients will be approved as having a functional health-related role in the diet; any such claims must still be backed up by appropriate scientific evidence.

Foods for Particular Nutritional Uses (PARNUTS)

Another consideration for functional foods is whether they will fall under the scope of Directive 89/398/EEC on foodstuffs for particular nutritional uses (the so-called PARNUTS Directive).[2] As a "framework" directive, it does not set standards but prescribes definitions of "foods for particular nutritional uses" and the categories of persons whose nutritional requirements must be met by such foods. PARNUTS foods are defined as foodstuffs that, owing to their special composition or manufacturing process, are clearly distinguishable from foodstuffs for normal consumption, are suitable for their claimed nutritional purposes, and are marketed in such a way as to indicate such suitability. These foods must also have a particular nutritional use that fulfills the nutritional requirements of (1) certain categories of persons whose digestive processes or metabolism are disturbed, (2) certain categories of persons who are in a special physiological condition and who are therefore able to obtain special benefit from controlled consumption of certain substances, or (3) infants or young children in good health. Therefore, it could be argued that functional foods should be considered as PARNUTS foods if their labeling refers to a suitability for a particular nutritional use.

However, it is more than likely that functional foods will differ from foods for a particular nutritional use in that they will be aimed at the general population rather than at those with special dietary or nutritional needs, such as diabetics. If this is indeed the case, then the rules on PARNUTS foods will not apply, but manufacturers must be aware of these rules if they intend to market a functional food as being suitable for a specific target group.

Labeling

There is no specific legislation on claims, including health claims, at one EC level. For some, health claims also cover medical claims; that is, claims that have a specific reference to disease.

Under the EC general labeling directive, 79/112/EEC, as amended[3] (implemented in all the member states), it is specifically prohibited to attribute to any foodstuff the property of preventing, treating, or curing a human disease, or refer to such properties, other than where this is specifically provided for in EC provisions (ie, on mineral waters and foods for particular nutritional uses). Medical claims are therefore not permitted except under these particular provisions.

Furthermore, directive 79/112/EEC requires that labeling generally must not be false or misleading to the purchaser with regard to the characteristics of a food, such as its nature, identity, properties, composition, quantity, durability, origin or provenance, or method of manufacture or production. Therefore, these general rules are applied in the absence of specific provisions on a particular labeling issue.

Health claims in the EU, therefore, generally consist of a reference to a possible disease risk factor, or the effect of a nutrient on the body or other general health claims, such as "good for you" or "a healthy food." In a number of member states, it is therefore acceptable to claim that a product may help to reduce cholesterol, where this can be shown to be justified; however, it is not acceptable to make a reference to heart disease. However, even the acceptability of such claims will vary from state to state, depending on the view of the authorities there, and in the absence of specific legislation it can be difficult for manufacturers to determine what claim is acceptable in each member state. Generally, it is advisable to consult with the enforcement authorities of the member state in question to check the acceptability of specific claims.

UNITED KINGDOM

General Provisions

Within the United Kingdom, there is no legal definition of a functional food, and there are no specific regulations to which such foods must conform. Since the majority of foodstuffs are not subject to specific compositional controls, it is unlikely that specific rules will be drawn up for a separate category of "functional foods."

In the absence of any specific regulations, foodstuffs in general must satisfy the general provisions of the Food Safety Act of 1990,[4] and it is widely accepted that the same system should be applicable to functional foods. The act requires that foods must be of the nature, substance, and quality demanded by the purchaser (Section 14), that they must not be injurious to health (Section 7), that they must comply with food safety requirements (Section 8), and that their labeling must not be false or likely to mislead a purchaser with regard to the nature, substance, or quality of the food (Section 15). In essence, foods must be safe and wholesome and must be labeled accurately and truthfully.

The provisions of the Trade Descriptions Act of 1968[5] are also applicable to foodstuffs, including functional foods, which make it an offense to apply a false or misleading trade description to goods.

Health Claims

Increased recognition of the role of diet in maintaining good health has led to the growth of the market for so-called "functional" foods and the use of health claims in their labeling. Therefore, the need to prevent the use of false, exaggerated, misleading, and prohibited health claims has become a priority issue for the food industry, law enforcement officers, and consumers, not only to protect the consumer but also to promote fair trade in commerce.

It is also quite clear that the use of medicinal claims on foods is not permitted. Unfortunately, there is little flexibility in the interpretation of this rule, which may restrict the use of genuine claims that are backed by generally accepted scientific data and that offer positive health messages to the consumer. For example, in the case of the link between spina bifida and folic acid, where there is clear and generally accepted scientific evidence that folic acid has a role to play in preventing spina bifida, such a claim is prohibited, whereas in the United States the practice is encouraged and there are specific health claims for such allowance.

Currently, there is no specific EC or UK legislation permitting the use of approved health claims (as in Japan and the United States). Manufacturers are obviously wary of making such claims, as substantiation of such claims is ambiguous and such products have been open to criticism from consumer watchdogs.

An example of how difficult it is to get the balance right is the case of the Ribena juice and fiber product that appeared on the United Kingdom market several years ago. The product contained soluble and insoluble fiber ingredients that the manufacturer claimed could help reduce excess blood cholesterol, one of the risk factors in heart disease. Unfortunately, the Advertising Standards Authority ruled that the advertising for the product "exaggerated the drink's likely health benefits" and asked the manufacturers to moderate the claims made.[6]

JHCI Code of Practice on Health Claims

In February 1996, the Ministry of Agriculture, Fisheries and Food's Food Advisory Committee (FAC) was asked to carry out a review of health claims, focusing particularly on functional foods. The FAC issued its conclusions and draft guidelines on health claims for comment in December 1996; however, following the 3-month consultation period, their progress was delayed owing to a general election.

As a result, in June 1997, the Food and Drink Federation, the National Food Alliance, and the Local Authorities Coordinating Body on Food and Trading Standards established the Joint Health Claims Initiative (JHCI) to further advance the work already done in this area and to issue a code of practice on health claims. The JHCI approach was that of a partnership between food manufacturers and retailers, enforcement officers, regulators, and consumers.

In the development of the code, a strong consensus emerged that the present legal and enforcement framework governing claims was both incomplete and inflexible. All members of JHCI believed that there was a need to clarify and strengthen the requirements for evidence to substantiate health claims. They also believed that once scientific validation had been established, claims should be able to express clearly and more directly the increasing variety of relationships between foods or food ingredients and human health now being documented by research. The code attempts to add both rigor and flexibility to the existing system. In particular, the provisions for "generic" and "innovative" claims build in the capacity to adapt to emerging scientific research. Nonetheless, all partners of JHCI believed that there was a need for the government to review the existing law on food claims in light of scientific advances, in particular because the prohibition on foods claiming to prevent, treat, or cure diseases can inhibit communication of the role of a healthy diet in maintaining good health and reducing the risk of disease. Accordingly, a recommendation has been made to ministers to reassess the existing law and its interpretation in order to facilitate the communication of beneficial scientifically valid information to consumers in terms they can understand.

At the time of writing, a final draft of the JHCI Code of Practice on Health Claims on Foods had been circulated during August 1998.[7] The code outlines the general principles to be applied when making a health claim and contains details on the substantiation of health claims and examples of the types of health claim that may be considered acceptable.

The objectives of the code include

- protecting public health
- providing accurate and responsible information relating to food to enable consumers to make informed choices
- promoting fair trade and innovation in the food industry
- promoting consistency in the use of health claims in the United Kingdom, Europe, and internationally

Status of the Code

Although the code has no direct legal status, it aims to clarify and augment the existing legislation and to complement existing codes of practice and guidelines. Compliance with the code should assist companies to establish a defense of due diligence if prosecuted for a health claim under the Food Safety Act of 1990. Instead, it is proposed that to meet these objectives a code administrative body be established to assist enforcement authorities in deciding whether a health claim is legal and scientifically justified.

General Principles for Making a Health Claim

The Code of Practice on Health Claims on Foods allows for the making of acceptable health claims, which are defined as any claims, whether direct, indirect, or implied in food labeling, advertising, and promotion, that a food has specific health benefits. These include nutrient function claims describing the physiological role of the nutrients in growth, development, and normal functions of the body (eg, "Calcium aids in the development of strong bones and teeth"), but do not include nutrient content claims (eg, that a food is low fat, reduced cholesterol, or high fiber).

Under the code, the use of "generic health claims" would be acceptable. A generic health claim is defined as a health claim based on well-established, generally accepted knowledge and/or evidence in scientific literature and/or recommendations from national or international public health bodies, such as the Committee on Medical Aspects of Food and Nutrition Policy (COMA), the US Food and Drug Administration, and the EU Scientific Committee for Foods.

The general principle in making health claims outlined in the code is that health claims should assist consumers to make informed choices. Consumers expect that health claims are substantiated and have been checked for accuracy by independent experts prior to use and will continue to be controlled by the enforcement authorities. Therefore, the code should be applied in the spirit as well as by the letter of the law. Certain factors must be taken into consideration when deciding upon the acceptability of a health claim, including marketing imagery, careful use of words, and literature given out with the product.

Principles of the Code

The principles of the Code are as follows:

1. Health claims must be truthful and must not mislead, exaggerate, or deceive, either directly or by implication. (Section 6.2.1)
2. A claim must be consistent with the nature and scope of the evidence. If data are collected from a representative cross section of the population, then the claim should extend to the whole population. However, if the evidence is collected from a specific target group, then the health claim should refer to a benefit only for that target group. (Section 6.2.2)
3. The use of medicinal claims is prohibited. References to specific diseases should be avoided, as this is likely to be regarded by the consumer as implying that the food has a medicinal effect

in relation to that disease. Nonspecific references to disease may imply that the food will have a medicinal effect in preventing disease in general. (Section 6.2.3)

4. It is acceptable to refer to the maintenance of good health in general or to a specific part or organ of the body; for example, "Food 'x' helps to maintain a healthy heart" or "Food 'x' helps to keep your body healthy." (Section 6.2.4)

5. It is also acceptable to refer to risk factors that may adversely affect good health; for example, "Food 'x' helps to keep your cholesterol levels healthy, which helps to maintain a healthy heart." However, any such reference must indicate that the overall benefit is within the context of a healthy diet and lifestyle, with the aim of reducing the risk of disease rather than having any preventative effect on the development of the disease. (Section 6.2.5)

6. Health claims must not encourage or condone excessive consumption of any food or disparage good dietary practice. The health claim must be made in the context of the role of the food in relation to the overall diet or other lifestyle factors (eg, "if eaten as part of a low-fat diet") unless the evidence indicates that this is inappropriate or unnecessary—for example, in the case of the effect of folic acid. (Section 6.2.6)

7. Health claims should not denigrate any other food or imply that normal foods cannot provide a healthy diet. (Section 6.2.7)

8. The benefit from the health claim must be gained from the food as intended to be consumed and as recommended by the manufacturer or the instructions for use given on the label. (Section 6.2.8)

9. If the health claim relates to any general properties of the food or any of its ingredients or components, this must be made clear to the consumer. (Section 6.2.9)

10. The benefit from the health claim must be derived wholly from the food for which the health claim is made and must not rely on any benefit derived from consuming the food with other foods, even if this may be normal practice or the intended mode of consumption, as in the case of breakfast cereal with milk. (Section 6.2.10)

11. Health claims of synergistic benefits are acceptable; for example, the bioavailability of iron improving when taken in conjunction with vitamin C. (Section 6.2.11)

12. Health claims relating to ingredients or components of the food must be based on good evidence of a likelihood of benefit in the target population or of a clear benefit from either reducing or increasing the intake of a particular substance. (Section 6.2.12)

13. The health claim must be fulfilled in the target population when the food is consumed in the quantities that can reasonably be expected to be consumed in one day. (Section 6.2.13)

14. Health claims must be communicated in such a way as to promote consumer understanding of the relationship between diets, specific nutrients or components, and the physiological benefits, to allow people to make informed and appropriate food choices. (Section 6.2.14)

15. Vulnerable sectors of the population, such as pregnant women, lactating mothers, children, and the elderly may have particular nutritional requirements, and companies must take care to ensure that any health claims do not mislead these sectors. Likewise, health claims directed at specific sectors of the population should not be presented in such a way as to mislead the general population. General health claims that may apply differently to specific sectors of the population must be explained. (Section 6.2.15)

16. Health claims that could encourage high levels of consumption must not be made for any substance where there is evidence that high intakes of the food or substance could be harmful or unlikely to contribute to a healthy diet. (Section 6.2.16)

17. A food for which a health claim is made should have a nutritional profile at least equivalent to that of other foods typical of that group. (Section 6.2.17)

Labeling

In addition to the labeling particulars required by the Food Labeling Regulations of 1996, as amended, the code recommends that where a health claim is made, manufacturers should provide additional labeling information to assist consumers' understanding of the significance of the claim. The code suggests inclusion of the following information:

- full nutrition labeling information
- details of a quantified serving size
- details of the target population and/or any persons who should avoid consumption of the food
- a statement indicating the quantity of the food and pattern of consumption required to achieve the beneficial effect
- where there is evidence that high intakes could be harmful, a safe maximum intake
- wherever a food component provides the basis for a health claim, a declaration of the amount of this functional ingredient in 100 g and a serving of the food

Substantiation

The code makes a distinction between well-established health claims (for the purposes of the code, called "generic" claims) and "innovative" (new) health claims. In the case of generic health claims, no specific substantiation would be required, and the manufacturer could use the claim without restriction or further documentation, although all other parts of the code would still apply. It is intended that a list of generic health claims will be drawn up by the code administrative body, with participation from industry, and will be regularly reviewed and updated as necessary in light of any new scientific evidence or consensus. In the case of innovative health claims, substantiation or scientific evaluation of the health claim is considered essential. Companies must show that the health claim is true and that scientific evidence in support of the claim outweighs opposing evidence or opinion.

When making an innovative health claim, companies must be able to demonstrate

1. that the food (or components) in question will cause or contribute to a significant and positive physiological benefit when consumed by the target population as part of their normal diet
2. that the claimed effect can be achieved by consuming a reasonable amount of the food on a regular basis or by the food's making a reasonable contribution to the diet
3. that the effect is maintained over a reasonable period of time and is not a short-term response to which the body adjusts, unless the resultant health claim is relevant for a short- or medium-term benefit (eg, folic acid in neural tube defects)
4. the minimum or maximum amount and the frequency of consumption required to achieve the effect or that the food provides a reasonable dietary contribution to the amount required to achieve the effect
5. who can benefit from the effect—for example, whether this is the entire population, "at-risk" groups, or target groups
6. how the effect is brought about, although the exact biological mechanism(s) need not be fully understood or explained
7. any potential effects on vulnerable groups, such as the elderly, pregnant women, or children

Sources of Evidence

The health claim must be based on a systematic review of all the available scientific evidence relating to the validity of the health claim. Substantiation of the claim must be based on the totality of the evidence (not just data in support of the claim) and on studies or evidence in humans. The code accepts that gathering full clinical evidence on foods can be difficult, so evidence of the effects of the food on established biomarkers is considered acceptable; for example, if a food reduces cholesterol, this can also be taken to reduce the risk of heart disease.

Documentation

All evidence must be able to stand up to peer review and be fully documented. The documentation should include summaries of the evidence on which the health claim is based and should also include a description of the food and its ingredient, its intended use, any warnings (potential allergies, etc), a copy of the product label, and any advertising documentation. The evidence should be made available on request, although there will be exceptions for commercially sensitive information.

Code Administrative Body

It is proposed that the code be supported by a code administrative body consisting of a council and a secretariat. The council would monitor the code and decide on policy matters, and the secretariat would administer the code. An "expert authority" would also be established who would provide consistent, independent, and credible scientific advice to enforcement agencies and regulators and would give premarket advice to companies. Three options have been put forward as to how the expert authority should be formed. One option is to set up a new expert committee, another is to contract out the work to a reputable nutrition or academic unit; and still another is to use an existing expert committee, such as COMA or FAC. Each option has its advantages and disadvantages, but it is clear that the expert authority must be seen to be both credible and independent and must be acceptable to all parties of JHCI, as it will be the key to the successful operation of the code.

FRANCE

General Provisions

Like most of the EU member states, France does not currently have any specific regulations on functional foods or health claims. General provisions that foodstuffs must be safe and suitable for consumption apply to the sale of functional foods, and provisions on false or misleading labeling are applicable in the area of health claims.

Health Claims

CEDAP Opinion on Health Claims

In 1997, the Inter-Ministerial Commission for the Study of Products Intended for a Particular Diet (CEDAP), an advisory body to the French authorities, published an opinion on health claims.[8] This opinion, dated December 1996, aimed to assist the authorities in appreciating the nonmisleading character of functional nutritional claims.

Among other issues, the opinion called for guidelines to be defined that would set conditions under which the role of a nutrient or substance might be expressed without misleading the consumer and suggested provisions allowing for nonmisleading claims to be made. It was concluded that, when consumers were informed of the presence of a nutrient or substance in a food, they could also be informed of the role of the nutrient or substance whose presence was emphasized in the labeling. In concurrence with Swedish opinion, it was considered consistent for such action in labeling to be taken in terms of providing nutrition information to the public. According to the commission, the choice of exact wording could be left to the manufacturer, provided properties that were scientifically established were used. Expressions such as "contributes to" or "is necessary for"—nonmisleading concepts rather than so-called "word-for-word" statements—were suggested. Specific examples discussed included the role of calcium in bone construction and the mineralization of bone tissue, the role of iron in the functioning of hemoglobin and the red blood cells, and the role of vitamin A in vision.

CNA Opinion on Health Claims

Furthermore, the National Food Council (CNA) issued an opinion relating to functional foods and health claims in June 1998.[9] CNA proposed that various conditions be followed for making claims relating to diet and health, including functional claims, so that such claims would be beneficial to consumers. It was recognized that scientifically sound claims can be objectively made for certain food products under precise consumption conditions.

CNA believes that the following types of health claims may be acceptable, under certain conditions:

- *Claims that a food may help in reducing the risk of disease*, provided that the manufacturer obtains prior authorization from the authority (express authorization), unless these appear on a positive regulatory list
- *Claims that a food makes a positive contribution to the state of health*. CNA was unable to reach a consensus on how such claims should be regulated but proposed several different options, including express authorization for each claim, passive authorization (ie, notification and a delay period before marketing), and notification at the time of marketing with consequent evaluation only (this later option was favored by the Board of Food Processing Industries and was also accepted by a number of experts, provided that a trial period would be included)
- *Functional nutritional claims*, which describe the positive role of a nutrient in the normal functions of the body, provided that they comply with the regulation on nonmisleading advertising and are approved either by reference to a reference list or by an independent reference body

Furthermore, CNA concluded that the following conditions should be applied when making health claims:

- Therapeutic (medicinal claims) on foods are prohibited.
- The same conditions for assessing health claims must be applied to all products intended for human consumption, whatever their nature or origin.
- It may be necessary to allow a health claim initially for a brief trial period, which will allow assessment of the impact of the health claim, how it might affect dietary behavior, and so on.
- Claims should be made only where there is knowledge of the link between the nutrient and health and must be placed in the context of the total diet (ie, the benefit of a healthy, balanced diet), the benefit of exercise, and so on.
- There is need for clarification of the regulatory system, which must be effective and proportional to the level of presentation of the health claim, to the extent that the current legislative framework requires adaptation in order to ensure the protection of consumers, their safety, and the fairness of exchanges.
- The scientific evaluation of health claims must be carried out by independent, expert bodies.
- Good practice guidelines for claims should be developed within the general framework of fair and nondeceptive advertising.
- Responsibility should be imposed on all parties involved in the food chain using health claims.

CNA recognized the need for improvement of consumer knowledge of the link between diet and health and proposed improvements in the current system for educating consumers and the development of a nutritional policy.

SWEDEN

Health Claims

In Sweden, an Industry Guide on Health Claims in the Labeling and Marketing of Food Products[10] was drawn up in 1990. The authorities considered that an increasing number of Swedish consumers were interested in matters relating to health and diet, including "added-value" nutritional products such as functional foods. This led to the desire of manufacturers to use health claims in the marketing of their food products, so, to protect consumers from misleading health claims, a clear distinction had to be made between nutrition, health, and medical claims.

The guide defines a health claim as an assessment of the positive health effects of a foodstuff—that is, a claim that the nutritional composition of a product can be connected with prophylactic effects or the reduced risk of a diet-related disease. The health claim must be consistent with official Swedish dietary and nutrition recommendations; it must be based on scientific facts generally accepted in Sweden and must be formulated to take into account the need for a balanced diet containing all different nutrients. In addition, marketing and information material containing health claims should be designed to give the consumer good insight into the connection between diet and health.

The general rules state that health claims in marketing should be used in such a way as to create confidence in food and the food industry, thus ensuring that health claims are not used to damage consumer confidence in specific food products or food in general. The claim must consist of two parts: information on a diet-health relationship, followed by information on the composition of the product (known as the two-step principle); for example, "To prevent osteoporosis in later life, it is important to eat calcium-rich food and to exercise. X is a calcium-rich food." Although not an exhaustive list, examples of approved and nonapproved health claims are shown in Table 17–2). It is interesting to note that claims relating to the suggested health benefits of probiotic microorganisms are not permitted, as the Swedish authorities believe there is no consensus on their validity. Nor are general claims referring to diet and cancer allowed.

Furthermore, the Swedish authorities have determined that products are being marketed with health claims given in accordance with these self-regulatory guidelines, examples of which are shown in Table 17–3.

Proposal for Product-Specific Health Claims

The guidelines currently prohibit the making of any health claim based on the statement that an individual product has a specific health effect. Such claims may be made only on application to the National Food Administration for approval to market the product as a PARNUTS or to the Medical Products Agency with an application for registration as a natural remedy. However, this approach is now considered to be quite restrictive, and manufacturers may wish to make product-specific health claims. At the time of writing, a proposal to extend the permitted health claims to include product-specific physiological claims had been published.[11] Under this proposal, a product-specific physiological claim is defined as a claim concerning the health-promoting effects of the product itself, which has been manipulated in a special manner to provide a specific effect. The product must be shown to make a positive contribution to a nutritionally adequate diet. Such a claim must be based on

Table 17–2 Examples of Approved and Nonapproved Health Claims (Sweden)

Category of Health Claim	Example of Approved Health Claim	Example of Nonapproved Health Claim
Obesity: Obesity, or over-weight, can be prevented (and treated) through a diet that has a low or reduced energy content. Energy reduction through lower amounts of fat is especially important.	According to the Swedish recommendations, intake of food containing a moderate amount of fat, combined with exercise, is the basis for maintaining a healthy weight. The product X has a low fat content and therefore contains less energy (fewer calories) than ordinary butter or margarine.	X—the perfect slimming aid.[1]
Blood cholesterol level: Reduction in saturated fat intake, by lowering total fat or by replacing saturated fat by mo-nounsaturated or polyunsaturated fat, may help to reduce the blood cholesterol level. As *trans*-fatty acids also increase the level of cholesterol in the blood, health claims should not be made for products with high levels of *trans*-fatty acids. Some types of soluble, gel-forming dietary fiber can contribute to a reduction in the cholesterol level.	Saturated fatty acids increase the level of cholesterol in the blood. The product X contains low levels of saturated fat and total fat.	Eating X will improve your blood cholesterol level.[t] Y will help you reduce your blood cholesterol level.[t]
Blood pressure: A reduction in salt (sodium chloride) consumption can counteract high blood pressure. The use of mineral salts containing potassium can contribute to a reduction in sodium intake, but there is no other documented positive effect of mineral salts on blood pressure.	In high quantities, ordinary salt may increase the risk of high blood pressure. X has a low salt content.	Salt X will lower your blood pressure.[t]
Atherosclerosis: High levels of blood cholesterol and high blood pressure are diet-related risk factors for atherosclerosis. Measures that lead to a reduction in these risk factors thus reduce the risk of atherosclerosis and associated cardiovascular diseases.	High levels of cholesterol in the blood increase the risk of cardiovascular disease. The intake of saturated fat contributes to an increase in the level of cholesterol in the blood. X contains a low amount of saturated fatty acids.	X provides excellent protection against cardiovascular disease through its low content of saturated fatty acids.[t] X has a low salt content and therefore protects against cardiovascular disease.[t]

continues

Table 17–2 continued

Category of Health Claim	Example of Approved Health Claim	Example of Nonapproved Health Claim
Naturally occurring omega-3 fatty acids, as found in abundant levels in fish oils and fish products, can afford some degree of protection against cardiovascular diseases.	High blood pressure is a risk factor in cardiovascular disease. By restricting your intake of salt, it is possible to reduce the risk of high blood pressure. X has a low salt content.	
	Omega-3 fatty acids have a positive effect on blood lipids and can therefore help protect against cardiovascular disease. Fish X or the fish product X is rich in omega-3 fatty acids.	Omega-3 fatty acids have a positive effect on blood lipids and can therefore help protect against cardiovascular disease. Bread X is (Y's eggs are) rich in omega-3 fatty acids.
Constipation: Dietary fiber speeds up the passage of food through the intestinal tract and counteracts constipation caused by low fiber intake.	It is important to eat sufficient amounts of dietary fiber to prevent constipation. X is high in dietary fiber.	X prevents constipation, thanks to its high dietary fiber content.[†]
Osteoporosis: Osteoporosis can be counteracted by an intake of calcium equivalent to the recommended daily intake, physical exercise, and abstention from smoking.	To prevent osteoporosis in later life, it is important to eat calcium-rich food and to exercise. X is a calcium-rich food.	X gives strong bones.[†]
Caries: The absence of sugars and other easily fermentable carbohydrates in food that is eaten often and between meals reduces the risk of caries.	When you eat food containing sugar or other carbohydrates that are easily broken down by bacteria in the mouth, acid is formed on the teeth. The more often this happens, the greater the risk of caries. X chewing gum is sugar-free.	Because X chewing gum contains xylitol, it offers excellent protection against caries.[†]
Iron deficiency: Food that contains significant amounts of iron is important in preventing iron deficiency. Components that stimulate and inhibit iron absorption should also be taken into consideration.	Iron deficiency is common among women but can be prevented by good dietary habits. X is an important source of the type of iron that is readily absorbed in the body.	X is the best medicine for iron deficiency.[†]

*Low-energy and energy-reduced foods intended for weight control must be registered as foodstuffs intended for particular nutritional uses (PARNUTS) (SLVFS 1991:12) Statens Livsmedelsverk Forfattningssamling [National Food Administration Ordinance]).

[†]The claim that a product has specific effects may not be made under the food industry's self-regulating program. If health-promoting claims are to be used in the marketing of a product as such, and if these can be documented, it is possible to report the product or apply for registration as a foodstuff intended as a natural remedy (the Medical Products Agency).

Source: Adapted from *Health Claims in the Labeling and Marketing of Food Products: The Food Industry's Rules* (Self-Regulating Programme), pp. 10–11, 28/8/96, Sweden.

Table 17–3 Types of Health Claims Currently Used in Accordance with the Swedish Self-Regulatory Guidelines

Category of Health Claim	Number of Products on Market in Sweden Bearing a Health Claim
Obesity	5
Blood cholesterol levels	6
Blood pressure	1
Atherosclerosis	1
Constipation	21
Osteoporosis	0
Caries	0 (excluding chewing gum)
Iron deficiency	1

sound scientific evidence that is unobjectionable and subject to peer review; the data must be relevant to the consumption of the food in humans as part of a normal diet; and effects must be shown to be lasting. The claim must also take into account any target groups.

CODEX ALIMENTARIUS

The general principles outlined in this chapter form the basis for the provisions laid down by Codex Alimentarius, the international body responsible for the execution of the Joint Foods Standards Programme of the Food and Agricultural Organization (FAO) and the World Health Organization (WHO). Owing to the increasing influence of Codex under the World Trade Agreement, Codex provisions will continue to gain importance in the international arena, so it is important to look at Codex controls on functional claims. General Codex principles on claims require that no food be described or presented in a manner that is false, misleading, or deceptive or is likely to create an erroneous impression regarding its character in any respect. In addition, the person marketing the food must be able to justify the claims made.

Guidelines on Nutrition Claims

Codex Alimentarius Guidelines for the Use of Nutrition Claims were adopted by the Codex Alimentarius Commission in 1997 (CAC/GL 23-1997).[12] These guidelines have been sent to all member nations and associate members of the FAO and WHO as an advisory text, and it is up to the individual governments to decide how they make use of these guidelines. In addition to the requirements on nutrient content claims, they include provisions on nutrient function claims—for example, "Protein helps build and repair body tissues" and "Calcium aids in the development of strong bones and teeth"—and on claims related to dietary guidelines or healthy diets.

Proposed Recommendations on Health Claims

A Codex proposal was issued in April 1997 that, if and when adopted, would replace the above provisions on nutrient function claims. The proposed draft "Recommendation for the Use of Health Claims" (at step 3 of the Codex adoption procedure)[13] defined a health claim as any representation that states, suggests, or implies that a relationship exists between a food or a nutrient or other substance contained in a food and a disease or health-related condition. Examples of the type of health claim that may be permitted under these recommendations are listed in Table 17–4. Under these current proposed recommendations, a health claim that a food or nutrient or substance contained in a food has an effect on an adverse health-related condition in the body (in other words, a medicinal claim) is not permitted. However, a health claim that the consumption, or reduced consumption, of a food, nutrient, or substance contained in a food, as part of a total dietary pattern, may have an effect on a disease or health-related condition is permitted if it meets certain conditions:

- There is scientific consensus supported by the competent authority that a relationship exists between the food, nutrient, or substance and the disease or adverse health-related condition.
- The wording of the claim is within the context of total dietary patterns.
- The food for which the claim is made must be a significant source of the nutrient or substance if increased consumption is recommended, or "low" in or "free" from the nutrient or substance if reduced consumption is recommended.
- The claim must not state or imply that consumption of a particular food can cure, prevent, or treat a disease.
- The claim must not be made if the consumption of the food would result in the intake of a nutrient or substance in an amount that would increase the risk of a disease or health-related condition. The kind and amount of the nutrient or substance should be clearly specified.

As part of the discussions on the proposal, several delegations, including the EC observer, pointed out that their national legislation did not currently allow the use of claims related to the prevention, cure, or treatment of diseases but that there had been debate on the relationship between health and diet. Some of the delegations, including consumer organizations, were not in favor of health claims in general; however, as these were appearing on food found in the market, there was a need to consider

Table 17–4 Types of Health Claims Permitted in the Proposed Draft Codex Recommendations for the Use of Health Claims on Foods

Type of Health Claim	Examples of Health Claims
Health-related effects on the body directly attributed to a food or nutrient or substance	X fish oil lowers serum triglycerides and increases clotting times.
	X bran lowers blood cholesterol levels.
	X vegetable oil is low in saturated fat and will help reduce blood cholesterol levels.
	Contains soluble fiber that lowers blood cholesterol.
	Contains sorbitol. Polyols are more slowly absorbed than sugars and decrease the insulin response.
Disease prevention attributed to a nutrient or substance contained in a food	X contains soluble fiber that reduces the risk of heart disease.
	X is low in saturated fat, which reduces the risk of heart disease.
Disease prevention or health-related effects related to a food	A low-fat diet will reduce the risk of cancer. X is a low-fat food.
	Saturated fat raises blood cholesterol levels. A diet low in saturated fat will reduce blood cholesterol levels and reduce the risk of cardiovascular disease. X is low in saturated fat.

Source: Data from *Report of the Twenty-Sixth Session of the Codex Committee on Food Labeling,* ALINORM 99/22, Appendix X (Proposed Draft Recommendations for the Use of Health Claims (Step 3)).

this issue further in terms of consumer information and education in health and nutrition matters. The committee noted that the scientific basis of health claims must be considered in more detail and that it would be useful to refer this matter for advice to the Codex Committee on Nutrition and Foods for Special Dietary Uses.

The committee generally agreed that health claims should not refer to one single food but should be placed in the context of the total diet and that they should be substantiated by scientific evidence. The committee noted that there was need for further discussion on the issue of health claims, and agreed to return to the text to step 3 for further comments and consideration at its next session and to consult the Committee on Nutrition and Foods for Special Dietary Uses for advice regarding the scientific basis of health claims.

NETHERLANDS

In the Netherlands, there are no specific regulations on functional foods or health claims, although various codes of practice with respect to labeling and advertising deception apply.

Advertising Code of Practice

The Inspection Board for the Public Promotion of Medicines (*Keuringsraad Openlijke Aanprijzing Geneesmiddelen* KOAG) and the Inspection Board for the Promotion of Health Products (*Keuringsraad Aanprijzing Gezondheidsproducten* KAG) have published two codes of practice, one on the advertising of medicines and the other on the promotion of health products.[14] The main body of interest here is KAG, which is responsible for the supervision of the public advertising, including labeling, of health products. It is important to be aware that all advertising of health products must be submitted to KAG for assessment and authorization prior to marketing of the product.

The Code Governing the Promotion of Health Products is to a large extent based on Articles 19 and 20 of the Consumer Goods Act, which stipulates that it is prohibited to state or imply that consumer goods have properties to prevent, treat, or cure diseases.

Health products are defined under the code as products in a pharmaceutical form that have a pharmaceutical appearance or for which a primary health-related function is claimed. Health products are not intended to be used as medicines and are covered by the Consumer Goods Act. Products that consumers can clearly distinguish from medicines, such as foods, are not considered "health products," even if a health-related claim is made. Therefore, a food product, the primary function of which is the provision of nutrients, is not considered a "health product," even if a health claim is made.

A list of the claims permitted for both medicines and health products forms an important part of this code; each possible claim is listed as to whether it is permitted or prohibited for use as a health product. The lists are categorized into claims for body organs and systems (eg, heart and blood circulation), by physiological functions (eg, menstruation), and by adverse effects, such as allergies.

Code of Practice on Scientific Evaluation of Health Claims

A code of practice assessing the scientific evidence for health benefits stated in health claims on food and drink products was drawn up by the Netherlands Nutrition Center (Voedingscentrum) in 1998.[15] Although the code is voluntary, it gives manufacturers, importers, and distributors of food and drink products a standard against which the validity of health claims can be measured. One aim of the code is to increase consumer confidence in the use of health claims on food and drink products.

This code was drawn up in conjunction with industry, consumer organizations, advisory services, scientific institutes, and the Dutch government and took about 3 years to complete.

The code defines a health claim as a direct, indirect, or implied claim, which may be made in the labeling, advertising, or presentation of a food, that a food has special qualities that can improve or maintain the consumer's health. Imported products are also subject to the provisions of the code.

The code does not apply to medicinal claims, which, according to the Dutch Commodities Act, are defined as any statement, depiction, or suggestion that attributes preventative or curative effects to a food or drink product. Any such medicinal statements or depiction on foods or drinks are prohibited. Nor does the code apply to nutrient content claims.

The scientific evidence necessary to substantiate any health claim is assessed by an independent panel of experts at the request of the person wishing to make the health claim in the marketing of a product. The expert panel considers only the scientific evidence for the health claim and not the literal text in which the evidence is used (ie, whether the perceived health benefit is correctly presented in the health claim), which is considered the province of other organizations, such as the Advertising Code Foundation and its appeals tribunal.

The scientific evidence is assessed using the following criteria:

1. The scientific evidence must be of high quality.
    - It must be based on relevant scientific data on human subjects.
    - It must apply to a specific product or product group; evidence relating to the properties of an ingredient or component is not acceptable. If the functionality of an ingredient or substance has been proven in prior research, it must also be shown that its properties are not reduced when used in the product concerned.
    - The evidence must be reproduced.
2. The data must relate to the normal use (consumed quantities) by the target population, and the health claim must relate to that target group.
3. The health benefit must not conflict with dietary guidelines.

There is no distinction in the Dutch Code between "generic" and "new" health claims, as is the case in the United Kingdom's code of practice.

The applicant must submit a dossier of evidence to the panel of experts, who have 3 months in which to consider the application for approval of the health claim. All information relating to the application remains confidential until a final assessment report has been issued (and is not subject to appeal) and the product has been placed on the market bearing the health claim approved under the code of practice.

BELGIUM

At the time of writing of this chapter, Belgian authorities were in the initial stages of considering the need for controls on health claims on foods. Three options for action were proposed: (1) Do nothing (existing controls would be sufficient); (2) Issue strict legislation to control health claims that would require prior approval for any health claim and would include both positive and negative wordings of permitted health claims; and (3) Opt for a self-regulation scheme.

Following an informal consultation, it was decided that a voluntary code of practice on health claims would be drawn up. The Belgian code of practice seems likely to be very similar in content to both the British and Dutch codes in that an administrative body would be formed to consider the acceptability of an applicant's health claim on the basis of the criteria for substantiation outlined in the code. Again, scientifically evaluated and generally agreed-on evidence would be required to substantiate any health claims made on foods. The final agreed-on code of practice was due to be issued during 1998.

JAPAN

Foods for Specified Health Use (FOSHU)

It is well recognized that the concept of functional foods originated in Japan. The Japanese authorities had been concerned for some time about changes in the national diet, which meant that certain nutrients, such as calcium and fiber, were deficient. Because the Japanese felt that the link between some nutrients and health was clear, a new category of foods was introduced in 1991. The aim was to raise consumer awareness and provide an easy way for consumers to increase their intake of the components that were felt to be most beneficial. To achieve this, the Japanese Ministry of Health and Welfare introduced a new category of FOSHU to add to those already regulated by its Nutrition Improvement Law.[16]

Foods for specified health use are defined as "foods, based on the knowledge concerning the relationship between foods or food components and health, that are expected to have certain health benefits and have been licensed to bear labeling claiming that a person using them may expect to obtain that health use through the consumption of these foods." Many foods for specified health use contain "effective ingredients" added to help in the maintenance of a healthy body, such as dietary fiber, sugar alcohols, oligosaccharides, proteins, lacto- or bifido-bacilli, and chitosan, which have been approved by the authorities as having a specific health benefit.

Once a food has been approved as being a food for specified health use, it may bear an approved statement indicating its specific health benefits and must bear the Ministry of Health and Welfare's permission/approval mark for FOSHU.

By 1998, approximately 100 different products had been permitted as FOSHU foods by the Japanese authorities. Examples of foods approved for specified health use and the types of health claims allowed are outlined in Table 17–5. However, in many cases manufacturers chose not to make claims about their products and relied instead on the Japanese consumer's awareness of the beneficial effects of the particular ingredient. If manufacturers do not make a claim, they do not need to have their foods licensed.

Licensing of FOSHU

Licensing of foods for specified health use is carried out on a product-by-product basis, and a license will be issued to food products only when they have undergone a comprehensive clinical and nutritional evaluation of the health benefits of the food, conducted by a committee of experienced specialists, and when the product's safety has been confirmed. For a food for specified health use to be approved/permitted, it must satisfy the following criteria:

1. The food must be expected to provide an increased maintenance of health by improvement of the diet.
2. The health benefits of the food or its relevant component must have a clear medical and nutritional basis.
3. An appropriate level of consumption of the food or its relevant component should be defined, based on medical and nutritional knowledge.
4. The food or its relevant components must be safe as judged from experience.
5. The relevant components should be well defined in terms of
   • Their physical and chemical properties and analysis methods
   • The methods of qualitative and quantitative analysis
6. There should be no significant loss in the nutritional composition of the product compared to the composition of nutrients that are normally contained in similar types of foods.

Table 17–5 Examples of Foods Approved for Specified Health Use in Japan

Type of Food	Functional Ingredient	Permitted Health Claim
Lactic acid bacteria drink	Xylo-oligosaccharide	This product is made with xylo-oligosaccharides, which promote the correct growth of *Bifidus* bacteria in the intestines. This helps to keep the environment in the intestines satisfactory and regulates the gut. Warning required: excessive intake causes a loosening of the bowels.
Soft drink	CCM (citric acid, malic acid, and calcium)	This drink is suitable to supplement calcium (which may be present in an insignificant amount in a normal diet), since it contains a calcium ingredient with enhanced bioavailability.
Chewing gum	Palatinose, maltitol	This gum is nice to teeth.
Tofu	Soya protein	This product contains isolated soya protein, which prevents the absorption of cholesterol and is useful for the improvement of dietary patterns of people with high blood cholesterol levels.
Frankfurter	Soya protein	This sausage is designed to reduce absorption of cholesterol and helps the improvement of the diet of people who are fond of meat but are concerned about high levels of cholesterol.
All-bran cereal	Wheat bran	Since this product is made from wheat bran that is rich in dietary fiber, it helps keep your stomach in good condition.
Soft drink	Casein dodecapeptide	This food contains casein dodecapeptide and is a neutraceutical beneficial for people with mild hypertension.
Biscuit	Chitosan	This biscuit contains chitosan, which suppresses the absorption of cholesterol and is useful for people with high blood cholesterol levels or those who are careful about their intake of cholesterol.
Soft drink	Heme iron	This product is good for people who often suffer from anemia and who need a supplement of iron.
Chocolate	Green tea polyphenols, palatinose	Green tea polyphenols and palatinose do not cause dental caries. This chocolate is unlikely to cause dental caries.
Yogurt	*Lactobacillus acidophilus, Bifidobacterium longum*	*Acidophilus* bacteria and *Bifidus* bacteria improve the environment in the intestines and help to regulate the intestines.

Source: Reprinted from Ministry of Health and Welfare, Japan.

7. The product should be a type of food that is normally consumed as part of an ordinary diet, rather than a food that may be consumed occasionally.
8. The product should be in the form of ordinary foods and not, for example, as a pill or capsule.
9. The food or its relevant component must not be used exclusively as a medicine.

The applicant must submit specific information to the Ministry of Health and Welfare certifying the product's efficacy and safety. This information should include a sample of the labeling of the product and documents outlining the medical and nutritional health benefits of the food or its ingredients, including details of the scientific basis of such health benefits and documents outlining the safety of the food or its ingredients.

However, the original approval process that has been in operation since 1991 has proved to be both lengthy and costly. The Ministry of Health and Welfare felt that consumers had an interest in encouraging more manufacturers to submit products for approval, so in November 1997 the ministry revised the FOSHU approval system to reduce the burden on the applicant and to shorten the process. The most important of the changes to the system, which came into effect in early 1998, included shortening of the application procedure to permit the manufacturer's own analytical testing of the product, where previously testing by the National Health and Nutrition Laboratory had been required. In addition, there was reduction in the documentation required in support of an application. Other changes included removal of the time limit on the approval (previously approval had been granted only for a period of 4 years) and removal of the need for separate application for each variation of a product, so as to allow changes in the scope of the FOSHU-approved product after initial approval had been granted.

Labeling

Once a product is licensed as a food for specified health use, it must be labeled with details of the brand name of the food, a list of ingredients, name and address of the manufacturer, permitted/approved health claim, reasons for permission/approval, indication that the product is a FOSHU, and nutrition information, including energy value, list of ingredients, shelf life, recommended daily consumption, and instructions for use. However, the authorities may require other specific details to be declared as a condition of the approval, such as warning statements.

CONCLUSION

In summary, it can be seen that in Europe there is no harmonized regulatory framework under which functional foods must be marketed, and hence, the general controls on marketing and labeling of foods apply. This is the exact opposite of the strict regulatory framework for FOSHU foods that has been adopted in Japan.

In the absence of specific EC legislation controlling the composition of functional foods or the use of health claims, several of the EU member states are going down the route of labeling controls and industry self-regulating guidelines to control functional foods. In this climate, the future of any functional food product depends on exactly how it is marketed and how the health benefits of the product are conveyed to the consumer by way of labeling. The main role of the regulators is to ensure that consumer protection and confidence are maintained, bearing in mind that the industry must not be stifled by over-regulation to the detriment of consumer choice. Health claims, therefore, need to be carefully worded to avoid any reference to disease. In theory, claims that relate to risk factors, such as cholesterol levels, should be acceptable provided they can be justified; however, there may be differences in the tolerance of such claims between the EU member states. However, because most member states simply prohibit medical claims and do not have legislation on the use of health claims, the acceptability of any specific health claim must be checked with the authorities of each member state.

Unfortunately, the absence of a level regulatory 'playing field' does not aid the manufacturer who wishes to market a functional food throughout the whole of Europe. There may be light at the end of the tunnel; however, it is interesting to see that the individual approaches to regulating functional foods and health claims that have been introduced at a national level in several of the member states are very similar in approach. Hopefully, this will allow the market for functional foods to grow and develop in Europe, rather than being stifled by excessive legal controls.

REFERENCES

1. Regulation (EC) No. 258/97 of the European Parliament and of the Council of 27 January 1997 concerning novel foods and novel food ingredients. *Off J Eur Communities.* February 1997;40(L43):1–6.

2. Council Directive 89/398/EEC of 3 May 1989 on the approximation of the laws of the member states relating to foodstuffs for particular nutritional uses, as amended. *Off J Eur Communities.* June 1989;L186:27–32.

3. Council Directive 79/112/EEC of 18 December 1978 on the approximation of the laws of the member states relating to the labelling, presentation and advertising of foodstuffs for sale to the ultimate consumer, as amended. *Off J Eur Communities.* February 1979;22(L33):1–14.

4. Food Safety Act 1990, Chapter 16. HMSO, London.

5. Trade Descriptions Act of 1968, Chapter 29. HMSO, London.

6. Advertising Standards Authority. http//www.asa.org.uk/

7. Joint Health Claims Initiative Draft Code of Practice on Health Claims on Foods, HCI/15/98, 27/8/98. Food and Drink Federation, London.

8. Opinion of the Inter-Ministerial Commission for the Study of Products Intended for a Particular Diet. *Bull Off Concurrence, Comsommation Répression Fraudes.* 1997;57(17).

9. Opinion of the National Food Council. *Bull Off Concurrence, Consommation Repression Fraudes.* 1998;58(16):452–455.

10. *Health Claims in the Labelling and Marketing of Food Products: The Food Industry's Rules (Self-Regulating Programme).* Federation of Swedish Food Industries, August 1996.

11. National Food Administration, Press release. December 1998.

12. Codex Alimentarius Guidelines for the Use of Nutrition Claims, CAC/GL 23–1997. Codex Alimentarius Commission; 1997.

13. Report of the Twenty-Sixth Session of the Codex Committee on Food Labelling, ALINORM 99/22, Appendix X (Proposed Draft Recommendations for the Use of Health Claims (Step 3)). Codex Alimentarius Commission; 1998.

14. Guidelines issued by *Keuringsraad Aanprijzing Gezondheidsproducten.* 9th ed. May 1997.

15. *Code of Practice Assessing the Scientific Evidence for Health Benefits Stated in Health Claims on Food and Drink Products.* The Hague: Netherlands Nutrition Center (Voedingscentrum); 1998.

16. *Notification on the Establishment of a System of Licensing Foods for Specified Health Use.* Japanese Ministry of Health and Welfare; 1991.

Index